SPACE, GENDER, KNOWLEDGE

FEMINIST READINGS

Edited by
Linda McDowell
and
Joanne P. Sharp

A member of the Hodder Headline Group
LONDON • NEW YORK • SYDNEY • AUCKLAND

First published in Great Britain 1997 by
Arnold, a member of the Hodder Headline Group
338 Euston Road, London NW1 3BH

Copublished in the US, Central and South America by
John Wiley & Sons, Inc. 605 Third Avenue,
New York, NY 10158-0012

British Library Cataloguing in Publication Data
A catalogue record for this book is available from the British Library

Library of Congress Cataloging-in-Publication Data
A catalog record for this book is available from the Library of Congress

ISBN 0 340 677929 (pb)
ISBN 0 470 236698 (Wiley)
ISBN 0 340 70019 X (hb)
ISBN 0 470 23668 X (Wiley)

Typeset in 10/12pt Times by J&L Composition Ltd, Filey, North Yorkshire
Printed and bound in Great Britain by J W Arrowsmith Ltd, Bristol

CONTENTS

PREFACE

Producing a reader is always a terrifying experience and when the aim is to try and reflect the diversity of feminist scholarship by geographers and others interested in the variety in gender relations across space, then the task is even more intimidating. Choosing the papers to include was both a nightmare and a joy. How could we possibly squeeze the remarkable flowering of feminist research and commentary into a small number of representative sections/issues? Should we choose well- or lesser known authors and papers. By geographers or not?

What we have included here reflects our own predilections but also our certainty that gender matters for geographical scholarship. It was impossible to produce a representative selection, nor did we wish to be authoritative. What we hope we do is stimulate you to read on.

We also had fun, though, putting together this collection, reading and re-reading the remarkable feminist literature together. We can truly say that it was a joint production. We chose the papers together as well as commenting on and redrafting the introductions together. Part of the work was done in Cambridge and part in Glasgow and in each city we sampled a range of restaurants as well as the libraries. Our collaboration has brought each of us a new friendship and the opportunity to explore the similarities and differences in our feminist understanding and practices.

We were supported by a great editor: Laura McKelvie, who knew when to cajole and when to push. The result owes a lot to her commitment to us. Our wider debt is to the expanding community of feminist geographers and to the pleasure of exploring feminism in such company. Hugh McDowell was an excellent photocopier, for which we give thanks. Jo would also like to thank Lorraine Dowler for her stimulating discussions of feminist geography during the early stages of this project. Thanks are also due to the authors, publishers and journals who gave us permission to include their work, usually promptly and often with helpful advice or words of encouragement. And to Elizabeth Grosz, thanks for the title for Section Four. Finally in the dedication we acknowledge our own joint and particular debt to Lucy Adrian, who taught us both and set us the best example of feminist geography in practise that it is possible to find.

(1988), and ' "Gender for a Marxist Dictionary: the Sexual Politics of a Word' from *Simians, Cyborgs and Women: the Reinvention of Nature* (Free Association Books, 1991); the author and W.W. Norton and Company, Inc. for 'Housing and American Life' from *Redesigning the American Dream: the Future of Housing, Work and Family Life.* Copyright © 1984 by Delores Hayden. Reprinted by permission of W.W. Norton and Company, Inc.; the author and Pion Ltd. for 'Femininity, Post-Fordism and the "New Traditionalism" ' by D.A. Leslie from *Environment and Planning D: Society and Space* **11** (1993); the author and Carfax Publishing Company for 'Industrial Restructuring as Class Restructuring: Production Decentralization and Local Uniqueness' by Doreen Massey from *Regional Studies* **17** (1983); the author and Routledge, Inc. for 'No Longer in a Future Heaven: Gender, Race and Nationalism' from *Imperial Leather: Race, Gender and Sexuality in the Colonial Contest* by Anne McClintock (1994). Reprinted by permission of the publisher, Routledge: New York and London; the Royal Geographical Society for 'Doing Gender: Feminism, Feminists and Research Methods in Human Geography' by Linda McDowell from *Transactions of the Institute of British Geographers* **17** (1992); the authors and Economic Geography for 'Missing Subjects: Gender, Sexuality and Power in Merchant Banking' by Linda McDowell and Gill Court from *Economic Geography* **70** (1994); the author, Blackwell Publishers Ltd., and Stanford University Press for 'Feminist Encounters: Locating the Politics of Experience' by Chandra Talpade Mohanty from M. Barrett and A. Phillips (eds) *Destabilising Theory: Contemporary Feminist Debates* (Polity, 1992); the author and *New Left Review* for 'Man Bad, Woman Good? Essentialisms and Ecofeminisms' by Caroline New from *New Left Review* **216** (1996); the author, Blackwell Publishers Ltd., and the University of Minnesota Press for 'Looking at Landscape: the Uneasy Pleasures of Power' from *Feminism and Geography* by Gillian Rose (Polity, 1993), and The National Gallery, London, for permission to reproduce 'Mr and Mrs Andrews' by Thomas Gainsborough; the author and Routledge Inc. for 'The Earth is Not Your Mother' from *Earth Follies* by Joni Seager. Reprinted by permission of the publisher, Routledge: New York and London (1994); the author and Lynne Rienner Publishers for 'Middle East Politics through Feminist Lenses: Toward Theorizing International Relations from Women's Struggles' by Simona Sharoni from *Alternatives: Social Transformation and Humane Governance* **18**(1) winter 1993. Copyright © 1993 by Alternatives. Reprinted with permission of Lynne Rienner Publishers, Inc.; the author and Zed Books Ltd. for 'Women in Nature' from *Staying Alive: Women, Ecology and Development* by Vandana Shiva (1989); the author and Elsevier Science Ltd. for 'Can there be a Feminist Ethnography?' by Judith Stacey from *Women's Studies International Forum* **11**(1) (1988); the author and Macmillan Press Ltd. for 'Women, Employment and the Family' by Sharon Stichter from S. Stichter and J.L. Parpart (eds) *Women, Employment and the Family in the International Division of Labour* (1988); the author and Pion Ltd. for '(Hetero)Sexing Space: Lesbian Perceptions and Experiences of Everyday Spaces' by Gill Valentine from *Environment and Planning D: Society and Space* **11** (1993); the authors and Routledge Ltd. for 'Why Study

Feminist Geography?' by the Women and Geography Study Group from *Geography and Gender: An Introduction to Feminist Geography* (Hutchinson, in association with the Explorations in Feminism Collective, 1984); the author for 'Into the Labyrinth' from *The Sphinx in the City: Urban Life, the Control of Disorder and Women* by Elizabeth Wilson (Virago, 1991); the author and Princeton University Press for 'The Scaling of Bodies and the Politics of Identity' from *Justice and the Politics of Difference* by Iris Marion Young. Copyright © 1990 by Princeton University Press. Reprinted by permission of Princeton University Press; the author and Routledge Ltd. for 'Gender and Nation' by Nira Yuval-Davis from *Ethnic and Racial Studies* **16**(4) 1993.

Every effort has been made to trace copyright holders of material, and those who hold the rights to the reproduction of this material. Any rights not acknowledged here will be acknowledged in subsequent printings if notice is given to the publishers.

INTRODUCTION

It is a tall order to outline the history of feminist scholarship and its relationship to geography in a few brief pages of an introduction. Add to that a justification of how we have chosen and organised the papers that we have included and the difficulty is magnified. But these are the two aims of our introduction. We want to sketch in the main lines of change in feminist scholarship in general, in geographical work in particular and also try to link these threads to feminist politics.

It is, of course, hard to separate these aims. The papers that we have included were chosen to given some flavour of the short history of feminist geographical scholarship – its key debates and its shifting focus – as it developed from a tiny number of papers published in the 1970s, to its currently remarkably diverse and provocative status. While we hope to capture some of this diversity, any collection is, of course, a compromise. While we have tried here to be reasonably comprehensive and to range across geographical scales and parts of the world, it is inevitable that we have had to leave out more than we have included. We make no claims to be an 'essential anthology', as recent editors of collections in both geography and womens' studies have done (see for example Agnew *et al.*, 1996 and Jackson, 1993). Instead our aim is to be provocative: to prompt the reader to ask for more; to stimulate you to go on and search out all those things we have had to leave out. We have listed the papers that ideally we would have included in the reader, and many others, in our guides to further reading at the end of each section. And then there are the new papers published since we went to press in a wide range of journals and books, some within the bounds of our discipline and many, many more outside.

We have tried to include papers that differ in their style and aims: some are theoretical pieces, others are reviews of literature, or progress in an aspect of feminist scholarship. Some papers are empirically-based, reporting on research, in others the voices of women (and some men) are heard, discussing their experiences of gendered spaces, gendered social relations and strategies of resistance. In some of these papers the voices of the oppressed speak loud, others are testaments to the joy of being a woman and celebrate female solidarity. But, we (like most women?) are not romantics clinging to notions of international sisterhood. We recognise the complexities of the relations between gender, class, sexuality, ethnicity and other socially-constructed divisions. These are shifting, complex and often mutually contradictory dimensions of selfhood and group membership, which vary in their significance in different circumstances. Gender is only one aspect of our identities as women and men, its significance varies across space and time, as does its very constitution and meaning, but we do not see this as reducing the materiality of these divisions in the daily lives of women across the globe. Increasingly, however, feminist scholars have come to recognise that there are multiple ways of 'doing gender', of being male and female, masculine and feminine that are more or less appropriate, more or less socially sanctioned in particular spaces and at different times. Interestingly, this complexity and the idea of the positionality and contextual nature of gendered identities has, in our view, placed geographical research, a spatial imagination, right at the centre of current feminist scholarship.

In what follows, we have tried to include 'classics': some of the early 'landmark' statements about the aims, purpose and scope of feminist geography, or geographies as we now recognise. These are accompanied by more recent work by geographers, and a range of other papers, that have both influenced and stimulated geographers, or that seems to us to be amongst the best and clearest discussions of what we claim are essentially geographical issues by feminist scholars working in other disciplines. Therefore, we present this collection as a reader *for* feminist geographers rather than as a summary *of* feminist geography. In their early work, we think that feminist geographers have had especially close links with the concerns of feminist anthropologists. Indeed, some of the earliest papers about gender and the salience of the nature/culture division were by anthropologists. The links between the two disciplines are clear. Both geographers and anthropologists are interested in difference and diversity in social relations and meanings and symbols across space and over time. (And both too have their social and 'natural' sides.) Indeed, when asked for a definition of the scope of feminist geography, we both find Henrietta Moore's definition of feminist anthropology hard to beat. Feminist anthropology she argues is 'what it is to be a woman, how cultural understandings of the category "woman" vary through space and time, and how these understandings relate to the position of women in different societies' (Moore, 1988 p. 12). Defining the category 'woman', of course, demands a definition of gender, and here too Moore is helpful. Gender, she suggests, 'may be seen either as a symbolic construction or as a social relationship' (p. 13).

Let us first pursue the question of why space matters in the analysis and explanation of the gendered social relations. Spatial relations and layout, the differences between and within places, the nature and form of the built environment, images and representations of this environment and of the 'natural' world, ways of writing about it, as well as our bodily place within it, are all part and parcel of the social constitution of gendered social relations and the structure and meaning of place. The spaces in which social practices occur affect the nature of those practices, who is 'in place', who is 'out of place' and even who is allowed to be there at all. But the spaces themselves in turn are constructed and given meaning through the social practices that define men and women as different and unequal. Physical and social boundaries reinforce each other and spatial relations act to socialise people into the acceptance of gendered power relations – they reinforce power, privileges and oppression and literally keep women in their place. But one can also 'push against oppressive boundaries' to 'invent spaces of radical openness' (hooks 1990, pp. 145 and 148) within which to challenge dominant power, taking it on from the margins.

Interestingly, 'space', or sometimes 'spatiality' which gives more of a sense of its active agency, is an increasingly ubiquitous term in critical thinking, where it is used in both metaphorical and material ways. In addition to spatial relations at a societal level, gendered spatiality has been theorised at the level of individual bodies – the body as a surface to be mapped, a surface for inscription, as a boundary between the individual subject and that which is Other to it, as the container of individual identity, but also as a permeable boundary which leaks and bleeds and is penetrable.

In pursuing understandings of these diverse social constructions of femininity and womanhood, their cultural meanings and the sets of social relationships that hold these meanings in place, the relative emphases of work by feminist geographers in the English-speaking world have varied over time, and indeed between the different nations. It is possible to see a distinctively US, UK and Antipodean set of emphases (Peake, 1989; Monk, 1994), as well as to map the move from remedying women's absence through detailed descriptions of the different space-time coordinates of their lives (Tivers, 1979, 1985; Palm and Pred, 1976), to a sustained challenge to the key assumptions of western enlightenment thought that feminist geographers have made in association with other feminist theorists (Duncan, 1996; McDowell, 1992a and b; Rose, 1993; Women and Geography Study Group, 1997). In an earlier discussion of the changing emphases of feminist geography one of us (McDowell, 1992a and b) argued that there has been a move from what was termed feminist empiricism to a focus on questions about identity and difference, of the mapping of complex and multiple feminist geographies. In this move, as we shall show briefly below and return to in our introductions to each of the parts of this reader, a greater awareness of the multiplicity of ways of being a woman, of 'doing gender' has become dominant. Further, the social construction of multiple ways of doing masculinity has also become a subject of geographical attention.

Despite these changes in the theorisation of gender over time, it is possible (though perhaps rash!) to identify a key aim and constant of a feminist geography or geographies: it is to demonstrate the ways in which hierarchical gender relations are both affected by and reflected in the spatial structure of societies, as well as in the theories that purport to explain these relationships and in the methods used to investigate them. But we also believe that something more than that is involved. A commitment to a different and less inequitable society is also implied in the term 'feminist', although what these societies may look like is a disputed issue. In our view, feminist geography, or as we prefer to term it, a scholarship 'for feminists', is more than the analysis of gender relations. It also implies a progressive view of the future. In this sense our endeavours remain close to politically progressive modernist projects. Feminism and feminist politics have always been a complicated endeavour, embodying many different positions and perspectives on the ways in which being a woman and a man is achieved and the consequences that result. But in addition to acknowledging and investigating these diversities, a commitment to change is important. As we shall show below, this commitment has often led to (we believe misunderstood) accusations of political bias being levelled at a feminist scholarship.

In the beginning there were two positions . . .

From the 1960s, when second wave feminism (so-called to distinguish it from the suffrage movement or first wave) became both a visible political movement and began to influence academic practices and scholarship, there has been huge debate about the social construction of gender divisions and the explanations for them. Radical feminists, who in the main were more influential outside the academy, emphasised women's difference from men, demanding a separatist politics and celebrating women's strength in poetry, art and direct action. Women's nurturing abilities, their/our passivity and pacifism, closeness to nature and ability to give birth were regarded positively and as a unifying set of characteristics. In the academy, a more dour and serious but perhaps less exlusive feminism dominated. Often termed socialist feminism, it considered the ways in which capitalist society as a whole and individual men within it, appropriated women's domestic labour to ensure the daily reproduction of men as wage labourers, of children as future workers and of the system as a whole. Women were socialised into their gender roles and the social relations between men and women were unequal.

Getting complicated: race, class and gender

These two positions began to lose their distinctiveness throughout the 1970s and 1980s for a whole set of reasons, connected both to changes in feminist alliances and politics and to retheorising in the academy.

These two strands came together in powerful criticisms from many women of colour, who argued that the concerns of feminist politics, in the US and the UK at least, reflected issues that were dominant in the lives of white, middle class women – rights to employment outside their homes, the right to equal pay, the right to control fertility, for example. For other women – women who may be poor, or refugees – a different set of political agendas were important – the right to remain at home, for example, to have children rather than to control their fertility. Third world women in 'western' societies found that the calls to international sisterhood and solidarity were often exclusive rather than inclusive, and a more sophisticated understanding of the ways in which race, class and gender divisions were mutually constituted, in ways which lead to cleavages between women as well as unity among them, was gradually forged. The old claims about threefold oppression – as a woman, as a member of the working class and as a person of colour – were recast into analyses that emphasised interconnections and difference (Sanders, 1990). A series of theorists have argued that an individual cannot be represented by a single identity (Spelman, 1986). The effect of different subjectivities is transformative rather than additive. Race cannot simply be added to gender: it is different being a woman of colour and a white woman, just as class divisions force women apart. In a moving account of a Caribbean holiday, US writer and academic June Jordan (1989) explained her pain in being unable to connect with the life and experiences of the woman who waited on her in her holiday hotel even though they were both black women from the Caribbean.

Decentering 'woman'

The feminist recognition of difference coincided with a series of interesting debates by a diverse group of theorists, who now tended to be subsumed into two groups: the first is poststructural or postmodern theorists; the second postcolonial theorists. The recognition of political differences between women and the diverse ways in which they found themselves aligned on particular questions was paralleled by a theoretical exploration of difference. As feminists had long pointed out, western enlightenment knowledge, with its liberal assumption of a disembodied, rational and universal point of view, in fact reflected a very particular set of ideas about power and knowledge, about truth and humanity. A growing band of 'Others': women, post-imperial subjects, the 'deviant', people with 'alternative sexual identities' or 'alternative' ways of living questioned both their silencing in what passed for inclusive theories and the right of the powerful and dominant to speak on their behalf. Marxism and feminism, which believed that they spoke for the powerless – the working class and women respectively – were not immune from the criticisms, imbued as they were with a still singular notion of the truth and a progressive politics. Poststructuralism, postmodernism and postcolonialism emphasised the breakup of the unitary subject, be it man, woman or worker, and argued instead for alternative notions of identity, power and knowledge. Postcolonial critics have offered thorough critiques of western feminism for its silencing of differences other than those based around gender. Writers including Gayatri C. Spivak, Chandra T. Mohanty, and bell hooks (to name just a few of the most influential) have forced white feminists to re-examine the female subject of their writings and to theorise the complexities of gendered and raced subjectivities. Similarly, lesbian women have pointed to the implicit heterosexuality of much feminist work, and have insisted that questions of sexuality be brought into feminist writings.

This recognition and diversity, however, was accompanied by a more threatening challenge to the notion of the subject 'woman' that had hitherto been taken for granted. Feminist scholars, and other theorists influenced by psychoanalysis and by the work of Foucault, began to retheorise the subject as relational and contingent. The subject, rather than being a fixed entity which entered into social relations with its gender (and class and race) fixed in place, is always fluid, in the process of becoming, anxious to create and hold on to an identity which is constructed through discourse and everyday actions. In some ways this was a great step forward in theorising gender relations as instead of taking binary divisions for granted it opened up to interrogation the ways in which the attributes of femininity and masculinity are constructed in particular circumstances and times. Thus what it meant to be a woman (or a man) was seen to be context dependent, relational, complex and variable. Gender, as Linda Alcoff argues, 'is not a point to start from in the sense of being a given thing but is, instead, a posit or construct, formalizable in a nonarbitrary way through a matrix of habits, practices, and discourses' (Alcoff, 1988 p. 431).

A 'cultural turn'? Performing gender/writing gender

One of the consequences of these new debates has been a shift in the focus of feminist scholarship towards the practices and discourse identified by Alcoff. One of the most influential shifts has been towards theorising gender as a performance, influenced in particular by the work of Judith Butler, an extract of which we include in Section Four. Butler's interest is in drag as a performance which destabilises conventional heterosexuality, and her approach has influenced many geographers looking at gay and lesbian politics (see Bell and Valentine, 1995 for examples) but it has also influenced understandings of femininity more generally. As Elizabeth Grosz argues

> the practices of femininity can readily function in certain contexts that are difficult to ascertain in advance, as modes of guerrilla subversion of

> patriarchal codes, although the line between compliance and subversion is always a fine one, difficult to draw with any certainty. All of us, men as much as women, are caught up in modes of self-production and self-observation; these modes may entwine us in various networks of power, but never do they render us merely passive and compliant. They are constitutive of both bodies and subjects.
>
> (Grosz, 1994 p. 144)

Indeed, we might argue that it may be possible to be both compliant and subversive at the same time depending on the reading of particular acts.

This raises some interesting questions about how to explain what Coward (1984) has termed 'the lure of those identities which also oppress us'; about our own investments in femininity and the pleasures we gain from them. These questions about the interstices of power, pleasure and desire take us a long way from what some regard as suitable subjects for geographical treatment!

The emphasis on the social construction of power relationships in discursive practices has also led to a shift towards the analysis of representational practices: art, performance, writing and to a new awareness of our own practices as authors. Gillian Rose (1993), for example, has played a critical role for geographers in revealing the significance of the masculine gaze in high art as well as in our discipline itself. In this work, links between geographical scholarship and the humanities are becoming as significant as the connections with economics, sociology and anthropology were in the 1980s. Feminist geographers, in common with others in our discipline, have been influenced by the 'cultural turn' which has both critiqued and complemented earlier more economistic accounts of gender relations. Although Moore in the definition outlined above emphasised the importance of both social process and symbolic structures of gender, feminist geographers tended initially to focus almost exclusively on the social relations that constructed and maintained gender divisions. More recent writing has a greater concern with meaning and symbolism (see for example Jacobs, 1996) and the complex interdependencies between image and materiality (Enloe, 1989, 1993; Seager, 1993; Sharp, 1996). In particular there has been a move to study the metaphors invoked in writing about gender: most significantly, the intertextualities of the parallel dualisms in western thought, woman/man, nature/culture (see Blunt and Rose, 1994).

As a result of this cultural turn, feminist geographers have become intrigued by the spatial significance of non-geographical writing – writing that has inspired geographers through perhaps incidental invocation of spatial concepts, especially the importance of the rise of spatial metaphors (Bondi, 1993; Grewal and Kaplan, 1994; Kirby, 1996). Feminists have seen women silenced from the narratives of history and so have sought a position, a political stance, not in the rhetoric of time and progress but from a spatialised position – women have adopted the margins, challenged the centre, transgressed borders, written over maps. . . . While geographers have obviously been excited by such work (Katz and Smith, 1993), we have perhaps also been cautious of too ready an acceptance of the significance of spatial metaphors, for they too cannot avoid the exclusivity, the power/knowledge of history, but are as political and strategic as any other terms.

The possibility of feminist politics

These changes in the understanding of the social construction of femininity and masculinity, of the various possible subject positions, have had an exciting and in many ways liberating effect. To be able to challenge the centred male subject, in whose image knowledge was constructed, has been very important in building new understandings of what is often termed identity politics. Rather than a singular understanding of 'woman' the new positional and relational definition of the subject has led to political analyses that stress coalition building around particular issues. Women of colour, for example, may feel their allegiances lie with men in anti-racist struggles, some women may have green allegiances – but this is no longer taken for granted as it was when all women were assumed to have a pro-environmental stance because of their 'closeness to nature'.

The deconstruction of categorical analyses of the subject, however, has led to fears of the disablement of a specifically feminist politics in some quarters. Whilst holding a great deal of sympathy for arguments about the provisional and positional construction of identity, Linda Alcoff also fears the political implications of losing the identity of 'woman':

> What can we demand in the name of women if 'women' do not exist and demands in their name simply reinforce the myth that they do? How can we speak out against sexism as detrimental to the interests of women if the category is a fiction? How can we demand legal abortions, adequate child care, or wages based on comparative worth without invoking a concept of 'woman'?
>
> (Alcoff, 1988 p. 420)

Her fear parallels a similar anxiety about the theoretical implications of the decentring of identity and the key place this notion holds in postmodern thought – a group of theoretical positions, which, although close to feminism in many ways, is still very much dominated by male theorists. As Fox-Genovese has speculated:

> Surely it is no coincidence that the western white male elite proclaimed the death of the subject at precisely the moment at which it might have had to share that status with women and peoples of other races and classes who were beginning to challenge its supremacy.
>
> (Fox-Genovese, 1986 p. 121)

Similar anxieties have been expressed by feminist anthropologists and geographers (see for example, Mascia-Lees *et al.*, 1989; McDowell, 1991). Nancy Fraser, however, believes that the current debates are caught in a series of unacceptable series of antitheses and oppositions: structuralism versus poststructuralism; difference versus essentialism. She wants to try and move forward (a reminder of the progressive nature of a feminist agenda), seeing counterposing arguments about women's identity as fluid, multiple and uncertain – as problematic as earlier views about a fixed and essential femininity. She argues

> that the current proliferation of identity-dereifying, fungible, commodified images and significations constitutes as great a threat to women's liberation as do fixed, fundamentalist identities. In fact, dereifying processes and reifying processes are two sides of the same post-Fordist coin. They demand a two-sided response. Feminists need both deconstruction and reconstruction, destablisation of meaning and projection of utopian hope.
>
> (Fraser, 1991 p. 175)

So perhaps through this two-sided response, we may be able to choose an appropriate politics according to the circumstances. At times, women's interests may be divided and an awareness of and sensitivity to the differences between us is crucial for alliance building. But at other times, it seems vital to speak as women. In societies where women are denied full educational opportunities, other civil rights or the ability to earn their living, then surely we must be able to speak as women against their oppression. We cannot evade the responsibility to establish criteria to distinguish progressive from reactionary forms of identity politics, despite the differences between us.

The structure of the reader

We have organised the papers in this reader into seven sections. The substantive divisions are based on a combination of different aspects of life and on geographic scale. We want to give an indication of the ubiquity of gendered ideas and processes in different places and through different activities (both 'publicly' and 'privately' enacted), in addition to a range of interdependent scales, linking the gendering of the body to the nation, the workplace, and to nature. Further, in each section, we have tried to give an indication of the most significant issues and debates. Thus, after two sections which examine the development of feminist scholarship, in general and in our discipline (Section One) and look at questions about 'practising' feminist geography (Section Two), we focus in turn on nature, the body, everyday life and waged work, nationality and international politics. In almost all of the papers and extracts that are included, a common set of concepts are found. The authors struggle over the distinctions between man and woman, masculine and feminine, the mind and the body, showing how these binary distinctions construct woman as man's 'Other' and as inferior. The challenge and success of feminist scholarship has been to deconstruct the 'naturalness', the common sense or taken-for-granted nature of these concepts, and reveal their social construction, and to make space for a more complex and multiple understanding of the character of gender divisions.

References and further reading

Agnew, J, Livingstone, D and Rodgers, A (eds) 1996: *Human Geography: An Essential Anthology*. Blackwell: Oxford.

Alcoff, L 1988: Cultural feminism versus post-structuralism, *Signs* **3**, 404–36.

Ardener, S (ed.) 1981: *Women and Space*. Croom Helm: London.

Barrett, M and Phillips, A (eds) 1992: *Destabilizing Theory: Contemporary Feminist Debates*. Polity: Cambridge.

Bell, D and Valentine G (eds) 1995: *Mapping Desire*. Routledge: London.

Blunt, A and Rose, G (eds) 1994: *Writing Women and Space*. The Guildford Press: London.

Coward, R 1984: *Female Desire: Women's Sexuality Today*. Granada Books: London.

Duncan, N (ed.) 1996: *BodySpace*. Routledge: London.

Enloe, C 1989: *Bananas, Beaches and Bases*. University of California Press: Berkeley and Los Angeles.

Enloe, C 1993: *The Morning After*. University of California Press: Berkeley and Los Angeles.

Fox-Genovese, E 1986: The claims of a common culture: gender, race, class and the canon. *Salmagundi* **72**, 119–32.

Fraser, N 1991: False antitheses: a response to Seyla Benhabib and Judith Butler. *Praxis International* **11**, 166–77.

Grewal, I and Kaplan, C (eds) 1994: *Scattered Hegemonies*. University of Minnesota Press: Minneapolis.

Grosz, E 1994: *Volatile Bodies: Towards a Corporeal Feminism*. Indiana University Press: Indiana.

hooks, b 1990: Choosing the margin as a space of radical openness. In *Yearning: Race Gender and Cultural Politics*. Turnaround Books: London. 145–53.

Jacobs, J 1996: *Edge of the Empire*. Routledge: London.

Jackson, S 1993: *Women's Studies: Essential Readings*. New York University Press: New York.

Jordan, J 1989: *Moving Towards Home*. Virago: London.

Katz, C and Smith, N 1993: Spatializing metaphors: towards a spatialized politics. In Keith, M and Pile, S (eds) *Place and the Politics of Identity*. Routledge: London. 67–83.

Kirby, K 1996: *Indifferent Boundaries: Spatial Concepts of Human Subjectivity*. Guildford Press: London.

Mascia-Lees, F, Sharp, P and Cohen, C B 1989: The postmodern turn in anthropology: cautions from a feminist perspective. *Signs* **15**, 7–33.

McDowell, L 1991: The baby and the bathwater: deconstruction, diversity and feminist theory in geography. *Geoforum* **22**, 123–34.

McDowell, L 1992a: Space, place and gender relations: part 1. Feminist empiricism and the geography of social relations. *Progress in Human Geography* **17**, 157–79.

McDowell, L 1992b: Space, place and gender relations: part 2. Identity, difference, feminist geometries and geographies. *Progress in Human Geography* **17**, 305–18.

Mohanty, CT, Russo, A and Torres, L 1991: *Third World Women and the Politics of Feminism*. Indiana University Press: Indiana.

Monk, J 1994: *Contextualising Feminist Geography: International Perspectives*. IGU Working Paper 27 Commission on Gender and Geography.

Moore, H 1988: *Feminism and Anthropology*. Polity: Cambridge.

Nicholson, L (ed.) 1990: *Feminism/Postmodernism*. Routledge: London.

Peake, L (ed.) 1989: The challenge of feminist geography. *Journal of Geography in Higher Education* **13**, 85–121.

Palm, R and Pred, A 1976: *A Time-Geographic Approach on Problems of Inequality for Women*. Working Paper 236 Institute for Urban and Regional Development. University of California: Berkeley.

Rose, G 1993: *Feminism and Geography*. Polity: Cambridge.

Sanders, R 1990: Integrating race and ethnicity into geographic gender studies. *Professional Geographer* **42**, 228–31.

Seager, J 1993: *Earth Follies*. Routledge: London.

Sharp, J 1996: Gendering nationhood: a feminist engagement with national identity. In Duncan, N (ed.) *BodySpace*. Routledge: London, 97–108.

Spelman, E 1986: *Inessential Woman*. Beacon Press: Boston.

Tivers, J 1979: How the other half lives. *Area* **11**, 302–6.

Tivers, J 1985: *Women Attached: The Daily Lives of Women with Young Children*. Croom Helm: London.

Women and Geography Study Group 1997: *Feminist Geographies: Explorations in Diversity and Difference*. Longman: London.

SECTION ONE

THINKING THROUGH GENDER

Editors' introduction

As we have argued, the debates about space, gender and knowledge have become immeasurably more sophisticated in the short history of feminist and gender scholarship. In Section One of the reader we capture some of this complexity by focusing on the changing ways in which gender relations have been theorised. Our aim is to reproduce the 'development' of work in feminist geography, from the first statements insisting upon the inclusion of women and feminist methodology in the discipline, to more recent 'ambivalent' reflections upon the utility of gender as an explanatory device in a critical geography.

The first reading is the initial statement of the British geographers who as a collective enterprise produced *Geography and Gender* in 1984. Twelve years on this is a fascinating reminder of the concerns of the time when feminist scholarship in geography was beginning to establish itself 'against the odds'. The institutional climate of individual departments and the professional associations of the time were relatively hostile. It had taken what now looks like a surprisingly long time to appear since the publication of what we think was the very first paper in a geographical journal raising specifically feminist issues – a paper by Pat Burnett in *Antipode* 1973 about women's position in the city, with a response by Irene Bruegel in the same issue. *Antipode* carried another paper the next year and then in *Area* between 1978 and 1980 Jackie Tivers (1979), Linda McDowell, (1979), and Suzanne MacKenzie and Jo Foord (1980) had an exchange of views about geographic research on and by women. In the United States, the earliest papers were published at about the same time. Zelinsky (1973), Zelinsky, Monk and Hanson (1982) and then Monk and Hanson (1982) lamented the absence of women from US geography, both as teachers and as the subjects of human geographical research. A comparison with the situation in the mid 1990s when almost all the major geographic journals take gender issues seriously, and there has been an expansion of feminist scholarship published by both mainstream and specialist feminist presses (notice the recent launch of *Gender, Place and Culture*) is indicative of how the discipline is being changed by feminist scholarship. In these years there has also been a marked shift in the reaction to feminist scholarship. At first, feminist scholars had to become used to defending their academic work against an accusation of 'too political' or biased. This has always been a strange criticism. Class theorists, documenting the divisions of labour in contemporary capitalist societies and the resulting inequalities, or scholars interested in race relations were seldom attacked in quite the same vituperative and personal way, although marxists may have experienced the same accusation of political bias. The reasons for the antagonism are not hard to find, however, as feminist theorists began to challenge those social divisions that were the most taken-for-granted, assumed to have a biological basis, as being 'only natural'. To problematise the social divisions between the sexes, the very meaning of being a man or a woman, and then to turn to sexuality as well, meant that the most personal questions about identity and our sense of ourselves had to be addressed. Feminist politics has been expanded into the 'private' sphere to encompass those aspects of our lives that we have learnt to consider most defining of our character. This may explain the antagonism of anti-feminists, especially the many men who thought that they were bound to lose authority if the social construction of the bases of their power was made visible. Gradually, however, feminism has gained academic respectability (although some fear that this may have blunted the radical edge afforded by intellectual autonomy). Research that uncovered the crucial significance of gender divisions to both material social relations and in structures of representation, meaning, symbols, and beliefs led to the wide recognition that gender relations could not be ignored as a central element in the humanities and social sciences.

But yet, lest these signs be read too optimistically, there remains only a tiny number of specifically feminist options in undergraduate geography courses, and too many 'mainstream' courses are untouched by feminist perspectives. It was this latter 'integrationist' approach that Monk and Hanson (1982) advocated over twenty years ago, worried by the prospect of feminist teaching being sidelined or seen as 'women's issues' without relevance to male teachers and students. This is too common an attitude now, as we have both found in our own teaching, and as many other feminists verify. Despite their evident presence, women and feminists remain a tiny minority in most geography departments in English-speaking universities, and elsewhere (see Momsen, 1980; McDowell and Peake, 1990). It is also worrying that the commitment to equal opportunities policies and programmes in many institutions is little more than window-dressing and elsewhere, notably in California, is actually being withdrawn.

There have been a number of important early statements regarding the need for a feminist agenda, including the statement by Monk and Hanson and the *Antipode* pieces mentioned above. To represent initial feminist agendas we have chosen the introduction to *Geography and Gender* from 1984 by the **Women and Geography Study Group** of the

Institute of British Geographers, not because it was first, or a necessarily better statement of a way forward than these others but because it can be seen to be a key agenda-setting document for the directions a feminist geography might take. In what now seems an idealistic move in a far more competitive and individualised academy at the end of the century, the book was authored and edited by a group who chose not to claim individual authorship. However, a second book also published under the collective umbrella of the Women and Geography Study Group appeared in 1997 bravely repeating the 1984 gesture at individual authority. A much larger group of feminist geographers were involved this time: an indication of the growing influence of feminist perspectives in British universities.

The next pair of papers are by Donna **Haraway** and Bob **Connell**. A feminist reader without a contribution from Donna Haraway is unthinkable. Her influence on feminist scholarship as a whole has been remarkable, and her range, erudition and sparkling style has dazzled, enthralled and inspired many feminist geographers, as is evident in the number of times she is cited. She became known to a wider geographic audience through her participation in a debate at an AAG conference with David Harvey, since published in *Environment and Planning D: Society and Space* in 1995, as well as through her voluminous work published elsewhere. We were spoilt for choice about which papers of hers to include. Her work has been important in the debates about situated knowledge, as well as about the social construction of nature (Haraway, 1989). Her cyborgs paper is insightful about trends in the contemporary economy of the USA (Haraway, 1991). In the end we chose to include two pieces. The first is her 'definition of "gender" for a marxist dictionary' which provides a good summary of the ways in which gender has been defined in different feminist discourses. This long paper is an extremely thought-provoking survey of the links between sex and gender and the historical shifts in the theorisation of what Haraway terms the 'sex–gender system' drawing on Gayle Rubin's early paper (Rubin, 1975). For those readers new to feminist scholarship, Haraway's paper provides an exceptionally clear summary of different theoretical approaches to the understanding of gender divisions.

We have paired Haraway's first contribution with Bob Connell's clear outline of gender as social practice, illustrated in this instance, by the case of masculinity, which, we are aware, does not receive a great deal of explicit attention in our selection of readings. Connell, an Australian theorist now working in the USA, is one of the scholars who has done most to advance thinking about the ways in which gender is a socially constituted, and historically and geographically differentiated, set of relations. His notion of a gender regime which is 'lumpy' is close to the ideas of the regulation theorists, Lipietz and Aglietta, which are perhaps more familiar to many geographers. Like these theorists, Connell argues that a hegemonic gender regime is established at any particular period, which becomes destabilised by social change and is subject to sudden changes. Hegemonic versions of appropriate ways of being masculine and feminine – the heterosexual male breadwinner and the feminine home maker of the Fordist era in western societies is a good example – are destabilised by changes in accepted norms and mores which loosen the bounds of these hegemonic versions. Connell has written widely on the multiple ways of being masculine, including a book published in 1995 which provides a sweeping approach to the variants of masculinity across space and time. Among geographers, Peter Jackson (1991; 1994) in the UK has been at the forefront of efforts to establish geographical variations in heterosexual masculinity, whereas in the USA perhaps Larrie Knopp (1992, 1994, 1995) has been the key scholar in the analysis of gay masculinity.

The second paper by **Haraway** is one of three papers which turn to more recent feminist theorising and the complex questions raised by the theoretical interrelationships between postmodernism, post-colonial critiques and feminism. We will pursue the implications for geographic practice in the set of papers in Section Two. As we indicated above, the encounters between these three positions on the nature and status of knowledge have been vexed, and, at times, heated. Haraway's paper about 'situated knowledges' is an important and extremely influential paper that concerns feminist attempts to reclaim an 'objectivity' for their work. She achieves this by insisting upon the embodiment of *all* knowledge, so directly challenging the scientific orthodoxy of objectivity that assumes a disembodied, disengaged and placeless gaze of knowledge. A number of feminist theorists have indicated the importance of decentring knowledge production to those most often silenced by dominant positions, but Haraway offers a warning. She cautions us not to romanticise the standpoints of the marginalised and the repressed. She insists that they not be seen as somehow 'innocent' but that they should always be taken as partial perspectives, offering situated accounts, as do all the figures involved in knowledge production. However, Haraway argues that because the poor and oppressed have a vested interest in revealing the structures of power that maintain their oppression, their view of the world is a clearer one than that of the unlocated 'view from above'.

In its various guises postmodernism has posed a challenge to geography that has presented feminism with interesting opportunities. Postmodernism, like feminism, insists upon the need to situate knowledge production, highlights the instability of subjectivity, and warns of the exclusions of (historical) narrative. And yet, as many feminists have observed, the enthusiasm with which geographers have embraced postmodernism should not be confused with a commitment to feminist goals. Although since the late 1980s, sessions and discussions at IBG and AAG meetings have 'changed out of all recognition' as far as an attentiveness to difference and subjectivity are concerned, it appears that 'this recognition is based not on an understanding of feminist arguments, and more certainly not on a commitment to feminism's revolutionary project, but rather on an eclectic reading of postmodernism' (McDowell, 1992a p. 58). In other words, the rush to accept the polyphony inherent in an acceptance of postmodernism may have more to do with attempts to achieve an 'academic therapy' than an active engagement with important excluded voices. The influence of postmodernism in geography has not brought feminism so much into the 'heart' of the discipline as might have been expected. Some observers have gone further still, indicating that just as previously marginalised voices start to be heard, just as they begin to achieve a recognised subjectivity, masculinist theories negate the existence of both of these concepts.

Liz **Bondi**, in her paper included in Section One, addresses the relationship between postmodernism and feminism in geography. She, like many others, some of whom are included in later sections of this collection, is grappling with the problem of on the one hand, adopting a political project of women's issues/feminism as a powerful and united political group, and on the other, allowing the differences of women's experiences and identities to survive, to avoid being repressed under the necessities of unity of 'women' or 'feminists'. (see also Bondi, 1993 and Bondi and Domosh, 1992). In the final paper here, Chandra Talpade **Mohanty** addresses this issue more directly, and with more anger. She fears the imposition of white feminist agendas onto women's struggles around the globe. As a result, other categories of identification and repression – most importantly to Mohanty, of race – dissolve, as if 'the categories of race and class have to be invisible for gender to be visible'. Rather than a 'sisterhood', Mohanty offers the possibility of a feminist 'coalition' recognising that not everyone necessarily wants to join the coalition but sees it as the best option, a 'strategic essentialism' that facilitates both a unified front in the face of the multiple effects of patriarchal structures and heterogeneous expression of womanhood.

References and further reading

Aglietta, M 1979: *A Theory of Capitalist Regulation.* New Left Books: London.

Bondi, L 1993: Locating identity politics. In Keith, M and Pile, S (eds) *Place and the Politics of Identity.* Routledge: London. 84–101

Bondi, L and Domosh, M 1992: Other figures in other places: on feminism, postmodernism and geography, *Environment and Planning D: Society and Space* **10**, 199–213.

Bruegel, I 1973: Cities, women and social class: a comment. *Antipode* **5**, 62–3.

Burnett, P 1973: Social change, the status of women and models of city form and development. *Antipode* **5**, 57–61.

Connell, R W 1995: *Masculinities.* Blackwell: Oxford.

Haraway, D 1989: *Primate Visions: Gender, Race and Nature in Modern Societies.* Routledge: London.

Haraway, D and Harvey, D 1995: Nature, politics and possibilities: a debate and discussion with David Harvey and Donna Haraway. *Environment and Planning D: Society and Space* **13**, 507–27.

Jackson, P 1991: The cultural politics of masculinity. *Transactions, Institute of British Geographers* **16**, 199–213.

Jackson, P 1994: Black Male: advertizing and the cultural politics of masculinity. *Gender, Place and Culture* **1**, 49–60.

Knopp, L 1992: Sexuality and the spatial dynamics of capitalism. *Environment and Planning D: Society and Space* **10**, 651–69.

Knopp, L 1994: Social justice, sexuality and the city. *Urban Geography* **7**, 644–60.

Knopp, L 1995: Sexuality and urban space: a framework for analysis. In Bell, D and Valentine, G (eds) *Mapping Desire*. Routledge: London. 149–61.

Lipietz, A 1987: *Mirages and Miracles: the crises of global Fordism*. Verso: London.

Mackenzie, S, Foord, J and McDowell, L 1980: Women's place, women's space. Comments arising from papers in Area, *Area* **12**, 47–51.

McDowell, L 1979: Women in British Geography. *Area* **11**, 151–55.

McDowell, L 1992a: Space, place and gender relations: Part 1. Feminist empiricism and the geography of social relations. *Progress in Human Geography* **17**, 157–79.

McDowell, L and Peake, L 1990: Women in geography revisited. *Journal of Geography in Higher Education* **14**, 19–30.

Momsen J 1980: Women in Canadian geography. *Canadian Geography* **24**, 177–83.

Monk, J and Hanson, S 1982: On not excluding half the human from human geography. *Professional Geographer* **34**, 11–23.

Rubin, G 1975: The Traffic in Women. In Reiter, R. (ed.) *Towards an Anthropology of Women*. Monthly Review Press: New York.

Tivers, J 1979: How the other half lives. *Area* **11**, 302–6.

Women and Geography Study Group 1997: *Feminist Geographies: Explorations in Diversity and Difference*. Longman: London.

Zelinsky, W 1973: The strange case of the missing female geographer. *The Professional Geographer*. **25**, 101–5.

Zelinsky, W, Monk, J and Hanson, S 1982: Women and geography: a review and a prospectus. *Progress in Human Geography*. **6**, 357–66.

1 Women and Geography Study Group

'Why Study Feminist Geography?'

Excerpt from: *Geography and Gender: An Introduction to Feminist Geography*, pp. 19–23. London: Hutchinson, in association with the Explorations in Feminism Collective (1984)

A 'geography of women' or 'feminist geography'?

Looking through a representative selection of geography books on a library or bookshop shelf, it appears as if most geography is concerned with 'man'. We are confronted by 'man and his physical environment', 'man and culture' or simply 'man and environment'. The authors of such texts may not intend to portray humanity as being entirely *male*. Nevertheless, it is this image which is created in the mind of the reader and which persists when the subject matter of the books is consulted. We are presented, for example, with *man* as the agent of change in agricultural landscapes, *men* digging for coal (or being made redundant by the closure of coal-mines), and the results of surveys in which *men*, as heads of households, have been asked for their opinions on recreation resources, transport needs or housing. We might, in fact, be forgiven for thinking that *women* simply do not exist in the spatial world.

If we turn to the historical study of human activities, we find the same situation. One of the features of human history appears to be that groups which are disadvantaged, being in some way dominated or oppressed by other groups – slaves, the common people, non-whites, women – do not figure as prominently in historical accounts as their social significance or sheer weight of numbers would suggest. Often, such groups are barely mentioned at all. In the majority of historical studies, the researcher is concerned with the way in which *men* gradually increase their control over nature, with the exploits of warriors, princes and kings and, more recently, of various states*men*, business moguls, inventors and scientists. The history of dominated groups has been taken as more or less incidental to the lives of those characters who, although few in number, occupy the centre of the historical stage and exercise control over society's wealth and resources. If the activities or experiences of subordinate groups are considered, they are often portrayed as part of some natural, even divinely-inspired, order, represented in such a way as to suggest that members of such groups are happy with their lot, or they are simply denigrated.

Rowbotham (1979), challenging such distortions, has suggested that women are 'Hidden from History'. We contend that women are also 'hidden from geography'. Hayford (1974) clearly makes this point when examining the situation in geography in North America. She argues that, although geographers recognize the existence of women, they make little effort to investigate the role they play in society other than in terms of adjustment to a male-dominated and male-determined order. Later in the 1970s Tivers (1978) published an article suggesting that geographers should consider 'How the other half [i.e. women] lives', and drop their patronizing attitude, common to workers in the other social sciences, to the study of women's perceptions and behaviour. During the late 1970s and early 1980s, a variety of work has appeared in which women's spatial behaviour and environmental perceptions have been highlighted, and the papers by Monk and Hanson (1982) and by Zelinsky, Monk and Hanson (1982) provide numerous examples in many fields of geography. This work has quite clearly demonstrated the existence and effects of women's inferior economic and social status, and shown how traditional geographical work is deficient in this respect.

Despite the increase of interest shown by some (although still not many) researchers within geography in women's activities and problems, the writers of introductory texts seem to have remained largely unaffected by such a change of attitude. One of the aims of this book is, therefore, to redress the balance of introductory geographical material in favour of *women*.

. . . Encouraging geographers to consider women's daily lives and problems as legitimate, sensible and important areas for research and teaching is, however, not the principal aim of this book. We are concerned that there should be a geography of women as well as of men. Nevertheless, we do not wish this to be seen as merely 'adding' women to existing types of geographical analyses without any alteration to the theoretical assumptions underlying these analyses. Making women visible is simply not enough. The very fact that differences between male and female spatial behaviour patterns can be so readily identified does not in itself guarantee that the geographer will do anything other than conclude: 'Well, men and women are different, and it's interesting to see how this comes out in their behaviour'! In common with other approaches in geography which are critical of mainstream work, we must analyse and understand why women remain in a subordinate position.

What we argue for in this book is not, therefore, an increase in the number of studies of women *per se* in geography, but an entirely different approach to geography as a whole. Consequently we consider that the implications of *gender* in the study of geography are at least as important as the implications of any other social or economic factor which transforms society and space. We use the term 'gender' to refer to *socially created* distinctions between femininity and masculinity, while the term 'sex' is used to refer to biological differences between men and women. Hence, we are concerned to introduce the idea of *feminist geography* – a geography which explicitly takes into account the socially created gender structure of society; and in which a commitment both towards the alleviation of gender inequality in the short term and towards its removal, through social change towards real equality, in the longer term, is expressed.

Why bother?

So far we have outlined the general approach of this book. But, before any further consideration of feminist geography, we should answer one very obvious question which may well be in the mind of the reader – why bother?

Why should we, as geographers, be worried about sexism or the 'invisibility' of women? Surely things are different these days and women have equal rights with men? Why should we any longer need to stress the existence of inequality based on gender?

To take the last point first, it is perfectly clear that inequalities between the sexes have not been eradicated within advanced capitalist societies. For example, it is true that in Britain, as in many other similar countries, a certain amount of *legislative* change occurred in the 1970s. The Equal Pay Act was introduced to prevent the payment of unequal wages to men and women for doing the same or similar work. However, employers have found many ways to evade the legislation, and their task was made easier by a five year easing-in period. It is extremely difficult to compare the wages paid to men and women for similar work because, in so many instances, workforces have been increasingly segregated so that the scope for job comparison is limited. The Sex Discrimination Act which was intended to provide equal opportunities and which covers education, training, employment, housing and the provision to the public of goods, facilities and services, appears far-reaching, but is extremely complex and full of loopholes. In 1977, the Employment Protection Act also came into force, giving women the legal right, given two years' full-time service, to return to their job after having a baby. Women were also granted the right to maternity pay under certain conditions. Such measures do represent a small step towards providing a degree of financial independence for women. However, two areas which have as yet remained resistant to legislative change but in which men and women are treated unequally, are the tax and social security systems.

However, legislative change does not necessarily herald a fundamental change in societal organization or norms and values. This has been the case partly because of the very limited nature of the legislation. For example, while 'equal pay' and 'equal opportunity' have been officially supported as worthy *concepts* to be worked towards, no attention has been paid to the practical need of working mothers for child care, in order to allow them to work in the first place. In addition, even the limited though positive gains made on behalf of women during the 1970s have begun to be eroded as the current economic recession has deepened. For example, it is now much more difficult for working women to qualify for maternity pay, and the Conservative government attempted (although unsuccessfully) in 1980 to 'rationalize' various forms of maternity benefit, to the detriment of working women. In many ways, this situation is not surprising. Someone has to foot the bill for child care if women no longer provide it, for the increased wages (if women truly obtain equal pay and opportunities), and for a whole range of maternity benefits. The money can be raised directly from private or public employers if they provide free child care and other benefits for their employees, or indirectly via various forms of taxation. During times of economic prosperity, profits are higher and 'society' can more readily afford to accede to at least some of the demands made by women (bearing in mind the fact that many other groups will also be struggling to obtain 'a larger share of the cake'). During times of recession, not only is there likely to be a squeeze on wages in many sectors of the economy, but also public expenditure is more likely to be cut. Both these trends have been only too evident in Britain since the late 1970s.

Inequality of employment opportunity is not the only area in which women remain subordinate to men. Of particular importance is the difference in educational attainment between men and women. At all levels above GCE 'O'-level men predominate. There are more male graduates, and therefore more men than women entering managerial and professional occupations. Women are clearly *not* inferior to men in terms of intelligence or application to study. As in the case of employment, there are social factors which intervene to prevent women from achieving their full potential in many cases. In an interesting article by Linda McDowell (1979) on the gender structure of British geography departments, statistics were presented to show that, whereas 42 per cent of full-time student geographers were women, this was true of only 7 per cent of full-time university geography teachers. However, 44 per cent of part-time teachers were women. Part-time employment, offering lower status and less security than full-time service, is the most 'convenient' way to utilize the talents of women, without at the same time disturbing the societal 'status quo' to any significant extent.

The vast preponderance of full-time male academics in geography departments in institutions of further and higher education, therefore, both reflects and perpetuates a pattern of social relationships which are male dominated. This brings us back to the first of our original points – why should *geographers* be concerned about the gender structure of society? The answer to this question is not only political. In recent years geographers have been active in research into questions of societal inequality based on class, income or race, not only because of a growing belief in the need to expose injustice and work towards a better and fairer future, but also because of the importance of an accurate analysis of human geographical phenomena for the development of our understanding of society and space. We cannot interpret human behaviour unless we take into account all the societal patterns and structures which exist. The same is clearly true of a feminist perspective in geography. Without an understanding of the gender roles which underlie the workings of society, we cannot hope to present a reasonable analysis of the spatial behaviour of women and men, nor of the institutions both dependent on and influencing that behaviour.

It is because of our belief in the importance of a feminist approach to geography that the Women and Geography Study Group of the Institute of British Geographers has been formed, and, through this book, seeks to put forward an introductory explanation of such an approach.

References

Hayford, A 1974: The geography of women: an historical introduction. *Antipode* **6**, 1–19.

McDowell, L 1979: Women in British geography. *Area* **11**, 151–55.

Monk, J and Hanson, S 1982: On not excluding half the human from human geography. *Professional Geographer* **34**, 11–23.
Rowbotham, S 1977: *Hidden from History*. Pluto: London.
Tivers, J 1978: How the other half lives. *Area* **11**, 302–6.
Zelinsky, W, Monk, J and Hanson, S 1982: Women and geography: a review and a prospectus. *Progress in Human Geography* **6**, 357–66.

2 Donna Haraway

'"Gender" for a Marxist Dictionary: the Sexual Politics of a Word'

Excerpts from: *Simians, Cyborgs and Women: the Reinvention of Nature*, pp. 127–48. London: Free Association Books (1991)

In 1983, Nora Räthzel from the autonomous women's collective of the West German independent Marxist journal, *Das Argument*, wrote to ask me to write a 'keyword' entry for a new Marxist dictionary. As Räthzel expressed it, 'We, that is the women's editorial group, are going to suggest some keywords which are missing, and we want some others rewritten because the women do not appear where they should' (personal communication, 2 December 1983). This gentle understatement identified a major arena of feminist struggle – the canonization of language, politics, and historical narratives in publishing practices, including standard reference works.

'The women do not appear where they should.' The ambiguities of the statement were potent and tempting. Here was an opportunity to participate in producing a reference text. I had up to five typed pages for my assignment: sex/gender. Foolhardy, I wrote to accept the task.

There was an immediate problem: I am anglophone, with variously workable but troubled German, French, and Spanish. This crippled language accomplishment reflects my political location in a social world distorted by US hegemonic projects and the culpable ignorance of white, especially, US citizens. English, especially American English, distinguishes between sex and gender. That distinction has cost blood in struggle in many social arenas, as the reader will see in the discussion that follows. German has a single word, *Geschlecht*, which is not really the same as either the English *sex* or *gender*. Further, the dictionary project, translating foreign contributors' entries into German, proposed to give each keyword in German, Chinese (both ideogram and transcription), English, French, Russian (in transcription only), and Spanish. The commingled histories of Marxism and of imperialism loomed large in that list. Each keyword would inherit those histories.

At least I knew that what was happening to *sex* and *gender* in English was not the same as what was going on with *género*, *genre*, and *Geschlecht*. The specific histories of women's movement in the vast global areas where these languages were part of living politics were principal reasons for the differences. The old hegemonic grammarians – including the sexologists – had lost control of gender and its proliferating siblings. Europe and North America could not begin to discipline the twentieth-century fate of its imperializing languages. However, I did not have a clue what to make of my sex/gender problem in Russian or Chinese. Progressively, it became clear to me that I had rather few clues what to make of sex/gender in *English*, even in the United States, much less in the anglophone world. There are so many Englishes in the United States alone, and all of them suddenly seemed germane to this promised five-page text for a German Marxist dictionary. My English was marked by race, generation, gender (!), region, class, education, and political history. How could *that* English be my matrix for sex/gender *in general*? Was there any such thing, even as words, much less as anything else, as 'sex/gender in general'? Obviously not.

. . . I was moderately cheered to learn that the editorial group really wanted an entry on the sex/gender *system*. That helped; there was a specific textual locus for the first use of the term – Gayle Rubin's (1975) stunning essay written when she was a graduate student at the University of Michigan, 'The traffic in women: notes on the political economy of sex'. I could just trace the fate of the 'sex/gender system' in the explosion of socialist and Marxist feminist writing indebted to Rubin. That thought provided very brief consolation. First, the editors directed that each keyword had to locate itself in relation to the corpus of Marx and Engels, whether or not they used the precise words. I think Marx would have been amused at the dead hand guiding the living cursor on the video display terminal. Second, those who adopted Rubin's formulation did so out of many histories, including academic and political interests. US white socialist feminists generated the most obvious body of writing for tracing the 'sex/gender system' narrowly considered. That fact itself was a complex problem, not a solution. Much of the most provocative feminist theory in the last twenty years has insisted on the ties of sex and *race* in ways that problematized the birth pangs of the sex/gender system in a discourse more focused on the interweaving of gender and *class*.[1] It has seemed very rare for feminist theory to hold race, sex/gender, and class analytically together – all the best intentions, hues of authors, and remarks in prefaces notwithstanding. In addition, there is as much reason for feminists to argue for a race/gender system as for a sex/gender system, and the two are not the same *kind* of analytical move. And, again, what happened to class? The evidence is building of a need for a theory of 'difference' whose geometries, paradigms, and logics break out of binaries, dialectics, and nature/culture models of any kind. Otherwise, threes will always reduce to twos, which quickly become lonely ones in the vanguard. And no one learns to count to four. These things matter politically.

Also, even though Marx and Engels – or Gayle Rubin, for that matter – had not ventured into sexology, medicine, or biology for their discussions of sex/gender or the woman question, I knew I would have to do so. At the same time, it was clear that other BIG currents of modern feminist writing on sex, sexuality, and gender interlaced constantly with even the most modest interpretation of my assignment. *Most* of those, perhaps especially the French and British feminist psychoanalytic and literary currents, do not appear in my entry on *Geschlecht*. In general, the entry below focuses on writing by US feminists. That is not a trivial scandal.[2]

So, what follows shows the odd jumps of continual reconstructions over six years.

Keyword

Gender (English), Geschlecht (German), Genre (French), Género (Spanish)

The root of the English, French, and Spanish words is the Latin verb, *generare*, to beget, and the Latin stem *gener-*, race or kind. An obsolete English meaning of 'to gender' is 'to copulate' (*Oxford English Dictionary*). The substantives 'Geschlecht', 'gender', 'genre', and 'género' refer to the notion of sort, kind, and class. In English, 'gender' has been used in this 'generic' sense continuously since at least the fourteenth century. In French, German, Spanish, and English, words for 'gender' refer to grammatical and literary categories. The modern English and German words, 'gender' and 'Geschlecht', adhere closely to concepts of sex, sexuality, sexual difference, generation, engendering, and so on, while the French and Spanish seem not to carry those meanings as readily. Words close to 'gender' are implicated in concepts of kinship, race, biological taxonomy, language, and nationality. The substantive 'Geschlecht' carries the meanings of sex, stock, race, and family, while the adjectival form 'geschlechtlich' means in English translation both sexual and generic. Gender' is at the heart of constructions and classifications of systems of difference. Complex differentiation and merging of terms for 'sex' and 'gender' are part of the political history of the words. Medical meanings related to 'sex' accrue to 'gender' in English progressively through the twentieth century. Medical, zoological, grammatical, and literary meanings have all been contested in modern feminisms. The shared categorical racial and sexual meanings of gender point to the interwoven modern histories of colonial, racist, and sexual oppressions in systems of bodily production and inscription and their consequent liberatory and oppositional discourses. The difficulty of accommodating racial and sexual oppressions in Marxist theories of class is paralleled in the history of the words themselves. This background is essential to understanding the resonances of the theoretical concept of the 'sex-gender system' constructed by Western anglophone feminists in the 1970s.[3] In all their versions, feminist gender theories attempt to articulate the specificity of the oppressions of women in the context of cultures which make a distinction between sex and gender salient. That salience depends on a related system of meanings clustered around a family of binary pairs: nature/culture, nature/history, natural/human, resource/product. This interdependence on a key Western political-philosophical field of binary oppositions – whether understood functionally, dialectically, structurally, or psychoanalytically – problematizes claims to the universal applicability of the concepts around sex and gender; this issue is part of the debate about the cross-cultural relevance of Euro-American versions of feminist theory (Strathern, 1988).

History

Articulation of the problem area in the writings of Marx and Engels. In a critical, political sense, the concept of gender was articulated and progressively contested and theorized in the context of the post-Second World War, feminist women's movements. The modern feminist concept for gender is not found in the writings of Marx and Engels, although their writings and other practice, and those of others in the Marxist tradition, have provided crucial tools for, as well as barriers against, the later politicization and theorization of gender. Despite important differences, all the modern feminist meanings of gender have roots in Simone de Beauvoir's claim that 'one is not born a woman' (de Beauvoir, 1949; 1952, p. 249) and in post-Second World War social conditions that have enabled constructions of women as a collective historical subject-in-process. Gender is a concept developed to contest the naturalization of sexual difference in multiple arenas of struggle. Feminist theory and practice around gender seek to explain and change historical systems of sexual difference, whereby 'men' and 'women' are socially constituted and positioned in relations of hierarchy and antagonism. Since the concept of gender is so closely related to the Western distinction between nature and society or nature and history, via the distinction between sex and gender, the relation of feminist gender theories to Marxism is tied to the fate of the concepts of nature and labour in the Marxist canon and in Western philosophy more broadly.

Traditional Marxist approaches did not lead to a political concept of gender for two major reasons: first, women, as well as 'tribal' peoples, existed unstably at the boundary of the natural and social in the seminal writings of Marx and Engels, such that their efforts to account for the subordinate position of women were undercut by the category of the natural sexual division of labour, with its ground in an unexaminable natural heterosexuality; and second, Marx and Engels theorized the economic property relation as the ground of the oppression of women in marriage, such that women's subordination could be examined in terms of the capitalist relations of class, but not in terms of a specific sexual politics between men and women. The classical location of this argument is Engels' *The Origins of the Family, Private Property and the State* (1884). Engels' analytic priority of the family as a mediating formation between classes and the state 'subsumed any separate consideration of the division of the sexes as an antagonistic division' (Coward, 1983, p. 160).[4] Despite their insistence on the historical variability of family forms and the importance of the question of the subordination of women, Marx and Engels could not historicize sex and gender from a base of natural heterosexuality.

The German Ideology (Part I, Theses on Feuerbach) is the major locus for Marx and Engels' naturalization of the sexual division of labour, in their assumption of a pre-social division of labour in the sex act (heterosexual intercourse), its supposed natural corollaries in the reproductive activities of men and women in the family, and the consequent inability to place women in their relations to men unambiguously on the side of history and of the fully social. In *The Economic and Philosophic Manuscripts of 1844*, Marx refers to the relation of man and woman as the 'most natural relation of human being to human being' (Marx, 1964b, p. 134). This assumption persists in volume one of *Capital* (Marx, 1964a, p. 351). This inability fully to historicize women's labour is paradoxical in view of the purpose of *The German Ideology* and subsequent work to place the family centrally in history as the place where social divisions arise. The root difficulty was an inability to historicize sex itself; like nature, sex functioned analytically as a prime matter or raw material for the work of history. Relying on Marx's research on ethnographic writings (1972), Engels' *Origins* (1884) systematized Marx's views about the linked transitions of family, forms of property, the organization of the division of labour, and the state. Engels almost laid a basis for theorizing the specific oppressions of women in his brief assertion that a fully materialist analysis of the production and reproduction of immediate life reveals a twofold character: the production of the means of existence and '*the production of human beings themselves*' (1884; 1972, p. 71). An exploration of this latter character has been the starting point for many Euro-American Marxist-feminists in their theories of the sex/gender division of labour.[5]

The 'woman question' was widely debated in the many European Marxist parties in the late nineteenth and early twentieth centuries. In the context of the German Social Democratic Party the other of the two most influential Marxist treatments of the position of women was written, August Bebel's *Woman under Socialism* (1883; orig. *Women in the Past, Present and Future*, 1878). Alexandra Kollontai drew on Bebel in her struggles for women's emancipation in Russia and the Soviet Union; and within German social democracy, Clara Zetkin, a leader of the International Socialist Women's Movement, developed Bebel's position in her 1889 'The Question of Women Workers and Women at the Present Time'.[6]

Current problematic

The gender identity paradigm. The story of the political reformulations of gender by post-1960s Western feminists must pass through the construction of meanings and technologies of sex and gender in normalizing, liberal, interventionist-therapeutic, empiricist, and functionalist life sciences, principally in the United States, including psychology, psychoanalysis, medicine, biology, and sociology. Gender was located firmly in an individualist problematic within the broad 'incitement to discourse' (Foucault, 1976) on sexuality characteristic of bourgeois, male-dominant, and racist society. The concepts and technologies of 'gender identity' were crafted from several components: an instinctualist reading of Freud; the focus on sexual somatic- and psychopathology by the great nineteenth-century sexologists (Krafft-Ebing, Havelock Ellis) and their followers; the ongoing development of biochemical and physiological endocrinology from the 1920s; the psychobiology of sex differences growing out of comparative psychology; proliferating hypotheses of hormonal, chromosomal, and neural sexual dimorphism converging in the 1950s; and the first gender reassignment surgeries around 1960 (Linden, 1981). 'Second-wave' feminist politics around 'biological determinism' *vs.* 'social constructionism' and the biopolitics of sex/gender differences occur within discursive fields pre-structured by the gender identity paradigm crystallized in the 1950s and 60s. The gender identity paradigm was a functionalist and essentializing version of Simone de Beauvoir's 1940s insight that one is not born a woman. Significantly, the construction of what could count as a woman (or a man) became a problem for bourgeois functionalists and pre-feminist existentialists in the same historical post-war period in which the social foundations of women's lives in a world capitalist, male-dominant system were undergoing basic reformulations.

. . . The version of the nature/culture distinction in the gender identity paradigm was part of a broad liberal reformulation of life and social sciences in the post-Second World War, Western, professional and governing élites' divestment of pre-war renditions of biological racism. These reformulations failed to interrogate the political-social history of binary categories like nature/culture, and so sex/gender, in colonialist Western discourse. This discourse structured the world as an object of knowledge in terms of the appropriation by culture of the resources of nature. Many recent oppositional, liberatory literatures have criticized this ethnocentric epistemological and linguistic dimension of the domination of those inhabiting 'natural' categories or living at the mediating boundaries of the binarisms (women, people of colour, animals, the non-human environment) (Harding, 1986, pp. 163–96; Fee, 1986). Second-wave feminists early criticized the binary logics of the nature/culture pair, including dialectical versions of the Marxist-humanist story of the domination, appropriation, or mediation of 'nature' by 'man' through 'labour'. But these efforts hesitated to extend their criticism fully to the derivative sex/gender distinction. That distinction was too useful in combating the pervasive biological determinisms constantly deployed against feminists in urgent 'sex differences' political struggles in schools, publishing houses, clinics, and so on. Fatally, in this constrained political climate, these early critiques did not focus on historicizing and culturally relativizing the 'passive' categories of sex or nature. Thus, formulations of an essential identity as a woman or a man were left analytically untouched and politically dangerous.

In the political and epistemological effort to remove women from the category of nature and to place them in culture as constructed and self-constructing social subjects in history, the concept of gender has tended to be quarantined from the infections of biological sex. Consequently, the ongoing constructions of what counts as sex or as female have been hard to theorize, except as 'bad science' where the female emerges as naturally subordinate. 'Biology' has tended to denote the body itself, rather than a social discourse open to intervention. Thus, feminists have argued against 'biological determinism' and for 'social constructionism' and in the process have been less powerful in deconstructing how bodies, including sexualized and racialized bodies, appear as objects of knowledge and sites of intervention in 'biology'. Alternatively, feminists have sometimes affirmed the categories of nature and the body as sites of resistance to the dominations of history, but the affirmations have tended to obscure the *categorical* and overdetermined aspect of 'nature' or the 'female body' as an oppositional ideological resource. Instead, nature has seemed simply there, a reserve to be preserved from the violations of civilization in general. Rather than marking a categorically determined pole, 'nature' or 'woman's body' too easily then means the saving core of reality distinguishable from the social impositions of patriarchy, imperialism, capitalism, racism, history, language. That repression of the *construction* of the category 'nature' can be and has been both used by and used against feminist efforts to theorize women's agency and status as social subjects.

Judith Butler (1989) argued that gender identity discourse is intrinsic to the fictions of heterosexual coherence, and that feminists need to learn to produce narrative legitimacy for a whole array of non-coherent genders. Gender identity discourse is also intrinsic to feminist racism, which insists on the non-reducibility and antagonistic relation of coherent women and men. The task is to 'disqualify' the analytic categories, like sex or nature, that lead to univocity. This move would expose the illusion of an interior organizing gender core and produce a field of race and gender difference open to resignification. Many feminists have resisted moves like those Butler recommends, for fear of losing a concept of agency for women as the concept of the subject withers under the attack on core identities and their constitutive fictions. Butler, however, argued that agency is an instituted practice in a field of enabling constraints. A concept of a coherent inner self, achieved (cultural) or innate (biological), is a regulatory fiction that is unnecessary – indeed, inhibitory – for feminist projects of producing and affirming complex agency and responsibility.

A related 'regulatory fiction' basic to Western concepts of gender insists that motherhood is natural and fatherhood is cultural: mothers make babies naturally, biologically. Motherhood is known on sight; fatherhood is inferred. Analysing gender concepts and practices among Melanesians, Strathern (1988, pp. 311–39) went to great pains to show both the ethnocentric quality of the self-evident Western assertion that 'women make babies' and the inferential character of *all* vision. She showed the productionist core of the belief that women make babies (and its pair, that man makes himself), which is intrinsic to Western formulations of sex and gender. Strathern argued that Hagen men and women do not exist in permanent states as subjects and objects within Aristotelian, Hegelian, Marxist, or Freudian frames. Hagen agency has a different dynamic and geometry. For Westerners, it is a central consequence of concepts of gender difference that a person may be turned by another person into an object and robbed of her or his status as subject. The proper state for a Western person is to have ownership of the self, to have and hold a core identity as if it were a possession. That possession may be made from various raw materials over time, that is, it may be a cultural production, or one may be born with it. Gender identity is such a possession. Not to have property in the self is not to be a subject, and so not to have agency. Agency follows different pathways for the Hagen, who as persons 'are composed of multiple gendered parts, or multiple gendered persons, who are interacting with one another as donors and recipients in maintaining the flow of elements throught the body' (Douglas, 1989, p. 17). Sexist domination between persons can and does systematically occur, but it cannot be traced or addressed by the same analytical moves that would be appropriate for many Western social fields of meaning (Strathern, 1988, pp. 334–9). Butler could – cautiously – use Strathern's ethnographic arguments to illustrate one way to disperse the coherence of gender without losing the power of agency.

So, the ongoing tactical usefulness of the sex/gender distinction in life and social sciences has had dire consequences for much feminist theory, tying it to a liberal and functionalist paradigm despite repeated efforts to transcend those limits in a fully politicized and historicized concept of gender. The failure lay partly in not historicizing and relativizing sex and the historical-epistemological roots of the logic of analysis implied in the sex/gender distinction and in each member of the pair. At this level, the modern feminist limitation in theorizing and struggling for the empirical life and social sciences is similar to Marx and Engels' inability to extricate themselves from the natural sexual division of labour in heterosexuality despite their admirable project of historicizing the family.

Sex/gender differences discourse exploded in US sociological and psychological literature in the 1970s and 80s. The explosion is part of a vigorous political and scientific contestation over the construction of sex and gender, as categories and as emergent historical realities, in which feminist writing becomes prominent about the mid-1970s, primarily in criticisms of 'biological determinism' and of sexist science and technology, especially biology and medicine. Set up within the epistemological binary framework of nature/culture and sex/gender, many feminists (including socialist and Marxist feminists) appropriated the sex/gender distinction and the interactionist paradigm to argue for the primacy of culture-gender over biology-sex in a panoply of debates in Europe and the United States. These debates have ranged from genetic differences in mathematics ability of boys and girls, the presence and significance of sex differences in neural organization, the relevance of animal research to human behaviour, the causes of male dominance in the organization of scientific research, sexist structures and use patterns in language, sociobiology debates, struggles over the meanings of sex chromosomal abnormalities, to the similarities of racism and sexism. By the mid-1980s, a growing suspicion of the category of gender and the binarism sex/gender entered the feminist literature in these debates. That scepticism was partly an outgrowth of challenges to racism in the Euro-American women's movements, such that some of the colonial and racist roots of the framework became clearer.[7]

The sex-gender system. Another stream of feminist sex/gender theory and politics came through appropriations of Marx and Freud read through Lacan and Lévi-Strauss in an influential formulation by Gayle Rubin (1975) of the 'sex-gender system'. Rubin and those indebted to her theorization adopted a version of the nature/culture distinction, but one flowing less out of US empiricist life and social science, and more from French psychoanalysis and structuralism. Rubin examined the 'domestication of women', in which human females were the raw materials for the social production of women, through the exchange systems of kinship controlled by men in the institution of human culture. She defined the sex-gender system as the system of social relations that transformed biological sexuality into products of human activity and in which the resulting historically specific sexual needs are met. She then called for a Marxian analysis of sex/gender systems as products of human activity which are changeable through political struggle. Rubin viewed the sexual division of labour and the psychological construction of desire (especially the oedipal formation) as the foundations of a system of production of human beings vesting men with rights in women which they do not have in themselves. To survive materially where men and women cannot perform the other's work and to satisfy deep structures of desire in the sex/gender system in which men exchange women, heterosexuality is obligatory. Obligatory heterosexuality is therefore central to the oppression of women.

Adrienne Rich (1980) also theorized compulsory heterosexuality to be at the root of the oppression of women. Rich figured 'the lesbian continuum' as a potent metaphor for grounding a new sisterhood. For Rich, marriage resistance in a cross-historical sweep was a defining practice constituting the lesbian continuum. Monique Wittig (1981) developed an independent argument that also foregrounded the centrality of obligatory heterosexuality in the oppression of women. In a formulation which its authors saw as the explanation for the decisive break with traditional Marxism of the Movement pour la Libération des Femmes (MLF) in France, the group associated with Wittig argued that all women belong to a class constituted by the hierarchical social relation of sexual difference that gives men ideological, political and economic power over women (Editors of *Questions féministes*, 1980).[8] What *makes* a woman is a specific relation of appropriation by a man. Like race, sex is an 'imaginary' formation of the kind that produces reality, including bodies then perceived as prior to all construction. 'Woman' only exists as this kind of imaginary being, while women are the product of a social relation of appropriation, naturalized as sex. A feminist is one who fights for women as a class and for the disappearance of that class. The key struggle is for the destruction of the social system of heterosexuality, because 'sex' is the naturalized political category that founds society as heterosexual. All the social sciences based on the category of 'sex' (most of them) must be overthrown. In this view, lesbians are not 'women' because they are outside the political economy of heterosexuality. Lesbian society destroys women as a natural group (Wittig, 1981).

Thus, theorized in three different frames, withdrawal from marriage was central to Rubin's, Rich's, and Wittig's political visions in the 1970s and early 80s. Marriage encapsulated and reproduced the antagonistic relation of the two coherent social groups, men and women. In all three formulations both the binary of nature/culture and the dynamic of productionism enabled the further analysis. Withdrawal of women from the marriage economy was a potent figure and politics for withdrawal from men, and therefore for the self-constitution of women as personal and historical subjects outside the institution of culture by men in the exchange and appropriation of the products (including babies) of women. To be a subject in the Western sense meant reconstituting women outside the relations of objectification (as gift, commodity, object of desire) and appropriation (of babies, sex, services). The category-defining relation of men and women in objectification, exchange, and appropriation, which was the theoretical key to the category 'gender' in major bodies of feminist theory by white women in this period, was one of the moves that made an understanding of the race/gender or race/sex system and the barriers to cross-racial 'sisterhood' hard for white feminists analytically to grasp.

However, these formulations had the powerful virtue of foregrounding and legitimating lesbianism at the heart of feminism. The figure of the lesbian has been repeatedly at the contentious, generative centre of feminist debate (King, 1986). Audre Lorde put the black lesbian at the heart of her understanding of the 'house of difference':

> Being women together was not enough. We were different. Being gay-girls together was not enough. We were different. Being Black together was not enough. We were different. Being Black women together was not enough. We were different. Being Black dykes together was not enough. We were different . . . It was a while before we came to realize that our place was the very house of difference rather than the security of any one particular difference.
>
> (Lorde, 1982, p. 226)

This concept of difference grounded much US multi-cultural feminist theorizing on gender in the late 1980s.

There have been many uses and criticisms of Rubin's sex-gender system. In an article at the centre of much Euro-American Marxist and socialist-feminist debate, Hartmann (1981) insisted that patriarchy was not simply an ideology, as Juliet Mitchell seemed to argue in her seminal 'Women: the Longest Revolution' (1966) and its expansion in *Women's Estate* (1971), but a material system that could be defined 'as a set of social relations between men, which have a material base, and which, though hierarchical, establish or create interdependence and solidarity among men that enable them to dominate women' (Hartmann, 1981, p. 14). Within this frame, Hartmann attempted to explain the partnership of patriarchy and capital and the failure of male-dominated socialist labour movements to prioritize sexism. Hartmann used Rubin's concept of the sex-gender system to call for an understanding of the mode of production of human beings in patriarchal social relations through male control of women's labour power.

In the debate stimulated by Hartmann's thesis, Iris Young (1981) criticized the 'dual systems' approach to capital and patriarchy, which were then allied in the oppressions of class and gender. Note how race, including an interrogation of white racial positioning, remained an unexplored system in these formulations. Young argued that 'patriarchal relations are internally related to production relations as a whole' (1981, p. 49), such that a focus on the gender division of labour could reveal the dynamics of a single system of oppression. In addition to waged labour, the gender division of labour also included the excluded and unhistoricized labour categories in Marx and Engels, that is, bearing and rearing children, caring for the sick, cooking, housework, and sex-work like prostitution, in order to bring gender and women's specific situation to the centre of historical materialist analysis. In this theory, since the gender division of labour was also the first division of labour, one must give an account of the emergence of class society out of changes in the gender division of labour. Such an analysis does not posit that all women have a common, unified situation; but it makes the historically differentiated positions of women central. If capitalism and patriarchy are a single system, called capitalist patriarchy, then the struggle against class and gender oppressions must be unified. The struggle is the obligation of men and women, although autonomous women's organization would remain a practical necessity. This theory is a good example of strongly rationalist modern approaches, for which the 'postmodern' moves of the disaggregation of metaphors of single systems in favour of complex open fields of criss-crossing plays of domination, privilege, and difference appeared very threatening. Young's 1981 work was also a good example of the power of modernist approaches in specific circumstances to provide political direction.

In exploring the epistemological consequences of a feminist historical materialism, Nancy Hartsock (1983a,b) also concentrated on the categories that Marxism had been unable to historicize: (1) women's sensuous labour in the making of human beings through child-bearing and raising; and (2) women's nurturing and subsistence labour of all kinds. But Hartsock rejected the terminology of the *gender* division of labour in favour of the *sexual* division of labour, in order to emphasize the bodily dimensions of women's activity. Hartsock was also critical of Rubin's formulation of the sex-gender system because it emphasized the exchange system of kinship at the expense of a materialist analysis of the labour process that grounded women's potential construction of a revolutionary standpoint. Hartsock relied on versions of Marxist humanism embedded in the story of human self-formation in the sensuous mediations of nature and humanity through labour. In showing how women's lives differed systematically from men's, she aimed to establish the ground for a feminist materialist standpoint, which would be an engaged position and vision, from which the real relations of domination could be unmasked and a liberatory reality struggled for. She called for exploration of the relations between the exchange abstraction and abstract masculinity in the hostile systems of power characterizing phallocratic worlds. Several other Marxist feminists have contributed to intertwined and independent versions of feminist standpoint theory, where the debate on the sex/gender division of labour is a central issue. Fundamental to the debate is a progressive problematization of the *category* labour, or its extensions in Marxist-feminist meanings of reproduction, for efforts to theorize women's active agency and status as subjects in history.[9] Collins (1989a) adapted standpoint theory to characterize the foundations of black feminist thought in the self-defined perspective of black women on their own oppression.

Sandra Harding (1983) took account of the feminist theoretical flowering as a reflection of a heightening of lived contradictions in the sex-gender system, such that fundamental change can now be struggled for. In extending her approach to the sex-gender system to *The Science Question in Feminism* (1986), Harding stressed three variously interrelated elements of gender: (1) a fundamental category through which meaning is ascribed to everything, (2) a way of organizing social relations, and (3) a structure of personal identity. Disaggregating these three elements has been part of coming to understand the complexity and problematic value of politics based on gender identities. Using the sex-gender system to explore post-Second World War politics of sexual identity in gay movements, Jeffrey Escoffier (1985) argued for a need to theorize the emergence and limitations of new forms of political subjectivity, in order to develop a committed, positioned politics without metaphysical identity closures. Haraway's (1985) 'Manifesto for Cyborgs' developed similar arguments in order to explore Marxist-feminist politics addressed to women's positionings in multi-national science- and technology-mediated social, cultural, and technical systems.

In another theoretical development indebted to Marxism, while critical of both it and of the language of gender, Catherine MacKinnon (1982, p. 515) argued that

> Sexuality is to feminism what work is to marxism: that which is most one's own, yet most taken away . . . Sexuality is that social process which creates, organizes, expresses, and directs desire, creating the social beings we know as women and men, as their relations create society . . . As the organized expropriation of the work of some for the benefit of others defines a class – workers – the organized expropriation of the sexuality of some for the use of other defines the sex, woman.

MacKinnon's position has been central to controversial approaches to political action in much of the US movement against pornography, defined as violence against women and/or as a violation of women's civil rights; that is, a refusal to women, via their construction as woman, of the status of citizen. MacKinnon saw the construction of woman as the material and ideological construction of the object of another's desire. Thus women are not simply alienated from the product of their labour; in so far as they exist as 'woman', that is to say, sex objects, they are not even potentially historical subjects. 'For women, there is no distinction between objectification and alienation because women have not authored objectifications, we have been them' (1982, pp. 253–4). The epistemological and political consequences of this position are far reaching and have been extremely controversial. For MacKinnon, the production of women is the production of a very material illusion, 'woman'. Unpacking this material illusion, which is women's lived reality, requires a politics of consciousness-raising, the specific form of feminist politics in MacKinnon's frame. 'Sexuality determines gender', and 'women's sexuality is its use, just as our femaleness *is* its alterity' (p. 243). Like independent formulations in Lacanian feminisms, MacKinnon's position has been fruitful in theorizing processes of representation, in which 'power to create the world from one's point of view is power in its male form' (p. 249).

In an analysis of the gendering of violence sympathetic to MacKinnon's, but drawing on different theoretical and political resources, Teresa de Lauretis's (1984, 1985) approaches to representation led her to view gender as the unexamined tragic flaw of modern and postmodern theories of culture, whose faultline is the heterosexual contract. De Lauretis defined gender as the social construction of 'woman' and 'man' and the semiotic production of subjectivity; gender has to do with 'the history, practices, and imbrication of meaning and experience'; that is, with the 'mutually constitutive effects in semiosis of the outer world of social reality with the inner world of subjectivity' (1984, pp. 158–86). De Lauretis drew on Charles Peirce's theories of semiosis to develop an approach to 'experience', one of the most problematic notions in modern feminism, that takes account both of experience's intimate embodiment and its mediation through signifying practices. Experience is never *im*-mediately accessible. Her efforts have been particularly helpful in understanding and contesting inscriptions of gender in cinema and other areas where the idea that gender is an embodied semiotic difference is crucial and empowering. Differentiating technologies of gender from Foucault's formulation of technologies of sex, de Lauretis identified a specific feminist gendered subject position within sex/gender systems. Her formulation echoed Lorde's understanding of the inhabitant of the house of difference: 'The female subject of feminism is one constructed across a multiplicity of discourses, positions, and meanings, which are often in conflict with one another and inherently (historically) contradictory' (de Lauretis, 1987, pp. ix–x).

Offering a very different theory of consciousness and the production of meanings from MacKinnon or de Lauretis, Hartsock's (1983a) exploration of the sexual division of labour drew on anglophone versions of psychoanalysis that were particularly important in US feminist theory, that is, object relations theory as developed especially by Nancy Chodorow (1978). Without adopting Rubin's Lacanian theories of always fragmentary sexed subjectivity, Chodorow adopted the concept of the sex-gender system in her study of the social organization of parenting, which produced women more capable of non-hostile relationality than men, but which also perpetuated the subordinate position of women through their production as people who are structured for mothering in patriarchy. Preferring an object relations psychoanalysis over a Lacanian version is related to neighbouring concepts like 'gender identity', with its empirical social science web of meanings, over 'acquisition of positions of sexed subjectivity', with this concept's immersion in Continental cultural/textual theory. Although criticized as an essentializing of woman-as-relational, Chodorow's feminist object relations theory has been immensely influential, having been adapted to explore a wide range of social phenomena. Drawing on and criticizing Lawrence Kohlberg's neo-Kantian theories, Gilligan (1982) also argued for women's greater contextual consciousness and resistance to universalizing abstractions, for example in moral reasoning.

Chodorow's early work was developed in the context of a related series of sociological and anthropological papers theorizing a key role for the public/private division in the subordination of women (Rosaldo and Lamphere, 1974). In that collection, Rosaldo argued the universal salience of the limitation of women to the domestic realm, while power was vested in the space men inhabit, called public. Sherry Ortner connected that approach to her structuralist analysis of the proposition that women are to nature as men are to culture. Many Euro-American feminist efforts to articulate the social positioning of women that followed *Woman, Culture, and Society* and *Toward an Anthropology of Women* (Reiter, 1975), both strategically published in the mid-1970s, were deeply influenced by the universalizing and powerful theories of sex and gender of those early collections. In anthropology as a discipline, criticisms and other outgrowths of the early formulations were rich, leading both to extensive cross-cultural study of gender symbolisms and to fundamental rejection of the universal applicability of the nature/culture pair. Within the disciplines, there was growing criticism of universalizing explanations as an instance of mistaking the analytical tool for the reality (MacCormack and Strathern, 1980; Rosaldo, 1980; Ortner and Whitehead 1981; Rubin, 1984). As feminist anthropology moved away from its early formulations, they none the less persisted in much feminist discourse outside anthropological disciplinary circles, as if the mid-1970s positions were permanently authoritative feminist anthropological theory, rather than a discursive node in a specific political-historical-disciplinary moment.

The universalizing power of the sex-gender system and the analytical split between public and private were also sharply criticized politically, especially by women of colour, as part of the ethnocentric and imperializing tendencies of European and Euro-American feminisms. The category of gender obscured or subordinated all the other 'others'. Efforts to use Western or 'white' concepts of gender to characterize a 'Third World Woman' often resulted in reproducing orientalist, racist, and colonialist discourse (Mohanty, 1984; Amos *et al.*, 1984). Furthermore, US 'women of colour', itself a complex and contested political construction of sexed identities, produced critical theory about the production of systems of hierarchical differences, in which race, nationality, sex, and class were intertwined, both in the nineteenth and early twentieth centuries and from the earliest days of the women's movements that emerged from the 1960s civil rights and anti-war movements.[10] These theories of the social positioning of women ground and organize 'generic' feminist theory, in which concepts like 'the house of difference' (Lorde) 'oppositional consciousness' (Sandoval), 'womanism' (Walker), 'shuttle from center to margin' (Spivak), 'Third World feminism' (Moraga and Smith), 'el mundo zurdo' (Anzaldúa and Moraga), 'la mestiza' (Anzaldúa), 'racially-structured patriarchal capitalism' (Bhavnani and Coulson, 1986), and 'inappropriate/d other' (Trinh, 1986–7, 1989) structure the field of feminist discourse, as it decodes what counts as a 'woman' within as well as outside 'feminism'. Complexly related figures have emerged also in feminist writing by 'white' women: 'sex-political classes' (Sofoulis, 1987); 'cyborg' (Haraway, 1985); the female subject of feminism (de Lauretis, 1987).

Rubin's 1975 theory of the sex/gender system explained the complementarity of the sexes (obligatory heterosexuality) and the oppression of women by men through the central premise of the exchange of women in the founding of culture through kinship. But what happens to this approach when women are not positioned in similar ways in the institution of kinship? In particular, what happens to the idea of gender if whole groups of women and men are positioned *outside* the *institution* of kinship altogether, but in relation to the kinship systems of another, dominant group? Carby (1987), Spillers (1987), and Hurtado (1989) interrogated the concept of gender through an exploration of the history and consequences of these matters.

Carby clarified how in the New World, and specifically in the United States, black women were not constituted as 'woman', as white women were. Instead, black women were constituted simultaneously racially and sexually – as marked female (animal, sexualized and without rights), but not as woman (human, potential wife, conduit for the name of the father) – in a specific institution, slavery, that excluded them from 'culture' defined as the circulation of signs through the system of marriage. If kinship vested men with rights in women that they did not have in themselves, slavery abolished kinship for one group in a legal discourse that produced whole groups of people as alienable property (Spillers, 1987). MacKinnon (1982, 1987) defined woman as an imaginary figure, the object of another's desire, made real. The 'imaginary' figures made real in slave discourse were objects in another sense that made them different from either the Marxist figure of the alienated labourer or the 'unmodified' feminist figure of the object of desire. Free women in US white patriarchy were exchanged in a system that oppressed them, but white women *inherited* black women and men. As Hurtado (1989, p. 841) noted, in the nineteenth century prominent white feminists were *married* to white men, while black feminists were *owned* by white men. In a racist patriarchy, white men's 'need' for racially pure offspring positioned free and unfree women in incompatible, asymmetrical symbolic and social spaces.

The female slave was marked with these differences in a most literal fashion – the flesh was turned inside out, 'add[ing] a lexical dimension to the narratives of woman in culture and society' (Spillers, 1987, pp. 67–8). These differences did not end with formal emancipation; they have had definitive consequences into the late twentieth century and will continue to do so until racism as a founding institution of the New World is ended. Spillers called these founding relations of captivity and literal mutilation 'an American grammar' (p. 68). Under conditions of the New World conquest, of slavery, and of their consequences up to the present, 'the lexis of reproduction, desire, naming, mothering, fathering, etc. [are] all thrown into extreme crisis' (p. 76). 'Gendering, in its coeval reference to African-American women, *insinuates* an implicit and unresolved puzzle both within current feminist discourse *and* within those discursive communities that investigate the problematics of culture' (p. 78).

Spillers foregrounded the point that free men and women inherited their *name* from the father, who in turn had rights in his minor children and wife that they did not have in themselves, but he did not own them in the full sense of alienable property. Unfree men and women inherited their *condition* from their mother, who in turn specifically did not control their children. They had no *name* in the sense theorized by Lévi-Strauss or Lacan. Slave mothers could not transmit a name; they could not be wives; they were outside the system of marriage exchange. Slaves were unpositioned, unfixed, in a system of names; they were, specifically, unlocated and so disposable. In these discursive frames, white women were not legally or symbolically *fully* human; slaves were not legally or symbolically human *at all*. 'In this absence from a subject position, the captured sexualities provide a physical and biological expression of "otherness"' (Spillers, 1987, p. 67). To give birth (unfreely) to the heirs of property is not the same thing as to give birth (unfreely) to property (Carby, 1987, p. 53).

This little difference is part of the reason that 'reproductive rights' for women of colour in the US prominently hinge on comprehensive control of children – for example, their freedom from destruction through lynching, imprisonment, infant mortality, forced pregnancy, coercive sterilization, inadequate housing, racist education, or drug addiction (Hurtado, 1989, p. 853). For white women the concept of property in the self, the ownership of one's own body, in relation to reproductive freedom has more readily focused on the field of events around conception, pregnancy, abortion, and birth, because the system of white patriarchy turned on the control of legitimate children and the consequent constitution of white females as woman. To have or not have children then becomes literally a subject-defining choice for women. Black women specifically – and the women subjected to the conquest of the New World in general – faced a broader social field of reproductive unfreedom, in which their children did not inherit the status of human in the founding hegemonic discourses of US society. The problem of the black mother in this context is not simply her own status as subject, but also the status of her children and her sexual partners, male and female. Small wonder that the image of uplifting the race and the refusal of the categorical separation of men and women – without flinching from an analysis of coloured and white sexist oppression – have been prominent in New World black feminist discourse (Carby, 1987, pp. 6–7; hooks, 1981, 1984).

The positionings of African-American women are not the same as those of other women of colour; each condition of oppression requires specific analysis that refuses the separations but insists on the non-identities of race, sex, and class. These matters make starkly clear why an adequate feminist theory of gender must *simultaneously* be a theory of racial difference in specific historical conditions of production and reproduction. They also make clear why a theory and practice of sisterhood cannot be grounded in shared positionings in a system of sexual difference and the cross-cultural structural antagonism between coherent categories called women and men. Finally, they make clear why feminist theory produced by women of colour has constructed alternative discourses of womanhood that disrupt the humanisms of many Western discursive traditions.

While contributing fundamentally to the breakup of any master subject location, the politics of 'difference' emerging from this and other complex reconstructings of concepts of social subjectivity and their associated writing practices is deeply opposed to levelling relativisms. Non-feminist theory in the human sciences has tended to identify the breakup of 'coherent' or masterful subjectivity as the 'death of the subject'. Like others in newly *unstably* subjugated positions, many feminists resist this formulation of the project and question its emergence at just the moment when raced/sexed/colonized speakers begin 'for the first time', that is, they claim an originary authority to represent themselves in institutionalized publishing practices and other kinds of self-constituting practice. Feminist deconstructions of the 'subject' have been fundamental, and they are not nostalgic for masterful coherence. Instead, necessarily political accounts of constructed embodiments, like feminist theories of gendered racial subjectivities, have to take affirmative *and* critical account of emergent, differentiating, self-representing, contradictory social subjectivities, with their claims on action, knowledge, and belief. The point involves the commitment to transformative social change, the moment of hope embedded in feminist theories of gender and other emergent discourses about the breakup of masterful subjectivity and the emergence of inappropriate/d others (Trinh, 1986–7, 1989).

The multiple academic and other institutional roots of the literal (written) category 'gender', feminist and otherwise, sketched in this entry have been part of the race-hierarchical system of relations that obscures the publications by women of colour because of their origin, language, genre – in short, 'marginality', 'alterity', and 'difference' as seen from the 'unmarked' positions of hegemonic and imperializing ('white') theory. But 'alterity' and 'difference' are precisely what 'gender' is 'grammatically' about, a fact that constitutes feminism as a politics defined by its fields of contestation and repeated refusals of master theories. 'Gender' was developed as a category to explore what counts as a 'woman', to problematize the previously taken-for-granted. If feminist theories of gender followed from Simone de Beauvoir's thesis that one is not born a woman, with all the consequences of that insight, in the light of Marxism and psychoanalysis, for understanding that any finally coherent subject is a fantasy, and that personal and collective identity is precariously and constantly socially reconstituted (Coward, 1983, p. 265), then the title of bell hooks's provocative book, echoing the great nineteenth-century black feminist and abolitionist, Sojourner Truth, *Ain't I a Woman* (1981), bristles with irony, as the identity of 'woman' is both claimed and deconstructed simultaneously. Struggle over the agents, memories, and terms of these reconstitutions is at the heart of feminist sex/gender politics.

The refusal to become or to remain a 'gendered' man or a woman, then, is an eminently political insistence on emerging from the nightmare of the all-too-real, imaginary narrative of sex and race. Finally and ironically, the political and explanatory power of the 'social' category of gender depends upon historicizing the categories of sex, flesh, body, biology, race, and nature in such a way that the binary, universalizing opposition that spawned the concept of the sex/gender system at a particular time and place in feminist theory implodes into articulated, differentiated, accountable, located, and consequential theories of embodiment, where nature is no longer imagined and enacted as resource to culture or sex to gender. Here is my location for a utopian intersection of heterogeneous, multi-cultural, 'Western' (coloured, white, European, American, Asian, African, Pacific) feminist theories of gender hatched in odd siblingship with contradictory, hostile, fruitful, inherited binary dualisms. Phallogocentrism was the egg ovulated by the master subject, the brooding hen to the permanent chickens of history. But into the nest with that literal-minded egg has been placed the germ of a phoenix that will speak in all the tongues of a world turned upside down.

Notes

1 A curious linguistic point shows itself here: there is no marker to distinguish (biological) race and (cultural) race, as there is for (biological) sex and (cultural) gender, even though the nature/culture and biology/society binarisms pervade Western race discourse. The linguistic situation highlights the very recent and uneven entry of gender into the political, as opposed to the grammatical, lexicon. The non-naturalness of race – it is always and totally an arbitrary, cultural construction – can be emphasized from the lack of a linguistic marker. But, as easily, the total collapse of the category of race into biologism is linguistically invited. All these matters continue to hinge on unexamined functioning of the productionist, Aristotelian logic fundamental to so much Western discourse. In this linguistic, political, and historical matrix, matter and form, act and potency, raw material and achieved product play out their escalating dramas of production and appropriation. Here is where subjects and objects get born and endlessly reincarnated.

2 Although not mutually exclusive, the language of 'gender' in Euro-American feminist discourse usually is the language of 'sexed subject position' and 'sexual difference' in European writing. For British Marxist feminism on the 'sexed subject in patriarchy', see Kuhn and Wolpe (1978), Marxist-Feminist Literature Collective (1978), Brown and Adams (1979), the journal *m/f*; Barrett (1980). German socialist-feminist positions on sexualization have stressed the dialectic of women's self-constructing agency, already structured social determinations, and partial restructurings. This literature examines how women construct themselves into existing structures, in order to find the point where change might be possible. If women are theorized as passive victims of sex and gender as a system of domination, no theory of liberation will be possible. So social constructionism on the question of gender must not be allowed to become a theory of closed determinism (Haug, 1980, 1982; Haug *et al.* 1983, 1987; Mouffe, 1983). Looking for a theory of experience, of how women actively embody themselves, the women in the collective writing the *Frauenformen* pubications insisted on a descriptive/theoretical practice showing 'the ways we live ourselves in bodily terms' (Haug *et al.*, 1987, p. 30). They evolved a method called 'memory work' that emphasizes collectively criticized, written narratives about 'a stranger', a past 'remembered' self, while problematizing the self-deluding assumptions of autobiography and other causal accounts. The problem is to account for the emergence of 'the sexual itself as the process that produces the insertion of women into, and their subordination within, determinate social practices' (p. 33). Ironically, self-constituted as sexualized, as woman, women cannot be accountable for themselves or society (p. 27). Like all the theories of sex, sexuality, and gender surveyed in this effort to write for a standard reference work that inevitably functions to canonize some meanings over others, the *Frauenformen* versions insist on gender as a gerund or a verb, rather than a finished noun, a

substantive. For feminists, gender means making and unmaking 'bodies' in a contestable world; an account of gender is a theory of experience as signifying and significant embodiment.

3 Joan Scott (1988, pp. 28–50) wrote an incisive treatment of the development of gender as a theoretical category in the discipline of history. She noted the long history of play on the grammatical gender difference for making figurative allusions to sex or character (p. 28). Scott quoted as her epigram *Fowler's Dictionary of Modern English Usage*'s insistence that to use gender to mean the male or female sex was either a mistake or a joke. The ironies in this injunction abound. One benefit of the inheritance of feminist uses of gender from grammar is that, in that domain, 'gender is understood to be a way of classifying phenomena, a socially agreed-upon system of distinctions, rather than an objective description of inherent traits' (p. 29).

4 See Coward (1983, chs 5 and 6) for a thorough discussion of the concepts of the family and the woman question in Marxist thought from 1848 to about 1930.

5 Rubin (1975), Young and Levidow (1981), Harding (1983, 1986), Hartsock (1983 a, b), Hartmann (1981), O'Brien (1981), Chodorow (1978), Jaggar (1983).

6 See *The Woman Question* (1951); Marx and Aveling (1885–6); Kollantai (1977).

7 To sample the uses and criticisms, see Sayers (1982), Hubbard *et al.* (1982), Bleier (1984, 1986), Fausto-Sterling (1985), Kessler and McKenna (1978), Thorne and Henley (1975), West and Zimmermann (1987), Morawski (1987), Brighton Women and Science Group (1980), Lowe and Hubbard (1983), Lewontin *et al.* (1984).

8 Several streams of European feminisms (some disavowing the name) were born after the events of May '68. The stream drawing from Simone de Beauvoir's formulations, especially work by Monique Wittig, Monique Plaza, Colette Guillaumin, and Christine Delphy, published in *Questions féministes, Nouvelles questions féministes*, and *Feminist Issues*, and the stream associated complexly with the group 'Psychanalyse et Politique' and/or with Julia Kristeva, Luce Irigaray, Sarah Kofman, and Hélène Cixous have been particularly influential in international feminist development on issues of sexual difference. (For introductory summaries, see Marks and de Courtivron, 1980; Gallop, 1982; Moi, 1985; Duchen, 1986). These streams deserve large, separate treatments; but in the context of this entry two contributions to theories of 'gender' from these writers, who are deeply opposed among themselves on precisely these issues, must be signalled. First, there are Wittig's and Delphy's arguments for a materialist feminism, which insist that the issue is 'domination', not 'difference'. Second, there are Irigaray's, Kristeva's, and Cixous's various ways (intertextually positioned in relation to Derrida, Lacan and others) of insisting that the subject, which is perhaps best approached through writing and textuality, is always in process, always disrupted, that the idea of woman remains finally unclosed and multiple. Despite their important opposition between and within the francophone streams, all these theorists are possessed with flawed, contradictory, and critical projects of denaturalization of 'woman'.

9 Smith (1974), Flax (1983), O'Brien (1981), Rose, H. (1983, 1986), Harding (1983).

10 See, for example, Ware (1970); Combahee River Collective (1979); Bethel and Smith (1979); Joseph and Lewis (1981); hooks (1981, 1984); Moraga and Anzaldúa (1981); Davis (1982); Hull *et al.* (1982); Lorde (1982, 1984); Aptheker (1982); Moraga (1983); Walker (1983); Smith (1983); Bulkin *et al.* (1984); Sandoval (n.d.); Christian (1985); Giddings (1985); Anzaldúa (1987); Carby (1987); Spillers (1987); Collins (1989, 1989b); Hurtado (1989).

References

Anzaldúa, Gloria (1987) *Borderlands/La Frontera*. San Francisco: Spinsters/Aunt Lute.

Aptheker, Betina (1982) *Woman's Legacy: Essays on Race, Sex, and Class in American History*. Amherst: University of Massachusetts Press.

Barrett, Michèle (1980) *Women's Oppression Today*. London: Verso.

Bethel, Lorraine and Smith, Barbara, eds (1979) *The Black Women's Issue, Conditions* 5.

Bleier, Ruth (1984) *Science and Gender: A Critique of Biology and Its Themes on Women*. New York: Pergamon.

——, ed. (1986) *Feminist Approaches to Science*. New York: Pergamon.

Brighton Women and Science Group (1980) *Alice through the Microscope*. London: Virago.

Brown, Beverley and Adams, Parveen (1979) 'The feminine body and feminist politics', *m/f* 3: 35–57.

Bulkin, Elly, Pratt, Minnie Bruce, and Smith, Barbara (1984) *Yours in Struggle: Three Feminist Perspectives on Racism and Anti-Semitism*. New York: Long Haul.

Carby, Hazel (1987) *Reconstructing Womanhood: The Emergence of the Afro-American Woman Novelist*. New York: Oxford University Press.

Chodorow, Nancy (1978) *The Reproduction of Mothering: Psychoanalysis and the Sociology of Gender*. Los Angeles: University of California Press.

Christian, Barbara (1985) *Black Feminist Criticism: Perspectives on Black Women Writers*. New York: Pergamon.

Collins, Patricia Hill (1989) 'The social construction of Black feminist thought', *Signs* 14(4): 745–73.

—— (1989a) 'A comparison of two works on Black family life', *Signs* 14(4): 875–84.

Combaheee River Collective (1979) 'A Black feminist statement', in Zillah Eisenstein, ed. *Capitalist Patriarchy and the Case for Socialist Feminism*. New York: Monthly Review.

Coward, Rosalind (1983) *Patriarchal Precedents: Sexuality and Social Relations*. London: Routledge & Kegan Paul.

Davis, Angela (1982) *Women, Race, and Class*. London: Women's Press.

Duchen, Claire (1986) *Feminism in France from May '68 to Mitterrand*. London: Routledge & Kegan Paul.

Fausto-Sterling, Anne (1985) *Myths of Gender: Biological Theories about Women and Men*. New York: Basic.

Flax, Jane (1983) 'Political philosophy and the patriarchal unconscious: a psychoanalytic perspective on epistemology and metaphysics', in Harding and Hintikka eds pp. 245–82.

Frankenberg, Ruth (1988) 'The social construction of whiteness', University of California at Santa Cruz, PhD thesis.

Gallop, Jane (1982) *The Daughter's Seduction: Feminism and Psychoanalysis*. New York: Macmillan.

Giddings, Paula (1985) *When and Where I Enter: The Impact of Black Women on Race and Sex in America*. Toronto: Bantam.

Haraway, Donna (1989) *Primate Visions: Gender, Race, and Nature in the World of Modern Science*. New York: Routledge.

Harding, Sandra (1983) 'Why has the sex/gender system become visible only now?', in Harding and Hintikka, eds *Discovering Reality. Feminist Perspectives on Epistemology, Metaphysics, Methodology, and Philosophy of Science*. Dordrecht: Reidel, pp. 311–24.

—— (1986) *The Science Question in Feminism*. Ithaca: Cornell University Press.

Hartmann, Heidi (1981) 'The unhappy marriage of marxism and feminism', in Sargent, L. *Women and Revolution*. Boston: South End, pp. 1–41.

Hartsock, Nancy (1983a) 'The feminist standpoint: developing the ground for a specifically feminist historical materialism', in Harding and Hintikka (1983), pp. 283–310.

—— (1983b) *Money, Sex, and Power*. New York: Longman; Boston: Northeastern University Press, 1984.

Haug, Frigga, ed. (1980) *Frauenformen: Alltagsgeschichten und Entwurf einer Theorie weiblicher Sozialisation*. Berlin: Argument Sonderband 45.

—— (1982) 'Frauen und Theorie', *Das Argument* 136(11/12).

——, *et al.* (1983) *Sexualisierung: Frauenformen 2*. Berlin: Argument-Verlag.

——, *et al.* (1987) *Female Sexualization: A Collective Work of Memory*. London: Verso.

hooks, bell (1981) *Ain't I a Woman*. Boston: South End.

—— (1984) *Feminist Theory: From Margin to Center*. Boston: South End.

Hubbard, Ruth, Henifin, Mary Sue, and Fried, Barbara, eds (1982) *Biological Woman, the Convenient Myth*. Cambridge, MA: Schenkman.

Hull, Gloria, Scott, Patricia Bell, and Smith, Barbara, eds (1982) *All the Women Are White, All the Men Are Black, But Some of Us Are Brave*. Old Westbury: The Feminist Press.

Hurtado, Aida (1989) 'Relating to privilege: seduction and rejection in the subordination of white women and women of color', *Signs* 14(4): 833–55.

Jaggar, Alison (1983) *Feminist Politics and Human Nature*. Totowa, NJ: Roman & Allenheld.

Joseph, Gloria and Lewis, Jill (1981) *Common Differences*. New York: Anchor.

Kessler, Suzanne and McKenna, Wendy (1978) *Gender: An Ethnomethodological Approach*. Chicago: University of Chicago Press.

Kollontai, Alexandra (1977) *Selected Writings*. London: Allison & Busby.

Kuhn, Annette and Wolpe, AnnMarie, eds (1978) *Feminism and Materialism*. London: Routledge & Kegan Paul.

Lewontin, R.C., Rose, Steven, and Kamin, Leon J. (1984) *Not in Our Genes: Biology, Ideology, and Human Nature*. New York: Pantheon.

Lorde, Audre (1982) *Zami, a New Spelling of My Name*. Trumansberg, NY: Crossing, 1983.

—— (1984) *Sister Outsider*. Trumansberg, NY: Crossing.

Lowe, Marian and Hubbard, Ruth, eds (1983) *Woman's Nature: Rationalizations of Inequality*. New York: Pergamon.

Marks, Elaine and de Courtivron, Isabelle, eds (1980) *New French Feminisms*. Amherst: University of Massachusetts Press.

Marx, Eleanor and Aveling, E. (1885–6) *The Woman Question*. London: Swann & Sonnenschein.

Marx, Karl (1964a) *Capital* vol. 1. New York: International.

—— and Engels, Frederick (1970) *The German Ideology*. London: Lawrence & Wishart.

Marxist-Feminist Literature Collective (1978) 'Women's writing', *Ideology and Consciousness* 1(3): 27–48.

Moi, Toril (1985) *Sexual/Textual Politics*. New York: Methuen.

Moraga, Cherríe (1983) *Loving in the War Years: lo que nunca pasó por sus labios*. Boston: South End.

—— and Anzaldúa, Gloria, eds (1981) *This Bridge Called My Back: Writings by Radical Women of Color*. Watertown: Persephone.

Morawski, J.G. (1987) 'The troubled quest for masculinity, femininity and androgyny', *Review of Personality and Social Psychology* 7: 44–69.
Mouffe, Chantal (1983) 'The sex-gender system and the discursive construction of women's subordination', *Rethinking Ideology*. Berlin: Argument Sonderband 84.
O'Brien, Mary (1981) *The Politics of Reproduction*. New York: Routledge & Kegan Paul.
Rose, Hilary (1983) 'Hand, brain, and heart: a feminist epistemology for the natural sciences', *Signs* 9(1): 73–90.
—— (1986) 'Women's work: women's knowledge', in Juliet Mitchell and Ann Oakley, eds, *What Is Feminism? A Re-Examination*. New York: Pantheon, pp. 161–83.
Rubin, Gayle (1975) 'The traffic in women: notes on the political economy of sex', in Rayna Rapp Reiter (1975), pp. 157–210.
Sandoval, Chela (n.d.) *Yours in Struggle: Women Respond to Racism, a Report on the National Women's Studies Association*. Oakland, CA: Center for Third World Organizing.
Sayers, Janet (1982) *Biological Politics: Feminist and Anti-Feminist Perspectives*. London: Tavistock.
Scott, Joan Wallach (1988) *Gender and the Politics of History*. New York: Columbia University Press.
Smith, Barbara ed. (1983) *Home Girls: A Black Feminist Anthology*. New York: Kitchen Table, Women of Color Press.
Smith, Dorothy (1974) 'Women's perspectives as a radical critique of sociology', *Sociological Inquiry* 44.
Spillers, Hortense (1987) 'Mama's baby, papa's maybe: an American grammar book', *Diacritics* 17(2): 65–81.
Thorne, Barrie and Henley, Nancy, eds (1975) *Language and Sex: Difference and Dominance*. Rowley, MA: Newbury.
Walker, Alice (1983) *In Search of Our Mothers' Gardens*. New York: Harcourt Brace Jovanovitch.
Ware, Celestine (1970) *Woman Power*. New York: Tower.
West, Candance and Zimmermann, D.H. (1987) 'Doing gender', *Gender and Society* 1(2): 125–51.
Young, Robert M. and Levidow, Les, eds (1981, 1985) *Science, Technology and the Labour Process*, 2 vols. London: CSE and Free Association Books.

---

3 R. W. Connell

'Gender as a Structure of Social Practice'

Excerpt from: *Masculinities*, pp. 71–81. Cambridge: Polity (1995)

Gender as a structure of social practice

Gender is a way in which social practice is ordered. In gender processes, the everyday conduct of life is organized in relation to a reproductive arena, defined by the bodily structures and processes of human reproduction. This arena includes sexual arousal and intercourse, childbirth and infant care, bodily sex difference and similarity.

I call this a 'reproductive arena' not a 'biological base' to emphasize that we are talking about a historical process involving the body, not a fixed set of biological determinants. Gender is social practice that constantly refers to bodies and what bodies do, it is not social practice reduced to the body. Indeed reductionism presents the exact reverse of the real situation. Gender exists precisely to the extent that biology does *not* determine the social. It marks one of those points of transition where historical process supersedes biological evolution as the form of change. Gender is a scandal, an outrage, from the point of view of essentialism.

Social practice is creative and inventive, but not inchoate. It responds to particular situations and is generated within definite structures of social relations. Gender relations, the relations among people and groups organized through the reproductive arena, form one of the major structures of all documented societies.

Practice that relates to this structure, generated as people and groups grapple with their historical situations, does not consist of isolated acts. Actions are configured in larger units, and when we speak of masculinity and femininity we are naming configurations of gender practice.

'Configuration' is perhaps too static a term. The important thing is the *process* of configuring practice. (Jean-Paul Sartre speaks in *Search for a Method* of the 'unification of the means in action'.) Taking a dynamic view of the organization of practice, we arrive at an understanding of masculinity and femininity as *gender projects*. These are processes of configuring practice through time, which transform their starting-points in gender structures.

We find the gender configuring of practice however we slice the social world, whatever unit of analysis we choose. The most familiar is the individual life course, the basis of the common-sense notions of masculinity and femininity. The configuration of practice here is what psychologists have traditionally called 'personality' or 'character'.

Such a focus is liable to exaggerate the coherence of practice that can be achieved at any one site. It is thus not surprising that psychoanalysis, originally stressing contradiction, drifted towards the concept of 'identity'. Post-structuralist critics of psychology such as Wendy Hollway have emphasized that gender identities are fractured and shifting, because multiple discourses intersect in any individual life.[1] This argument highlights another site, that of discourse, ideology or culture. Here gender is organized in symbolic practices that may continue much longer than the individual life (for instance: the construction of heroic masculinities in epics; the construction of 'gender dysphorias' or 'perversions' in medical theory).

Social science has come to recognize a third site of gender configuration, institutions such as the state, the workplace and the school. Many find it difficult to accept that institutions are substantively, not just metaphorically, gendered. This is, nevertheless, a key point.

The state, for instance, is a masculine institution. To say this is not to imply that the personalities of top male office-holders somehow seep through and stain the institution. It is to say something much stronger: that state organizational practices are structured in relation to the reproductive arena. The overwhelming majority of top office-holders are men because there is a gender configuring of recruitment and promotion, a gender configuring of the internal division of labour and systems of control, a gender configuring of policymaking, practical routines, and ways of mobilizing pleasure and consent.[2]

The gender structuring of practice need have nothing biologically to do with reproduction. The link with the reproductive arena is social. This becomes clear when it is challenged. An example is the recent struggle within the state over 'gays in the military', i.e., the rules excluding soldiers and sailors because of the gender of their sexual object-choice. In the United States, where this struggle was most severe, critics made the case for change in terms of civil liberties and military efficiency, arguing in effect that object-choice has little to do with the capacity to kill. The admirals and generals defended the status quo on a variety of spurious grounds. The unadmitted reason was the cultural importance of a particular definition of masculinity in maintaining the fragile cohesion of modern armed forces.

It has been clear since the work of Juliet Mitchell and Gayle Rubin in the 1970s that gender is an internally complex structure, where a number of different logics are superimposed. This is a fact of great importance for the analysis of masculinities. Any one masculinity, as a configuration of practice, is simultaneously positioned in a number of structures of relationship, which may be following different historical trajectories. Accordingly masculinity, like femininity, is always liable to internal contradiction and historical disruption.

We need at least a three-fold model of the structure of gender, distinguishing relations of (a) power, (b) production and (c) cathexis (emotional attachment). This is a provisional model, but it gives some purchase on issues about masculinity.[3]

(a) *Power relations* The main axis of power in the contemporary European/American gender order is the overall subordination of women and dominance of men – the structure Women's Liberation named 'patriarchy'. This general structure exists despite many local reversals (e.g., woman-headed households, female teachers with male students). It persists despite resistance of many kinds, now articulated in feminism. These reversals and resistances mean continuing difficulties for patriarchal power. They define a problem of legitimacy which has great importance for the politics of masculinity.

(b) *Production relations* Gender divisions of labour are familiar in the form of the allocation of tasks, sometimes reaching extraordinarily fine detail. (In the English village studied by the sociologist Pauline Hunt, for instance, it was customary for women to wash the inside of windows, men to wash the outside.) Equal attention should be paid to the economic consequences of gender divisions of labour, the dividend accruing to men from unequal shares of the products of social labour. This is most often discussed in terms of unequal wage rates, but the gendered character of capital should also be noted. A capitalist economy working through a gender division of labour is, necessarily, a gendered accumulation process. So it is not a statistical accident, but a part of the social construction of masculinity, that men and not women control the major corporations and the great private fortunes. Implausible as it sounds, the accumulation of wealth has become firmly linked to the reproductive arena, through the social relations of gender.[4]

(c) *Cathexis* Sexual desire is so often seen as natural that it is commonly excluded from social theory. Yet when we consider desire in Freudian terms, as emotional energy being attached to an object, its gendered character is clear. This is true both for heterosexual and homosexual desire. (It is striking that in our culture the non-gendered object choice, 'bisexual' desire, is ill-defined and unstable.) The practices that shape and realize desire are thus an aspect of the gender order. Accordingly we can ask political questions about the relationships involved: whether they are consensual or coercive, whether pleasure is equally given and received. In feminist analyses of sexuality these have become sharp questions about the connection of heterosexuality with men's position of social dominance.[5]

Because gender is a way of structuring social practice in general, not a special type of practice, it is unavoidably involved with other social structures. It is now common to say that gender 'intersects' – better, interacts – with race and class. We might add that it constantly interacts with nationality or position in the world order.

This fact also has strong implications for the analysis of masculinity. White men's masculinities, for instance, are constructed not only in relation to white women but also in relation to black men. Paul Hoch in *White Hero, Black Beast* more than a decade ago pointed to the pervasiveness of racial imagery in Western discourses of masculinity. White fears of black men's violence have a long history of colonial and post-colonial situations. Black fears of white men's terrorism, founded in the history of colonialism, have a continuing basis in white men's control of police, courts and prisons in metropolitan countries. African-American men are massively over-represented in American prisons, as Aboriginal men are in Australian prisons. This situation is strikingly condensed in the American black expression 'The Man', fusing white masculinity and institutional power. As the black rap singer Ice-T put it,

> It makes no difference whether you're in or out. The ghetto, the Pen, it's all institutionalized. It's being controlled by the Man . . . Ever since 1976, they stop trying to rehabilitate Brothers. Now it's strictly punishment. The Man's answer to the problem is not more education – it's more prisons. They're saying let's not educate them, let's lock them the fuck up. So when you come outta there you're all braindead, so yeah it's a cycle.[6]

Similarly, it is impossible to understand the shaping of working-class masculinities without giving full weight to their class as well as their gender politics. This is vividly shown in historical work such as Sonya Rose's

Limited Livelihoods, on industrial England in the nineteenth century. An ideal of working-class manliness and self-respect was constructed in response to class deprivation and paternalist strategies of management, at the same time and through the same gestures as it was defined against working-class women. The strategy of the 'family wage', which long depressed women's wages in twentieth-century economies, grew out of this interplay.[7]

To understand gender, then, we must constantly go beyond gender. The same applies in reverse. We cannot understand class, race or global inequality without constantly moving towards gender. Gender relations are a major component of social structure as a whole, and gender politics are among the main determinants of our collective fate.

Relations among masculinities: hegemony, subordination, complicity, marginalization

With growing recognition of the interplay between gender, race and class it has become common to recognize multiple masculinities: black as well as white, working-class as well as middle-class. This is welcome, but it risks another kind of oversimplification. It is easy in this framework to think that there is *a* black masculinity or *a* working-class masculinity.

To recognize more than one kind of masculinity is only a first step. We have to examine the relations between them. Further, we have to unpack the milieux of class and race and scrutinize the gender relations operating within them. There are, after all, gay black men and effeminate factory hands, not to mention middle-class rapists and cross-dressing bourgeois.

A focus on the gender relations among men is necessary to keep the analysis dynamic, to prevent the acknowledgement of multiple masculinities collapsing into a character typology, as happened with Fromm and the *Authoritarian Personality* research. 'Hegemonic masculinity' is not a fixed character type, always and everywhere the same. It is, rather, the masculinity that occupies the hegemonic position in a given pattern of gender relations, a position always contestable.

A focus on relations also offers a gain in realism. Recognizing multiple masculinities, especially in an individualist culture such as the United States, risks taking them for alternative lifestyles, a matter of consumer choice. A relational approach makes it easier to recognize the hard compulsions under which gender configurations are formed, the bitterness as well as the pleasure in gendered experience.

With these guidelines, let us consider the practices and relations that construct the main patterns of masculinity in the current Western gender order.

Hegemony

The concept of 'hegemony', deriving from Antonio Gramsci's analysis of class relations, refers to the cultural dynamic by which a group claims and sustains a leading position in social life. At any given time, one form of masculinity rather than others is culturally exalted. Hegemonic masculinity can be defined as the configuration of gender practice which embodies the currently accepted answer to the problem of the legitimacy of patriarchy, which guarantees (or is taken to guarantee) the dominant position of men and the subordination of women.[8]

This is not to say that the most visible bearers of hegemonic masculinity are always the most powerful people. They may be exemplars, such as film actors, or even fantasy figures, such as film characters. Individual holders of institutional power or great wealth may be far from the hegemonic pattern in their personal lives. (Thus a male member of a prominent business dynasty was a key figure in the gay/transvestite social scene in Sydney in the 1950s, because of his wealth and the protection this gave in the cold-war climate of political and police harassment.)[9]

Nevertheless, hegemony is likely to be established only if there is some correspondence between cultural ideal and institutional power, collective if not individual. So the top levels of business, the military and government provide a fairly convincing *corporate* display of masculinity, still very little shaken by feminist women or dissenting men. It is the successful claim to authority, more than direct violence, that is the mark of hegemony (though violence often underpins or supports authority).

I stress that hegemonic masculinity embodies a 'currently accepted' strategy. When conditions for the defence of patriarchy change, the bases for the dominance of a particular masculinity are eroded. New groups may challenge old solutions and construct a new hegemony. The dominance of *any* group of men may be challenged by women. Hegemony, then, is a historically mobile relation. Its ebb and flow is a key element of the picture of masculinity proposed in this book.

Subordination

Hegemony relates to cultural dominance in the society as a whole. Within that overall framework there are specific gender relations of dominance and subordination between groups of men.

The most important case in contemporary European/American society is the dominance of heterosexual men and the subordination of homosexual men. This is much more than a cultural stigmatization of homosexuality or gay identity. Gay men are subordinated to straight men by an array of quite material practices.

These practices were listed in early Gay Liberation texts such as Dennis Altman's *Homosexual: Oppression and Liberation*. They have been documented at length in studies such as the NSW Anti-Discrimination Board's 1982 report *Discrimination and Homosexuality*. They are still a matter of everyday experience for homosexual men. They include political and cultural exclusion, cultural abuse (in the United States gay men have now become the main symbolic target of the religious right), legal violence (such as imprisonment under sodomy statutes), street violence (ranging from intimidation to murder), economic discrimination and personal boycotts. It is not surprising that an Australian working-class man, reflecting on his experience of coming out in a homophobic culture, would remark:

> You know, I didn't totally realize what it was to be gay. I mean it's a bastard of a life.[10]

Oppression positions homosexual masculinities at the bottom of a gender hierarchy among men. Gayness, in patriarchal ideology, is the repository of whatever is symbolically expelled from hegemonic masculinity, the items ranging from fastidious taste in home decoration to receptive anal pleasure. Hence, from the point of view of hegemonic masculinity, gayness is easily assimilated to femininity. And hence – in the view of some gay theorists – the ferocity of homophobic attacks.

Gay masculinity is the most conspicuous, but it is not the only subordinated masculinity. Some heterosexual men and boys too are expelled from the circle of legitimacy. The process is marked by a rich vocabulary of abuse: wimp, milksop, nerd, turkey, sissy, lily liver, jellyfish, yellowbelly, candy ass, ladyfinger, pushover, cookie pusher, creampuff, motherfucker, pantywaist, mother's boy, four-eyes, ear-'ole, dweeb, geek, Milquetoast, Cedric, and so on. Here too the symbolic blurring with femininity is obvious.

Complicity

Normative definitions of masculinity, as I have noted, face the problem that not many men actually meet the normative standards. This point applies to hegemonic masculinity. The number of men rigorously practising the hegemonic pattern in its entirety may be quite small. Yet the majority of men gain from its hegemony, since they benefit from the patriarchal dividend, the advantage men in general gain from the overall subordination of women.

Accounts of masculinity have generally concerned themselves with syndromes and types, not with numbers. Yet in thinking about the dynamics of society as a whole, numbers matter. Sexual politics is mass politics, and strategic thinking needs to be concerned with where the masses of people are. If a large number of men have some connection with the hegemonic project but do not embody hegemonic masculinity, we need a way of theorizing their specific situation.

This can be done by recognizing another relationship among groups of men, the relationship of complicity with the hegemonic project. Masculinities constructed in ways that realize the patriarchal dividend, without the tensions or risks of being the frontline troops of patriarchy, are complicit in this sense.

It is tempting to treat them simply as slacker versions of hegemonic masculinity – the difference between the men who cheer football matches on TV and those who run out into the mud and the tackles themselves. But there is often something more definite and carefully crafted than that. Marriage, fatherhood and community life often involve extensive compromises with women rather than naked domination or an uncontested display of authority.[11] A great many men who draw the patriarchal dividend also respect their wives and mothers, are never violent towards women, do their accustomed share of the housework, bring home the family wage, and can easily convince themselves that feminists must be bra-burning extremists.

Marginalization

Hegemony, subordination and complicity, as just defined, are relations internal to the gender order. The interplay of gender with other structures such as class and race creates further relationships between masculinities.

New information technology became a vehicle for redefining middle-class masculinities at a time when the meaning of labour for working-class men was in contention. This is not a question of a fixed middle-class masculinity confronting a fixed working-class masculinity. Both are being reshaped, by a social dynamic in which class and gender relations are simultaneously in play.

Race relations may also become an integral part of the dynamic between masculinities. In a white-supremacist context, black masculinities play symbolic roles for white gender construction. For instance, black sporting stars become exemplars of masculine toughness, while the fantasy figure of the black rapist plays an important role in sexual politics among whites, a role much exploited by right-wing politics in the United Sates. Conversely, hegemonic masculinity among whites sustains the institutional oppression and physical terror that have framed the making of masculinities in black communities.

Robert Staples's discussion of internal colonialism in *Black Masculinity* shows the effect of class and race relations at the same time. As he argues, the level of violence among black men in the United States can only be understood through the changing place of the black labour force in American capitalism and the violent means used to control it. Massive unemployment and urban poverty now powerfully interact with institutional racism in the shaping of black masculinity.[12]

Though the term is not ideal, I cannot improve on 'marginalization' to refer to the relations between the masculinities in dominant and subordinated classes or ethnic groups. Marginalization is always relative to the *authorization* of the hegemonic masculinity of the dominant group. Thus, in the United States, particular black athletes may be exemplars for hegemonic masculinity. But the fame and wealth of individual stars has no trickle-down effect; it does not yield social authority to black men generally.

The relation of marginalization and authorization may also exist between subordinated masculinities. A striking example is the arrest and conviction of Oscar Wilde, one of the first men caught in the net of modern anti-homosexual legislation. Wilde was trapped because of his connections with homosexual working-class youths, a practice unchallenged until his legal battle with a wealthy aristocrat, the Marquess of Queensberry, made him vulnerable.[13]

These two types of relationship – hegemony, domination/subordination and complicity on the one hand, marginalization/authorization on the other – provide a framework in which we can analyse specific masculinities. (This is a sparse framework, but social theory should be hardworking.) I emphasize that terms such as 'hegemonic masculinity' and 'marginalized masculinities' name not fixed character types but configurations of practice generated in particular situations in a changing structure of relationships. Any theory of masculinity worth having must give an account of this process of change.

Notes

1 Hollway, 1984.
2 Franzway *et al.* 1989; Grant and Tancred, 1992.
3 Mitchell, 1971; Rubin, 1975. The three-fold model is spelt out in Connell, 1987.
4 Hunt, 1980. Feminist political economy is, however, under way, and these notes draw on Mies, 1986; Waring, 1988; Armstrong and Armstrong, 1990.
5 Some of the best writing on the politics of heterosexuality comes from Canada: Valverde, 1985; Buchbinder *et al.* 1987. The conceptual approach here is developed in Connell and Dowsett, 1992.
6 Interview with Ice-T in *City on a Hill Press* (Santa Cruz, CA), 21 Jan 1993; Hoch, 1979.
7 Rose, 1992, ch. 6 especially.
8 I would emphasize the dynamic character of Gramsci's concept of hegemony, which is not the functionalist theory of cultural reproduction often portrayed. Gramsci always had in mind a social struggle for leadership in historical change.
9 Wotherspoon, 1991 (chapter 3) describes this climate, and discreetly does not mention individuals.
10 Altman, 1972; Anti-Discrimination Board, 1982. Quotation from Connell, Davis and Dowsett 1993: 122.
11 See, for instance, the white US families described by Rubin, 1976.
12 Staples, 1982. The more recent United States literature on black masculinity, e.g., Majors and Gordon 1994, has made a worrying retreat from Staples's structural analysis towards sex role theory; its favoured political strategy, not surprisingly, is counselling programs to resocialize black youth.
13 Ellmann, 1987.

References

Altman, Dennis. 1972. *Homosexual: Oppression and Liberation.* Sydney: Angus & Robertson.
Anti-Discrimination Board, New South Wales, 1982. *Discrimination and Homosexuality.* Sydney: Anti-Discrimination Board.
Armstrong, Pat and Hugh Armstrong. 1990. *Theorizing Women's Work.* Toronto: Garamond Press.
Buchbinder, Howard, Varda Burstyn, Dinah Forbes and Mercedes Steedman. 1987. *Who's On Top? The Politics of Heterosexuality.* Toronto: Garamond Press.
Connell, R. W. 1987. *Gender and Power: Society, the Person and Sexual Politics.* Cambridge: Polity Press.
Connell, R. W., M. Davis and G. W. Dowsett. 1993. 'A bastard of a life: homosexual desire and practice among men in working-class milieux'. *Australian and New Zealand Journal of Sociology* **29**: 112–35.
Connell, R. W. and G. W. Dowsett, (eds) 1992. *Rethinking Sex: Social Theory and Sexuality Research.* Melbourne: Melbourne University Press.
Ellmann, Richard. 1987. *Oscar Wilde.* London: Hamish Hamilton.

Franzway, Suzanne, Dianne Court and R. W. Connell. 1989. *Staking a Claim: Feminism, Bureaucracy and the State.* Sydney: Allen & Unwin; Cambridge: Polity Press.
Grant, Judith and Peta Tancred. 1992. 'A feminist perspective on state bureaucracy'. In *Gendering Organization Analysis*, Albert J. Mills and Peta Tancred (eds). Newbury Park, CA. Sage. 112–28.
Hoch, Paul, 1979. *White Hero, Black Beast: Racism, Sexism and the Mask of Masculinity.* London: Pluto Press.
Hollway, Wendy. 1984. 'Gender difference and the production of subjectivity'. In *Changing the Subject*, J. Henriques *et al.* (eds). London: Methuen. 227–63.
Hunt, Pauline. 1980. *Gender and Class Consciousness.* London: Macmillan.
Majors, Richard G. and Jacob U. Gordon. 1994. *The American Black Male: His Present Status and his Future.* Chicago: Nelson-Hall.
Mies, Maria. 1986. *Patriarchy and Accumulation on a World Scale: Women in the International Division of Labour.* London: Zed Books.
Mitchell, Juliet. 1971. *Woman's Estate.* Harmondsworth: Penguin.
Rose, Sonya O. 1992. *Limited Livelihoods: Gender and Class in Nineteenth-Century England.* Berkeley: University of California Press.
Rubin, Gayle. 1975. 'The traffic in women: notes on the "political economy of sex"'. In *Toward an Anthropology of Women*, Rayna R. Reiter (ed.). New York: Monthly Review Press. 157–210.
Sartre, Jean Paul. 1968 [1960]. *Search for a Method.* New York: Vintage.
Staples, Robert. 1982. *Black Masculinity: The Black Male's Role in American Society.* San Francisco: Black Scholar Press.
Valverde, Mariana. 1985. *Sex, Power and Pleasure.* Toronto: Women's Press.
Waring, Marilyn. 1988. *Counting for Nothing: What Men Value and What Women are Worth.* Wellington: Allen & Unwin and Port Nicholson Press.
Wotherspoon, Gary. 1991. *City of the Plain: History of a Gay Sub-culture.* Sydney: Hale & Iremonger.

4 Donna Haraway

'Situated Knowledges: the Science Question in Feminism and the Privilege of Partial Perspective'

From: *Feminist Studies* **14,** 575–99 (1988)

Academic and activist feminist inquiry has repeatedly tried to come to terms with the question of what we might mean by the curious and inescapable term "objectivity." We have used a lot of toxic ink and trees processed into paper decrying what *they* have meant and how it hurts *us*. The imagined "they" constitute a kind of invisible conspiracy of masculinist scientists and philosophers replete with grants and laboratories. The imagined "we" are the embodied others, who are not allowed *not* to have a body, a finite point of view, and so an inevitably disqualifying and polluting bias in any discussion of consequence outside our own little circles, where a "mass"-subscription journal might reach a few thousand readers composed mostly of science haters. At least, I confess to these paranoid fantasies and academic resentments lurking underneath some convoluted reflections in print under my name in the feminist literature in the history and philosophy of science.

It has seemed to me that feminists have both selectively and flexibly used and been trapped by two poles of a tempting dichotomy on the question of objectivity. Certainly I speak for myself here, and I offer the speculation that there is a collective discourse on these matters. Recent social studies of science and technology, for example, have made available a very strong social constructionist argument for *all* forms of knowledge claims, most certainly and especially scientific ones.[1] According to these tempting views, no insider's perspective is privileged, because all drawings of inside-outside boundaries in knowledge are theorized as power moves, not moves toward truth. So, from the strong social constructionist perspective, why should we be cowed by scientist's descriptions of their activity and accomplishments; they and their patrons have stakes in throwing sand in our eyes. They tell parables about objectivity and scientific method to students in the first years of their initiation, but no practitioner of the high scientific arts would be caught dead *acting on* the textbook versions. Social constructionists make clear that official ideologies about objectivity and scientific method are particularly bad guides to how scientific knowledge is actually *made*. Just as for the rest of us, what scientists believe or say they do and what they really do have a very loose fit.

The only people who end up actually *believing* and, goddess forbid, acting on the ideological doctrines of disembodied scientific objectivity – enshrined in elementary textbooks and technoscience booster literature – are nonscientists, including a few very trusting philosophers. Of course, my designation of this last group is probably just a reflection of a residual disciplinary chauvinism acquired from identifying with historians of science and from spending too much time with a microscope in early adulthood in a kind of disciplinary preoedipal and modernist poetic moment when cells seemed to be cells and organisms, organisms. *Pace*, Gertrude Stein. But then came the law of the father and its resolution of the problem of objectivity, a problem solved by always already absent referents, deferred signifieds, split subjects, and the endless play of signifiers. Who wouldn't grow up warped? Gender, race, the world itself – all seem the effects of warp speeds in the play of signifiers in a cosmic force field.

In any case, social constructionists might maintain that the ideological doctrine of scientific method and all the philosophical verbiage about epistemology were cooked up to distract our attention from getting to know the world *effectively* by practicing the sciences. From this point of view, science – the real game in town – is rhetoric, a series of efforts to persuade relevant social actors that one's manufactured knowledge is a route to a desired form of very objective power. Such persuasions must take account of the structure of facts and artifacts, as well as of language mediated actors in the knowledge game. Here, artifacts and facts are parts of the powerful art of rhetoric.

Practice is persuasion, and the focus is very much on practice. All knowledge is a condensed node in an agonistic power field. The strong program in the sociology of knowledge joins with the lovely and nasty tools of semiology and deconstruction to insist on the rhetorical nature of truth, including scientific truth. History is a story Western culture buffs tell each other; science is a contestable text and a power field; the content is the form.[2] Period.

So much for those of us who would still like to talk about *reality* with more confidence than we allow to the Christian Right when they discuss the Second Coming and their being raptured out of the final destruction of the world. We would like to think our appeals to real worlds are more than a desperate lurch away from cynicism and an act of faith like any other cult's, no matter how much space we generously give to all the rich and always historically specific mediations through which we and everybody else must know the world. But the further I get in describing the radical social constructionist program and a particular version of postmodernism, coupled with the acid tools of critical discourse in the human sciences, the more nervous I get. The imagery of force fields, of moves in a fully textualized and coded world, which is the working metaphor in many arguments about socially negotiated reality for the postmodern subject is, just for starters, an imagery of high-tech military fields, of automated academic battlefields, where blips of light called players disintegrate (what a metaphor!) each other in order to stay in the knowledge and power game. Technoscience and science fiction collapse into the sun or their radiant (ir)reality – war.[3] It shouldn't take decades of feminist theory to sense the enemy here. Nancy Hartsock got all this crystal clear in her concept of abstract masculinity.[4]

I, and others, started out wanting a strong tool for deconstructing the truth claims of hostile science by showing the radical historical specificity, and so contestability, of *every* layer of the onion of scientific and technological constructions, and we end up with a kind of epistemological electroshock therapy, which far from ushering us into the high stakes tables of the game of contesting public truths, lays us out on the table with self-induced multiple personality disorder. We wanted a way to go beyond showing bias in science (that proved too easy anyhow) and beyond separating the good scientific sheep from the bad goats of bias and misuse. It seemed promising to do this by the strongest possible constructionist argument that left no cracks for reducing the issues to bias versus objectivity, use versus misuse, science versus pseudo-science. We unmasked the doctrines of objectivity because they threatened our budding sense of collective historical subjectivity and agency and our "embodied" accounts of the truth, and we ended up with one more excuse for not learning any post-Newtonian physics and one more reason to drop the old feminist self-help practices of repairing our own cars. They're just texts anyway, so let the boys have them back.

Some of us tried to stay sane in these disassembled and dissembling times by holding out for a feminist version of objectivity. Here, motivated by many of the same political desires, is the other seductive end of the objectivity problem. Humanistic Marxism was polluted at the source by its structuring theory about the domination of nature in the self-construction of man and by its closely related impotence in relation to historicizing anything women did that didn't qualify for a wage. But Marxism was still a promising resource as a kind of epistemological feminist mental hygiene that sought our own doctrines of objective vision. Marxist starting points offered a way to get to our own versions of standpoint theories, insistent embodiment, a rich tradition of critiquing hegemony without disempowering positivisms and relativisms and a way to get to nuanced theories of mediation. Some versions of psychoanalysis are of aid in this approach, especially anglophone object relations theory, which maybe did more for US socialist feminism for a time than anything from the pen of Marx or Engels, much less Althusser or any of the late pretenders to sonship treating the subject of ideology and science.[5]

Another approach, "feminist empiricism," also converges with feminist uses of Marxian resources to get a theory of science which continues to insist on legitimate meanings of objectivity and which remains leery of a radical constructivism conjugated with semiology and narratology.[6] Feminists have to insist on a better account of the world; it is not enough to show radical historical contingency and modes of construction for everything. Here, we, as feminists, find ourselves perversely conjoined with the discourse of many practicing scientists, who, when all is said and done, mostly believe they are describing and discovering things *by means of* all their constructing and arguing. Evelyn Fox Keller has been particularly insistent on this fundamental matter, and Sandra Harding calls the goal of these approaches a "successor science." Feminists have stakes in a successor science project that offers a more adequate, richer, better account of a world, in order to live in it well and in critical, reflexive relation to our own as well as others' practices of domination and the unequal parts of privilege and oppression that make up all positions. In traditional philosophical categories, the issue is ethics and politics perhaps more than epistemology.

So, I think my problem, and "our" problem, is how to have *simultaneously* an account of radical historical contingency for all knowledge claims and knowing subjects, a critical practice for recognizing our own "semiotic technologies" for making meanings, *and* a no-nonsense commitment to faithful accounts of a "real" world, one that can be partially shared and that is friendly to earthwide projects of finite freedom, adequate material abundance, modest meaning in suffering, and limited happiness. Harding calls this necessary multiple desire a need for a successor science project and a postmodern insistence on irreducible difference and radical multiplicity of local knowledges. *All* components of the desire are paradoxical and dangerous, and their combination is both contradictory and necessary. Feminists don't need a doctrine of objectivity that promises transcendence, a story that loses track of its mediations just where someone might be held responsible for something, and unlimited instrumental power. We don't want a theory of innocent powers to represent the world, where language and bodies both fall into the bliss of organic symbiosis. We also don't want to theorize the world, much less act within it, in terms of Global Systems, but we do need an earthwide network of connections, including the ability partially to translate knowledges among very different – and power-differentiated – communities.

We need the power of modern critical theories of how meanings and bodies get made, not in order to deny meanings and bodies, but in order to build meanings and bodies that have a chance for life.

Natural, social, and human sciences have always been implicated in hopes like these. Science has been about a search for translation, convertibility, mobility of meanings, and universality – which I call reductionism only when one language (guess whose?) must be enforced as the standard for all the translations and conversions. What money does in the exchange orders of capitalism, reductionism does in the powerful mental orders of global sciences. There is, finally, only one equation. That is the deadly fantasy that feminists and others have identified in some versions of objectivity, those in the service of hierarchical and positivist orderings of what can count as knowledge. That is one of the reasons the debates about objectivity matter, metaphorically and otherwise. Immortality and omnipotence are not our goals. But we could use some enforceable, reliable accounts of things not reducible to power moves and agonistic, high-status games of rhetoric or to scientistic, positivist arrogance. This point applies whether we are talking about genes, social classes, elementary particles, genders, races, or texts; the point applies to the exact, natural, social and human sciences, despite the slippery ambiguities of the words "objectivity" and "science" as we slide around the discursive terrain. In our efforts to climb the greased pole leading to a usable doctrine of objectivity, I and most other feminists in the objectivity debates have alternatively, or even simultaneously, held on to both ends of the dichotomy, a dichotomy which Harding describes in terms of successor science projects versus postmodernist accounts of difference and which I have sketched in this essay as radical constructivism versus feminist critical empiricism. It is, of course, hard to climb when you are holding on to both ends of a pole, simultaneously or alternatively. It is, therefore, time to switch metaphors.

The persistence of vision

I would like to proceed by placing metaphorical reliance on a much maligned sensory system in feminist discourse: vision.[7] Vision can be good for avoiding binary oppositions. I would like to insist on the embodied nature of all vision and so reclaim the sensory system that has been used to signify a leap out of the marked body and into a conquering gaze from nowhere. This is the gaze that mythically inscribes all the marked bodies, that makes the unmarked category claim the power to see and not be seen, to represent while escaping representation. This gaze signifies the unmarked positions of Man and White, one of the many nasty tones of the word "objectivity" to feminist ears in scientific and technological, late-industrial, militarized, racist, and male-dominant societies, that is, here, in the belly of the monster, in the United States in the late 1980s. I would like a doctrine of embodied objectivity that accommodates paradoxical and critical feminist science projects: Feminist objectivity means quite simply *situated knowledges*.

The eyes have been used to signify a perverse capacity – honed to perfection in the history of science tied to militarism, capitalism, colonialism, and male supremacy – to distance the knowing subject from everybody and everything in the interests of unfettered power. The instruments of visualization in multinationalist, postmodernist culture have compounded these meanings of disembodiment. The visualizing technologies are without apparent limit. The eye of any ordinary primate like us can be endlessly enhanced by sonography systems, magnetic reasonance imaging, artifical intelligence-linked graphic manipulation systems, scanning electron microscopes, computed tomography scanners, color-enhancement techniques, satellite surveillance systems, home and office video display terminals, cameras for every purpose from filming the mucous membrane lining the gut cavity of a marine worm living in the vent gases on a fault between continental plates to mapping a planetary hemisphere elsewhere in the solar system. Vision in this technological feast becomes unregulated gluttony; all seems not just mythically about the god trick of seeing everything from nowhere, but to have put the myth into ordinary practice. And like the god trick, this eye fucks the world to make techno-monsters. Zoe Sofoulis calls this the cannibaleye of masculinist extra-terrestrial projects for excremental second birthing.

A tribute to this ideology of direct, devouring, generative, and unrestricted vision, whose technological mediations are simultaneously celebrated and presented as utterly transparent, can be found in the volume celebrating the 100th anniversary of the National Geographic Society. The volume closes its survey of the magazine's quest literature, effected through its amazing photography, with two juxtaposed chapters. The first is on "Space," introduced by the epigraph. "The choice is the universe – or nothing."[8] This chapter recounts the exploits of the space race and displays the color-enhanced "snapshots" of the outer planets reassembled from digitalized signals transmitted across vast space to let the viewer "experience" the moment of discovery in immediate vision of the "object."[9] These fabulous objects come to us simultaneously as indubitable recordings of what is simply there and as heroic feats of technoscientific production. The next chapter, is the twin of outer space: "Inner Space," introduced by the epigraph, "The stuff of stars has come alive."[10] Here, the reader is brought into the realm of the infinitesimal, objectified by means of radiation outside the wave lengths that are "normally" perceived by hominid primates, that is, the beams of lasers and scanning electron microscopes, whose signals are processed into the wonderful full-color snapshots of defending T cells and invading viruses.

But, of course, that view of infinite visions is an illusion, a god trick. I would like to suggest how our insisting metaphorically on the particularity and embodiment of all vision (although not necessarily organic embodiment and including technological mediation), and not giving in to the tempting myths of vision as a route to disembodiment and second-birthing allows us to construct a usable, but not an innocent, doctrine of objectivity. I want a feminist writing of the body that metaphorically emphasizes vision again, because we need to reclaim that sense to find our way through all the visualizing tricks and powers of modern sciences and technologies that have transformed the objectivity debates. We need to learn in our bodies, endowed with primate color and stereoscopic vision, how to attach the objective to our theoretical and political scanners in order to name where we are and are not, in dimensions of mental and physical space we hardly know how to name. So, not so perversely, objectivity turns out to be about particular and specific embodiment and definitely not about the false vision promising transcendence of all limits and responsibility. The moral is simple: only partial perspective promises objective vision. All Western cultural narratives about objectivity are allegories of the ideologies governing the relations of what we call mind and body, distance and responsibility. Feminist objectivity is about limited location and situated knowledge, not about transcendence and splitting of subject and object. It allows us to become answerable for what we learn how to see.

These are lessons that I learned in part walking with my dogs and wondering how the world looks without a fovea and very few retinal cells for color vision but with a huge neural processing and sensory area for smells. It is a lesson available from photographs of how the world looks to the compound eyes of an insect or even from the camera eye of a spy satellite or the digitally transmitted signals of space probe-perceived differences "near" Jupiter that have been transformed into coffee table color photographs. The "eyes" made available in modern technological sciences shatter any idea of passive vision; these prosthetic devices show us that all eyes, including our own organic ones, are active perceptual systems, building on translations and specific ways of seeing, that is, ways of life. There is no unmediated photograph or passive camera obscura in scientific accounts of bodies and machines; there are only highly specific visual possibilities, each with a wonderfully detailed, active, partial way of organizing worlds. All these pictures of the world should not be allegories of infinite mobility and interchangeability but of elaborate specificity and difference and the loving care people might take to learn how to see faithfully from another's point of view, even when the other is our own machine. That's not alienating distance; that's a *possible* allegory for feminist versions of objectivity. Understanding how these visual systems work, technically, socially, and psychically ought to be a way of embodying feminist objectivity.

Many currents in feminism attempt to theorize grounds for trusting especially the vantage points of the subjugated; there is good reason to believe vision is better from below the brilliant space platforms of the powerful.[11] Building on that suspicion, this essay is an argument for situated and embodied knowledges and an argument against various forms of unlocatable, and so irresponsible knowledge claims. Irresponsible means unable to be called into account. There is a premium on establishing the capacity to see from the peripheries and the depths. But here there also lies the serious danger of romanticizing and/or appropriating the vision of the less powerful while claiming to see from their positions. To see from below is neither easily learned nor unproblematic, even if "we" "naturally" inhabit the great underground terrain of subjugated knowledges. The positionings of the subjugated are not exempt from critical reexamination, decoding, deconstruction, and interpretation; that is, from both semiological and hermeneutic modes of critical inquiry. The standpoints of the subjugated are not "innocent" positions. On the contrary, they are preferred because in principle they are least likely to allow denial of the critical and interpretive core of all knowledge. They are knowledgeable of modes of denial through repression, forgetting, and disappearing acts – ways of being nowhere while claiming to see comprehensively. The subjugated have a decent chance to be on to the god trick and all its dazzling – and, therefore, blinding – illuminations. "Subjugated" standpoints are preferred because they seem to promise more adequate, sustained, objective, transforming accounts of the world. But *how* to see from below is a problem requiring at least as much skill with bodies and language, with the mediations of vision, as the "highest" technoscientific visualizations.

Such preferred positioning is as hostile to various forms of relativism as to the most explicitly totalizing versions of claims to scientific authority. But the alternative to relativism is not totalization and single vision, which is always finally the unmarked category whose power depends on systematic narrowing and obscuring. The alternative to relativism is partial, locatable, critical knowledges sustaining the possibility of webs of connections called solidarity in politics and shared conversations in epistemology. Relativism is a way of being nowhere while claiming to be everywhere equally. The "equality" of positioning is a denial of responsibility and critical inquiry. Relativism is the perfect mirror twin of totalization in the ideologies of objectivity; both deny the stakes in location, embodiment, and partial perspective: both make it impossible to see well. Relativism and totalization are both "god tricks" promising vision from everywhere and nowhere equally and fully, common myths in rhetorics surrounding Science. But it is precisely in the politics and epistemology of partial perspectives that the possibility of sustained, rational, objective inquiry rests.

So, with many other feminists, I want to argue for a doctrine and practice of objectivity that privileges contestation, deconstruction, passionate construction, webbed connections, and hope for transformation of systems of knowledge and ways of seeing. But not just any partial perspective will do; we must be hostile to easy relativisms and holisms built out of summing and subsuming parts. "Passionate detachment"[12] requires more than acknowledged and self-critical partiality. We are also bound to seek perspective from those points of view, which can never be known in advance, that promise something quite extraordinary, that is, knowledge potent for constructing worlds less organized by axes of domination. From such a viewpoint, the unmarked category would *really* disappear – quite a difference from simply repeating a disappearing act. The imaginary and the rational – the visionary and objective vision – hover close together. I think Harding's plea for a successor science and for postmodern sensibilities must be read as an argument for the idea that the fantastic element of hope for transformative knowledge and the severe check and stimulus of sustained critical inquiry are jointly the ground of any believable claim to objectivity or rationality not riddled with breathtaking denials and repressions. It is even possible to read the record of scientific revolutions in terms of this feminist doctrine of rationality and objectivity. Science has been utopian and visionary from the start; that is one reason "we" need it.

A commitment to mobile positioning and to passionate detachment is dependent on the impossibility of entertaining innocent "identity" politics and epistemologies as strategies for seeing from the standpoints of the subjugated in order to see well. One cannot "be" either a cell or molecule – or a woman, colonized person, laborer, and so on – if one intends to see and see from these positions critically. "Being" is much more problematic and contingent. Also, one cannot relocate in any possible vantage point without being accountable for that movement. Vision is *always* a question of the power to see – and perhaps of the violence implicit in our visualizing practices. With whose blood were my eyes crafted? These points also apply to testimony from the position of "oneself." We are not immediately present to ourselves. Self-knowledge requires a semiotic-material technology to link meanings and bodies. Self-identity is a bad visual system. Fusion is a bad strategy of positioning. The boys in the human sciences have called this doubt about self-presence the "death of the subject" defined as a single ordering point of will and consciousness. That judgement seems bizarre to me. I prefer to call this doubt the opening of nonisomorphic subjects, agents, and territories of stories unimaginable from the vantage point of the cyclopean; self-satisfied eye of the master subject. The Western eye has fundamentally been a wandering eye, a traveling lens. These peregrinations have often been violent and insistent on having mirrors for a conquering self – but not always. Western feminists also *inherit* some skill in learning to participate in revisualizing worlds turned upside down in earth-transforming challenges to the views of the masters. All is not to be done from scratch.

The split and contradictory self is the one who can interrogate positionings and be accountable, the one who can construct and join rational conversations and fantastic imaginings that change history.[13] Splitting, not being, is the privileged image for feminist epistemologies of scientific knowledge. "Splitting" in this context should be about heterogeneous multiplicities that are simultaneously salient and incapable of being squashed into isomorphic slots or cumulative lists. This geometry pertains within and among subjects. Subjectivity is multidimensional: so, therefore, is vision. The knowing self is partial in all its guises, never finished, whole, simply there and original; it is always constructed and stitched together imperfectly, and *therefore* able to join with another, to see together without claiming to be another. Here is the promise of objectivity: a scientific knower seeks the subject position, not of identity, but of objectivity, that is, partial connection. There is no way to "be" simultaneously in all, or wholly in any, of the privileged (i.e., subjugated) positions structured by gender, race, nation, and class. And that is a short list of critical positions. The search for such a "full" and total position is the search for the fetishized perfect subject of oppositional history, sometimes appearing in feminist theory as the essentialized Third World Woman.[14] Subjugation is not grounds for an ontology; it might be a visual clue. Vision requires instruments of vision; an optics is a politics of positioning. Instruments of vision mediate standpoints; there is no immediate vision from the standpoints of the subjugated. Identity, including self-identity, does not produce science; critical positioning does, that is, objectivity. Only those occupying the positions of the dominators are self-identical, unmarked, disembodied, unmediated, transcendent, born again. It is unfortunately possible for the subjugated to lust for and even scramble into that subject position – and then disappear from view. Knowledge from the point of view of the unmarked is truly fantastic, distorted, and irrational. The only position from which objectivity could not possibly be practiced and honored is the standpoint of the master, the Man, the One God, whose Eye produces, appropriates, and orders all difference. No one ever accused this God of monotheism of objectivity; only of indifference. The god trick is self-identical, and we have mistaken that for creativity and knowledge, omniscience even.

Positioning is, therefore, the key practice in grounding knowledge organized around the imagery of vision, and much Western scientific and philosophic discourse is organized in this way. Positioning implies responsibility for our enabling practices. It follows that politics and ethics ground struggles for and contexts over what may count as rational knowledge. That is, admitted or not, politics and ethics ground struggles over knowledge projects in the exact natural, social, and human sciences. Otherwise, rationality is simply impossible, an optical illusion projected from nowhere comprehensively. Histories of science may be powerfully told as histories of the technologies. These technologies are ways of life, social orders, practices of visualization. Technologies are skilled practices. How to see? Where to see from? What limits to vision. What to see for? Whom to see with? Who gets to have more than one point of view? Who gets blinded? Who wears blinders? Who interprets the visual field? What other sensory powers do we want to cultivate besides vision? Moral and political discourse should be the paradigm for rational discourse about the imagery and technologies of vision. Sandra Harding's claim, or observation, that moments of social revolution have most contributed to improvements in science might be read as a claim about the knowledge consequences of new technologies of positioning. But I wish Harding had spent more time remembering that social and scientific revolutions have not always been liberatory, even if they have always been visionary. Perhaps this point could be captured in another phrase: the science question in the military. Struggles over what will count as rational accounts of the world are struggles over *how* to see. The terms of vision: the science question of colonialism, the science question in exterminism,[15] the science question in feminism.

The issue of politically engaged attacks on various empiricisms, reductionisms, or other versions of scientific authority should not be relativism – but location. A dichotomous chart expressing this point might look like this:

universal rationality	ethnophilosophies
common language	heteroglossia
new organon	deconstruction
unified field theory	oppositional positioning
world system	local knowledges
master theory	webbed accounts

But a dichotomous chart misrepresents in a critical way the positions of embodied objectivity that I am trying to sketch. The primary distortion is the illusion of symmetry in the chart's dichotomy, making any position appear, first, simply alternative and, second, mutually exclusive. A map of tensions and reasonances between the fixed ends of a charged dichotomy better represents the potent politics and epistemologies of embodied, therefore accountable, objectivity. For example, local knowledges have also to be in tension with the productive structurings that force unequal translations and exchanges – material and semiotic – within the webs of knowledge and power. Webs *can* have the property of being systemic, even of being centrally structured global systems with deep filaments and tenacious tendrils into time, space, and consciousness, which are the dimensions of world history. Feminist accountability requires a knowledge tuned to reasonance, not to dichotomy. Gender is a field of structured and structuring difference, in which the tones of extreme localization, of the intimately personal and individualized body, vibrate in the same field with global high-tension emissions. Feminist embodiment, then, is not about fixed location in a reified body, female or otherwise, but about nodes in fields, inflections in orientations, and responsibility for difference in material-semiotic fields of meaning. Embodiment is significant prosthesis; objectivity cannot be about fixed vision when what counts as an object is precisely what world history turns out to be about.

How should one be positioned in order to see, in this situation of tensions, reasonances, transformations, resistances, and complicities? Here, primate vision is not immediately a very powerful metaphor or technology for feminist political-epistemological clarification, because it seems to present to consciousness already processed and objectified fields; things seem already fixed and distanced. But the visual metaphor allows one to go beyond fixed appearances, which are only the end products. The metaphor invites us to investigate the varied apparatuses of visual production including the prosthetic technologies interfaced with our biological eyes and brains. And here we find highly particular machineries for processing regions of the electromagnetic spectrum into our pictures of the world. It is in the intricacies of these visualizations technologies in which we are embedded that we will find metaphors and means for understanding and intervening in the patterns of objectification in the world – that is, the patterns of reality for which we must be accountable. In these metaphors, we find means for appreciating simultaneously *both* the concrete, "real" aspect and the aspect of semiosis and production in what we call scientific knowledge.

I am arguing for politics and epistemologies of location, positioning, and situating, where partiality and not universality is the condition of being heard to make rational knowledge claims. These are claims on people's lives. I am arguing for the view from a body always a complex, contradictory, structuring, and structuring body, versus the view from above, from nowhere, from simple city. Only the god trick is forbidden. Here is a criterion for deciding the science question in militarism, that dream science technology of perfect language, perfect communication, future order.

Feminism loves another science: the sciences and politics of interpretation, translation, stuttering, and the partly understood. Feminism is about the sciences of the multiple subject with at least double vision. Feminism is about a critical vision consequence upon a critical positioning in unhomogeneous gendered social space.[16] Translation is always interpretative, critical, and partial. Here is a ground for conversation, rationality, and objectivity – which is power-sensitive, not pluralist, "conversation." It is not even the mythic cartoons of physics and mathematics – incorrectly caricatured in antiscience ideology as exact, hypersimple knowledges – that have come to represent the hostile other to feminist paradigmatic models of scientific knowledge, but the dream of the perfectly known in high-technology, permanently militarized scientific productions and positionings, the god trick of a Star Wars paradigm of rational knowledge. So location is about vulnerability; location resists the politics of closure, finality, or to borrow from Althusser, feminist objectivity resists "simplification in the last instance." That is because feminist embodiment resists fixation and is insatiably curious about the webs of differential positioning. There is no single feminist standpoint because our maps require too many dimensions for that metaphor to ground our visions. But the feminist standpoint theorists' goal of an epistemology and politics of engaged, accountable positioning remains eminently potent. The goal is better accounts of the world, that is, "science."

Above all, rational knowledge does not pretend to disengagement: to be from everywhere and so nowhere, to be free from interpretation, from being represented, to be fully self-contained or fully formalizable. Rational knowledge is a process of ongoing critical interpretation among "fields" of interpreters and decoders. Rational knowledge is power-sensitive conversation.[17] Decoding and transcoding plus translation and criticism; all are necessary. So science becomes the paradigmatic model, not of closure, but of that which is contestable and contested. Science becomes the myth, not of what escapes human agency and responsibility in a realm above the fray, but, rather, of accountability and responsibility for translations and solidarities linking the cacophonous visions and visionary voices that characterize the knowledges of the subjugated. A splitting of senses, a confusion of voice and sight, rather than clear and distinct ideas, becomes the metaphor for the ground of the rational. We seek not the knowledges ruled by phallogocentrism (nostalgia for the presence of the one true Word) and disembodied vision. We seek those ruled by partial sight and limited voice – not partiality for its own sake but, rather, for the sake of the connections and unexpected openings situated knowledges make possible. Situated knowledges are about communities, not about isolated individuals. The only way to find a larger vision is to be somewhere in particular. The science question in feminism is about objectivity as positioned rationality. Its images are not the products of escape and transcendence of limits (the view from above) but the joining of partial views and halting voices into a collective subject position that promises a vision of the means of ongoing finite embodiment of living within limits and contradictions – of views from somewhere.

Objects as actors: the apparatus of bodily production

Throughout this reflection on "objectivity," I have refused to resolve the ambiguities built into referring to science without differentiating its extraordinary range of contexts. Through the insistent ambiguity, I have foregrounded a field of commonalities binding exact, physical, natural, social, political, biological, and human sciences; and I have tied this whole heterogeneous field of academically (and industrially, e.g., in publishing, the weapons trade, and pharmaceuticals) institutionalized knowledge production to a meaning of science that insists on its potency in ideological struggles. But, partly in order to give play to both the specificities and the highly permeable boundaries of meanings in discourse on science, I would like to suggest a resolution to one ambiguity. Throughout the field of meanings constituting science, one of the commonalities concerns the status of any object of knowledge and of related claims about the faithfulness of our accounts to a "real world," no matter how mediated for us and no matter how complex and contradictory these worlds may be. Feminists, and others who have been most active as critics of the sciences and their claims or associated ideologies, have shied away from doctrines of scientific objectivity in part because of the suspicion that an "object" of knowledge is a passive and inert thing. Accounts of such objects can seem to be either appropriations of a fixed and determined world reduced to resource for instrumentalist projects of destructive Western societies, or they can be seen as masks for interests, usually dominating interests.

For example, "sex" as an object of biological knowledge appears regularly in the guise of biological determinism, threatening the fragile space for social constructionism and critical theory, with their attendant possibilities for active and transformative intervention, which were called into being by feminist concepts of gender as socially, historically, and semiotically positioned difference. And yet, to lose authoritative biological accounts of sex, which set up productive tensions with gender, seems to be to lose too much. It seems to be to lose not just analytic power within a particular Western tradition but also the body itself as anything but a blank page for social inscriptions, including those of biological discourse. The same problem of loss attends the radical "reduction" of the objects of physics or of any other science to the ephemera of discursive production and social construction.[18]

But the difficulty and loss are not necessary. They derive partly from the analytic tradition, deeply indebted to Aristotle and to the transformative history of "White Capitalist Patriarchy" (how may we name this scandalous Thing?) that turns everything into a resource for appropriation, in which an object of knowledge is finally itself only matter for the seminal power, the act, of the knower. Here, the object both guarantees and refreshes the power of the knower, but any status as *agent* in the productions of knowledge must be denied the object. It – the world – must, in short, be objectified as a thing, not as an agent; it must be matter for the self-formation of the only social being in the productions of knowledge, the human knower. Zoe Sofoulis[19] identified the structure of this mode of knowing in technoscience as "resourcing" – as the second birthing of Man through the homogenizing of all the world's body into resource for his perverse projects. Nature is only the raw material of culture, appropriated, preserved, enslaved, exalted, or otherwise made flexible for disposal by culture in the logic of capitalist colonialism. Similarly, sex is only matter to the act of gender: the productionist logic seems inescapable in traditions of Western binary oppositions. This analytical and historical narrative logic accounts for my nervousness about the sex/gender distinction in the recent history of feminist theory. Sex is "resourced" for its representation as gender, which "we" can control. It has seemed all but impossible to avoid the trap of an appropriationist logic of domination built into the nature/culture opposition and its generative lineage, including the sex/gender distinction.

It seems clear that feminist accounts of objectivity and embodiment – that is, of a world – of the kind sketched in this essay require a deceptively simple maneuver within inherited Western analytical traditions, a maneuver begun in dialectics but stopping short of the needed revisions. Situated knowledges require that the object of knowledge be pictured as an actor and agent, not as a screen or a ground or a resource, never finally as slave to the master that closes off the dialectic in his unique agency and his authorship of "objective" knowledge. The point is paradigmatically clear in critical approaches to the social and human sciences, where the agency of people studied itself transforms the entire project of producing social theory. Indeed, coming to terms with the agency of the "objects" studied is the only way to avoid gross error and false knowledge of many kinds in these sciences. But the same point must apply to the other knowledge projects called sciences. A corollary of the insistence that ethics and politics covertly or overtly provide the bases for objectivity in the sciences as a heterogeneous whole, and not just in the social sciences, is granting the status of agent/actor to the "objects" of the world. Actors come in many and wonderful forms. Accounts of a "real" world do not, then, depend on a logic of "discovery" but on a power-charged social relation of "conversation." The world neither speaks itself nor disappears in favor of a master decoder. The codes of the world are not still waiting only to be read. The world is not raw material for humanization; the thorough attacks on humanism, another branch of "death of the subject" discourse have made this point quite clear. In some critical sense that is crudely hinted at by the clumsy category of the social or of agency the world encountered in knowledge projects is an active entity. Insofar as a scientific account has been able to engage this dimension of the world as object of knowledge, faithful knowledge can be imagined and can make claims on us. But no particular doctrine of representation or decoding or discovery guarantees anything. The approach I am recommending is not a version of "realism," which has proved a rather poor way of engaging with the world's active agency.

My simple, perhaps simple-minded, maneuver is obviously not new in Western philosophy, but it has a special feminist edge to it in relation to the science question in feminism and to the linked question of gender as situated difference and the question of female embodiment. Ecofeminists have perhaps been most insistent on some version of the world as active subject, not as resource to be mapped and appropriated in bourgeois, Marxist, or masculinist projects. Acknowledging the agency of the world in knowledge makes room for some unsettling possibilities, including a sense of the world's independent sense of humor. Such a sense of humor is not comfortable for humanists and others committed to the world as resource. There are, however, richly evocative figures to promote feminist visualizations of the world as witty agent. We need not lapse into appeals to a primal mother resisting her translation into resource. The Coyote or Trickster as embodied in Southwest native American accounts, suggests the situation we are in when we give up mastery but keep searching for fidelity, knowing all the while that we will be hoodwinked. I think these are useful myths for scientists who might be our allies. Feminist objectivity makes room for surprises and ironies at the heart of all knowledge production; we are not in charge of the world. We must live here and try to strike up noninnocent conversations by means of our prosthetic devices, including our visualization technologies. No wonder science fiction has been such a rich writing practice in recent feminist theory. I like to see feminist theory as a reinvented coyote discourse obligated to its sources in many heterogeneous accounts of the world.

Another rich feminist practice in science in the last couple of decades illustrates particularly well the "activation" of the previously passive categories of objects of knowledge. This activation permanently problematizes binary distinctions like sex and gender, without eliminating their strategic utility. I refer to the reconstructions in primatology (especially, but not only, in women's practice as primatologists, evolutionary biologists, and behavioral ecologists) of what may count as sex, especially as female sex, in scientific accounts.[20] The *body*, the object of biological discourse, becomes a most engaging being. Claims of biological determinism can never be the same again. When female "sex" has been so thoroughly retheorized and revisualized that it emerges as practically indistinguishable from "mind," something basic has happened to the categories of biology. The biological female peopling current biological behavioral accounts has almost no passive properties left. She is structuring and active in every respect; the "body" is an agent, not a resource. Difference is theorized *biologically* as situational, not intrinsic, at every level from gene to foraging pattern, thereby fundamentally changing the biological politics of the body. The relations between sex and gender need to be categorically reworked within these frames of knowledge. I would like to suggest that this trend in explanatory strategies in biology is an allegory for interventions faithful to projects of feminist objectivity. The point is not that these new pictures of the biological female are simply true or not open to contestation and conversation – quite the opposite. But these pictures foreground knowledge as situated conversation at every level of its articulation. The boundary between animal and human is one of the stakes in this allegory, as is the boundary between machine and organism.

So I will close with a final category useful to a feminist theory of situated knowledges: the apparatus of bodily production. In her analysis of the production of the poem as an object of literary value, Katie King offers tools that clarify matters in the objectivity debates among feminists. King suggests the term "apparatus of literary production" to refer to the emergence of literature at the intersection of art, business, and technology. The apparatus of literary production is a matrix from which "literature" is born. Focusing on the potent object of value called the "poem," King applies her analytic framework to the relation of women and writing technologies.[21] I would like to adapt her work to understanding the generation – the actual production and reproduction – of bodies and other objects of value in scientific knowledge projects. At first glance, there is a limitation to using King's scheme inherent in the "facticity" of biological discourse that is absent from literary discourse and its knowledge claims. Are biological bodies "produced" or "generated" in the same strong sense as poems? From the early stirrings of Romanticism in the late eighteenth century, many poets and biologists have believed that poetry and organisms are siblings. *Frankenstein* may be read as a meditation on this proposition. I continue to believe in this potent proposition but in a postmodern and not a Romantic manner. I wish to translate the ideological dimensions of "facticity" and "the organic" into a cumbersome entity called a "material-semiotic actor." This unwieldy term is intended to portray the object of knowledge as an active, meaning-generating part of apparatus of bodily production without *ever* implying the immediate presence of such objects or, what is the same thing, their final or unique determination of what can count as objective knowledge at a particular historical juncture. Like "poems," which are sites of literary production where language too is an actor independent of intentions and authors, bodies as objects of knowledge are material-semiotic generative nodes. Their *boundaries* materialize in social interaction. Boundaries are drawn by mapping practices; "objects" do not preexist as such. Objects are boundary projects. But boundaries shift from within; boundaries are very tricky. What boundaries provisionally contain remains generative, productive of meanings and bodies. Siting (sighting) boundaries is a risky practice.

Objectivity is not about disengagement but about mutual *and* usually unequal structuring, about taking risks in a world whereas "we" are permanently mortal, that is, not in "final" control. We have, finally, no clear and distinct ideas. The various contending biological bodies emerge at the intersection of biological research and writing, medical and other business practices, and technology, such as the visualization technologies enlisted as metaphors in this essay. But also invited into that node of intersection is the analogue to the lively languages that actively intertwine in the production of literary value: the coyote and the protean embodiments of the word as witty agent and actor. Perhaps the world resists being reduced to mere resource because it is – not mother/matter/mutter – but coyote, a figure of the always problematic, always potent tie between meaning and bodies. Feminist embodiment, feminist hopes for partiality, objectivity, and situated knowledges, turn on conversations and codes at this potent node in fields of possible bodies and meanings. Here is where science, science fantasy and science fiction converge in the objectivity question in feminism. Perhaps our hopes for accountability, for politics, for ecofeminism, turn on revisioning the world as coding trickster with whom we must learn to converse.

Notes

This essay originated as a commentary on Sandra Harding's *The Science Question in Feminism*, at the Western Division meetings of the American Philosophical Association, San Francisco, March 1987. Support during the writing of this paper was generously provided by the Alpha Fund of the Institute for Advanced Study, Princeton, New Jersey. Thanks especially to Joan Scott, Judy Butler, Lila Abu-Lughod, and Dorinne Kondo.

1 For example, see Karin Knorr-Cetina and Michael Mulkay, eds., *Science Observed: Perspectives on the Social Study of Science* [London: Sage, 1983]; Wiebe E. Bijker, Thomas P. Hughes, and Trevor Pinch, eds., *The Social Construction of Technological Systems* (Cambridge: MIT Press, 1987); and esp. Bruno Latour's *Les microbes, guerre et paix, suivi de irréductions* (Paris: Métailié, 1984) and *The Pasteurization of France, Followed by Irreductions: A Politico-Scientific Essay* (Cambridge: Harvard University Press, 1988). Borrowing from Michel Tournier's *Vendredi* (Paris: Gallimard, 1967), *Les microbes* (p. 171), Latour's brilliant and maddening aphoristic polemic against all forms of reductionism, makes the essential point for feminists: "Méfiez-vous de la pureté; c'est le vitriol de l'ame" (Beware of purity; it is the vitriol of the soul). Latour is not otherwise a notable feminist theorist, but he might be made into one by readings as perverse as those he makes of the laboratory, that great machine for making significant mistakes faster than anyone else can, and so gaining world-changing power. The laboratory for Latour is the railroad industry of epistemology, where facts can only be made to run on the tracks laid down from the laboratory out. Those who control the railroads control the surrounding territory. How could we have forgotten? But now it's not so much the bankrupt railroads we need as the satellite network. Facts run on light beams these days.

2 For an elegant and very helpful elucidation of a noncartoon version of this argument see Hayden White, *The Content of the Form: Narrative Discourse and Historical Representation* (Baltimore: Johns Hopkins University Press, 1987). I still want more; and unfulfilled desire can be a powerful seed for changing the stories.

3 In "Through the Lumen: Frankenstein and the Optics of Re-Origination" (Ph.D. diss.) University of California at Santa Cruz, 1988), Zoe Sofoulis has produced a dazzling (she will forgive me the metaphor) theoretical treatment of technoscience, the psychoanalysis of science fiction culture, and the metaphorics of extraterrestrialism, including a wonderful focus on the ideologies and philosophies of light, illumination, and discovery in Western mythics of science and technology. My essay was revised in dialogue with Sofoulis's arguments and metaphors in her dissertation.

4 Nancy Hartsock, *Money, Sex, and Power: An Essay on Domination and Community* (Boston: Northeastern University Press, 1984).

5 Crucial to this discussion are Sandra Harding, *The Science Question in Feminism* (Ithaca: Cornell University Press, 1987); Evelyn Fox Keller, *Reflections on Gender and Science* (New Haven: Yale University Press, 1984); Nancy Hartsock, "The Feminist Standpoint: Developing the Ground for a Specifically Feminist Historical Materialism in *Discovering Reality: Feminist Perspectives on Epistemology, Metaphysics, and Philosophy of Science*, eds. Sandra Harding and Merrill B. Hintikka (Dordrecht, The Netherlands Reidel, 1983): 283–310; Jane Flax's "Political Philosophy and the Patriarchal Unconcscious," in *Discovering Reality*, 245–81; and "Postmodernism and Gender Relations in Feminist Theory," *Signs* 12 (Summer

1987): 621–43; Evelyn Fox Keller and Christine Grontkowski, "The Mind's Eye," in *Discovering Reality*, 207–24; Hilary Rose, "Women's Work, Women's Knowledge," in *What is Feminism? A Re-Examination*, eds. Juliet Mitchell and Ann Oakley (New York: Pantheon, 1986), 161–83; Donna Haraway, Manifesto for Cyborgs: Science, Technology, and Socialist Feminism in the 1980 *Socialist Review*, no. 80 (March–April 1985): 65–107; and Rosalind Pollack Petches. "Fetal Images: The Power of Visual Culture in the Politics of Reproduction," *Feminist Studies* 13 (Summer 1987): 263–92.

Aspects of the debates about modernism and postmodernism affect feminist analysis of the problem of "objectivity." Mapping the fault line between modernism and postmodernism in ethnography and anthropology – in which the high stakes are an authorization or prohibition to craft *comparative* knowledge across "cultures" – Marilyn Strathern made the crucial observation that it is not the written ethnography that's parallel to the work of art as object-of-knowledge, but the *culture*. The Romantic and modernist natural-technical objects of knowledge, in science and in other cultural practice, stand on one side of this divide. The postmodernist formation stands on the other side, with its "anti-aesthetic" of permanently split, problematized, always receding as deferred "objects" of knowledge and practice, including signs, organisms, systems, selves, and cultures. "Objectivity" in a postmodern framework cannot be about the problematic *objects*; it must be about specific prosthesis and always partial translation. At root, objectivity is about crafting *comparative* knowledge: How may a community name things to be stable and to be like each other? In postmodernism, this question translates into a question of the politics of redrawing of boundaries in order to have non-innocent conversations and connections. What is at stake in the debates about modernism and postmodernism is the pattern of relationships between and within bodies and language. This is a crucial matter for feminists. See Marilyn Strathern, "Out of Context: The Persuasive Fictions of Anthropology," *Current Anthropology* 28 (June 1987): 251–81, and "Partial Connections," Munro Lecture, University of Edinburgh, November 1987, unpublished manuscript.

6 Harding, 24–26, 161–62.

7 John Varley's science fiction short story, "The Persistence of Vision," in *The Persistence of Vision* (New York: Dell, 1978), 263–316, is part of the inspiration for this section. In the story, Varley constructs a utopian community designed and built by the deaf-blind. He then explores these people's technologies and other mediations of communication and their relations to sighted children and visitors. In the story, "Blue Champagne," in *Blue Champagne* (New York: Berkeley, 1986), 17–79, Varley transmutes the theme to interrogate the politics of intimacy and technology for a paraplegic young woman whose prosthetic device, the golden gypsy, allows her full mobility. But because the infinitely costly device is owned by an intergalactic communications and entertainment empire, for which she works as a media star making "feelies," she may keep her technological, intimate, enabling, other self only in exchange for her complicity in the commodification of all experience. What are her limits to the reinvention of experience for sale? Is the personal political under the sign of simulation? One way to read Varley's repeated investigations of finally always limited embodiments, differently abled beings, prosthetic technologies, and cyborgian encounters with their finitude, despite their extraordinary transcendence of "organic" orders, is to find an allegory for the personal and political in the historical mythic time of the late twentieth century, the era of techno-biopolitics. Prosthesis becomes a fundamental category for understanding our most intimate selves. Prosthesis is semiosis, the making of

meanings and bodies, not for transcendence, but for power-charged communication.

8 C.D.B. Bryan, *The National Geographic Society: 100 Years of Adventure and Discovery* (New York: Harry N. Abrams, 1987), 352.

9 I owe my understanding of the experience of these photographs to Jim Clifford, University of California at Santa Cruz, who identified their "land ho!" effect on the reader.

10 Bryan, 454.

11 See Hartsock, "The Feminist Standpoint: Developing the Ground for a Specifically Feminist Historical Materialism"; and Chela Sandoral, *Yours in Struggle: Women Respond to Racism* (Oakland: Center for Third World Organizing. n.d.); Harding; and Gloria Anzaldúa, *Borderlands/La Frontera* (San Francisco: Spinsters/Aunt Lute, 1987).

12 Annette Kuhn, *Women's Pictures: Feminism and Cinema* (London: Routledge & Kegan Paul, 1982), 3–18.

13 Joan Scott reminded me that Teresa de Lauretis put it like this:

> Differences among women may be better understood as differences within women. . . . But once understood in their constitutive power – once it is understood, that is, that these differences not only constitute each woman's consciousness and subjective limits but all together define the *female subject of feminism* in its very specificity, is inherent and at least for now irreconcilable contradiction – these differences, then, cannot be again collapsed into a fixed identity, a sameness of all women as Woman, or a representation of Feminism as a coherent and available image.

See Theresa de Lauretis, "Feminist Studies/Critical Studies: Issues, Terms, and Contexts," in her *Feminist Studies/Critical Studies* (Bloomington: Indian University Press, 1986), 14–15.

14 Chandra Mohanty, "Under Western Eyes," *Boundary* 2 and 3 (1984): 333–58.

15 See Sofoulis, unpublished manuscript.

16 In *The Science Question in Feminism* (p. 18), Harding suggests that gender has three dimensions, each historically specific: gender symbolism, the social-sexual division of labor, and processes of constructing individual gendered identity. I would enlarge her point to note that there is no reason to expect the three dimensions to covary or codetermine each other, at least not directly. That is, extremely steep gradients between contrasting terms in gender symbolism may very well not correlate with sharp social-sexual divisions of labor or social power, but they may be closely related to sharp racial stratification or something else. Similarly, the processes of gendered subject formation may not be directly illuminated by knowledge of the sexual division of labor or the gender symbolism in the particular historical situation under examination. On the other hand, we should expect mediated relations among the dimensions. The mediations might move through quite different social axes of organization of both symbols, practice, and identity, such as race – and vice versa. I would suggest also that science, as well as gender or race, might be usefully broken up into such a multipart scheme of symbolism, social practice, and subject position. More than three dimensions suggest themselves when the parallels are drawn. The different dimensions of, for example, gender, race and science might mediate relations among dimensions on a parallel chart. That is, racial divisions of labor might mediate the patterns of connection between symbolic connections and formation of individual subject positions on the science or gender chart. Or formations of gendered or racial subjectivity might

mediate the relations between scientific social division of labor and scientific symbolic patterns.

The chart below begins an analysis by parallel dissections. In the chart (and in reality?), both gender and science are analytically asymmetrical; that is, each term contains and obscures a structuring hierarchicalized binary opposition, sex/gender and nature/science. Each binary opposition orders the silent term by a logic of appropriation, as resource to product, nature to culture, potential to actual. Both poles of the opposition are constructed and structure each other dialectically. Within each voiced or explicit term, further asymmetrical splittings can be excavated, as from gender, masculine to feminine, and from science, hard sciences to soft sciences. This is a point about remembering how a particular analytical tool works, willy-nilly, intended or not. The chart reflects common ideological aspects of discourse on science and gender and may help as an analytical tool to crack open mystified units like Science or Woman.

GENDER	SCIENCE
1) symbolic system	symbolic system
2) social division of labor (by sex, by race, etc.)	social division of labor (e.g., by craft or industrial logics)
3) individual identity/subject position (desiring/desired; autonomous relational)	individual identity/subject position (knower/known; scientist/other)
4) material culture (e.g., gender paraphernalia and daily gender technologies, the narrow tracks on which sexual difference runs)	material culture (e.g., laboratories, the narrow tracks on which facts run)
5) dialectic of construction and discovery	dialectic of construction and discovery

17 Katie King, "Canons without Innocence" (Ph.D. diss., University of California at Santa Cruz, 1987).

18 Evelyn Fox Keller, in "The Gender/Science System: Or, Is Sex to Gender As Nature Is to Science?" (*Hypatia* 2 [Fall 1987]: 37–49], has insisted on the important possibilities opened up by the construction of the intersection of the distinction between sex and gender, on the one hand, and nature and science, on the other. She also insists on the need to hold to some nondiscursive grounding in "sex" and "nature," perhaps what I am calling the "body" and "world."

19 See Sofoulis, chap. 3.

20 Donna Haraway, *Primate Visions: Gender, Race, and Nature in the World of Modern Science* (New York: Routledge & Kegan Paul), forthcoming Spring 1989.

21 Katie King, prospectus for "The Passing Dreams of Choice . . . Once Before and After: Audre Lorde and the Apparatus of Literary Production" (*MS*, University of Maryland, College Park, Maryland, 1987).

5 Liz Bondi

'Feminism, Postmodernism and Geography: Space for Women?'

From: *Antipode* **22**, 156–67 (1990)

According to its proponents, postmodernism seeks to recover that which existing cultural forms, social theories and epistemologies have excluded. 'Space' is numbered among those exclusions and geographers have noted that postmodernism appears to sensitise diverse traditions in social thought to geographical difference (Gregory, 1989). The reality behind that appearance is disputed, hence responses to postmodernism that vary from the enthusiastic (Cooke, 1989) through guarded excitement (Graham, 1988) to hostility (Harvey, 1987) shading into derision (Lovering, 1989). But what is notable is that the same issues and perspectives remain marginal: women, ethnic minorities and collected 'others' get tagged along as categories to be 'recovered', with or without the benefit of postmodernism. The possibility that we might be 'in there' already, that alternative perspectives on the dichotomy between 'self' and 'other' already exist, goes unnoticed (Morris, 1988, 11–16). In particular, the transformative potential of feminism is simply ignored; it remains outside 'the project' of radical geography as well as mainstream geography (Christopherson, 1989), as continued silence about the vigorous debate between feminism and postmodernism testifies.

This paper challenges such silence. In the first section I outline geographical discussions of postmodernism arguing that, whether critical or celebratory, these are characterised by a premature foreclosure of key issues raised, or at least highlighted, by postmodernism. Thus, postmodernism is debated, or in some cases assimilated, within existing theoretical frameworks in ways that resist any fundamental challenge to existing 'radical' geography. Second, I consider how feminists might respond to these discussions of postmodernism, and in so doing comment on the significance of postmodernism for feminist geography. In this section my comments focus on the gender coding of knowledge and on the question of difference.

Postmodernism: geographical encounters

Whereas poststructuralism insists on the instability of meanings, postmodernism exploits such instability, widening the gap between signifier and signified until images are liberated from any constant points of reference. Consequently, postmodernism defies definition and slips out in the myriad of spaces (absences) of my preceding sentence. But postmodernism leaves a trace, and I consider three elements of this trace, concerned with architectural style, cultural change and social theory respectively, that are apparent in geographical writings.

Postmodernism as architectural style provoked attention in the context of interpretations of the built environment (Dear, 1986; Knox, 1987). It has sometimes been welcomed as a humanising of the environment, in which value is attached to a human scale, to personal associations and to local histories (e.g. Ley, 1987). Others have been critical, interpreting it merely as another, more frenetic, phase of consumerism, commodification and profit-making (e.g. Harvey, 1987). Characterised by multiple allusions, juxtaposition and, above all, double codings, such different readings are of course intrinsic to postmodernism. But, more important for my argument is that, in different ways, these commentators interpret postmodernism as a cultural landscape, associated with a new urban middle-class. Analyses explore tensions within both postmodernism and the new class, drawing out, for example, both the oppositional potential of critiques of modernist styles and the self-satisfied, celebratory ransacking and packaging of history to provide heritage (e.g. Mills, 1988; Hutcheon, 1989). In this way, postmodernism is related to complex changes in the social order, and especially to reconfigurations of middle-class identity (see Dickens, 1989).

At this point the notion of postmodernism as architectural style converges with a broader notion of postmodernism as a 'sea change' in cultural experience and representation, or as a 'profound shift in the structure of feeling'. In this context, the key contribution of geographers has been to explore such claims in terms of the experiences and representation of time and space, and is best represented by David Harvey's account *The Condition of Postmodernity*. Developing Jameson's analysis of 'hyperspace', and recuperating earlier existential accounts of place and placelessness (e.g. Relph, 1976), Harvey (1989) offers an interpretation of four phases of cultural transformation in terms of what he calls 'time-space compression'. This phrase firmly roots his discussion in an historical-materialist analysis, in which postmodernism is essentially a manifestation of the transition from Fordism to flexible modes of accumulation. Extending his earlier analysis of the historical geography of capitalism, Harvey argues that crises of overaccumulation induce a search for spatial and temporal resolutions that create an overwhelming and disruptive sense of time-space compression. The current phase began in the late 1960s/early 1970s and prompted a series of responses that transform our sense of space and time through, for example, (a) the virtual collapse of spatial barriers as new technological and organisational forms permit the almost instantaneous transfer of information and capital around the world, and (b) the adoption of a wide variety of means of reducing turnover time from the switch to 'just-in-time' systems to the commodification of ephemeral images. Consequences include the fragmentation of time into 'a series of perpetual presents', and increased sensitivity to small spatial variations. For Harvey, the cultural shifts that follow from this involve attempts to resist, explain and represent the sense of time-space compression. The transformations effected in this way illustrate that space, place and scale are social constructs not external givens. Thus

> not only is the fragility and transitoriness of contemporary social relations expressed 'in space', the production of space increasingly constructs social difference.
>
> (Smith, 1989, 12)

Hence, time-space compression generates crises in representation as well as in experience.

Harvey's interpretation of postmodernism entails a two-fold claim about spatiality. First, in his dissection of modernity from the Enlightenment to the late twentieth century, he is unravelling ideas about time and space as inseparable albeit geographically and historically varying. He is not claiming the primacy of either space or time. Rather, his discussion of modernism stresses tensions between the immutable and the transient, an associated heightened awareness of time and a concomitant subduing of space. Consequently space is taken for granted; it is treated as 'natural' rather than socially created. Second, in his discussion of postmodernity, Harvey is arguing that this phase of time-space compression is greatly heightening awareness of, and is in effect 'de-naturalizing', geographical space. This heightened awareness is itself manifest within discourse on postmodernism and has led some geographers to talk of a 'reassertion of space within social theory' (see Cooke, 1989; Gregory, 1989; and especially Soja, 1989).

This brings me to the third strand, within which postmodernism is interpreted as a critique of the primacy accorded to time and history in social theory, and as expressing the recuperation of geography and spatiality. Inspiration is taken from an apparent geographical turn in the work of social and cultural theorists such as Foucault and Berger. This involves both recognition of past refusals to consider the potentialities of space, and forceful declarations of its current ascendancy, as the following citations exemplify.

> Space was treated as the dead, the fixed, the undialectical, the immobile. Time on the contrary was richness, fecundity, life, dialectic.
>
> (Foucault, 1980, cited by Soja, 1989, 10)

> The present epoch will perhaps be above all the epoch of space. We are in the epoch of simultaneity: we are in the epoch of juxtaposition, the epoch of the near and far, of the side-by-side, of the dispersed. We are at a moment, I believe, when our experience of the world is less that of a long life developing through time than that of a network that connects points and intersects with its own skein.
>
> (Foucault, 1986, cited by Soja, 1989, 10)

> Prophesy now involves a geographical rather than historical projection; it is space not time that hides consequences from us.
>
> (Berger, 1974, cited by Soja, 1989, 22)

This emphasis on space has prompted a geographical recasting of the postmodern critique of meta-narratives and grand theory such that the source of totalising tendencies in major bodies of social theory is traced to the unidimensionality and the unidirectionality of time. Space, by contrast encourages a shift away from universals, and greater sensitivity to difference, local discourses etc. Spatialising tendencies are therefore considered to be intrinsic to theoretical developments influenced by postmodernism. For geography, the claim is that what began as a tentative rapprochement with social theory has now gathered pace and promises a place at the heart of social theory. For some, this analysis is celebratory of both geography and postmodernism, and involves a reinterpretation of geographical research projects (embarked upon under different banners) as archetypal postmodern enterprises (Cooke, 1989). Others warn against the naive appropriation of spatial concepts, cautioning that celebration of geographical difference merely obscures relations of power, inequality and hierarchy (Smith, 1989).

These strands in the encounter between geography and postmodernism are neither exhaustive nor mutually exclusive. They serve to illustrate that postmodernism is difficult to limit: one can begin with a view of postmodernism as a relatively well-defined architectural style but consideration of its meaning and significance lead to broader questions about cultural experience and intellectual practice. Yet, in some ways, I think the geographical encounter with postmodernism displays a remarkable degree of containment. At one level, the response of geographers to postmodernism has been predictable and no different from responses to other intellectual innovations, namely to insist that the spatial and environmental context of social life be taken seriously. Thus, existing interest within human geography in the representation of social relations in the spaces we create and inhabit, and in the social consequences of those creations, has been extended rather than transformed. Further, to use a currently favoured metaphor, most geographers seem keen to 'ground' postmodernism by tying it to some kind of material social 'reality', thereby refusing the more radical claims that no 'reality' is material and that there is no basis upon which to choose between competing and free-floating images. I am not wholly averse to such attempts but I think they are undertaken prematurely. Reading geographical writings on postmodernism, I repeatedly have the feeling that the real import of cultural and intellectual developments is being evacuated in a rush to ensure containment within existing categories. Nowhere is this more apparent than in relation to feminism and gender issues.

Postmodernism and geography: space for women?

Although feminism has never achieved a high profile in geography, awareness of gender issues has increased over the last decade. The most blatant forms of linguistic sexism are disappearing as academic journals and publishers issue new guidelines encouraging the use of gender-neutral language. However, it seems to me that what amounts to a polite sanitisation of language is serving to obscure important issues. For example, the phrase 'master-narrative' is conspicuous by its absence in geographical discussions of postmodernism (see Owens, 1983). My suspicion is that careful avoidance of obviously gendered language is in practice a new strategy for avoiding to think about the importance of gender in intellectual practice. And in other ways, discussions of postmodernism reveal more familiar strategies of avoidance. The illustrations included in David Harvey's discussion of postmodern culture (pp. 39 *et seq.*) consist entirely of images of women and yet this goes unmentioned in his commentary. Slightly more subtly, both Harvey (1989) and Soja (1989) claim that historical materialism can and should 'recuperate' issues of race and gender, but no attempt is made to do so, and in any case the ultimate superiority of class politics is unambiguously reasserted in both accounts (see for example, Harvey, 1989, pp. 353 *et seq.*). Similarly, discussions of postmodernism as expressive of reconfigurations of middle-class identity ignore the centrality of gender differentiation and inequality in class relations (Phillips, 1987; Johnson, 1989; Walby, 1989).

What these and numerous other examples add up to is a refusal to open up the symbolic categories Woman/Man, or the sociological categories women/men, to scrutiny. In this there is little difference between postmodernism and other subjects on which geographers have written. But, since the claim that postmodernism is an intrinsically geographical project has counterparts in some feminist responses to postmodernism, geographers might do well to consider the ramifications of selective deafness. Further, if feminists are to do more than recycle existing critiques, the relationship between feminism and postmodernism must be explored.

Gender and space

Postmodernism is claimed by some to be symptomatic of feminist interventions in cultural and intellectual practice. For example, feminism has been highly critical of claims to universality in philosophy and political theory, on the grounds that they are rooted in a culturally specific conception of the individual as a masculine subject (Griffiths and Whitford, 1988; Pateman, 1988). Feminism can, therefore, be viewed as a particular version, and perhaps an instigator, of the postmodern attack on discourses that claim privileged access to truth. Conversely postmodernism would appear to endorse and broaden the basis of feminist critiques of patriarchal discourse, leading Flax (1987) to claim that feminism is a 'type of postmodern philosophy' (see also Weedon, 1987). To elaborate, postmodernism diagnoses a crisis in the authority of Western intellectual thought and culture, and responds by attempting to recover and recuperate that which the associated meta-narratives exclude. According to Alice Jardine (1985, 25)

> such thinking has involved above all, a reincorporation and reconceptualization of that which has been the master narratives' own 'nonknowledge', what has eluded them, what has engulfed them. This other-than-themselves is almost always a 'space' of some kind (over which the narrative has lost control), and this space has been coded as feminine, as woman.

This is crucial for geography because the coding of knowledge/non-knowledge or representable/unrepresentable as masculine/feminine is closely entwined with issues of time and space. The 'other' in Jardine's statement is represented as a space. Although used metaphorically, it is precisely this widespread appeal to spatial metaphors that has so excited geographers. After all, if everyone is now 'thinking spatially', what was once a private geographical obsession is rapidly gaining converts and surely, as Harvey argues, attempts to represent contemporary experiences are not entirely unrelated to their material context. Moreover, 'otherness' is not just conceived of as a different space or another country; it stands in opposition to time. It represents the polymorphous, the multi-dimensional, as opposed to the linear, the singular, the uni-directional. Harvey, Jardine and many others link together time and becoming on the one hand, space and being on the other. What the geographers then fail to do is consider the coding of the former as masculine and the latter as feminine. Some, to be fair, recognise the systematic silencing of collected 'others', including women, gays and blacks, within modernist intellectual practice, and suggest that postmodernism entails a shift from the masculine to the androgynous. But associations between femininity and both space and being remain studiously ignored. Thus, Michael Dear (1986, 367) quotes Jencks's assertion that '[d]efining our world today as Post-Modern is rather like defining women as non-men' and, like Jencks, appears to intend no irony: both seem to be saying 'I'm not so sexist as to consider women as negatives' while politely ignoring the sexism that enabled Western philosophy to do precisely that, i.e. to define women as non-men, as other, as unknowable. Alongside this polite avoidance, I am tempted to wonder whether the association between femininity and space might be just a bit too threatening: male geographers might be faced with deeply unsettling thoughts about their own predilections.

One possibility for feminist geographers might therefore be simply to insist on the relevance of this coding, step up the challenge to our male colleagues, and, in effect, recover geography for ourselves. This response, however, would seem to me to take neither postmodernism nor feminism far enough. Postmodernism, after all promises to dissolve, to go beyond, these dichotomies. It celebrates their instabilities and the possibility of playing with reversals and recodings. Anarchy is supposed to supersede familiar hierachies based on fixed categories, leading, at least in some interpretations, to the polarisation between femininity and masculinity dissolving into androgyny. But, at this point, feminism and postmodernism are, surely, deeply in conflict. Whatever the instabilities of feminine and masculine codings, a feminist analysis insists that gender relations and gender hierarchies cannot just, playfully, be wished away (Fraser and Nicholson, 1988). What postmodernism appears to do is to elide rather than deconstruct a dichotomy between ideas and materiality. Postmodernism may recognise the masculine bias of Western intellectual traditions, but it is accompanied by a preoccupation with gender symbolism at the expense of 'flesh and blood' women and men. In other words, the masculinity of ideas within these traditions is divorced from the maleness of their progenitors. This separation is carried over into postmodernism in an appropriation of the feminine that is as untarnished by 'real women' as the gender-neutral language of much contemporary human geography. Hence the refusal to acknowledge that gender difference is about systematic inequalities as well as different discourses; that it is about power and politics. This refusal leads Suzanne Moore (1988, 167) to characterise the postmodern venture as a 'new kind of gender tourism, whereby male theorists are able to take package trips into the world of femininity', in which they 'get a bit of the other' in the knowledge that they have return tickets to the safe, familiar and, above all, empowering terrain of masculinity.

For feminist geographers, therefore, it is important to maintain a distinction between the coding of space as feminine and the existence of women in geographical space. While the former suggests a convergence between feminism and postmodernism, the latter suggests divergence. Further, whereas gender symbolism entails a dichotomy, the notion of 'real women' introduces more complex forms of differentiation. To take this encounter further, it is necessary to consider concepts of difference employed by feminism and postmodernism.

Feminism and difference

The gender codings so far discussed invoke what Michèle Barrett terms a 'positional' concept of difference. This concept has, in a sense, relativised meaning and therefore made it possible

> to criticize and deconstruct the 'unified subject', whose appearance of universality disguised a constitution structured specifically around the subjectivity characteristic of the white, bourgeois male.
>
> (Barrett, 1987, 35)

Hence the convergence between postmodernism and feminism. But, postmodernism is radically anti-foundational and anti-essentialist, so that positional difference proliferates into a fragmentation of the subject and a differentiation between subjects so total that its effects are indistinguishable from the coherent, unified, stable conception of the subject it opposes. Thus,

> [f]or the liberal, race, class and gender are ultimately irrelevant to questions of justice and truth because 'underneath we are all the same'. For the post-structuralist [and postmodernist], race, class and gender are constructs and, therefore, incapable of decisively validating conceptions of justice and truth because underneath there lies no natural core to build on or liberate or maximize. Hence, once again, underneath we are all the same.
>
> (Alcoff, 1988, 420–1)

Feminism necessarily resists such a paralysing conception of difference, with its reactionary implications of 'post-feminism'. One response, associated with radical feminism, is to invoke a unified female subject as an alternative to the male subject of Enlightenment rationality. But here feminism comes into conflict not only with the anti-essentialism and de-centred subject of postmodernism but also with experiential, cultural and power-laden differences among women.

How to grapple with these differences has been a central issue for feminists in recent years and is beginning to be identified as a key theoretical issue among feminist geographers (McDowell, 1991). Socialist-feminist attempts to negotiate between class and gender differences have been extended within geography in the context of spatial variations in the impact and evolution of cultural and economic conditions (Bowlby, Lewis, McDowell and Foord, 1989). These formulations attempt to break down monolithic categories of masculinity and femininity, and to explore different constructions associated with different places and class positions. This kind of approach constructs differentiated human subjects from the 'outside', as occupants of positions in a kind of dynamic, multidimensional grid. It is an approach far removed from postmodernism in that it retains a commitment to the meta-narrative, albeit informed by positions understood to be other than universal. Thus, it insists on the authenticity of the experience of oppressed groups, although also, and in some ways contradictorily, accepting that experience and identity are socially constructed and capable of being transformed. But it implies that transformation proceeds from the outside in rather than the inside out.

Elsewhere feminists have drawn on poststructuralism to explore the construction of differentiated human subjects from the inside, especially via language, while refusing to wholly relativize the concept of gender (Weedon, 1987; Alcoff, 1988; Poovey, 1988; Scott, 1988). This entails retaining a notion of power relations. I read these accounts as negotiating a knife-edge between positional and experiential concepts of difference, as exploring the relationship between gender as a symbolic construct and gender as a set of social relations, and as attempting to reconcile femininity as a condition and as a process. Thus, feminist poststructuralists accept that women are caught within patriarchal definitions of femininity, and that the binary opposition of 'men' and 'women' is a patriarchal construct. There is, therefore, no "essential femininity' on which to base a feminist resistance. Rather, feminism must challenge the appeal to essentialism that underpins women's oppression in its 'endless variety and monotonous similarity' (Rubin, 1975, cited by Fraser and Nicholson, 1988, 383). In so doing, what unites women (falsely) will be undermined so that 'in the long run . . . feminists will need to write not only the history of women's oppression but also the future of gender difference(s)' (Poovey, 1988, 63). And, with or without postmodern spatial metaphors, that future is as geographical as its antecedents.

Feminist poststructuralism offers a more far-reaching critique of postmodernism in which convergences and divergences are again apparent. Hutcheon (1989, 11) characterises postmodernism as 'complicity and critique . . . that at once inscribes and subverts the conventions and ideologies of the dominant cultural and social forces of the twentieth century Western world.' Despite its critique of patriarchal discourse, postmodernism remains complicit in patriarchal practice. Feminism also inscribes and subverts: it struggles to move beyond the parameters of patriarchal practice but necessarily draws on resources bequeathed by patriarchy. But, whereas the radical relativism of postmodernism leads to political paralysis, the increasing sensitivity to difference within feminism is combined with an ideal of unity (Lovibund, 1989) that ensures political purpose is never eclipsed.

A number of feminists have pointed out that historical periodizations are rooted in gender-specific perspectives (Jardine, 1985; Christopherson, 1989). Consequently the trajectory of modernity/postmodernity has different implications for women and men. The notion that women have been excluded from modernity, to be recovered within postmodernity or by post-Enlightenment Marxism, merely reinscribes women's marginalisation within patriarchal culture. Without feminism radical geography reinforces this categorisation whether embracing or resisting postmodernism.

References

Alcoff, L. (1988) Cultural feminism versus poststructuralism: the identity crisis in feminist theory. *Signs: Journal of Women in Culture and Society* 13: 405–436.

Barrett, M. (1987) The concept of 'difference'. *Feminist Review* 26: 29–41.

Bowlby, S., J. Lewis, L. McDowell, and J. Foord (1989) The geography of gender. In R. Peet and N. Thrift (Eds) *New Models in Geography* (Volume II). London: Unwin Hyman, 157–175.

Christopherson, S. (1989) On being outside 'the project'. *Antipode* 21: 83–89.

Cooke, P. (1989) The contested terrain of locality studies. *Tijdschrift voor Economische en Sociale* 80: 14–29.

Dear, M. (1986) Postmodernism and planning. *Environment and Planning D: Society and Space* 4: 367–384.

Dickens, P. (1989) *Postmodernism, locality and the middle classes*. Paper presented at the Seventh Urban Change and Conflict Conference, Bristol, September 1989.

Flax, J. (1987) Postmodernism and gender relations in feminist theory. *Signs: Journal of Women in Culture and Society* 12: 621–643.

Fraser, N. and L. Nicholson (1988) Social criticism without philosophy: an encounter between feminism and postmodernism. *Theory, Culture and Society* 5: 373–394.

Graham, J. (1988) Post-modernism and marxism. *Antipode* 20: 60–66.

Gregory, D. (1989) Areal differentiation and post-modern human geography. In D. Gregory and R. Walford (Eds) *Horizons in Human Geography*. London: Macmillan, 67–96.

Griffiths, M. and M. Whitford (Eds) (1988) *Feminist Perspectives in Philosophy*. London: Macmillan.

Harvey, D. (1987) Flexible accumulation through urbanisation: reflection on 'postmodernism' in the American city. *Antipode* 19: 260–286.

Harvey, D. (1989) *The Condition of Postmodernity*. Oxford: Blackwell.

Hutcheon, L. (1989) *The Politics of Postmodernism*. London: Routledge.

Jardine, A. (1985) *Gynesis. Configurations of women and Modernity*. Ithaca: Cornell University Press.

Johnson, L. (1989) Weaving workplaces: sex, race and ethnicity in the Australian textile industry. *Environment and Planning A* 21: 681–4.

Knox, P. (1987) The social production of the built environment: architects, architecture and the postmodern city. *Progress in Human Geography* 11: 354–378.

Ley, D. (1987) Styles of the times: liberal and neo-conservative landscapes in inner Vancouver, 1968–1986. *Journal of Historical Geography* 13: 40–56.

Lovering, J. (1989) Postmodernism, marxism, and locality research: the contribution of critical realism to the debate. *Antipode* 21: 1–12.

Lovibund, S. (1989) Feminism and postmodernism. *New Left Review* 178: 5–28.

McDowell, L. (1991) The baby and the bathwater: diversity, deconstruction and feminist theory in geography. *Geoforum* 22: 123–34.

Mills, C. (1988) 'Life on the upslope': the postmodern landscape of gentrification. *Environment and Planning D: Society and Space* 6: 169–189.

Moore, S. (1988) Getting a bit of the other – the pimps of postmodernism. In R. Chapman and J. Rutherford (Eds) *Male Order*. London: Lawrence and Wishart, 165–192.

Morris, M. (1988) *The Pirate's Fiancee*. London: Verso.
Owens, C. (1983) The discourse of others: feminism and postmodernism. In H. Foster (Ed.) *Postmodern Culture*. London: Pluto, 57–82.
Pateman, C. (1988) *The Sexual Contract*. Cambridge: Polity.
Phillips, A. (1987) *Divided Loyalties*. London: Virago.
Poovey, M. (1988) Feminism and deconstruction. *Feminist Studies* 14: 51–65.
Relph, E. (1976) *Place and Placelessness*. London: Pion.
Scott, J.W. (1988) Deconstructing equality-versus-difference: or, the uses of post-structuralist theory for feminism. *Feminist Studies* 14: 33–50.
Smith, N. (1989) *Geography, difference and the politics of scale*. Paper presented at Conference on Postmodernism and the Social Science, St Andrews, August, 1989.
Soja, E. (1989) *Postmodern Geographies*. London: Verso.
Walby, S. (1989) *Theorising Patriarchy*. Oxford: Blackwell.
Weedon, C. (1987) *Feminist Practice and Poststructuralist Theory*. Oxford: Blackwell.

6 Chandra Talpade Mohanty

'Feminist Encounters: Locating the Politics of Experience'

Reprinted from: M. Barrett and A. Phillips (eds) *Destabilising Theory: Contemporary Feminist Debates*, pp. 74–92. Cambridge: Polity (1992)

Feminist and anti-racist struggles in the 1990s face some of the same urgent questions encountered in the 1970s. After two decades of engagement in feminist political activism and scholarship in a variety of socio-political and geographical locations, questions of difference (sex, race, class, nation), experience and history remain at the centre of feminist analysis. Only, at least in the US academy, feminists no longer have to contend as they did in the 1970s with phallocentric denials of the legitimacy of gender as a category of analysis. Instead, the crucial questions in the 1990s concern the construction, examination and, most significantly, the institutionalization of difference *within* feminist discourses. It is this institutionalization of difference that concerns me here. Specifically, I ask the following question: how does the politics of location in the contemporary USA determine and produce experience and difference as analytical and political categories in feminist 'cross-cultural' work? By the term 'politics of location' I refer to the historical, geographical, cultural, psychic and imaginative boundaries which provide the ground for political definition and self-definition for contemporary US feminists.[1]

Since the 1970s, there have been key paradigm shifts in western feminist theory. These shifts can be traced to political, historical, methodological and philosophical developments in our understanding of questions of power, struggle and social transformation. Feminists have drawn on decolonization movements around the world, on movements for racial equality, on peasant struggles and gay and lesbian movements, as well as on the methodologies of Marxism, psychoanalysis, deconstruction and post-structuralism to situate our thinking in the 1990s. While these developments have often led to progressive, indeed radical analyses of sexual difference, the focus on questions of subjectivity and identity which is a hallmark of contemporary feminist theory has also had some problematic effects in the area of race and Third World/post-colonial studies. One problematic effect of the post-modern critique of essentialist notions of identity has been the dissolution of the category of race – however, this is often accomplished at the expense of a recognition of racism. Another effect has been the generation of discourses of diversity and pluralism which are grounded in an apolitical, often individualized identity politics.[2] Here, questions of *historical interconnection* are transformed into questions of discrete and separate histories (or even herstories) and into questions of identity politics.[3] While I cannot deal with such effects in detail here, I work through them in a limited way by suggesting the importance of analysing and theorizing difference in the context of feminist cross-cultural work. Through this theorization of experience, I suggest that historicizing and locating political agency is a necessary alternative to formulations of the 'universality' of gendered oppression of struggles. This universality of gender oppression is problematic, based as it is on the assumption that the categories of race and class have to be invisible for gender to be visible. In the 1990s, the challenges posed by black and Third World feminists can point the way towards a more precise, transformative feminist politics. Thus, the juncture of feminist and anti-racist/Third World/post-colonial studies is of great significance, materially as well as methodologically.[4]

Feminist analyses which attempt to cross national, racial and ethnic boundaries produce and reproduce difference in particular ways. This codification of difference occurs through the naturalization of analytic categories which are supposed to have cross-cultural validity. I attempt an analysis of two recent feminist texts which address the turn of the century directly. Both texts also foreground analytic categories which address questions of cross-cultural, cross-national differences among women. Robin Morgan's 'Planetary Feminism: The Politics of the 21st Century' and Bernice Johnson Reagon's 'Coalition Politics: Turning the Century' are both *movement* texts and are written for diverse mass audiences. Morgan's essay forms the introduction to her 1984 book, *Sisterhood is Global: The International Women's Movement Anthology*, while Reagon's piece was first given as a talk at the West Coast Women's Music Festival in 1981, and has since been published in Barbara Smith's 1983 anthology, *Home Girls: A Black Feminist Anthology*.[5] Both essays construct contesting notions of experience, difference and struggle within and across cultural boundaries. I stage an encounter between these texts because they represent for me, despite their differences from each other, an alternative presence – a thought, an idea, a record of activism and struggle – which can help me both locate and position myself in relation to 'history'. Through this presence, and with these texts, I can hope to approach the end of the century and not be overwhelmed.

The status of 'female' or 'woman/women's' experience has always been a central concern in feminist discourse. After all, it is on the basis of shared experience that feminists of different political persuasions have argued for unity or identity among women. Teresa de Lauretis, in fact, gives this question a sort of foundational status: 'The relation of experience to discourse, finally, is what is at issue in the definition of feminism.'[6] Feminist discourses, critical and liberatory in intent, are not thereby exempt from inscription in their internal power relations. Thus, the recent definition, classification and assimilation of categories of experientially based notions of 'woman' (or analogously, in some analyses, 'lesbian') to forge political unity require our attention and careful analysis. Gender is *produced* as well as uncovered in feminist discourse, and definitions of experience, with attendant notions of unity and difference, form the very basis of this production. For instance, gender inscribed within a purely male/female framework reinforces what Monique Wittig has called the heterosexual contract.[7] Here difference is constructed along male/female lines, and it is being female (as opposed to male) which is at the centre of the analysis. Identity is seen as either male or female. A similar definition of experience can also be used to craft lesbian identity. King[8] criticizes feminist analyses in which difference is inscribed simply within a lesbian/heterosexual framework, with 'experience' functioning as an unexamined, catch-all category. This is similar to the female/male framework Wittig calls attention to, for although the terms of the equation are different, the status and definition of 'experience' are the same. The politics of being 'woman' or 'lesbian' are deduced from the *experience* of being woman or lesbian. Being female is thus seen as *naturally* related to being feminist, where the experience of being female transforms us into feminists through osmosis. Feminism is not defined as a highly contested political terrain; it is the mere effect of being female.[9] This is what one might call the feminist osmosis thesis: females are feminists by association and identification with the experiences which constitute us as female.

The problem is, however, we cannot avoid the challenge of *theorizing* experience. For most of us would not want to ignore the range and scope of the feminist political arena, one characterized quite succinctly by de Lauretis:

> feminism defines itself as a political instance, not merely a sexual politics but a politics of everyday life, which later . . . enters the public sphere of expression and creative practice, displacing aesthetic hierarchies and generic categories, and . . . thus establishes the semiotic ground for a different production of reference and meaning.[10]

It is this recognition that leads me to an analysis of the status of experience and difference, and the relation of this to political praxis in Robin Morgan's and Bernice Reagon's texts.

'A place on the map is also a place in history'[11]

The last decade has witnessed the publication of numerous feminist writings on what is generally referred to as an international women's movement, and we have its concrete embodiment in *Sisterhood is Global*, a text which in fact describes itself as '*The* International Women's Movement Anthology'. There is considerable difference between international feminist networks organized around specific issues like sex-tourism and multinational exploitation of women's work, and the notion of *an* international women's movement which, as I attempt to demonstrate, implicitly *assumes* global or universal sisterhood. But it is best to begin by recognizing the significance and value of the publication of an anthology such as this. The value of documenting the indigenous histories of women's struggles is unquestionable. Morgan states that the book took twelve years in conception and development, five years in actual work, and innumerable hours in networking and fundraising. It is obvious that without Morgan's vision and perseverance this anthology would not have been published. The range of writing represented is truly impressive. At a time when most of the globe seems to be taken over by religious fundamentalism and big business, and the colonization of space takes precedence over survival concerns, an anthology that documents women's organized resistances has significant value in helping us envision a better future. In fact, it is because I recognize the value and importance of this anthology that I am concerned about the political implications of Morgan's framework for cross-cultural comparison. Thus my comments and criticisms are intended to encourage a greater internal self-consciousness within feminist politics and writing, not to lay blame or induce guilt.

Universal sisterhood is produced in Morgan's text through specific assumptions about women as a cross-culturally singular, homogeneous group with the same interests, perspectives and goals and similar experiences. Morgan's definitions of 'women's experience' and history lead to a particular self-presentation of western women, a specific codification of differences among women, and eventually to what I consider to be problematic suggestions for political strategy.[12] Since feminist discourse is productive of analytic categories and strategic decisions which have material effects, the construction of the category of universal sisterhood in a text which is widely read deserves attention. In addition, *Sisterhood is Global* is still the only text which proclaims itself as the anthology of *the* international women's movement. It has had world-wide distribution, and Robin Morgan herself has earned the respect of feminists everywhere. And since authority is always charged with responsibility the discursive production and dissemination of notions of universal sisterhood is a significant political event which perhaps solicits its own analysis.

Morgan's explicit intent is 'to further the dialogue between and solidarity of women everywhere' (p. 8). This is a valid and admirable project to the extent that one is willing to assume, if not the reality, then at least the possibility, of universal sisterhood on the basis of shared good will. But the moment we attempt to articulate the operation of contemporary imperialism with the notion of an international women's movement based on global sisterhood, the awkward political implications of Morgan's task become clear. Her particular notion of universal sisterhood seems predicated on the erasure of the history and effects of contemporary imperialism. Robin Morgan seems to situate *all* women (including herself) outside contemporary world history, leading to what I see as her ultimate suggestion that transcendence rather than engagement is the model for future social change. And this, I think, is a model which can have dangerous implications for women who do not and cannot speak from a location of white, western, middle-class privilege. A place on the map (New York City) is, after all, also a locatable place in history.

What is the relation between experience and politics in Robin Morgan's text? In 'Planetary Feminism' the category of 'women's experience' is constructed within two parameters: woman as victim, and woman as truth-teller. Morgan suggests that it is not mystical or biological commonalities which characterize women across cultures and histories, but rather a common condition and world view:

> The quality of feminist political philosophy (in all its myriad forms) makes possible a totally new way of viewing international affairs, one less concerned with diplomatic postures and abstractions, but focused instead on concrete, *unifying* realities of priority importance to the survival and betterment of living beings. For example, the historical, cross-cultural opposition women express to war and our healthy skepticism of certain technological advances (by which most men seem overly impressed at first and disillusioned at last) are only two instances of shared attitudes among women which seem basic to a common world view. Nor is there anything mystical or biologically deterministic about this commonality. It is the result of a *common condition* which, despite variations in degree, is experienced by all human beings who are born female.
>
> (p. 4)

This may be convincing up to a point, but the political analysis that underlies this characterization of the commonality among women is shaky at best. At various points in the essay, this "common condition' that women share is referred to as the suffering inflicted by a universal 'patriarchal mentality' (p. 1), women's opposition to male power and androcentrism, and the experience of rape, battery, labour and childbirth. For Morgan, the magnitude of suffering experienced by most of the women in the world leads to their potential power as a world political force, a force constituted in opposition to Big Brother in the US, Western and Eastern Europe, Moscow, China, Africa, the Middle East and Latin America. The assertion that women constitute a potential world political force is suggestive; however, Big Brother is *not exactly the same* even in, say, the US and Latin America. Despite the similarity of power interests and location, the two contexts present significant differences in the manifestations of power and hence of the possibility of struggles against it. I part company with Morgan when she seems to believe that Big Brother is the same the world over because 'he' simply represents male interests, notwithstanding particular imperial histories or the role of monopoly capital in different countries.

In Morgan's analysis, women are unified by their shared perspective (for example, opposition to war), shared goals (betterment of human beings) and shared experience of oppression. Here the homogeneity of women as a group is produced not on the basis of biological essentials (Morgan offers a rich, layered critique of biological materialism), but rather through the psychologization of complex and contradictory historical and cultural realities. This leads in turn to the assumption of women as a unified group on the basis of secondary sociological universals. What binds women together is an ahistorical notion of the sameness of their oppression and, consequently, the sameness of their struggles. Therefore in Morgan's text cross-cultural comparisons are based on the assumption of the singularity and homogeneity of women as a *group*. This homogeneity of women as a group, is, in turn, predicated on a definition of the *experience of oppression* where difference can only be understood as male/female. Morgan assumes universal sisterhood on the basis of women's shared opposition to androcentrism, an opposition which, according to her, grows directly out of women's shared status as its victims. The analytic elision between the *experience* of oppression and the *opposition* to it illustrates an aspect of what I referred to earlier as the feminist osmosis thesis: being female and being feminist are one and the same, we are *all* oppressed and hence we *all* resist. Politics and ideology as self-conscious struggles and choices necessarily get written out of such an analysis.

Assumptions pertaining to the relation of experience to history are evident in Morgan's discussion of another aspect of women's experience: woman as truth-teller. According to her, women speak of the 'real' unsullied by 'rhetoric' or 'diplomatic abstractions'. They, as opposed to men (also a coherent singular group in this analytic economy), are authentic human beings whose 'freedom of choice' has been taken away from them: 'Our emphasis is on the individual voice of a woman speaking not as an official representative of her country, but rather as a truth-teller, with an emphasis on reality as opposed to rhetoric' (p. xvi). In addition, Morgan asserts that women social scientists are 'freer of androcentric bias' and 'more likely to elicit more trust and . . . more honest responses from female respondents of their studies' (p. xvii). There is an argument to be made for women interviewing women, but I do not think this is it. The assumptions underlying these statements indicate to me that Morgan thinks women have some kind of privileged access to the 'real', the 'truth', and can elicit 'trust' from other women purely on the basis of their being not-male. There is a problematic conflation here of the biological and the psychological with the discursive and the ideological. 'Women' are collapsed into the 'suppressed feminine' and men into the dominant ideology.

These oppositions are possible only because Morgan implicitly erases from her account the possibility that women might have *acted*, that they were anything but pure victims. For Morgan, history is a male construction; what women need is herstory, separate and outside of his-story. The writing of history (the discursive and the representational) is confused with women as historical actors. The fact that women are representationally absent from history does not mean that they are/were not significant social actors in history. However, Morgan's focus on herstory as separate and outside history not only hands over all of world history to the boys, but potentially suggests that women have been universally duped, not allowed to 'tell the truth', and robbed of all *agency*. The implication of this is that women as a group seem to have forfeited any kind of material referentiality.

What, then, does this analysis suggest about the status of experience in this text? In Morgan's account, women have a sort of cross-cultural coherence as distinct from men. The status or position of women is assumed to be self-evident. However, this focus on the position of women whereby women are seen as a coherent group in *all* contexts, regardless of class or ethnicity, structures the world in ultimately Manichaean terms, where women are always seen in opposition to men, patriarchy is always essentially the invariable phenomenon of male domination, and the religious, legal, economic and familial systems are implicitly assumed to be constructed by men. Here, men and women are seen as whole groups with *already constituted* experiences as groups, and questions of history, conflict and difference are formulated from what can only be this privileged location of knowledge.

I am bothered, then, by the fact that Morgan can see contemporary imperialism only in terms of a 'patriarchal mentality' which is enforced by men as a *group*. Women across class, race and national boundaries are participants to the extent that we are 'caught up in political webs not of our making which we are powerless to unravel' (p. 25). Since women as a unified group are seen as unimplicated in the process of history and contemporary imperialism, the logical strategic response for Morgan appears to be political transcendence: 'To fight back in solidarity, however, as a real political force requires that women transcend the patriarchal barriers of class and race, and furthermore, transcend even the solutions the Big Brothers propose to the problems they themselves created' (p. 18). Morgan's emphasis on women's transcendence is evident in her discussions of (1) women's deep opposition to nationalism as practised in patriarchal society, and (2) women's involvement in peace and disarmament movements across the world, because, in her opinion, they desire peace (as opposed to men who cause war). Thus, the concrete reality of women's involvement in peace movements is substituted by an abstract 'desire' for peace which is supposed to transcend race, class and national conflicts among women. Tangible responsibility and credit for organizing peace movements is replaced by an essentialist and psychological unifying desire. The problem is that in this case women are not seen as political agents; they are merely allowed to be well intentioned. Although Morgan does offer some specific suggestions for political strategy which require resisting 'the system', her fundamental suggestion is that women transcend the left, the right, and the centre, the law of the father, God, and the system. Since women have been analytically constituted outside real politics or history, progress for them can only be seen in terms of transcendence.

The *experience* of struggle is thus defined as both personal and ahistorical. In other words, the political is *limited to* the personal and all conflicts among and within women are flattened. If sisterhood itself is defined on the basis of personal intentions, attitudes or desires, conflict is also automatically constructed on only the psychological level. Experience is thus written in as simultaneously individual (that is, located in the individual body/psyche of woman) and general (located in women as a preconstituted collective). There seem to be two problems with this definition. First, experience is seen as being immediately accessible, understood and named. The complex relationships between behaviour and its representation are either ignored or made irrelevant; experience is collapsed into discourse and vice versa. Second, since experience has a fundamentally psychological status, questions of history and collectivity are formulated on the level of attitude and intention. In effect, the sociality of collective struggles is understood in terms of something like individual–group relations, relations which are common-sensically seen as detached from history. If the assumption of the *sameness* of experience is what ties woman (individual) to women (group), regardless of class, race, nation and sexualities, the notion of experience is anchored firmly in the notion of the individual self, a determined and specifiable constituent of European modernity. However, this notion of the individual needs to be self-consciously historicized if as feminists we wish to go beyond the limited bourgeois ideology of individualism, especially as we attempt to understand what cross-cultural sisterhood might be made to mean.

Towards the end of 'Planetary Feminism' Morgan talks about feminist diplomacy:

> What if feminist diplomacy turned out to be simply another form of the feminist aphorism 'the personal is political'? Danda writes here of her own feminist epiphany, Amanda of her moments of despair, La Silenciada of personally bearing witness to the death of a revolution's ideals. Tinne confides her fears. Nawal addresses us in a voice direct from prison, Hilkla tells us about her family and childhood; Ama Ata confesses the anguish of the woman artist, Stella shares her mourning with us, Mahnaz communicates her grief and her hope, Nell her daring balance of irony and lyricism, Paola the story of her origins and girlhood. Manjula isn't afraid to speak of pain, Corrine traces her own political evolution alongside that of her movement. Maria de Lourdes declares the personal and the political inseparable. Motlalepula still remembers the burning of a particular maroon dress, Ingrid and Renate invite us into their private correspondence, Marielouise opens herself in a poem, Elena appeals personally to us for help, Gwendoline testifies about her private life as a public figure . . .
>
> And do we not, after all recognize one another?
>
> (pp. 35–6)

It is this passage more than any other that encapsulates Morgan's individualized and essentially equalizing notion of universal sisterhood, and its corresponding political implications. The lyricism, the use of first names (the one and only time this is done), and the insistence that we must easily 'recognize one another' indicate what is left unsaid: we must identify with *all* women. But it is difficult to imagine such a generalized identification predicated on the commonality of women's interests and goals across very real divisive class and ethnic lines – especially, for example, in the context of the mass proletarianization of Third World women by corporate capital based in the US, Europe and Japan.

Universal sisterhood, defined as the transcendence of the 'male' world, thus ends up being a middle-class, psychologized notion which effectively erases material and ideological power differences within and among groups of women, especially between First and Third World women (and, paradoxically, removes us all as actors from history and politics). It is in this erasure of difference as inequality and dependence that the privilege of Morgan's political 'location' might be visible. Ultimately in this reductive utopian vision, men *participate* in politics while women can only hope to *transcend* them. Morgan's notion of universal sisterhood *does* construct a unity. However, for me, the real challenge arises in being able to craft a notion of political unity without relying on the logic of appropriation and incorporation and, just as significantly, a denial of *agency*. For me the unity of women is best understood not as *given*, on the basis of a natural/psychological commonality; it is something that has to be worked for, struggled towards – *in history*. What we need to do is articulate ways in which the historical forms of oppression relate to the category 'women', and not to try to deduce one from the other. In other words, it is Morgan's formulation of the relation of synchronous, alternative histories (herstories) to a diachronic, dominant historical narrative (History) that is problematic. One of the tasks of feminist analysis is uncovering alternative, non-identical histories which challenge and disrupt the spatial and temporal location of a hegemonic history. However, sometimes attempts to uncover and locate alternative histories code these very histories as either totally dependent on and determined by a dominant narrative, or as isolated and autonomous narratives, untouched in their essence by the dominant figurations. In these rewritings, what is lost is the recognition that it is the very co-implication of histories with History which helps us situate and understand oppositional agency. In Morgan's text, it is the move to characterize alternative herstories as separate and different from history that results in a denial of feminist agency. And it is this potential repositioning of the relation of oppositional histories/spaces to a dominant historical narrative that I find valuable in Bernice Reagon's discussion of coalition politics.

'It ain't home no more': rethinking unity

While Morgan uses the notion of sisterhood to construct a cross-cultural unity of women and speaks of 'planetary feminism as the politics of the 21st century', Bernice Johnson Reagon uses *coalition* as the basis to talk about the cross-cultural commonality of struggles, identifying *survival*, rather than *shared oppression*, as the ground for coalition. She begins with this valuable political reminder: 'You don't go into coalition because you *like* it. The only reason you would consider trying to team up with somebody who could possibly kill you, is because that's the only way you can figure you can stay alive' (p. 357).

The governing metaphor Reagon uses to speak of coalition, difference and struggle is that of a 'barred room'. However, whereas Morgan's barred room might be owned and controlled by the Big Brothers in different countries, Reagon's internal critique of the contemporary left focuses on the barred rooms constructed by oppositional political movements such as feminist, civil rights, gay and lesbian, and chicano political organizations. She maintains that these barred rooms may provide a 'nurturing space' for a little while, but they ultimately provide an illusion of community based on isolation and the freezing of difference. Thus, while sameness of experience, oppression, culture, etc. may be adequate to construct this space, the moment we 'get ready to clean house' this very sameness in community is exposed as having been built on a debilitating ossification of difference.

Reagon is concerned with differences *within* political struggles, and the negative effects, in the long run, of a nurturing, 'nationalist' perspective:

> At a certain stage nationalism is crucial to a people if you are going to ever impact as a group in your own interest. Nationalism at another point becomes reactionary because it is totally inadequate for surviving in the world with many peoples.
> (p. 358)

This is similar to Gramsci's analysis of oppositional political strategy in terms of the difference between wars of manoeuvre (separation and consolidation) and wars of position (re-entry into the mainstream in order to challenge it on its own terms). Reagon's insistence on breaking out of barred rooms and struggling for coalition is a recognition of the importance – indeed the inevitable necessity – of wars of position. It is based, I think, on a recognition of the need to resist the imperatives of an expansionist US state, and of imperial History. It is also, however, a recognition of the limits of identity politics. For once you open the door and let others in, 'the room don't feel like the room no more. And it ain't home no more' (p. 359).

The relation of coalition to home is a central metaphor for Reagon. She speaks of coalition as opposed, by definition, to home.[13] In fact, the confusion of home with coalition is what concerns her as an urgent problem, and it is here that the status of experience in her text becomes clear. She criticizes the idea of enforcing 'women-only' or 'woman-identified' space by using an 'in-house' definition of woman. What concerns her is not a sameness which allows us to identify with each other as women, but the exclusions particular normative definitions of 'woman' enforce. It is the exercise of violence in creating a legitimate *inside* and an illegitimate *outside* in the name of identity that is significant to her – or, in other words, the exercise of violence when unity or coalition is confused with home and used to enforce a premature sisterhood or solidarity. According to her this 'comes from taking a word like "women" and using it as a code' (p. 360). The experience of being woman can create an illusory unity, for it is not the experience of being woman, but the meanings attached to gender, race, class and age at various historical moments that is of strategic significance.

Thus, by calling into question the term 'woman' as the automatic basis of unity, Bernice Reagon would want to splinter the notion of experience suggested by Robin Morgan. Her critique of nationalist and culturalist positions, which after an initial necessary period of consolidation work in harmful and exclusionary ways, provides us with a fundamentally political analytic space for an understanding of experience. By always insisting on an analysis of the operations and effects of power in our attempts to create alternative communities, Reagon foregrounds our *strategic* locations and positionings. Instead of separating experience and politics and basing the latter on the former, she emphasizes the politics that always define and inform experience (in particular, in left, anti-racist and feminist communities). By examining the differences and potential divisions *within* political subjects as well as collectives, Reagon offers an implicit critique of totalizing theories of history and social change. She underscores the significance of the traditions of political struggle, what she calls an 'old-age perspective' – and this is, I would add, a global perspective. What is significant, however, is that the global is forged on the basis of memories and counter-narratives, not on an ahistorical universalism. For Reagon, global, old-age perspectives are founded on humility, the gradual chipping away of our assumed, often ethnocentric centres of self/other definitions.

Thus, her particular location and political priorities lead her to emphasize a politics of engagement (a war of position), and to interrogate totalizing notions of difference and the identification of exclusive spaces as 'homes'. Perhaps it is partly also her insistence on the urgency and difficult nature of political struggle that leads Reagon to talk about difference in terms of racism, while Morgan often formulates difference in terms of cultural pluralism. This is Bernice Reagon's way of 'throwing yourself into the next century':

> Most of us think that the space we live in is the most important space there is, and that the condition we find ourselves in is the condition that must be changed or else. That is only partially the case. If you analyze the situation properly, you will know that there might be a few things you can do in your personal, individual interest so that you can experience and enjoy change. But most of the things that you do, if you do them right, are for people who live long after you are forgotten. That will happen if you give it away . . . The only way you can take yourself seriously is if you can throw yourself into the next period beyond your little meager human-body-mouth-talking all the time.
>
> (p. 365)

We take ourselves seriously only when we go 'beyond' ourselves, valuing not just the plurality of the differences among us but also the massive presence of the Difference that our recent planetary history has installed. This 'Difference' is what we see only through the lenses of our present moment, our present struggles.

I have looked at two recent feminist texts and argued that feminist discourse must be self-conscious in its production of notions of experience and difference. The rationale for staging an encounter between the two texts, written by a white and black activist respectively, was not to identify 'good' and 'bad' feminist texts. Instead, I was interested in foregrounding questions of cross-cultural analysis which permeate 'movement' or popular (not just academic) feminist texts, and in indicating the significance of a politics of location in the

US of the 1980s and the 1990s. Instead of privileging a certain limited version of identity politics, it is the current *intersection* of anti-racist, anti-imperialist and gay and lesbian struggles which we need to understand to map the ground for feminist political strategy and critical analysis.[14] A reading of these texts also opens up for me a temporality of *struggle*, which disrupts and challenges the logic of linearity, development and progress which are the hallmarks of European modernity.

But why focus on a temporality of struggle? And how do I define *my* place on the map? For me, the notion of a temporality of struggle defies and subverts the logic of European modernity and the 'law of identical temporality'. It suggests an insistent, simultaneous, non-synchronous process characterized by multiple locations, rather than a search for origins and endings which, as Adrienne Rich says, 'seems a way of stopping time in its tracks'.[15] The year 2000 is the end of the Christian millennium, and Christianity is certainly an indelible part of post-colonial history. But we cannot afford to forget those alternative, resistant spaces occupied by oppositional histories and memories. By not insisting on *a* history or *a* geography but focusing on a temporality of struggle, I create the historical ground from which I can define myself in the USA of the 1990s, a place from which I can speak to the future – not the end of an era but the promise of many.

The USA of the 1990s: a geopolitical power seemingly unbounded in its effects, peopled with 'natives' struggling for land and legal rights, and 'immigrants' with their own histories and memories. Alicia Dujovne Ortiz writes about Buenos Aires as 'the very image of expansiveness'.[16] This is also how I visualize the USA of the 1990s. Ortiz writes of Buenos Aires:

> A city without doors. Or rather, a port city, a gateway which never closes. I have always been astonished by those great cities of the world which have such precise boundaries that one can say exactly where they end. Buenos Aires has no end. One wants to ring it with a beltway, as if to point an index finger, trembling with uncertainty, and say: 'You end there. Up to this point you are you. Beyond that, God alone knows!' . . . a city that is impossible to limit with the eye or the mind. So, what does it mean to say that one is a native of Buenos Aires? To belong to Buenos Aires, to be *Porteno* – to come from this Port? What does this mean? What or who can we hang onto? Usually we cling to history or geography. In this case, what are we to do? Here geography is merely an abstract line that marks the separation of the earth and sky.[17]

If the logic of imperialism and the logic of modernity share a notion of time, they also share a notion of space as territory. In the North America of the 1990s geography seems more and more like 'an abstract line that marks the separation of the earth and sky'. Witness the contemporary struggle for control over oil in the name of 'democracy and freedom' in Saudi Arabia. Even the boundaries between space and outer space are not binding any more. In this expansive and expanding continent, how does one locate oneself? And what does location as I have inherited it have to do with self-conscious, strategic location as I choose it now?

A National Public Radio news broadcast announces that all immigrants to the United States now have to undergo mandatory AIDS testing. I am reminded very sharply of my immigrant status in this country, of my plastic identification which is proof of my legitimate location in the US. But location, for feminists, necessarily implies self- as well as collective definition, since meanings of the self are inextricably bound up with our understanding of collectives as social agents. For me, a comparative reading of Morgan's and Reagon's documents of activism precipitates the recognition that experience of the self, which is often discontinuous and fragmented, must be historicized before it can be generalized into a collective vision. In other words, experience must be historically interpreted and theorized if it is to become the basis of feminist solidarity and struggle, and it is at this moment that an understanding of the politics of location proves crucial.

In this country I am, for instance, subject to a number of legal/political definitions: 'post-colonial', 'immigrant', 'Third World'. These definitions, while in no way comprehensive, do trace an analytic and political space from which I can insist on a temporality of struggle. Movement *between* cultures, languages and complex configurations of meaning and power have always been the territory of the colonized. It is this *process*, what Caren Kaplan in her discussion of the reading and writing of home/exile has called 'a continual reterritorialization, with the proviso that one moves on',[18] that I am calling a temporality of struggle. It is this process, this reterritorialization through struggle, that allows me a paradoxical continuity of self, mapping and transforming my political location. It suggests a particular notion of political agency, since my location forces and enables specific modes of reading and knowing the dominant. The struggles I choose to engage in are then an intensification of these modes of knowing – an engagement on a different level of knowledge. There is, quite simply, no transcendental location possible in the USA of the 1990s.

I have argued for a politics of engagement rather than a politics of transcendence, for the present and the future. I *know* – in my own non-synchronous temporality – that by the year 2000 apartheid will be discussed as a nightmarish chapter in black South Africa's history, the resistance to and victory over the efforts of the US government and multinational mining conglomerates to relocate the Navajo and Hopi reservations from Big Mountain, Arizona, will be written into elementary-school textbooks, and the Palestinian homeland will no longer be referred to as the 'Middle East question' – it will be a reality. But that is my preferred history: what I hope and struggle for, I garner as *my* knowledge, create it as the place from where I seek to know. After all, it is the way in which I understand, define and engage in feminist, anti-imperialist and anti-racist collectives and movements that anchors my belief in the future and in the efficacy of struggles for social change.

Notes

This essay is a slightly revised, updated version of an essay published in the journal *Copyright* 1, Fall 1987, pp. 30–44. I develop the arguments raised here, especially the question of political agency, in greater detail in my book in progress on western feminist theory. Third World feminisms, and the problems of cross-cultural enquiry.

1 I am indebted to Adrienne Rich's essay, 'Notes Toward a Politics of Location (1984)', for the notion of the 'politics of location' (in her *Blood, Bread, and Poetry: Selected Prose 1979–1985* [W.W. Norton & Company, New York, 1986], pp. 210–31). In a number of essays in this collection, Rich writes eloquently and provocatively about the politics of her own location as a white, Jewish, lesbian feminist in North America. See especially 'North American Tunnel Vision (1983)', and 'Blood, Bread and Poetry: The Location of the Poet (1984)'.

While I attempt to modify and extend Rich's notion, I share her sense of urgency as she asks feminists to re-examine the politics of our location in North America:

> A natural extension of all this seemed to me the need to examine not only racial and ethnic identity, but location in the United States of North America. As a feminist in the United States it seemed necessary to examine how we participate in mainstream North American cultural chauvinism, the sometimes unconscious belief that white North Americans possess a superior right to judge, select, and ransack other cultures, that we are more 'advanced' than other peoples of this hemisphere . . . It was not enough to say 'As a woman I have no country; as a woman my country is the whole world.' Magnificent as that vision may be, we can't explode into breadth without a conscious grasp on the particular and concrete meaning of our location here and now, in the United States of America.
>
> ('North American Tunnel Vision', p. 162)

2 I address one version of this, the management of race and cultural pluralism in the US academy, in some depth in my essay 'On Race and Voice: Challenges for Liberal Education in the 1990s', *Cultural Critique*, 14 (1989–90), pp. 179–208.

3 Two recent essays develop the point I am trying to suggest here. Jenny Bourne identifies the problems with most forms of contemporary identity politics which equalize notions of oppression, thereby writing out of the picture any analysis of structural exploitation or domination. See her 'Jewish Feminism and Identity Politics', *Race and Class*, XXIX (1987), pp. 1–24.

In a similar vein, S.P. Mohanty uses the opposition between 'History' and 'histories' to criticize an implicit assumption in contemporary cultural theory that pluralism is an adequate substitute for political analyses of dependent relationships and larger historical configurations. For Mohanty, the ultimate target is the cultural and historical *relativism* which he identifies as the unexamined philosophical 'dogma' underlying political celebrations of pure difference. This is how he characterizes the initial issues involved:

> Plurality [is] thus a political ideal as much as it [is] methodological slogan. But . . . a nagging question [remains]: How do we negotiate between my history and yours? How would it be possible for us to recover our commonality, not the humanist myth of our shared human attributes which are meant to distinguish us all from animals, but, more significantly, the imbrication of our various pasts and presents, the ineluctable relationships of shared and contested meanings, values, material resources? It is necessary to assert our dense particularities, our lived and imagined differences. But could we afford to leave unexamined the question of how our differences are intertwined and indeed hierarchically organized? Could we, in other words, really afford to have *entirely* different

histories, to see ourselves as living – and having lived – in entirely heterogeneous and discrete spaces?

See his 'Us and Them: On the Philosophical Bases of Political Criticism', *The Yale Journal of Criticism*, 2 (1989), pp. 1–31; p. 13.

4 For instance, some of the questions which arise in feminist analyses and politics which are situated at the juncture of studies of race, colonialism and Third World political economy pertain to the systemic production, constitution, operation and reproduction of the institutional manifestations of power. How does power operate in the constitution of gendered and racial subjects? How do we talk about contemporary political praxis, collective consciousness and collective struggle in the context of an analysis of power? Other questions concern the discursive codification of sexual politics and the corresponding feminist political strategies these codifications engender. Why is sexual politics defined around particular issues? One might examine the cultural and historical processes and conditions under which sexuality is constructed during conditions of war. One might also ask under what historical conditions sexuality is defined as sexual violence, and investigate the emergence of gay and lesbian sexual identities. The discursive organization of these questions is significant because they help to chart and shape collective resistance. Some of these questions are addressed by contributors in a collection of essays I have co-edited with Ann Russo and Lourdes Torres, entitled *Third World Women and the Politics of Feminism* (Indiana University Press, Bloomington, Ind., and Indianapolis, 1991).

5 Robin Morgan, 'Planetary Feminism: The Politics of the 21st Century', in her *Sisterhood is Global: The International Women's Movement Anthology* (Anchor Press/Doubleday, New York, 1984), pp. 1–37; I also refer to the 'Prefatory Note and Methodology' section (pp. xiii–xxiii) of *Sisterhood is Global* in this essay. Bernice Johnson Reagon, 'Coalition Politics: Turning the Century', in Barbara Smith (ed.), *Home Girls: A Black Feminist Anthology* (Kitchen Table, Women of Color Press, New York, 1983), pp. 356–68.

6 Teresa de Lauretis, 'Feminist Studies/Critical Studies: Issues, Terms and Contexts', in de Lauretis (ed.), *Feminist Studies/Critical Studies* (Indiana University Press, Bloomington, Ind., 1986), pp. 1–19; p. 5.

7 Monique Wittig develops this idea in 'The Straight Mind', *Feminist Issues*, 1 (1980), pp. 103–10; p. 103.

8 Katie King, 'The Situation of Lesbianism as Feminism's Magical Sign: Contests for Meaning and the US Women's Movement, 1968–1972', *Communication*, 9 (1986), pp. 65–91; p. 85.

9 Linda Gordon discusses this relation of female to feminist in her 'What's New in Women's History', in de Lauretis, *Feminist Studies/Critical Studies*, pp. 20–31.

10 de Lauretis, 'Feminist Studies/Critical Studies: Issues, Terms and Contexts', p. 10.

11 Rich, 'Notes Toward a Politics of Location', p. 212.

12 Elsewhere I have attempted a detailed analysis of some recent western feminist social science texts about the Third World. Focusing on works which have appeared in an influential series published by Zed Press of London, I examine this discursive construction of women in the Third World and the resultant western feminist self-representations. See 'Under Western Eyes: Feminist Scholarship and Colonial Discourses', *Feminist Review*, 30 (1988), pp. 61–88.

13 For an extensive discussion of the appeal and contradictions of notions of home and identity in contemporary feminist politics, see Biddy Martin and Chandra

Talpade Mohanty, 'Feminist Politics: What's Home Got to Do With It?', in de Lauretis, *Feminist Studies/Critical Studies*, pp. 191–212.

14 For a rich and informative account of contemporary racial politics in the US, see Michael Omi and Howard Winant, *Racial Formation in the United States: From the 1960s to the 1980s* (Routledge and Kegan Paul, New York and London, 1986). Surprisingly, this text erases gender and gay politics altogether, leading me to wonder how we can talk about the 'racial state' without addressing questions of gender and sexual politics. A good companion text which in fact emphasizes such questions is G. Anzaldúa and C. Moraga (eds), *This Bridge Called My Back: Writings By Radical Women of Color* (Kitchen Table, Women of Color Press, New York, 1983). Another, more contemporary text which continues some of the discussions in *This Bridge*, also edited by Gloria Anzaldúa, is entitled *Making Face, Making Soul, Haciendo Caras, Creative and Critical Perspectives by Women of Color* (Aunt Lute, San Francisco, 1990).

15 Rich, 'Notes Toward a Politics of Location', p. 227.

16 Alicia Dujovne Ortiz, '*Buenos Aires* (an excerpt)', *Discourse*, 8 (1986–7), pp. 73–83; p. 76.

17 Alicia Dujovne Ortiz, '*Buenos Aires* (an excerpt)', *Discourse*, 8 (1986–7), p. 76.

18 Caren Kaplan, 'The Poetics of Displacement in *Buenos Aires*', *Discourse*, 8 (1986–7), pp. 94–102; p. 98.

SECTION TWO

PRACTISING FEMINIST GEOGRAPHIES

Editors' introduction

Feminists have not only critiqued the theoretical aspects of geographical research and knowledge but also the sexism inherent in many forms of academic practice. They have pointed to the gender bias inherent in the methodologies of sampling, experimentation and fieldwork, and worked through the relationships between researchers and their 'objects' of study. Feminists have highlighted the importance of allowing the voices of research subjects into the finished texts, and also into the development of research questions in the first place. Too often, feminists have suggested, researchers have forced their concerns and research questions onto particular situations, and in doing so have simply appropriated the voices of their subjects in the drive to produce a closed and coherent explanation.

Paralleling both the range and the development of feminist thought discussed in the first section, feminist methods have also been diverse. Feminist methodologies can be seen to be divided along the same–difference split already mentioned, in that some methodologies are based on the premise that women possess a distinctly different way of knowing, whereas others have regarded the gendering of knowledge as an effect of patriarchal power intent upon maintaining women's inequality. In the book *Women's Ways of Knowing* published in 1986, Mary Belenky, Blythe Clinchy, Nancy Goldberger and Jill Tarule argue that knowledge is tied to experience: as a result of their different experiences, women's and men's ways of knowing differ. It has been argued that women do less well in competitive environments where abstract knowledge is required, and excel in more supportive environments where anecdotal, personalised knowledge is valued (Maher and Thompson-Tetreault, 1994). The argument continues that given the current forms of valued knowledge – perhaps at the pinnacle being abstract, scientific knowledge – women's knowledge has been devalued: this explains why women attain few positions of status and power in contemporary society. They are in a society where the male gendering of knowledge is disguised – it appears gender neutral. Were women's knowledge to be equally valued, women would have more success in gaining influence in the running of society.

In a similar vein, Sandra Harding's 'feminist standpoint theory' attempts to develop a feminist epistemology that avoids the essentialism that is attached to women's ways of knowing. Women have a different standpoint – and following from Haraway's article printed in the last section, a different situated knowledge – not because of natural or biological diversity but as a result of material social relations that position us/them differently from men in regards to power and knowledge. Thus, 'Women's lives (our many different lives and different experiences!) can provide the starting point for asking new, critical questions about not only those women's lives but also about men's lives and, most importantly, the causal relations between them' (Harding, 1993 p. 55). Quite simply, women ask different questions and different types of questions because of social structures that ensure that their lives are different from men's. This knowledge is not inherently better than men's knowledge, it is simply a different partial perspective. Harding's insistence of knowledge as partial does not dissolve knowledge into a relativistic morass of equally valid opinions but recognises that in particular situations, some perspectives are more useful or appropriate: masculinist thought is removed from its illusory all-seeing throne.

It should be noted here that other feminists have questioned this approach given its essentialising of the difference between women's and men's knowledges (even if this is a social rather than natural essentialism). Some are concerned to underscore the ways in which different power structures involving, amongst others, race, ethnicity and sexuality, have positioned women and men in a multitude of different positions that cannot simply be accounted for by 'women's' and 'men's' knowledge.

Some feminists have sought to go further in their questioning of research methodology by challenging the division between the practise of geography and of the politics of our lives more widely. On one level, this would involve bringing feminist concerns into the workings of the academy, an institution that is statistically – and culturally – dominated by men. Questions arise over recruiting: 'should women and men students be recruited and treated in identical ways or are there strategies that might be particularly appropriate in recruiting and training women?' (McDowell, 1992 p. 402). In other words, in recruiting, in order to redress the balance of women and men in the academy, should women be treated differently? The question comes back to the same–difference debate discussed in the Introduction.

The first excerpt here is from Linda **McDowell**'s paper 'Doing gender'. McDowell discusses the inherent sexism of many geographical methodologies that are often simply regarded as gender neutral. She laments the fact that feminist methodologies are rarely applied, but illustrates her contentions with a discussion of two studies that have introduced feminist concerns successfully. The excerpt closes with a reflection on 'some difficult questions' about the ethics of research.

Our next author, Judith **Stacey** wonders why, when so many feminists have criticised orthodox social science for its deafness to Other voices, so few of them have adopted ethnographic methods in their research, in order to allow formerly marginalised voices of women to be heard. Ethnography, many feminists believe, allows the voices, experiences and concerns of respondents to be brought to the forefront of research. It also allows the researcher to position her- or himself in the research process: opening the written work with an autobiographical account to illustrate to the reader where she or he is coming from. However, Stacey explains that ethnographic work is problematic in that it offers an illusion of intimacy between researcher and researched, especially if there is the appearance of a shared identity or politics. Stacey argues that this false sense of unity has rendered her blind to the particular experiences of the respondent. As a result, she describes the ethnographic approach as a very difficult one that challenges personal ethics and beliefs. In the paper included here, Stacey comments upon some of her research dilemmas – when she has been split between fidelity to a research project, to the truth, or to a political belief on one hand, and the feelings and comfort of an interviewee on the other. Nevertheless, she valorises an ethnographic approach for the openness it can offer researchers to position themselves within the process of project definition and knowledge acquisition.

The next paper, by J.K. **Gibson-Graham**, develops the question of representation further. Gibson-Graham offers a longer account of an engagement with feminist and postmodern politics in the development of a research project, and illustrates the complexities surrounding the right to speak for or about other people. The paper details the ways in which the project moves from a concern with the experience of 'miners' wives' (described with metaphors suggesting the researchers' mining of information from the rich lives of their respondents) to one more concerned with how academic work can resonate with the needs and concerns of the women, responding to their differences, their experiences and their political concerns – rather than putting together the pieces of a stereotype. Gibson-Graham ends up describing how she managed to help women devise political activism, and through this exchange, learned more about the women's needs and concerns. In discovering their ability to help, Gibson–Graham realised that the research process can indeed involve a degree of exchange. Obviously power relations remain, yet the research process operated as a true dialogue between the researchers and the researched with both 'sides' offering skills and knowledge to the other.

As critical pedagogy is of central importance to the academy, feminist methods cannot be limited to research but also need to be brought into the classroom. It has been argued that different teaching methods facilitate or hinder female involvement. Some feminists have argued that men and women communicate differently, that some women are intimidated by more self-confident men who can present their own ideas better than they can listen to others. It is this form of self-expression that is characteristically encouraged in higher education. Graduate departments are characterised by a

> Process of one-upmanship characteristic of the masculinist tradition by which we learn to be critical thinkers. In graduate school we are taught that a measure of our intelligence is the extent to which we can show others to be wrong. Thus the best students are those who can offer the most masterful critique, pointing to methodological flaws, finding gaps in the argument, and using the most sophisticated language. One consequence is an enormous loss of self-confidence and self-esteem so that it is the unusual student who emerges from a graduate program as a confident scholar who feels good about herself or himself.
>
> (Anderson, quoted in McDowell 1992 p. 402)

Clearly this argument is too simple in that it tends to force caricatures of men's and women's ways of knowing and communicating into a dualism. Nevertheless, it points to important issues in the manner in which we teach, and the goals that we offer students as good academic practice. Alternative positions can expose the limitations of the dominant viewpoint, as Renato Rosaldo explained for the case of early American history and the terms 'Puritans, thirteen colonies, a nation of immigrants, westward expansion, new frontiers':

> Their students ask why, when they look for themselves in the mirror of traditional American history, they see nothing. Asian-American students say that their ancestors came eastward, not westward. African-American students explain that being shackled in a slave galley cannot fruitfully be called an immigrant experience. Native American and Chicano students affirm that, rather than immigrating, they stood still and the border moved.
>
> (Quoted in Maher and Thompson-Tetreault 1994 p. 5)

Critiques such as this have come in for heavy attack as trying to demolish western history, but Rosaldo continues that their 'goal is not to destroy Western culture but to historicize it'. And so this is what a feminist classroom should offer: the possibility for other voices and experiences, and a supportive environment for other ways of knowing. And it must be a place open to difference not simply the voices of women.

> Black students sometimes get the feeling that feminism is a private white cult. The Black students' relentless efforts to link all discussions of gender with race may be contested by white students, who see this as deflecting attention away from feminist concerns. And so suddenly the feminist classroom is no longer the safe haven many women students imagined. Instead it presents conflict, tension, hostility.
>
> (bell hooks, quoted in Maher and Thompson-Tetreault 1994 pp. 14–15)

Feminist challenges to geographical knowledge in journals and books and in the classroom have been of great import, but some feminists have suggested that although we need this discourse, 'we also need politics and practice' (Chouinard and Grant, 1995 p. 158). The last selection in this section is Vera **Chouinard** and Ali **Grant**'s 'On not being anywhere near the "project"'. This paper is a response to a debate in the journal *Antipode* concerning who should decide upon the agenda of radical geography. Susan Christopherson (1989) argued that a feminist approach would insist upon an interrogation of any project to assess whose interests it served and whom it might exclude, in addition to assessing what it sought to achieve. She feels that Geography continually constructs women and minorities as Other so keeping them outside of the discipline's 'project'. Two years later in *Antipode*, Linda McDowell (1992) reflected on Christopherson's paper to evaluate the impact of critical geography's embracing of postmodernism for the marginalisation of women. McDowell concludes that if postmodernism is not just to replicate Geography's fascination with exotic Otherness, it must meet feminist challenges to situate the representations being produced and foreground power relations within the research process.

Chouinard and Grant turn Christopherson and McDowell's feminist critique on some of the assumptions of 'the feminist project' in academic geography, to show that here too there are exclusions. They highlight the silences of both postmodern radicalism and feminism in geography by engaging with it from the position of a disabled geographer and a lesbian geographer, to let 'the reader "see" society and our discipline through others' eyes.' They insist that feminists need to practice in real life what they preach in their work if the discipline is to be changed:

> The next time someone calls you (or a woman you know) a dyke, or hints that you are, look at why this has happened. How have you stepped out of line, how have you moved beyond the heteropatriarchal definition of a 'woman?' More critically, look at your response: do you quickly mention your male partner, apologize for that behavior, try to justify what you are doing, and/or step back into line? Or do you challenge, disrupt and unsettle? Do you recognize lesbian-baiting for what it is and do you recognize the power of heterosexism to control women? Do you challenge it, and say, 'Why, thank you'?
>
> (Chouinard and Grant, 1995 pp. 157–58)

We placed this paper last as it leaves so many questions open; uncomfortable questions that force us to reflect upon not only our work, but also what we might hitherto have considered to be our private lives.

References and further reading

Alcoff, L 1991: The problem of speaking for others. *Cultural Critique,* Winter, 5–32.

Alcoff, L and Potter, E (eds) 1993: *Feminist Epistemologies.* Routledge: London.

Belenky, M, Clinchy, B, Goldberger, N and Tarule, J 1986: *Women's Ways of Knowing*. Basic Books: New York.

Bell, D, Kaplan, P and Jahan Karim, W (eds) 1993: *Gendered Fields: Women, Men and Ethnography*.

Bowles, G, and Duelli Klein, R (eds) 1983: *Theories of Women's Studies*. Routledge: London.

Christopherson, S 1989: On being outside 'the project'. *Antipode* **21**(1), 83–9.

Di Leonardo, M 1991: *Gender at the Crossroads of Knowledge: Feminist Anthropology in the Postmodern Era*. University of California Press: Berkeley.

Dyck, I 1993: Ethnography: a feminist method? *The Canadian Geographer* **37**, 52–7.

Fonow, M and Cook, J (eds) 1991: *Beyond Methodology: Feminist Scholarship as Lived Research*. Indiana University Press: Indiana.

Gluck, S and Patai, D (eds) 1991: *Women's Words: the Feminist Practice of Oral History*. Routledge: London.

Harding, S 1983: *Feminism and Methodology*. Open University Press: Milton Keynes.

Harding, S 1993: Rethinking standpoint epistemology. In Alcoff, L and Potter, E (eds) *Feminist Epistemologies*. Routledge: London.

Lather, P. 1988: Feminist perspectives on empowering research methodologies. *Women's Studies International Forum* 11, 569–81.

McDowell, L 1992: Multiple voices: speaking from inside and outside 'the project'. *Antipode* **24**(1): 56–72.

Maher, FA and Thompson-Tetreault MK 1994: *The Feminist Classroom*. Basic Books: London.

Pratt, G 1993: Reflections on poststructuralism and feminist empirics, theory and practice. *Antipode* **25**(1), 51–63.

Professional Geographer 1994: Special issue 'women in the field' **46**(1).

Roberts, H (ed.) 1981: *Doing Feminist Research*. Routledge and Kegan Paul: London.

Rose, D 1993: Of feminism, method and methods in human geography: an idiosyncratic overview. *The Canadian Geographer* **37**, 51–63.

Spivak, GC 1993: *Outside in the Teaching Machine*. Routledge: London.

Stacey, L (ed.) 1990: *Feminist Praxis*. Routledge: London.

Stanley, L and Wise, S 1983: *Breaking Out: Feminist Consciousness and Feminist Research*. Routledge and Kegan Paul: London.

Strathern, M 1987: An awkward relationship: the case of feminism and anthropology. *Signs* **12**(2), 276–92.

Waring, M 1989: *If Women Counted*. Macmillan: London.

7 Linda McDowell

'Doing Gender: Feminism, Feminists and Research Methods in Human Geography'

Excerpt from: *Transactions of the Institute of British Geographers* **17**, 399–416 (1992)

Sexist biases in research methods: critiques of conventional methods

One reason for the long exclusion of women's lives from geographical research, indeed from the social sciences more generally, lies in the methodological arena. Here a number of factors combine to inhibit the investigation of women's lives and gender relations. These include the absence of statistics that distinguish women from their family or that accurately record their waged work, let alone their unpaid labour; the choice of research methodology and/or, when using interview techniques, the choice of the subjects to interview. As these criticisms are now relatively well rehearsed (see Oakley and Oakley, 1979; Roberts, 1981; Waring, 1989) a few examples should suffice.

Many geographical studies do not question a focus on the household as the appropriate unit of analysis, nor the unproblematic defintion of the male partner (where there is one present) as the head of household. This means that internal power relations within the household remain unexamined. Other studies focus only on men and exclude women altogether. Here the long tradition of ethnographic work is a salutary example. From the early monographs of the Chicago School in the 1920s and 1930s through to later texts men are the subjects. A good (bad) example is William Whyte's *Street Corner Society* (1955), a project during the course of which he seemed unaware that he had interviewed only men. The contemporary classics of the 'youth' and 'popular culture' literature (Cohen, 1972; Corrigan, 1979; Hebdige, 1979; Willis, 1977) are no different. The respondents, 'the youth' on the street, are almost always male (although see McRobbie, 1991).

A different strategy has been to include women in the initial research design but to drop them from the analysis part way through. This is often because the differences between men and women seem to raise intractable problems of classification, analysis or comparison. If women are included initially, in for example, studies of occupational mobility or housing inheritance, they tend to be dropped from the analysis because of the severe problems that are raised by trying to assign a class location to women. Should a woman's class location be based on her own job, if she has one, on her father's if she is single and on her husband's if she is married? This conventional solution – to allocate women on the basis of a man's social class whenever possible – raises problems of cross-generational inconsistencies when comparing women's social mobility, as the occupation structure itself has changed over time. So what exactly are we measuring by adopting these conventions? Perhaps, anyway, it is the household that should be the unit of analysis, if questions about social status, or about regional variations in living standards, are the focus of research. These complex difficulties have been the subject of long debate by sociologists (for example, see the exchange between Goldthorpe (1983, 1984) and Stanworth (1984)) although neglected by many geographers. Too frequently, the resolution is to exclude women and girls altogether. Not one of the classic social mobility studies (for example by Glass (1954) in the 1950s or Halsey *et al.* (1980) or Goldthorpe *et al.* (1980) in the 1980s) included women's experiences. These problems are not unique to sociology or geography. An interesting paper by Beverly Thiele (1986) includes a number of further illustrations of methodological and theoretical 'vanishing acts' practised by the 'grand old men' of social and political theory that have resulted in women's exclusion from most of the classic texts.

Conventional research methods in human geography especially those involving interviewing, have also been criticized on the grounds of the gender blindness of those administering the survey or undertaking the interviews. For example, male researchers may privilege male respondents without considering whether the information so obtained is systematically biased. And in certain circumstances, of course, male researchers are precluded from gathering certain types of information because of their gender. To take an extreme example men are precluded from research involving Asian women in purdah. Of course the reverse argument also applies and the work of women, too, is influenced by their own position (their gender, race, class, age, etc.) and that of their respondents. One of the lessons that feminist critiques have taught us is to be aware of this. However, this still leaves difficult questions of *how* we include our own social location into the interpretation of our work. In what ways should we take it into account both in the conduct of the research and in the ways in which we write up our results? What are the appropriate methods for building on the advantages we possess as women researching women? These are questions that have received a great deal of critical scrutiny from feminist social researchers in the last few years.

Feminist methods

There is a lively debate between feminists about whether there is, or whether there should be, an accepted set of feminist research methods. Although there has been relatively little consideration of this question by geographers, a large literature by feminist scholars exists elsewhere (Bowles and Duelli Klein, 1983; Fuss, 1989; Oakley, 1981; Roberts, 1981; Smart, 1984; Stacey, 1988; Stanley, 1990). In general, there is broad agreement that feminists, within and outside our own discipline, are searching for methods that are consonant with their values and aims as feminists, and appropriate to feminist topics. However, beyond this broad axiomatic statement of aims, there is less agreement about whether there are particular methods that are peculiarly suited to feminist investigations or, indeed, whether 'conventional' research methods might be appropriate for feminist ends, albeit used critically. Where views have tended to coincide, however, has been on an insistence on collaborative methods – on methods in which the typically unequal power relations between a researcher and her informants are broken down. Thus there has been a challenge to the argument in conventional methods manuals that involvement with and participation in the lives of those who are being investigated 'biases' the results. In the collection of 'data', for example, it is not assumed that the researcher is objective or value-free, nor is she assumed to stay 'at a distance' from her subjects. As women interviewing women, commonalities of experience should be recognized and become part of a mutual exchange of views. Hence the assertion in the title of a paper by Ann Oakley – 'Interviewing women: a contradiction in terms' (1981). As Australian feminist Elizabeth Grosz (1986) suggests

> the conventional assumption that the researcher is a disembodied, rational, sexually indifferent subject – a mind unlocated in space, time or constitutive interrelationships with others, is a status normally attributed only to angels.
>
> (p. 199)

The most common strategy advocated by feminists in a search for a collaborative and non-exploitive relationship with the participants in their/our research project has been some variant of a qualitative methodology, either based on in-depth interviews or, less frequently, on participant observation and ethnographic research. Thus it is often argued that qualitative, detailed, small scale and case study work is ideally suited to women studying women. It is assumed that such a methodological approach draws on women's (purported) abilities to listen, to empathize, and to validate personal experiences as part of the research process. Further, it is suggested that this type of research allows the development of a less exploitative and more egalitarian relationship between a researcher and her participants than is possible in other methodological frameworks. Thus the interconnections and the relationships that might develop between an interviewer and her subjects are seen as a valid part of the research process, rather than something to be guarded against. Intersubjectivity rather than 'objectivity' characterizes the ideal relationship between a feminist researcher and her 'subjects' and many texts and articles discussing feminist research methodology have concentrated on forms of participant observation as the preferred method. As Duelli Klein (1983) has argued: 'a methodology that allows for women studying women in an interactive process will end the exploitation of women as research objects' (p. 95).

To what extent have these arguments about feminist methodology been influential in feminist research strategies throughout the 1980s? Have feminist sociologists, economists and geographers been able to draw on their reserves of empathy and concern to construct a different type of research from that undertaken in the mainstream? And has Duelli Klein's very positive belief in the end of exploitation any validity?

The evidence, from geography in particular, is so far rather limited[1] and further, it is becoming clear that participant observation may not be as immune from the power differentials that mark conventional methodologies as was once imagined. It seems that the acceptance of subjectivity, involvement and interpersonal relationships in the research process is as likely to raise difficult ethical questions for researchers as do conventional methodologies, as well as posing difficult questions about the particular experience of those committed to feminist approaches in their geographical research. To date, however, we know very little about the particular experiences of feminist geographers, despite recent, and welcome, attempts to demystify geographical research and to examine the problems, as well as the advantages, of a (re)focus on qualitative methods in geography.

In two recent collections that are widely used in undergraduate teaching, little attention was given either to feminist methods or to the particular position of women as researchers in geography. In Eyles' *Research in Human Geography* (1988) I was the sole woman contributor and I ducked the chance to write a personal piece, raising instead general issues about feminist approaches. In Eyles and Smith's collection *Qualitative Methods in Human Geography* (1988), despite a number of thoughtful contributions by women, there is a surprising absence of direct discussion of gender issues.[2] Smith, for example, positioned herself as a white East Midlands woman and made it clear that her geographical 'otherness' as an East Midlander studying the West Midlands was at least as significant as her gender. And indeed, she was, as a whole, curiously absent from the drama she described in her chapter despite her subtitle 'the analysis of self in everyday life'. Donovan (1988) mentioned in passing her race, but not her gender, and it was left to Cornwall (1988) to raise questions about how the research process is determined by the social relations of the academy and the field. She explained how her gender gave her privileged access to networks of women informants, but once so positioned she found it difficult to gain access to similar informal networks of men via women. Direct approaches to men, presenting herself as a powerful individual (i.e. as an academic) were more successful. Evans (1988) was the exceptional male contributor to this collection who raised questions about his social position as a researcher, explicitly locating himself as a male 'insider' studying his own community.

In other disciplines there is also surprisingly little discussion among feminists of the particular problems that are raised by the adoption of an explicit attempt to open up the research process, and to overcome the inherent inequalities of women in privileged academic positions studying women who usually, but not always, are in less privileged circumstances. In the mid-1980s an excellent stream of books and research monographs appeared by, in the main, feminist sociologists and economists who adopted some variant of a qualitative or ethnographic research methodology. These studies include Ruth Cavendish's (1982) work in a car components factory, published as *Women on the line*; Sallie Westwood's (1984) sensitive investigation of gender relations and ritual on the shop floor in a Midlands hosiery factory; Cynthia Cockburn's (1983) now classic study, *Brothers: male dominance and technological change*, addressing issues of masculinity and work practices among print workers; Rosemary Pringle's (1988) work on sexuality and power in the office published as *Secretaries talk*; and Judith Stacey's

(1990) ethnographic study of two working class families in Silicon Valley, California, *Brave new families*.

In all cases the research involved detailed interviews, participant observation or in Cavendish's study, actually taking a job in the factory and sharing the monotony of women's everyday work lives on an assembly line. Sallie Westwood, while not actually working for the textile firm that she examined, also spent many hours with her informants sharing their lives not only on the factory floor but also participating in extra-work activities such as a hen night. These two case studies seem to me to be among the best examples of how to do gender-aware research. In both of the resulting books the reader gains a vivid picture of the women's lives. But even here the authors remained relatively hidden. Whilst Cavendish writes movingly of the effects of the monotonous work on her abilities to combine intellectual and manual labour, neither author addresses the affects that their presences had on the interactions between the women workers and between themselves as a researcher/worker and the other women. It would have been interesting to know whether Cavendish (actually a psuedonymn for Myriam Glucksmann!) revealed her 'true' identity to her co-workers and what they felt about it, as it would in Westwood's case. Each author remains relatively absent from her text despite working and socializing with the women she studied.

The extent to which scholars are able to reduce the barriers between themselves and their subjects[3] and the ways in which this influences the power relations of the research process as well as the form of the final text is, perhaps, raised more clearly if the example of Westwood's more recent research is taken. Here, she is not a woman interviewing women, with all the baggage of assumed commonality that goes along with that, but a white, middle-aged, middle class academic interviewing 'the other'. Her subject is the social construction of black masculinity on the streets of a city in the Midlands (Westwood, 1990). It involved hanging out with young Asian and black men, talking about football, girls and so on. I wanted to know a lot more about their reactions to her than she revealed in the published results. How much difference did it make that she was so different from the young men she interviewed? And by what criteria should we judge what she reported in her paper? Similar questions are raised by the interesting recent work of Cindi Katz (1989; 1991), which is perhaps more familiar to geographers than Westwood's work. Katz is challenging conventional definitions of appropriate subjects for geographical research by investigating the everyday lives, rituals and resistances of children and young people in the Sudan and in New York City. But again in her work she tends to draw a veil over the implications of her own position and the effects of her involvement on the rituals she records. Yet she surely is less than invisible in both of these 'foreign' fields. We need to begin to turn our attention to such questions as the difference that it makes who does the research and how to 'speak for' (speak with? but it is the researcher who ultimately controls the text) our subjects who may be very different from us. Are we forced to abandon the notion of an empowering and equitable dialogue that feminists envisioned a decade or so ago?

Some difficult questions

Feminist-inspired notions of doing research 'with' or 'for' rather than 'about' women (or other 'others') seem admirable and are becoming widely accepted within human geography (at least by those who hold to a notion of emancipatory geographies). However, it is becoming clear that the adoption of qualitative or ethnographic methods alone does not release the scholar from exploitative relations, or even the betrayal of her subjects. As I read the feminist texts on methods (and attempted to adapt them to my own work on landlords, and more recently on merchant banking), certain feelings of unease grew stronger. These were partly clarified while reading Pile's recent *Transactions* paper (1991) on interpretative methods. There Pile, influenced by psychotherapy rather than feminism but coming to similar conclusions, argued for the construction of a 'research alliance' rather than the more conventional 'distant' or uninvolved relationship between researcher and subject(s). He argued that, in building an alliance, despite the unequal power relations inherent in most interview situations, 'both interviewer and interviewed try to come to an understanding of what is taking place around them' with the intention of 'developing the trust that allows people to share their experiences and feelings in a safe and supportive atmosphere' (p. 459). This seems to me a highly dubious general statement which ignores the importance of the context of the research and the differences in social status, power and resources between the interviewer and subject(s). In the (perhaps atypical) circumstances which Pile and most feminist inspired discussions of research methodology seem to ignore, that is where the inequalities of power and prestige favour the research subjects, it is the researcher who is vulnerable and open to exploitation. Are we therefore permitted here to use the 'tricks of the trade', including 'feminine' wiles, to persuade our informants into confidences that they would prefer not to reveal? How, as women, do we appropriate particular versions of femininity in our presentation of self in different circumstances? And which masks of conventional femininity are most suitable for which circumstances? Is it ethical to be 'honest' with the relatively powerless women respondents that we study in certain circumstances, while disguising our purpose from others (often powerful men) whom we know would refuse to speak to us if they could read our minds? In such cases, revealing our own values and judgements may make it less, rather than more, likely that our informants would trust us. Clearly Pile's idealized notions are not uniformally applicable. Erica Schoenberger and I have recently debated this and other issues involved in interviewing the powerful (Schoenberger, 1991, 1992; McDowell, 1992).

Even taking the (more typical?) example of relatively powerful academics interviewing less privileged informants, a number of questions continue to puzzle. Is it a realistic aim to endeavour to empower the subjects of our research or does this in itself reveal contestable notions of domination? (Lather, 1988). A more appropriate aim may be to provide the means towards empowerment, ensuring that as we do so we do not make public information or strategies that may compromise the less powerful. But this still leaves us in the position of the judge of the utility/validity of our findings. Sometimes we may not know the implications of our decisions. And what do we do with the knowledge that we gain from our respondents that we would much rather not have? Research guidelines are not always helpful on these difficult ethical issues. Judith Stacey (1988), for example, although self-identified as a feminist, found that, in her own recent study of family relationships in California's Silicon Valley, she came to question the advice of the many feminists, and others, who advocate interpretative methods to reduce the distance between the researchers and subjects. She found that in detailed ethnographic fieldwork work she was more, not less, likely to become bound to her informants in a network of exploitative relationships, abandonment and betrayal than in her earlier work. Thus Stacey argues that, 'Precisely because ethnographic research depends upon human relationship, engagement and attachment, it places research subjects at grave risk of manipulation and betrayal by the ethnographer' (pp. 22–3). She described several situations (a lesbian affair, a secret paternity case and illicit activities) which all place her 'in situations of inauthenticity, dissimilitude, and potential, perhaps inevitable betrayal, situations that I now believe *are inherent in the fieldwork method* (p. 23, emphasis added).

Patai, in a similarly thoughtful dissenting discussion of the possibility of 'empowering research methodologies' (1991), has identified similar problems. There are dangers when, as she puts it, 'feminist researchers are unconsciously seductive towards their research subjects, raising their expectations and inducing dependency' (p. 143). Women doing gender research, usually although not always involving interviewing other women, are quite likely to find themselves in circumstances where they are more powerful, more affluent and with greater access to a range of resources than their subjects. It is too easy inadvertently to generate expectations of positive intervention on behalf of the women being studied, sometimes leading, as Stacey warned, to feelings of disappointment or even betrayal. There is no obvious way to resolve these problems. As Daniels (1983) concluded in her discussion of self-deception and self-discovery in fieldwork:

> It is in the nature of ethical problems that they are not generally clear-cut, readily or finally resolvable. It is in the nature of fieldwork that you are likely to find yourself up to the waist in a morass of personal ties, intimate experiences and lofty and base sentiments as your own sense of decency, vanity or outrage is tried.
>
> (p. 213)

It may be that the ways in which these dilemmas arise and are resolved should become part of the training programme of aspirant scholars.

It is, however, becoming increasingly clear that the notion of non-exploitative research relations is a utopian ideal that is receding from our grasp. As Harding (1991) has argued recently 'knowledge is socially situated and scientific methods bind the knower and the known together in social relationships of domination and subordination typical of the race-, class-, and gender-stratified society in which science is produced'. Thus we are forced to recognize that knowledge is always situated; that, as Stuart Hall (1991) has argued 'enunciation comes from somewhere. It cannot be unplaced, it cannot be unpositioned, it is always positioned in a discourse' (p. 36). I take this to mean that we must recognize and take account of our own position, as well as that of our research participants, and *write this into our research practice* rather than continue to hanker after some idealized equality between us.

Recognition of the difficult issues that this recommendation raises – of positionality, ethics, disclosure, power and representation – has not, of course, taken place in isolation within feminist debates. Methodological reflection has been part of a wider demand for critical theorists of whatever complexion to rethink their claims to knowledge in relation to their own positionality. This demand has arisen from a widespread critique of western enlightenment thought. Feminists, post-structuralists, post-colonial and queer theorists have developed a coincidence of interests in their project to reveal how the 'unmarked subject' of history embodies male, bourgeois and heterosexist assumptions. While this project has perhaps been *the* central purpose of feminism/s whatever their theoretical stance, it has also opened up an enormous challenge, as the centrality and the stability of the notion 'woman' and the taken-for-granted commonalities between women are subject to question too.

Notes

1 A forthcoming issue of *Environment and Planning D: Society and Space* considers issues raised by the turn (back) to ethnography that is currently influential in geography.

2 The unsatisfactory association of gender with the women contributors while men escape as 'ungendered' should not have escaped the reader. This is too common a practice, relegating all discussion of gender and 'women's issues' to the ghetto of feminist approaches in geography. It is not my intention to continue this association, but at the present time, unfortunately, it tends to be female scholars in the discipline who are more attuned to questions raised by the embodiment of the researcher. For example, it was surely not fortuitous that it was women who pointed to the gendered assumptions in Harvey's book. However, male geographers are also – at last – turning their attention to the significance of their masculinity for their scholarship (see in particular Jackson, 1991 and Pile 1991).

3 The term 'subjects' for the people we study is an interesting one. I tend to use it in the sense of a subject/person, somebody with an identity or subjectivity, but, of course, it also has connotations of authority/colonialism – the subject of a regime. I find this double meaning provocative, and hesitate to replace it with the more egalitarian term, 'participant' as what is at question is just how much the people that we study to enter our work as participants.

Selected references

Bowles, G. and Duelli Klein, R. (eds) (1983) *Theories of women's studies* (Routledge and Kegan Paul, London).

Cavendish, R. (1982) *Women on the line* (Routledge and Kegan Paul, London).

Cockburn, C. (1983) *Brothers: male dominance and technological change* (Pluto, London).

Cohen, S. (1972) *Folk devils and moral panics: the creation of the Mods and Rockers* (MacGibbon and Kee, London).

Cornwall, J. (1988) 'A case-study approach to lay health beliefs: reconsidering the research process', in Eyles, J. and Smith, D. (eds) *Qualitative methods in human geography* (Polity, Cambridge) pp. 219–32.

Corrigan, P. (1979) *Schooling the Smash Street kids* (Macmillan, London).

Daniels, A. K. (1983) 'Self-deception and self-discovery in fieldwork'. *Qualitative Soc.* 6: 195–214.

Donovan, J. (1988) 'When you're ill you gotta carry it: health and illness in the lives of black people in London', in Eyles J. and Smith, D. (eds) *Qualitative methods in human geography* (Polity Press, Cambridge) pp. 180–96.

Duelli Klein, R. (1983) 'How to do what we want to do: thoughts on feminist methodology', in Bowles, G. and Duelli Klein, R. (eds) (1983) *Theories of women's studies* (Routledge and Kegan Paul, London) pp. 88–104.

Evans, M. (1988) 'Participant observation: the researcher as a research tool', in Eyles, J. and Smith, D. *Qualitative methods in human geography* (Polity Press, Cambridge) pp. 197–218.

Eyles, J. (1988) *Research in human geography* (Basil Blackwell, Oxford).

Eyles, J. and Smith, D. (1988) *Qualitative methods in human geography* (Polity Press, Cambridge).

Fuss, D. (1989) *Essentially speaking: feminism, Nature and difference* (Routledge, London).

Glass, D. (1954) *Social mobility in Britain* (Routledge and Kegan Paul, London).

Goldthorpe, J. (1983) 'Women and class analysis: in defence of the conventional view', *Sociology* 17: 465–88.

Goldthorpe, J. (1984) 'Women and class analysis: a reply to the replies'. *Sociology* 18: 491–9.

Goldthorpe, J., Llewellyn, C. and Payne, C. (1980) *Social mobility and class structure in modern Britain* (Clarendon Press, Oxford).

Grosz, E. (1986) 'What is feminist theory?' in Pateman, C. and Grosz, E. (eds) *Feminist challenges: social and political theory* (Northeastern University Press, Boston) pp. 190–204.

Hall, S. (1991) The local and the global: globalization and ethnicity', in King, A. *Culture, globalization and the world system* (Macmillan, London) pp. 19–39.

Halsey, A. H., Heath, A. F. and Ridge, J. M. (1980) *Origins and destinations: family, class and education in modern Britain* (Clarendon Press, Oxford).

Harding, S. (1991) *Whose science? Whose knowledge: thinking from women's lives* (Cornell University Press, Ithaca).

Hebdige, D. (1979) *Subculture: the meaning of style* (Methuen, London).

Jackson, P. (1991) 'The cultural politics of masculinity: towards a social geography', *Trans. Inst. Br. Geogr.* 16: 199–213.

Katz, C. (1989) 'You can't drive a Chevy through a post-Fordist landscape: everyday cultural practices of resistance and reproduction among youth in New York City'. Paper presented at the 'Marxism now: traditions and differences' Conference, Amherst, MA.

Katz, C. (1991) 'Sow what you know: the struggle for reproduction in rural Sudan', *Ann. Ass. Am. Geogr.* 81: 488–514.

Lather, P. (1988) 'Feminist perspectives on empowering research methodologies', *Women's Stud. Int. Forum* 11: 569–81.

McDowell, L. (1992) 'Valid games?' *The Professional Geogr.* 44: 219–22.
McRobbie, A. (1991) *Feminism and youth culture* (Macmillan, London).
Oakley, A. (1981) 'Interviewing women: a contradiction in terms', in Roberts, H. (ed.) *Doing feminist research* (Routledge and Kegan Paul, London) pp. 31–61.
Oakley, A. and Oakley, R. (1979) 'Sexism in official statistics'. In Irvine, J., Miles, I. and Evans, J. (eds) *Demystifying social statistics* (Pluto, London) pp. 172–89.
Patai, D. (1991) 'US academics and Third World women: is ethical research possible?', in Gluck, S. B. and Patai, D. (eds) *Women's words: the feminist practice of oral history* (Routledge, London) pp. 137–53.
Pile, S. (1991) 'Practising interpretative geography'. *Trans. Inst. Br. Geogr.* 16: 458–69.
Pringle, R. (1988) *Secretaries talk: sexuality, power and work* (Verso, London).
Roberts, H. (ed.) (1981) *Doing feminist research* (Routledge and Kegan Paul, London).
Schoenberger, E. (1991) 'The corporate interview as a research method in economic geography', *The Professional Geogr.* 43: 180–9.
Schoenberger, E. (1992) 'A reply to Linda McDowell', *The Professional Geogr.* 44: 222–5.
Smart, C. (1984) *The ties that bind: law, marriage and the reproduction of patriarchal relations* (Routledge and Kegan Paul, London).
Stacey, J. (1988) 'Can there be a feminist ethnography?', *Women's Stud. Int. Forum* 11: 21–7.
Stacey, J. (1990) *Brave new families* (Basic, New York).
Stanley, L. (ed.) (1990) *Feminist praxis* (Routledge, London).
Stanworth, M. (1984) 'Women and class analysis: a reply to Goldthorpe', *Sociology* 18: 159–70.
Thiele, B. (1986) 'Vanishing acts in social and political thought: tricks of the trade', in Pateman, C. and Grosz, E. (eds) *Feminist challenges* (Northwestern University Press, Boston) pp. 30–43; also in McDowell, L. and Pringle, R. *Defining women: social institutions and gender divisions* (Polity Press, Cambridge).
Waring, M. (1989) *If women counted* (Macmillan, London).
Westwood, S. (1984) *All day, every day* (Pluto, London).
Westwood, S. (1990) 'Racism, black masculinity and the politics of space', in Hearn, J. and Morgan, D. (eds) *Men, masculinities and social theory* (Unwin Hyman, London) pp. 55–71.
Whyte, W. F. (1955) *Street corner society* (University of Chicago Press, Chicago and London) 2nd edition.
Willis, P. (1977) *Learning to labour* (Saxon House, London).

8 Judith Stacey

'Can there be a Feminist Ethnography?'

From: *Women's Studies International Forum* **11**(1), 21–27 (1988)

Most feminist researchers, committed, at a minimum, to redressing the sexist imbalances of masculinist scholarship, appear to select their research projects on substantive grounds. Personal interests and skills meld, often mysteriously, with collective feminist concerns to determine a particular topic of research, which, in turn, appears to guide the research methods employed in its service. Indeed, in such a fashion, I chose my last project, a study of patriarchy and revolution in China designed to address major theoretical questions about Western feminism and socialism. The nature of this subject, compounded by limitations in my training, necessitated the macro-structural, abstract approach based almost exclusively on library research that I adopted. And, as a consequence, *Patriarchy and Social Revolution in China*, its textual product, offered an analysis of socialism and patriarchy which, as several reviewers justly complained, left out stories about actual women or patriarchs (Stacey, 1983). My dissatisfaction with that kind of research process and outcome led me to privilege methodological considerations over substantive interests when I selected my current research project, a fieldwork study of family and gender relationships in California's Silicon Valley. I was eager for a "hands on," face-to-face research experience, which I also believed was more compatible with feminist principles.

Although there is no uniform canon of feminist research principles, and many lively debates about whether there should be, and, if so, what one should contain, still it is possible to characterize a dominant conception of feminist research currently prevailing among feminist scholars. Most view feminist research as primarily research on, by, and especially *for* women and draw sharp distinctions between the goals and methods of mainstream and feminist scholarship.[1] Feminist scholars evince widespread disenchantment with the dualisms, abstractions, and detachment of positivism, rejecting the separations between subject and object, thought and feeling, knower and known, and political and personal as well as their reflections in the arbitrary boundaries of traditional academic disciplines. Instead most feminist scholars advocate an integrative, trans-disciplinary approach to knowledge which grounds theory contextually in the concrete realm of women's everyday lives. The "actual experience and language of women is the central agenda for feminist social science and scholarship," asserts Barbara Du Bois in an essay advocating "Passionate Scholarship," and only a minority of feminist scholars would dissent (Du Bois, 1983: 108). Indeed feminists tend to celebrate "feeling, belief, and experientially based knowledge," which draw upon such traditionally feminine capacities as intuition, empathy, and relationship

(Stanley and Wise, 1983a). Discussions of feminist methodology generally assault the hierarchical, exploitative relations of conventional research, urging feminist researchers to seek instead an egalitarian research process characterized by authenticity, reciprocity, and intersubjectivity between the researcher and her "subjects" (Duelli Klein, 1983; Du Bois, 1983; Mies, 1983; Reinharz, 1983; Stanley and Wise, 1983a, 1983b). "A methodology that allows for women studying women in an interactive process," Renate Duelli Klein argues, "will end the exploitation of women as research objects" (Duelli Klein, 1983: 95).

Judged by such criteria, the ethnographic method, by which I mean intensive participant-observation study which yields a synthetic cultural account, appears ideally suited to feminist research. That is why in "The Missing Feminist Revolution in Sociology," an essay reflecting on the limitations of feminist efforts to transform sociology, Barrie Thorne and I wondered with disappointment why so few feminist sociologists had turned to the ethnographic tradition of community studies within the discipline, a tradition that seemed to us far more compatible with feminist principles than are the more widely practiced positivist methods (Stacey and Thorne, 1985). Many other feminist scholars share the view that ethnography is particularly appropriate to feminist research (Duelli Klein, 1983; Mies, 1983; Reinharz, 1983; Stanley and Wise, 1983a, 1983b). Like a good deal of feminism, ethnography emphasizes the experiential. Its approach to knowledge is contextual and interpersonal, attentive like most women, therefore, to the concrete realm of everyday reality and human agency. Moreover, because in ethnographic studies the researcher herself is the primary medium, the "instrument" of research, this method draws on those resources of empathy, connection, and concern that many feminists consider to be women's special strengths and which they argue should be germinal in feminist research. Ethnographic method also appears to provide much greater respect for and power to one's research "subjects" who, some feminists proppose, can and should become full collaborators in feminist research (Duelli Klein, 1983; Mies, 1983; Stanley and Wise, 1983a).

This, at least, is how ethnography appeared to me as I found myself unintentionally but irresistibly drawn to it in a study originally intended to be based on more conventional interview methods. An ethnographic approach seemed to resolve the "contradiction in terms" involved in interviewing women that Anne Oakley had identified in her critique of classical sociological interview methods (Oakley, 1981). Oakley rejected the hierarchical, objectifying, and falsely "objective" stance of the neutral, impersonal interviewer as neither possible nor desirable, arguing that meaningful and feminist research depends instead on empathy and mutuality. And I was reassured by Shulamit Reinharz's assertion that the problems of experiential fieldwork methodology "seem minor in comparison with the quality of relations that I develop with people involved in the study and the quality of the understanding that emerges from those relations" (Reinharz, 1983: 185).

But now after two and half years of fieldwork experience, I am less sanguine and more focused on the difficult contradictions between feminist principles and ethnographic method I have encountered than on their compatibility. Hence the question in my title which is modelled (but with a twist) on the implicit question in Oakley's "Interviewing Women: A Contradiction in Terms." The twist is that I now perceive the opposite contradiction between feminist ethics and methods than the one that Oakley discusses. I find myself wondering whether the appearance of greater respect for and equality with research subjects in the ethnographic approach masks a deeper, more dangerous form of exploitation.

There are two major areas of contradiction I wish to discuss. The first involves the ethnographic research process, the second its product. Precisely because ethnographic research depends upon human relationship, engagement, and attachment, it places research subjects at grave risk of manipulation and betrayal by the ethnographer, as the following vignette from my fieldwork illustrates. One of my key informants, now a married, fundamentalist Christian, was involved in a closeted lesbian relationship at the time of her conversion. I first learned of this relationship from her spurned lesbian lover, and this only six months after working in the field. Of course, this immediately placed me in an extremely awkward situation ethically, a situation of triangulation and potential betrayal in relation to these two women and of inauthenticity toward the more secretive one. Several months later (partly, I believe, in response to her perception of my inauthenticity) this informant "came out" to me about this affair, but she asked me to respect the confidentiality of this knowledge when relating to her relatives, friends, and co-workers. Moreover, she and her rejected lover began to compete for my allegiance, sympathy, and ultimately for my view of their shared history.

I could give numerous other examples (such as the case of a secret of paternity, of an illicit affair, and of illicit activities). All placed me in situations of inauthenticity, dissimilitude, and potential, perhaps inevitable betrayal, situations that I now believe are inherent in fieldwork method. For no matter how welcome, even enjoyable the fieldworker's presence may appear to "natives", fieldwork represents an intrusion and intervention into a system of relationships, a system of relationships that the researcher is far freer than the researched to leave. The inequality and potential treacherousness of this relationship seems inescapable.

So too does the exploitative aspect of ethnographic process seem unavoidable. The lives, loves, and tragedies that fieldwork informants share with a researcher are ultimately data, grist for the ethnographic mill, a mill that has a truly grinding power. More times that I would have liked, this study has placed me in a ghoulish, and structurally conflictual relationship to tragedy, a feature of ethnographic process that became particularly graphic during the recent death of another one of my key informants. My ethnographic role consigned me to experience this death both as friend and as researcher, and it presented me with numerous delicate, confusing dilemmas, such as whether or not, and to whom, to make a gift of the precious, but potentially hurtful tapes of an oral history I had once conducted with the deceased. I was confronted as well with the discomforting awareness that as researcher I stood to benefit from this tragedy. Not only would the funeral and family grieving process serve as further research "opportunity," but also the death may free me to include more of this family's "truths" in my ethnographic account than would have been possible had he lived. This and other fieldwork experiences forced my recognition that conflicts of interest and emotion between the ethnographer as authentic, related person (i.e. participant), and as exploiting researcher (i.e. observer) are also an inescapable feature of ethnographic method.

The second major area of contradiction between feminist principles and ethnographic method involves the dissonance between fieldwork practice and ethnographic product. Despite the aspects of intervention and exploitation I have described, ethnographic method appears to (and often does) place the researcher and her informants in a collaborative, reciprocal quest for understanding, but the research product is ultimately that of the researcher, however modified or influenced by informants. With very rare exceptions it is the researcher who narrates, who "authors" the ethnography. In the last instance an ethnography is a written document structured primarily by a researcher's purposes, offering a researcher's interpretations, registered in a researcher's voice.[2]

Here too, therefore, elements of inequality, exploitation, and even betrayal are endemic to ethnography. Perhaps even more than ethnographic process, the published ethnography represents an intervention into the lives and relationships of its subjects. As author an ethnographer cannot (and, I believe, should not) escape tasks of interpretation, evaluation, and judgement. It is possible (and most feminists might claim it is crucial) to discuss and negotiate one's final presentation of narrative with informants, but this does not eliminate the problem of authority, and it can raise a host of new contradictions for the feminist ethnographer.[3] For example after more than a year and a half and scores of hours of mutual reflections on the meaning of the lesbian relationship mentioned above, this "research collaboration" has asked me to leave this part of her history out of my ethnographic account. What feminist ethical principles can I invoke to guide me here? Principles of respect for research subjects and for a collaborative, egalitarian research relationship would suggest compliance, but this forces me to collude with the homophobic silencing of lesbian experience, as well as to consciously distort what I consider a crucial component of the ethnographic "truth" in my study. Whatever we decide, my ethnography will betray a feminist principle.

Indeed, the irony I now perceive is that ethnographic method exposes subjects to far greater danger and exploitation than do more positivist, abstract, and "masculinist" research methods. The greater the intimacy, the apparent mutuality of the researcher/researched relationship, the greater is the danger.

The account I have just given of the paradoxes of feminist ethnography is itself falsely innocent. I have presented my methodological/ethical quandaries the way that I first conceptualized them as a feminist researcher, innocent as then I was of relevant methodological literature by ethnographers who long have grappled with related concerns. I am no longer so innocent and ignorant, but I retained this construction to help underscore a curious fact. There has been surprisingly little cross-fertilization between the discourses of feminist epistemology and methods and those of the critical traditions within anthropology and sociology.[4] Most pertinent is the dearth of dialogue between feminist scholarship and the contemporaneous developments in the literature referred to as the "new" or "postmodern" or "poststructuralist" ethnography.[5] This is curious, because the new or postmodern ethnography is concerned with quite similar issues as those that concern feminist scholars and, at first glance, it offers a potential resolution to the feminist ethnographic paradox.

Postmodern ethnography is critical and self-reflexive ethnography and a literature of meditation on the inherent, but often unacknowledged hierarchical and power-laden relations of ethnographic writing.[6] Like feminist scholars, critical ethnographers tear the veil from scientific pretensions of neutral observation or description. They attempt to bring to their research an awareness that ethnographic writing is not cultural reportage, but cultural construction, and always a construction of self as well as of the other. In James Clifford's words the "historical predicament of ethnography" is "the fact that it is always caught up in the intervention, not the representation of cultures" (Clifford, 1986: 2). And at rare moments, critical or "postmodern" ethnographers incorporate feminist insights into their reflexive critiques. Vincent Crapanzano, for example, suggests that "Interpretation has been understood as a phallic, a phallic-aggressive, a cruel and violent, a destructive act, and as a fertile, a fertilizing, a fruitful, and a creative one," and he self-consciously retains the male pronoun to refer to the ethnographer "despite his or her sexual identity, for I am writing of a stance and not of the person" (Crapanzano, 1986: 52).

As I understand it, the postmodern ethnographic solution to the anthropologist's predicament is to fully acknowledge the limitations of ethnographic process and product and to reduce their claims. Like feminists, critical ethnographers eschew a detached stance of neutral observation, and they perceive their subjects as collaborators in a project the researcher can never fully control. Moreover, they acknowledge the indispensably intrusive and unequal nature of their participation in the studied culture. Even more than most feminist scholars, I believe, critical ethnographers have been excrutiatingly self-conscious about the distortions and limitations of the textual products of their studies. Here they have attempted first to fully acknowledge and own the interpretive authorial self and second to experiment with dialogic forms of ethnographic representation that place more of the voices and perspectives of the researched into the narrative and that more authentically reflect the dissonance and particularity of the ethnographic research process.

Finally postmodern ethnographers, influenced by deconstructionist fashions, aim only for "Partial Truths" as James Clifford titled his introduction to a major collection of this genre:

> Ethnographic truths are thus inherently *partial* – committed and incomplete. This point is now widely asserted – and resisted at strategic points by those who fear the collapse of clear standards of verification. But once accepted and built into ethnographic art, a rigorous sense of partiality can be a source of representational tact.
> (Clifford, 1986: 7)

This reflexivity and self-critique of "postmodern" ethnographic literature parallels and has much to contribute to feminist methodological reflections. Perhaps it unwittingly exploits some of the latter as well, as feminist social scientists have published similar reflections on matters of the self, commitment, and partiality in research (Krieger, 1985; Mies, 1983; Rosaldo, 1983; Stanley and Wise, 1983a, 1983b;). At the least it could temper feminist celebrations of ethnographic methods with a salutary note of humility about the limitations of cross-cultural and interpersonal understanding and representation. Certainly I favor much more dialogue and exchange between the two than has taken place to date.

Recently feminist anthropologist Marilyn Strathern also noted the surprising paucity of engagement between feminism and the new ethnography and, in an important contribution to such dialogue, offered an analysis of the grounds for mutual resistance that undergird what she termed the "awkward relationship" between the two (Strathern, 1987). Feminism and critical anthropology, Strathern claims, are mutually "vulnerable on the ethical grounds they hold to be so important": "each has a potential for undermining the other" because they rest upon incompatible constructions of the relationship between self and "Other" (Strathern, 1987: 289). Feminism, Strathern argues, presumes an antagonistic relationship to the male Other, a presumption which grounds its acute sensitivity to power inequalities and has the power to undermine those anthropological pretensions of alliance and collaboration with the Other upon which new ethnographic strategies for multiple authorship reside. Anthropology, in turn, from its cross-cultural vantage-point, suggests the illusory nature of feminist pretensions of actual separation from men of their own culture.

I view the resistances somewhat differently. Feminism's keen sensitivity to structural inequalities in research and to the irreconcilability of Otherness applies primarily, I believe, to its critique of research by men, particularly to research *by* men, but *about* women. The majority of feminist claims about *feminist* ethnographic and other forms of qualitative research, however, presume that such research occurs almost exclusively woman-to-woman. As such feminist researchers are apt to suffer the delusion of alliance more than the delusion of separateness and to suffer it more, I believe, than do most post-structuralist ethnographers. Recall the claims about empathy and identification between feminist researchers and the women they study and the calls by feminist scholars for an egalitarian research process, full collaboration, and even multiple authorship with which this essay began. Hence, it strikes me that a fruitful dialogue between feminism and critical ethnography might address their complementary sensitivities and naivetes about the inherent inequalities and the possibilities for relationships in the definition, study, and representation of the Other.

While I hope to further such a dialogue, in the end, I agree with Strathern that the relationship between feminism and ethnography is unavoidably ambivalent. I am less convinced that she of the virtues of this awkwardness, but I agree that while it can be mitigated, it cannot be effaced. Even an exhaustive, mutually beneficial exchange cannot resolve the feminist ethnographer's dilemma. The postmodern strategy is an inadequate response to the ethical issues endemic to ethnographic process and product that I have encountered and described. It acknowleges, but does little to ameliorate the problems of intervention, triangulation, or inherently unequal reciprocity with informants; nor can it resolve the feminist reporting quandries. For example, acknowledging partiality and taking responsibility for authorial construction cannot reduce my handling of the lesbian affair into a matter of "representational tact."

My current response to the question in my title is that while there cannot be a fully feminist ethnography, there can be (indeed there are) ethnographies that are partially feminist, accounts of culture enhanced by the application of feminist perspectives. There also can and should be feminist research that is rigorously self-aware and therefore humble about the partiality of its ethnographic vision and its capacity to represent self and other. Moreover, even after my loss of ethnographic innocence I believe the potential benefits of "partially" feminist ethnography seem worth the serious moral costs involved.

Indeed, as Carole Joffe has suggested to me, my assault on the ethical foundations of fieldwork may have been unduly harsh, a fairer measure, perhaps, of my prior illusions about ethnographic virtue than of ethnographic vice (Joffe, 1986). Certainly, as she and Shulamit Reinharz assert, fieldworkers can and do form valuable relationships with many of those we study, and some of our unsolicited interventions into the lives of our informants are constructive and deeply appreciated. Just last week, for example, a daughter of the informant, whose death I mentioned above consoled me on the sudden death of my own father and thanked me for having allowed her to repair her hostile relationship with her father before he died by helping her to perceive his pride in and identification with her. In certain circumstances fieldwork research offers particular research subjects practical and emotional support and a form of loving attention, of comparatively non-judgmental acceptance, that they come to value deeply.

But then again, beneficiaries of such attention may also come to depend upon it, and this suggests another ethical quandary in fieldwork, the potential for, indeed the likelihood of desertion by the researcher.[7] Yet rigorous self-awareness of the ethical pitfalls in the method enables one to monitor and then to mitigate some of the dangers to which ethnographers expose their informants. I conclude in this Talmudic fashion to leave the dialogue open, believing that an uneasy fusion of feminist and critical ethnographic consciousness may allow us to construct cultural accounts that, however partial and idiosyncratic, can achieve the contextuality, depth, and nuance I consider to be unattainable through less dangerous, but more remote research methods.

Notes

1 Perhaps the most comprehensive summary of the characteristic distinctions between these approaches which feminists draw appears in several pages of tables detailing contrasts between the two in Reinharz (1983: 168–72).

2 For just this reason Duelli Klein (1983, Mies (1983), and, to a lesser extent, Stanley

and Wise (1983b) argue against this approach and for fuller collaboration between researcher and subjects, particularly for activist research in the tradition of Pablo Freire generated by and accountable to grass roots women's movement projects. But, as Carol Smart (n.d.) as well as Stanley and Wise recognize, such an approach places severe restraints on who and what can be studied and on what could be written, restraints that could seriously harm feminist interests.

3 Carol Smart (n.d.) offers important reflections on the adverse implications of this ethical principle when feminists study, as she believes we should, the powerful and the agents of social control rather than their targets.

4 Critical reflections on the ethics and politics of fieldwork have a long history in both disciplines, and by now the literature is vast. (For important examples from the past two decades, see Asad, 1973; Emerson, 1983; Haan, Bellah, Rabinow, and Sullivan, 1983; Hymes, 1974; Thorne, 1978, 1980.)

5 Howard Becker makes a similar point about the unfortunate paucity of exchange between critical traditions in sociology and poststructuralist anthropology in a review of *Writing Culture* (Clifford and Marcus, 1986), a major text on new ethnography (Becker, 1987).

6 A good sampler and bibliography of postmodern ethnographic criticism appears in Clifford and Marcus, 1986. Other important texts include Clifford (1983), Crapanzano (1977), and Marcus and Cushman (1982).

7 In her inimicable witty style, Arlene Kaplan Daniels discusses the etiquette of abandoning one's research subjects as well as other ethical questions in fieldwork (Daniels, 1983). I believe that the problem of desertion is more serious in longterm ethnographic studies than in those based on more limited contact characteristic of other forms of qualitative research.

References

Asad, Talal. 1973. *Anthropology and the Colonial Encounter*. Ithaca Press, London.

Becker, Howard. 1987. The writing of science. *Contemporary Sociology* **16**(1): 25–27.

Clifford, James. 1983. On ethnographic authority. *Representations* **1**(2): 118–46.

Clifford, James. 1986. Introduction: Partial truths. In Clifford, James and Marcus, George, (eds) *Writing Culture. The Poetics and Politics of Ethnography*. University of California Press, Berkeley.

Clifford, James, and Marcus, George, (eds) 1986. *Writing Culture. The Poetics and Politics of Ethnography*. University of California Press, Berkeley.

Crapanzano, Vincent. 1977. The writing of ethnography. *Dialectical Anthropology* **2**: 69–73.

Crapanzano, Vincent. 1986. Hermes' dilemma: The masking of subversion in ethnographic description. In Clifford, James and Marcus, George, (eds) *Writing Culture. The Poetics and Politics of Ethnography*. University of California Press, Berkeley.

Daniels, Arlene Kaplan. 1983. Self-deception and self-discovery in fieldwork. *Qualitative Sociology* **6**(3): 195–214.

Du Bois, Barbara. 1983. Passionate scholarship: Notes on values, knowing and method in feminist social science. In Bowles, Gloria and Duelli Klein, Renate, (eds) *Theories of Women's Studies*. Routledge and Kegan Paul, London.

Duelli Klein, Renate. 1983. How to do what we want to do: Thoughts about feminist methodology. In Bowles, Gloria and Duelli Klein, Renate, (eds) *Theories of Women's Studies*. Routledge and Kegan Paul, London.

Emerson, Robert M. 1983. *Contemporary Field Research: A Collection of Readings*. Little-Brown, Boston.

Freire, Paulo. 1970. *Pedagogy of the Oppressed*. Seabury Press, New York.

Haan, Norma, Robert N. Bellah, Paul Rabinow, and William M. Sullivan, (eds) 1983. *Social Science as Moral Inquiry*. Columbia University Press, New York.

Hymes, Dell, (ed.) 1974. *Reinventing Anthropology*. Vintage, New York.

Joffe, Carole. 1986. Personal Communication to Author.

Krieger, Susan. 1985. Beyond "subjectivity": The use of the self in social science. *Qualitative Sociology* **8**(4): 309–324.

Marcus, George E. and Cushman, Dick. 1982. Ethnographies as Texts. *Annual Reviews of Anthropology* **11**: 25–69.

Mies, Maria. 1983. Towards a methodology for feminist research. In Bowles, Gloria and Duelli Klein, Renate, (eds) 1983. *Theories of Women's Studies*. Routledge and Kegan Paul, London.

Oakley, Anne. 1981. Interviewing women: A contradiction in terms. In Roberts, Helen, (ed.) *Doing Feminist Research*. Routledge and Kegan Paul, London.

Reinharz, Shulamit. 1983. Experiential analysis: A contribution to feminist research. In Bowles, Gloria and Duelli Klein, Renate, (eds) *Theories of Women's Studies*. Routledge and Kegan Paul, London.

Rosaldo, Michelle Z. 1983. Moral/Analytic Dilemmas Posed By the Intersection of Feminism and Social Science. In Haan, Norma, Bellah, Robert N., Rabinow, Paul, and Sullivan, William M., (eds) *Social Science as Moral Inquiry*. Columbia University Press, New York.

Smart, Carol, n.d. *Researching Prostitution: Some Problems for Feminist Research*. Unpublished paper, Institute of Psychiatry, London.

Stacey, Judith. 1983. *Patriarchy and Socialist Revolution in China*. University of California Press, Berkeley.

Stacey, Judith and Thorne, Barrie. 1985. The missing feminist revolution in sociology. *Social Problems* **32**(4): 301–316.

Stanley, Liz and Wise, Sue. 1983a. "Back into the personal" or: Our attempt to construct "feminist research." In Bowles, Gloria and Duelli Klein, Renate, (eds) *Theories of Women's Studies*. Routledge and Kegan Paul, London.

Stanley, Liz and Wise, Sue. 1983b. *Breaking Out: Feminist Consciousness and Feminist Research*. Routledge and Kegan Paul, London.

Strathern, Marilyn. 1987. An awkward relationship: The case of feminism and anthropology. *Signs* **12**(2): 276–292.

Thorne, Barrie. 1978. Political activist as participant observer: Conflicts of commitment in a study of the Draft Resistance Movement of the 1960s. *Symbolic Interaction* **2**(1): 73–88.

Thorne, Barrie. 1980. "You still takin' notes?" Fieldwork and problems of informed consent. *Social Problems* **27**: 284–297.

9 J. K. Gibson-Graham

'"Stuffed if I Know!": Reflections on Post-modern Feminist Social Research'

From: *Gender, Place and Culture* **1**(2), 205–24 (1994)

Introduction

'Stuffed if I know!'[1] is the phrase that resounds in my head as I reflect on doing feminist research in the 1990s. It was the frequent reply made by Leanne, a woman participating in the research project I want to discuss here, to any questions asked directly about her life, community and political involvements in a mining town in central Queensland. Of course she did know, and she knew a lot, although what she knew was not what she initially thought I wanted to know. Her predictable answer became a group joke as the research participants and I began to feel comfortable enough to allow self-deprecating humour. The amusing thing was that after this initial denial of knowledge, Leanne characteristically became one of the most expansive discussants in the group.

Why does this phrase continue to resonate? Partly because as it was said by Leanne it problematised knowledge. It questioned what one woman knew and did not know; what I wanted to know and why it might be she who knew it; and what of her knowing would be considered knowledge for my project. Partly, also, because as Leanne said it, it struck a chord within myself. When I get asked 'What do you know?' I experience a similar reaction and a resistance to the implicit assumption that I am a centred and knowing subject who is present to myself and can be spoken for. Hence my attraction to contemporary developments in post-structuralist theory that challenge the concepts of self and knowledge implicit in the Cartesian understanding of subjectivity (Young, 1990, p. 310). Partly, as well, because 'stuffed if I know' reflects my initial reaction to questions about feminist methodology in the light of post-modern critiques of modernist epistemology and science.[2]

As debates within feminism rage over the scientific and political status of women's knowledge, at times I have felt at a loss to know why I was researching mining town women at all and whether what I was doing was in any way feminist research. And yet, like Leanne, I do know something and I could try to communicate my own emerging ideas about doing anti-essentialist post-modern feminist social research and negotiating the difficult questions of subjectivity and identity.

Until quite recently feminist empiricism and feminist standpoint theory have offered epistemological positions that have been the basis for a phenomenal growth in feminist social science (Harding, 1986). Empirical studies conducted from a range of theoretical perspectives (radical, socialist and liberal feminist) have all in some way affirmed the existence of women's experience as a source of privileged understandings, if not the basis of an alternative social science. Now, however, the deconstruction of 'women' is having profoundly destabilising effects upon feminist theorising and research (Barrett, 1991).[3]

Wendy Brown (1991) writes of the 'palpable feminist panic' that has arisen as the situated and subjective knowledge of 'women', gleaned, for example, from ethnography, oral history material or consciousness raising groups, has come under attack for its presumption of representing the 'hidden truth' of women or women's experience.[4] With the turn to post-modernism many of the certainties of a feminist research practice have been dislodged. This has liberated a plethora of exciting philosophical, political and cultural endeavours that tackle the essentialism around women embedded in both feminist and non-feminist texts.[5] At the same time, however, feminist social analysts find themselves confronting an ironic impasse as what have been seen as the unifying objects of our research dissolve before our eyes.

This paper tells the tale of a social research process which has been situated within and shaped by the flux of current feminist debates. It takes up some of the problems of 'doing gender' outlined by other feminist geographers such as McDowell (1992), Dyck (1993), Rose (1993) and Pratt (1993). Like them, I take seriously the challenges posed by post-modern theory to feminist social scientific research. If we are to accept that there is no unity, centre or actuality to discover for women, what is feminist research about? How can we speak of our experiences as women? Can we still use women's experiences as resources for social analysis? Is it still possible to do research *for* women? How can we negotiate the multiple and decentred identities of women? In this paper I try to reflect self-consciously upon these questions as I discuss the research direction taken and the methods employed in my own project.[6]

Mining and representation

In many ways mining and social research are parallel activities. Both intervene in and disturb a landscape by probing and digging for a rich lode of ore or layer of stratum that has hitherto lain covered, or unknown, perhaps until now unvalued. Like bauxite, women's knowledge has lain potentially accessible, part of the surficial landscape, for centuries untouched by the technical interventions of dynamite and front-end loader, tape recorder and questionnaire, simply because society has had no use for it. Today, bauxite is mined by aluminium can makers and women's knowledge is mined by feminist researchers. This paper is about how and why we might go about 'mining' women's knowledge in a post-modern vein.

During 1987 and 1990 when I was researching the development of new mining localities in central Queensland, I consciously tried to document stories about the activities of women along with those told by company managers, union representatives and community workers. My attempt to include the narratives of 'ordinary' women in mining towns marked my phase of what Stanley and Wise have called 'corrective feminist research' (1983, p. 25). I was concerned to 'add women in', to redress the imbalance I saw in the (very much) his-stories of mine and mining town development since the 1960s in the frontierland of Australia's 'deep north'.

The research methods I employed resembled the blast and shovel techniques of open cut mining. Into my front-end loader I blithely proceeded to shovel conversations and interviews with women around the issue of public involvements in political, industrial and community issues. At this time I was not concerned to problematise the discursive strategy of using 'miners' wives' or 'mining town women' as categories for exploration. These fictions occupied a secure place within modern social science discourse and presented themselves to me as 'natural' or obvious social designations.

Two 'identities' or 'constructions' were available in relation to the women I was concerned with. One was the representation purveyed by mainstream Australian social analysts and service providers influenced by the liberal community studies tradition. In this discourse 'mining town women' are a client group who have needs for better social, psychological and health services. They are portrayed as independent and self-reliant, but defensive, vulnerable, cautious of emotional commitment, lonely, isolated, stressed and traditional. This literature is haunted by the spectre of 'country women', the truly centred, community-minded and integrated rural women to whom mining town women are the lesser, the Other. In that their social networks are more unstable and their commitments to community-based activities more unreliable than those of country women,[7] mining town women are constructed as psychologically deformed.

The other representation was that purveyed by socialist and socialist-feminist analysts in Australia, the UK and North America. In this international discourse 'miners' wives' are situated as auxiliary members of the industrial proletariat, the feminine face of the politically cohesive working class mining community that holds such a hallowed place in left wing sociology. While researchers studying the newer mining communities of Australia and North America have emphasised the differences between women in these more affluent and isolated new towns and those in traditional mining communities of the British, Welsh and Appalachian coalfields (see for example, Williams, 1981; Kingsolver, 1989; Gibson, 1991a), miners' wives are still constituted within a discourse that brings to the foregound the functional overlap of capitalist exploitation and patriarchal oppression. Working class first and women second, miners' wives are subsumed to the fictional identity 'working class' and relegated to the status of Other within this totalising conception.[8]

The focus in the socialist-feminist literature upon industrial disputes, in which women are expected to express their real identities through solidarity with working-class men, situates this literature solidly within essential Marxism-feminism where consciousness is true or false and subjectivity structurally constructed and constrained. Most of the studies within this tradition highlight those occasions when miners' wives invert their Other status, come to recognise their true class alignment, and heroically join, lead, or hold their men to authentic working-class consciousness and action.

As I collected stories of women's political involvements in mining towns, these two discourses formed a contradictory backdrop of concern. Corrective research, it seemed, could not add women into the picture without situating them with respect to one or the other representation. But the available characterisations seemed to construct narrow, unidimensional identities and subject positions with negative and disciplinary overtones. Both the client/victim/pathologised individual representation of liberal discourse and the proletarian/militant/supporter-cum-leader representation of socialist discourse denied the potential for a multiplicity of political subjectivities to emerge. And both seemed actively to organise women out of any independent involvement in either industrial or gender politics.

As I began to write up stories of women in mining towns, it occurred to me that a discourse, somewhat like a mineral, is a substance that is socially constructed and valued on account of the uses to which it can be put. In the process of mining women's knowledge the researcher looks for 'evidence' that can fit a discourse she or he is engaged in influencing or constructing. These traces might be stories told by interviewees, they might be personal impressions or responses to formal questions. And like the miner, the researcher cannot just uncover and extract new mineral mass; she or he must also ensure that the product can be sold in the intellectual market-place.

I was wary of the ways in which the stories I collected could be incorporated into the liberal and socialist feminist discourses and I worried about the effects of reinforcing existing representations. A few examples might illustrate my dilemma.

Exploratory probes and precious glimmers

My interest in the political voices of women in mining communities led me to problematise the unpredictable, public political activities of miners' wives and to collect stories about the most obvious examples of women's political activism, the Miners' Women's Auxiliaries in each town.[9] These organisations, once commonly found in older underground mining communities, have traditionally been adjuncts to the men's union, run basically as welfare organisations by wives of miners. Auxiliaries usually come into being at the time of a major dispute or disaster when mining town women become community carers, dispensing food, shelter and comfort. What follows are some of the stories of past and present Miners' Women's Auxiliaries and other organisations I came across in three towns.

Collinsville

Miner's wives in their sixties and seventies remembered the Collinsville and Scottville Miners' Women's Auxiliaries as active organisations that met in the 1950s on a regular basis. These neighbouring communities had originally been formed around underground mines in the 1920s. In both, the Communist Party of Australia was strong and the Women's Auxiliaries took on a broad range of issues:

> We used to meet once a month and it was a whole social morning. We had morning tea. We had proper meetings, properly run. We discussed what was going on around the town, we discussed politics, we wrote letters of protest here, there and everywhere. . . . We lived in a little valley but we kept up with what was going on.
>
> (Collinsville woman, 1990)

> We started off as the Miners' Women's Auxiliary and then joined the Housewives Association which then became the Union of Australian Women. Then the mine closed down and we disbanded.
>
> (Collinsville woman, 1990)

> We set up our own greengrocer's supply. It started at a (Women's Auxiliary) meeting when people were complaining about not getting enough fruit and vegies because the prices were outrageous . . . We ordered supplies from the Brisbane market. We paid market prices and fruit and vegies came by train. We charged market prices plus freight and we took turns going and doing up the orders. It kept going for years.
>
> (Collinsville woman, 1990)

In addition to organising around welfare and consumption issues, Collinsville and Scottville women actively supported their men in work place-based struggles such as the 1952 stay-down strike about wages and conditions at the Collinsville State Mine and the closure of the state mine in 1961.

From their long perspective the older women in Collinsville and Scottville see the contemporary period of mining activity as very different. There is no Women's Auxiliary operating in this community now. As they see it, the welfare role of the Women's Auxiliary is unnecessary today:[10]

> There's no struggle any more. They're [we're?] a dying race.
>
> (Elderly Collinsville woman, 1990)

In this comment I heard many things. I heard the lament for a modernist socialist politics that had so readily created a community of exclusion centred on the privileged and conscious few who could lead the way toward revolution. I heard the disgruntlement of old timers unhappy with the way things are now. I heard judgements of younger women, their politics bought off by material comforts. I heard sentiments that could have been read as marking the absence of socialist or feminist politics in today's style of working-class community.

I wondered, were the new breed of Collinsville and Scottville women a less hardy race, whose political concerns had been swapped for microwave and video-recorder? Was the liberal 'mining town woman' discourse more appropriate to an understanding of Collinsville's new generation of miners' wives? Why had these radical older women not raised another generation of stirrers? There were no public manifestations of any organisation, but did this preclude activism? How might the socialist visions of struggle, to which these older women adhered, have actually precluded different styles of politics, particularly ones that were women-focused rather than class-focused? As the older women spoke of a past age of courage and solidarity with their men, I began to see the discipline their vision of unity could enforce and how it might have contributed to the political silencing of younger women.

Moura

Moura is a town based on the modern, globally-oriented coal industry, a site of underground mines and the first open cut mine developed explicitly for coal export in 1960. In 1990, Moura was the only town in central Queensland that still had an operating Women's Auxiliary. It had 16 paid up members (mostly in their forties or older) who met once a month in the local park. Perhaps this organisation was still in existence in Moura because the community had suffered two very bad underground mine disasters in recent times. In each case the Women's Auxiliary played a major role in servicing the rescue teams and supporting the families of the dead miners.

As an ongoing organisation the Moura Women's Auxiliary defines itself in rather narrow terms. Its primary role is to distribute the 'men's money' to the needy – the families of sick or injured miners or those on strike. It is not allowed by its constitution to be a fund-raising organisation, though it can and does raise money on behalf of other charities in the town. Up to now it has not acted as a forum for discussion of specifically feminist or industrial politics.

Indeed, the juxtaposition in Moura of an operating Women's Auxiliary and a spontaneous grassroots political organisation, the Concerned Women of Moura, is one of the more curious discoveries of my research. The Concerned Women of Moura came together to lobby state parliamentarians about new industrial legislation that the right wing Queensland state premier was intending to introduce in 1987. This legislation proposed to empower the state government to declare strikers law breakers and to seize their personal assets if employment contracts were opposed. In the lead-up to the 1987 federal election both the conservative National and Liberal parties in Queensland endorsed the proposed changes.

While in most of the mining communities organised labour mounted campaigns to oppose the introduction of this repressive legislation, in Moura the struggle came to be seen as a community issue, mainly through the activities of this spontaneous organisation of women. Like many women in Queensland mining towns, the two women who spearheaded the Concerned Women of Moura had not been Labor Party[11] voters prior to this moment. For these women, and the five friends who joined them in the initial organisation of the group, it was a big step to consider changing their vote in an election to Labor, let alone to take direct political action as they did.

The struggle was seen by the Moura women as pitting their community against the state, and while unions were obviously involved, the women did not rely upon union structures to include their voices. Their grassroots organisation was successful in influencing the federal election vote and the proposed changes to industrial legislation were dropped shortly before the state government crumbled under allegations of corruption. After the election the Concerned Women of Moura disbanded. When I again visited Moura in 1990 the memory of this fight was already beginning to fade and the experience had not entered the collective knowledge of mining town women outside the town.

I found it heartening that Moura women challenged the opposition posed between 'mining town women' and 'country women' but I was also provoked to think of the relationship between their activities and feminist politics. Although both organisations gave their participants a sense of greater individual and collective power, it was not power with respect to, or over, men but power channelled *for* men in some way. In this sense, both stayed within the socialist, feminist discourse of miners' wives.

Within this discourse there was, however, an interesting glimmer of difference. The Women's Auxiliary was based upon a sense of continued, shared and internal identity gained by its members' relationship to mineworkers. It appealed to an older group of women but faced the constant problem of maintaining membership and negotiating the constraints placed upon it by its fixed and limited identity. The Concerned Women of Moura was also an organisation grounded in women's identities as miners' wives but it was ephemeral and came into being as a result of many different women's relation to an external threat. The subjectivity it created was based upon momentary and partial identification; it allowed a fluid set of practices to be directed towards a political goal, and then dissolved.[12]

Moranbah

In the stories above women's organisations gained some form of legitimation by supporting men's struggles. In a sense, all these organisations were designed to challenge the power of capital as expressed in the actons of coal companies, the state and local merchants. In Moranbah, I encountered women organising around so called women's (or non-economic) issues and it was interesting to hear of the considerable barriers to their efforts that had been constructed by men. The Moranbah stories could not easily be fitted into either the liberal or socialist discourses.

A group of miners' wives who had been active in the formation of a (rather temporary) Miners' Women's Auxiliary during the 6 week housing strike in 1978 (see Gibson, 1991a and b) saw the desperate need in the community for women to get some respite from the full-time care of their pre-school age children and to obtain information and support in child rearing. They also felt that there was a need for the town to initiate community development projects that were not directed and instigated by either the mine management or union organisations. The women began to lobby the state government, local union branches and the mining company to set up a neighbourhood centre, with a community social worker, to act as a focus for community development, as well as a child care centre offering occasional care for babies and pre-school children.

The unions were reluctant to support the building of a child care centre, and the miners themselves could not see the need for child care as that was what their wives were supposed to do. When women from the organising committee were allowed to address the rank and file union members at the mine site, the meeting was set up in the men's shower block, a venue the women speakers felt was deliberately selected to intimidate and embarrass them. The Combined Mining Unions refused support for the child care centre but agreed to contribute to the salary of the social worker associated with the neighbourhood centre.

A new building was finally built to house the day care centre and the neighbourhood centre. In the minds of some, the combination of both functions into one building was a mistake. The whole complex has tended to be seen as a child care centre and as a place for women and children, and the neighbourhood centre has had difficulty reaching out to the wider community. With the appointment of a number of young social workers whose radical feminist style and projects offended the values of many male and female community leaders, the centre has become feared, particularly by men.

> That's a place to keep your wife away from if you don't want her to divorce you.
> (Moranbah miner, 1990)

This project illustrates the problems for a feminist politics in mining communities where 'legitimate' politics means union politics and 'legitimate' women's politics means supporting the men.

One of the projects initiated by a (radical feminist) community worker at the Moranbah Neighbourhood Centre was the production of a video by a group of local women. Some eight women conceived of the topic, wrote and researched the script, conducted interviews and filmed the video with technical assistance from James Cook University, located at Townsville on the coast. The title selected for the video was 'Moranbah – is it so different?' In it the group challenged what they saw to be prevalent, negative representations of modern-day mining communities as towns with transient populations and a higher than average incidence of social problems like strikes, alcoholism, divorce, family break-up, domestic violence and child abuse. The video faces these images head on, arguing that there is no evidence to show that these problems are worse than in any other part of Australia. It counters the negative representations with positive images of the town, its sporting facilities, young population and affluence.

What I found interesting about the video was its silences about aspects of the social identity of mining communities that most social scientists highlight, such as class divisions, corporate control and militant unionism. Though women individually would admit that these are important features of Moranbah, the composition of the video production group prevented discussion of these issues. The project involved eight women, three of whom were wives of salaried staffmen (identified with management) at the mine, the rest being wives of wage workers. I began to see glimmers of how, as a *women's* group, participants experienced some of the contradictions of attaining a voice based upon a singular identity. The effort to represent the town positively culminated in a censoring of certain aspects of identity and a resultant code of silence around some of the most marked dimensions of difference.

Yet, while it bypassed any discussion of class power, the video project did allow the participants to confront and escape their traditional roles and confinement to the household. It did enable them to challenge their usual lack of participation in public image-making about their town. Both the Moranbah projects appeared to counter the socialist and liberal discourses on mining town women. They showed the capacity of women of different backgrounds to organise together, and to organise in the face of opposition from men. Here were forms of political activism and subjectivity that escaped the logic of the 'rural women' identity in which 'mining town women' are constructed as Other to 'country women', as well as the logic of working-class identity in which the 'miner's wife' is positioned as Other to the 'proletarian miner' (in both subject positions women in mining towns are constructed as the less potent, the subordinated and the underdeveloped). Here was a glimmer of actions by and for women which united and empowered and, at the same time, silenced and disciplined their participants.[13]

Processing

The stories I collected undermined the representation of 'mining town women' as individualistic and depressed, unwilling to connect with other women or contribute to community activities. But the traditionalism of the gender relations in these stories reinforced mainstream representations of women as personally and politically dependent upon men. The stories also overlapped with the socialist-feminist discourse on 'miners' wives' as members of the working-class, but at the same time allowed glimpses of other processes by which different (non-working class) subjectivities were continually being crafted, and sometimes enunciated in action (Kondo, 1990). A deconstruction of 'mining town women' and 'miners' wives' had allowed me to see both as regulatory fictions masquerading as self-evident categories of analysis, each of which positioned women in mining towns in subjugated positions.

I suspected that the traditional representation of 'mining town women' had a use within the discursive space occupied by social workers and service providers. The image of mining town women as individualistic non-joiners was a handy categorisation which justified the interventionist activities of service providers and staff wives (often the same people in smaller towns) who acted as gatekeepers for all social activity and any gender-based politics in the towns.

At the same time I was aware that the left representation of 'miner's wife' had a use within the discursive space occupied by union leaders, labour historians and socialist feminists. The image of a solid supporter of 'the men' and upholder of hard-won conditions valued and romanticised the contribution of women, and established the nobility of the miners' class struggle. For the wives of some miners this representation was a welcome reward for toeing the line in an important but self-effacing way. Women were accorded the accolades befitting true working-class warriors when they willingly subordinated their lives to the cause of jobs (for men), wages (for men), political rights (for men) or lower taxes (for men). But it seemed clear to me that these accolades would not be forthcoming if women overstepped some invisible mark and, for instance, interfered in wage negotiations or shift work changes when major disputes were not in the offing, or led a movement for paid work (for women) or wages for housework.

This initial process of deconstructing the categories 'mining town woman' and 'miner's wife' had allowed me to identify their politically powerful disciplinary and exclusionary effects. At the same time I had begun to see a glimpse of alternative subject positions and political identities for women in mining towns with which I could interact. It occurred to me that producing alternative discourses of gender and mining town life was one way of liberating alternative subjectivities for mining town women.

The question that soon emerged was why create an alternative discourse, an alternative voicing, and where was its audience? Who was interested in new subject positions for women in mining towns? And, anyway, why construct alternative subjectivities for mining town women if 'women' in general had disappeared? The usual answer to such questions harks back to the political project of feminism and the central role that research plays in the 'liberation' of women. But the political project of post-modern feminism is now a matter of considerable debate.

Research as politics/politics of research

In dissolving the presumed unity of women's identity post-modern feminism has liberated knowledge and given rise to fruitful theoretical controversies as to who women 'are' and how to 'know' them. At the same time, however, Brown's 'palpable feminist panic' (1991, mentioned at the outset of this paper) seems to have migrated from the realm of theory into the political realm, where the identity 'woman' has usually been constituted as the *necessary* ground of feminist political action.[14] Feminists have historically claimed that as 'women' we are dominated and oppressed, and feminist politics has staked its legitimacy upon the assumption of this shared or common, but importantly, *subordinated* identity. To surrender epistemological claims about women's shared identity has signified, for many, giving up the structural and moral position from which to organise politically to overcome oppression (Brown, 1991, p. 75). Without unity of women's identity, many critics see post-modern feminism as opening the doors to fragmentation, factionalism and political disempowerment.[15] It was in the face of this possibility that Spivak's strategic essentialism was proposed (in an interview with Grosz, 1984).

It seems that, for many, a paradox has emerged – as knowledge has been liberated, politics has been shackled. While feminists may agree that in theory, difference empowers, when it comes to politics many still hold to the adage that 'united we stand, divided we fall'. Ferguson (1993) argues that feminists must accept this contradiction and learn to live with the inevitable tension between articulating 'women's experience' (or interpretation in her terms) and deconstructing the texts that represent and enforce this presumed commonality (or genealogy in her terms). Rather than insisting on a real, originative essence that defines all women, Ferguson advocates constant movement between the (strategically essentialised) representation of women's experience and the (strategically non-essentialised) deconstructivist practice of undermining fixed categories of identity and gender.

Interpretation and genealogy are 'postures toward power and knowledge that need one another' (Ferguson, 1993, p. 29). In support of this view, Pratt (1993) calls for an 'equal commitment' and 'continuing dialogue' between these two moments of research practice. Barrett (1991) similarly positions the two moments in opposition:

> So it is an issue of whether one wants, speaking as a feminist, to deconstruct or to inhabit the category of 'women'.
>
> (p. 166)

Ferguson's suggestion for negotiating the problems produced by this contradictory strategy is to keep the discursive stance of irony always at hand. But I see a danger in posing these moments as opposing practices in irreducible/ironic tension, the one associated with grounded commitment, the other with relativism. Surely all deconstruction, or the tracing of 'how we produce truths' (Spivak, 1989, p. 214), is done from a specific theoretical and political entry point from which further interpretation also proceeds. Seeing one posture as less political because it highlights difference, and the other as more political because it highlights collective identity seems to suggest that the politics of identity is the only viable political form (for feminists at least). It also implies residual loyalty to the modernist separation of theory and practice – that conception of knowledge/theory existing separate from and prior to change/politics (we understand the world *in order to* change it). Practice or politics, in this formulation, can only be enacted by a collectivity of subjects all identically positioned *vis à vis* the structure of power that has been rendered visible by theory. What this conception betrays is an interesting failure to see knowledge and its production as an *always already* political process.

What is needed, it seems, is a rethinking of the relationship between politics and research. Following Foucault, I would see post-modern feminist politics starting from the assumption that power is everywhere inscribed, in and by women, as well as by men, in theory as well as in practice, in difference as well as in unity. Thus the process of theoretical production is as much a political intervention in changing power relations as is self-consciously (identity-based) political organisation. There is no prior reality or unified identity to gain access to or to be created by research from which we can launch a programme of change. There are, however, existing discourses that position subjects in relations of empowerment and disempowerment. The ways in which theory and research interact with these discourses have concrete political effects.

In this project I was not only working with existing discourses that situated and identified mining town women, but attempting to create new discourses and subject positions. As a social researcher, I was interested in the circulation of alternative discourses on women in mining towns within communities of interest not readily touched by academic writing. One mode of circulation open to me was the research process itself. Encouraged by feminists such as Brown and Weedon to engage in conversation and public discourse,[16] I wanted to move beyond a purely literary discursive intervention into representations of women in mining towns and into 'the field'.[17] I embarked upon a research process that attempted to involve women as 'knowing subjects' in the 'always already' political nature of the research process.

Immediately I was confronted by the practical dilemmas of how to include women in mining towns in the process of discursive deconstruction and the circulation of new discourses. One alternative was to attempt direct intervention in power relations between men and women by embarking on a project of action and participatory research. But in formal action research models the process of researching one's own situation with others who are similarly situated is designed to have the effect of raising consciousness of shared conditions and oppression. This method of research relies upon an identification between researcher and researched and the discovery of a shared subject position from which political intervention can be discussed and enacted (Reinharz, 1992). Without an assumed basis of unity between women could these research methods still be employed? In pursuing my idea of social research as a public engagement in the construction of alternative discourses, I was forced to rethink methods of action research in terms of post-modern feminist social research practice. The last section of this paper tells the story of the research roller coaster that took off.

As luck would have it . . .

As luck would have it, in 1989 an industrial issue had emerged which was disturbing many women in mining communities and I realised as I continued with my interviews that a market for an alternative discourse on women in mining towns existed right there amongst mining town women themselves. Both of the established identities and subject positions discussed in the literature were actively preventing women from addressing the issue of their concern. I could see the potential to create a space where another identity and subjectivity might begin to grow – women as researchers of their own conditions and potentially as industrial activists in a male-dominated industry.

In 1988 a new Industry Award had been instituted in the black coal industry of Queensland and New South Wales. An important element of this award was the introduction of new work rosters which would allow the mining production process to continue 24 hours a day, 7 days a week, 52 weeks a year, thereby maximising the utilisation of machinery and increasing mine productivity. New workers were not, by and large, employed to maintain this intensified production regime, but existing work practices were restructured to facilitate greater output from the established work-force.

In the coal mines of central Queensland most mines introduced what is known as the 7-day roster, a work timetable that involves shift work rotating on a 7-day basis with 1 or 2 days off between the seven afternoon and day shifts and a longer period of 4 days off after the night shift at the end of the rotation cycle. Whereas in the past most miners had had the voluntary option of working overtime at weekends (and many people worked one weekend day), after the new industry award overtime work was made compulsory and most workers were entitled to only one consecutive Saturday and Sunday off per month.

While there have been countless studies of the effects of shift work and different roster arrangements, most of these studies focus upon the disruption of circadian rhythms and other physiological and psychological effects on the well-being of the shift worker.[18] The male-worker bias of much of this research is evident in the preoccupation with evaluating impacts on the availability of leisure (and for protagonists of different roster systems the maximisation of leisure time). The fact that increased leisure for men (in terms of days off during the week) coincides with an unchanged (or even an increased) workload for their partners is rarely acknowledged.[19]

In Queensland the change of work roster was having a number of impacts on workers, their families and communities that were causing concern, particularly among women. The new work roster was quickly dubbed the 'divorce roster' and in this naming the connection between work changes and home life was clearly established. From informal discussions with a wide range of people in central Queensland coal mining towns in 1990 it seemed there was a need for research that was explicitly designed to break with the individualistic approach of much research into shift work, placing the worker in his family and community context in order to guage the wider impacts of industry restructuring. I therefore embarked on a research project with a very practical aim of documenting these impacts from the perspectives of women in mining towns.

The situation appeared to offer interesting potential to begin a number of conversations in different contexts, about women's role and input into industry policy, and about the possibility of feminist politics and research in the light of theoretical debates over post-modernism and difference.

Identity, difference and post-modern feminist action research

As my own research and that of others had established, in mining towns women are marginalised by many processes (Sturmey, 1989; Gibson, 1992a). The male dominance of the mining work-force and lack of alternative employment for women contributes to the impoverishment of women and their utter economic dependence upon men. It was clear from the stories I had collected previously that the traditions of mining unionism excluded women from anything but the most peripheral involvement in industrial politics, and the prominence of traditional images of masculine and feminine identity (Connell, 1987) constrained even the most mundane efforts towards women's social independence.

I decided to employ a number of miners' wives as co-researchers in the project in an attempt to confront some aspects of women's marginalisation.[20] The women employed had to be experiencing life with a shift worker on a 7-

day roster. They also were selected on the basis of their stage in the lifecycle and family formation. In each town the aim was to employ one woman with pre-school children, one with school-age children and one with no children living at home. The 12 participants (three each from four different mining towns)[21] were actively involved in the research design and questionnaire formulation and were trained as interviewers at an initial 2-day workshop to conduct recorded in-depth discussions with their friends and acquaintances.[22] They were each responsible for conducting and recording six discussions with women in their own social networks who had roughly the same family structure. In this way it was hoped that an established rapport would exist between interviewer and respondent and would be the basis for a more relaxed and revealing interview experience. The women received payment for their interviewing work and all expenses were covered to allow them to attend two workshops at a location that was central to all the towns. At the second workshop held later in the year, preliminary results were analysed, qualitative results discussed and possible interventions outlined.

I saw this project as a modified or post-modern form of action research. While I actively involved women in the process of researching their own situations with respect to shift working partners, the project had no underlying agenda of consciousness raising and direct group action. The initial training workshop and the later feedback workshop incorporated discussions of consciousness and action but there was no expectation on my part or that of the participants that a feminist political programme would or should emerge. Instead, it was at these workshops with the 12 employed participants that some of the murkier issues to do with subjectivity, identity and politics began to be raised. In the workshops and over the kitchen tables where one-to-one interviews were conducted, the research project created and cultivated spaces in which a feminist politics (the transformation of gendered power relations) was performed in conversation and group discussion.

In the process of creating this political space, 'place' assumed some importance. The mining towns from which the participants were selected are places where homogeneity and mutual 'community' identification can easily be enforced by the dominant group. Unlike in cities where people 'live together in relations of mediation among strangers with whom they are not in community' (Young, 1990, p. 303), in small mining 'communities' people foster links with each other through 'being the same' and excluding and ostracising anyone who is different.[23] Removing women from their 'community' was the first step in creating a discursive space in which to construct new political subjectivities.

Within the research team, various barriers could have stood in the way of political conversations. Many of the differences which divide and structure our everyday social experience were present in the group that assembled at the first training workshop – urban–rural born, married–unmarried, educated–uneducated, older–younger, childless–mother, adopted–not adopted, wealthy–poor, traditional–feminist, spiritual–materialist, new age–mainstream, fat–thin, forthcoming–shy, amongst others. In this group the decentredness of women's 'collective identity' and the overdetermined nature of subjectivity were patently obvious.

Throughout the workshops members of the research team inadvertently explored their differences with each other and with the other women they had interviewed, and of course with Joanne (the freelance community worker I employed as a co-researcher and local facilitator) and myself. In this space away from family, friends and 'community' women felt liberated to air differences without forcing conformity. I heard many comments prefaced by 'My life's not like that . . . my husband's not like that . . . of course things were different in those days than now . . . well, you're younger than me . . . some women must live in a very different situation than I do'. Openness to otherness was more pronounced among women from different towns, while within groups from the same town conflict was more likely to emerge as particular representations of town life or company actions were discussed and challenged.

For me the first workshop marked a transition in my relationship to the women I was employing, which was initially dominated by the hierarchy and differential power of the academic/mother and employer/employee relations. By the end of the workshop, one of the women who was most into differentiating herself and her particular experience from that of the others (especially because she was quite happy with her life and felt that others' complaints didn't ring true for her) expressed the view that even though I was a doctor I was really just one of them. This moment of identification referred primarily to one of my many subject positions, that of being a mother of small children, someone who could share in tales of childbirth, sleepless nights and the irrational frustrations of mothering. On the basis of this dimension of similarity I was somehow legitimated in her eyes, my power defused and her acceptance of me granted. I was homogenised and accepted into the fictional but collective unity 'mother'. In a different way Joanne (who was not a mother) was identified as a 'local' (that is, non-metropolitan), someone who experienced isolation, car breakdowns on outback roads and harsh climatic conditions, and partially accepted on that basis.[24]

At the outset of the research the notion of 'partial identification' was undeveloped and for many of the participants the lack of identification with each other was uppermost in their minds. At the end of the second workshop both the idea and the act of partial identification had become more developed. Interestingly, what emerged was not what I would see as identification around the shared experience of women (the recognition that *as women* we shared a common 'problem'). In fact personal differences in gender experience widened on many fronts. What took place was identification with respect to common problems of a very specific kind (ones that many women would not share) – living with a shift worker (or in the case of Joanne and myself, living with a self-employed partner who worked long and irregular hours, often including weekends); particular place-specific forms of male discipline; union, company and university reluctance to consider family life in industrial relations.

In a sense my research process was constructing a partial but shared, externally related identity, and beginning to create a public knowledge about rosters and family life, about terror in the face of male power over women's ability to speak out, about women's mistrust of each other. The fiction of the 'mining shift worker's wife' I was encouraging or imposing became a momentary reality – a basis for communication about many of the contours of power affecting political activism in mining towns. These comments were made with acceptance and resignation:

> Women don't trust one another.

> We're powerless in the face of decisions about the roster – the men won't listen to just us.

Other comments were made with surprise and consternation:

> Many women confessed to hating the 7-day roster but they refused to be interviewed. Their husbands were forbidding them to be involved.

> Even women who are normally very strong and stroppy said they couldn't do it.

> The men are so suspicious, of them and us. They thought this survey would be used against them.

In the process of research the participants became open to otherness and aware of their own political capacities:

> This has made me realise that not all people can cope with the roster system, and just because things in my part of the world run smoothly, does not mean there is nothing wrong with the town I live in.

> The course has made me get off my bum and go out and talk to people. I usually stay at home in my little cocoon and don't go out and meet people. I find that having talked so much about what the people of this town need, instead of sitting back and waiting for everyone else to do the work and me reap the benefits, I now have to be a doer and not a talker. This is something new for me. I have never been one to go out and take action and motivate people. This will be a turning point in my life and, provided I follow it through, it should do me a lot of good.

> The ladies I have met and mixed with have given me the feeling that there is hope for the future, as they care about what is happening around them, and I thought we as a society had lost that ability.

> The research helped me rationalise my thoughts about the 7-day roster and was very helpful in coming to terms with many personal issues. It boosted my self-worth and gave me a sense of achievement and involvement in community and value other than as a housewife/mother.

In the political space created by this research project a new discourse of mine shift work and a new subjectivity of 'the mining shiftworker's wife' started to emerge. As women engaged in the myriad conversations that formed part of the research, they actively displaced the existing discourses of 'mining town women' and 'miner's wife' that confined their subjectivities. Out of this process a new subject position has developed – one that is focused on the gender division of labour and the impact of industrial conditions and disputes on relations within the home.

At present this research process has had a multitude of effects, many of which I am unaware. Those I do know about include the above mentioned emergence of a non-individualistic shift work discourse in mining communities in central Queensland. As companies discuss the possibility of introducing 12 hour shifts in the mining industry and long-distance commuting from the coast to new mines, the results of this research are circulating as an alternative way of thinking through the issues. A booklet that illustrates the research findings using cartoons and verbatim comments from women has been published by the union and distributed throughout the region, significantly aiding the circulation of the discourse of shift work and family life.[25]

A new (the first) occasional child care centre has been built in Moura after attempts by one of the research participants (the one who felt she was ready to get off her bum) to set up a baby-sitting club failed. This woman took the research report along to the meeting to discuss the need for such a service and was able to influence the decision to establish the centre.

One of the women interviewed asked for the tape of her interview back, sat her husband down and made him listen to it. She then was able to broach her anger with him for spending all his days off from mine work at the new farm they had just purchased. Their interviewer was pleased to report that 'Now they're like a pair of newly weds'.

At some level the research is challenging the established discourse of industry policy – its boundaries, the actors it legitimises and its social effects. As women in all their diversity voice newly developed concerns around an issue of industry restructuring they enter wittingly or unwittingly an arena from which they have long been marginalised and excluded.

Conclusion

While I share no fundamental identity with any other person (as I am a unique ensemble of contradictory and shifting subjectivities), I am situated by one of the most powerful and pervasive discourses in social life (that of the binary hierarchy of gender) in a shared subject position with others who are identified, or identify themselves, as women. This subject position influences my entrée into social interactions and the ways I can speak, listen and be heard. In this sense I am enabled as a woman, to research with other women the conditions of our discursive construction and its effects.

As a feminist researcher, I am coming to understand my political project as one of discursive destabilisation. One of my goals is to undermine the hegemony of the binary gender discourse and to promote alternative subject positions for gendered subjects. I see my research as (participating in) creating identity/subjectivity, and in that process as constituting alternative sites of power and places of political intervention. Whether in conversation with mining town women or with other feminist academic researchers. I understand my discursive interventions as constitutive rather than reflective, political as well as academic.

In my research I found the metaphors of 'conversation' and 'performance' much more useful in imagining a research strategy than the mining metaphors I had initially adopted. The mining metaphors constitute research as a process of discovery, of revelation: as researchers we reveal truths that are hidden from the untutored observer, contributing hitherto untapped resources to the permanent store of knowledge. By contrast, conversation and performance are metaphors of creation and interaction. Both processes are ephemeral, yet each may have long lasting effects upon thought and action. Conversations can produce alternative discourses that entail new subject positions, supplementing or supplanting those that currently exist. These new subject positions crystallise power in new sites, enabling novel performances – individual or group interventions in a variety of social locations. In this way the creation of alternative discourses subverts the power of existing discourses and contributes to their destabilisation.

This research process has provided insights for me into the practice of a new, post-modern feminist politics of difference. Action research need not focus upon the uncovering or construction of a unified consciousness upon which later interventions will be based. Action research can be a means by which we 'develop political conversation(s) among a complex and diverse "we"' (Brown, 1991, p. 81). Within these conversations we create the discursive spaces in which new subjectivities can emerge. As the centred subject with its historic political mission departs the social stage, there is now room to talk of the inescapability of difference and the only/ever partial nature of identification. Yet such talk does not precede or preclude politics. For the babble emanating from this discursive space is a political process without end, and without a (unified collective) subject. In an overdetermined world conversations are interventions/actions/changes in and of themselves, no matter whether they do or do not also give rise to further planned interventions.

Notes

1 This is a common Australian saying which pronounces in decidedly emphatic tones a lack of knowledge.

2 Following Nicholson (1990) I use the term 'post-modern' here and in the title of this paper to denote an intellectual movement within feminism and social theory more generally which is centred upon a critique of modernism in all its forms – from theory production to the production of buildings (see Watson and Gibson [1994] for an indication of the diversity of this movement). Following Barrett (1992) I use the term 'post-structural' to denote one theoretical strand within the post-modernism movement – that which focuses upon a critique of structural modes of thought, particularly in Marxian and socialist-feminist theory. The theoretical insights developed in this paper have been influenced most directly by this type of feminist post-modernism.

3 And the argument that the category 'woman' is but a fiction of coherence which

served the interests of the heterosexual contract (Butler, 1990) has shocked and confronted many.

4 She writes

> 'the world from women's point of view' and 'the feminist standpoint' attempt resolution of the postfoundational epistemology problem by deriving from within women's experience the grounding for women's accounts. But this resolution requires suspending recognition that women's 'experience' is both thoroughly constructed and interpreted without end. Within feminist standpoint theory as well as much other modernist feminist theory, then, consciousness raising operates as feminism's epistemologically positivist movement.
>
> (Brown, 1991, p. 72)

The discussion of 'women's experience' is, from this perspective, the creation of a discourse which imposes a fixed identity, rather than uncovering of an unmediated truth.

5 Barrett talks of the 'turn to culture' within feminism as the interest in post-structuralist theory has prompted a shift away from 'the social sciences' preoccupation with things' towards words and language. She claims that '[a]cademically, the social sciences have lost their purchase within feminism and the rising star lies with the arts, humanities and philosophy' (1992, pp. 204–205).

6 The paper addresses similar issues to those Pratt (1993) has thoughtfully broached in relation to her own empirical research with Susan Hanson. I am aware that by talking retrospectively about the contours of my own research process I am creating a narrative of coherence out of a very random and haphazard set of experiences. I have attempted in this 'temporary retrospective fixing' (Weedon, 1987, p. 25) to highlight the overdetermined nature of the research process and the significant role played by confusion and chance as well as design and planning. The research findings will not be reported here. They have been written up in a number of research reports and articles; see Gibson 1991a, b, 1992a, b, 1993, also Gibson-Graham, 1994.

7 In the years since the 1970s height of Australia's mining boom an interesting literature has grown up on the topic of women in (usually remote) mining towns. Much of the research has been conducted from a behavioural social scientific perspective and has been concerned with the adequate provision of services for women in these otherwise male dominated townships. Researchers have studied the general health and well-being of mining town women (see Sturmey, 1989, for a good summary of this literature), their mental health (Neil, *et al.*, 1983) and social networks (Cotterrell, 1984). The thrust of this work is an identification of the relative well-being and/or disadvantage of women living in these types of community where isolation, lack of family networks, lack of female employment possibilities, limited social services and company domination of social life prevail.

8 An interesting aspect of this discourse is the precarious nature of women's position with respect to the working class in the mining industry. Metcalf (1987) argues that the exclusion of women from the coal mining work-force in nineteenth-century Britain marked the active introduction (largely by the male union movement) of structured gender divisions within coal mining communities. These divisions were transplanted to the Australian industry. In the terms of this discussion, this historical precedent provided the conditions under which miners' wives could only ever be constituted as Other to the working-class miner.

9 My focus upon public political organisations and actions was adopted as a research strategy, not because I privilege public over private politics.

10 Now coal miners earn some of the highest wages of any occupational group in Australia and the bureaucratisation of dispute handling procedures and the imperative of meeting export contracts has reduced the number and average duration of strikes. Without what was accepted by themselves and their men as the 'legitimate' welfare role of the Women's Auxiliary, this organisation has disintegrated and along with it has gone one of the means by which women engaged in public, political discussion and activity for themselves, as well as for their men.

11 In rural areas of Australia both rich and poor have traditionally supported the National (previously the Country) Party which can only attain power at state or federal level in coalition with the Liberal (conservative) Party.

12 Up until quite recently I have found myself lamenting the demise of the Concerned Women of Moura. I saw this dissolution as an indication of the weakly developed political resources available to women in mining towns. Upon further reflection and reading around the idea of a politics of difference I have become less discouraged by the momentary, and rather conversational nature of this type of political activism in which alliances and targeted actions give rise to particular results but not necessarily ongoing organisations. Iris Young's work has been inspirational in this context (1990).

13 Whether the silencing was equally felt or not it is unclear. My discussions with the video makers were not extensive, so though it appears in the discussion that some of the concerns of 'working-class' women were silenced, there may have also been concerns shared by the 'management wives' which were not addressed.

14 This political panic has been expressed in different ways across the range of theoretical interests within feminism, from standpoint theorist Nancy Hartsock's conspiratorial: 'Why is it that just at the moment when so many of us who have been silenced begin to demand the right to name ourselves, to act as subjects rather than objects of history, that just then the concept of subjecthood becomes problematic?' (1990, p. 163), to post-Marxist Michele Barrett's cautious 'Feminists recognize that the 'naming' of women and men occurs within an opposition that one would want to challenge and transform. Yet political silencing can follow from rejecting these categories altogether' (1991, p. 166), to post-structuralist Elizabeth Grosz's tongue-in-cheek rhetoric:

> . . . if women cannot be characterised in any general way . . . then how can feminism be taken seriously? What justifies the assumption that women are oppressed as a sex? What, indeed, does it mean to talk about women as a category? If we are not justified in taking women as a category, then what political grounding does feminism have?
>
> (1990, p. 341)

15 Some, like Hartsock (1990, p. 163) have noted the unfair timing of the post-modern critique and called foul. Others, such as Hekman (1992) and Weedon (1987), accept the embarrassment and ambiguity of occupying a discursive space carved out by a political movement which was itself born of the emancipatory liberal humanism of modernist theory, and from that space of enunciation damning the very politics and epistemology which are its foundations.

16 Brown (1991) has argued that in countering post-modern social fragmentation feminists need to orient their political conversations 'towards diversity and the common, toward world rather than self' and she encourages us to engage in a 'conversion of one's knowledge of the world from a situated (subject) position into a public idiom' (1991, pp. 80–81). In a similar vein Weedon emphasises the public realm, arguing that 'in order for a discourse to have a social effect, [it] must at least be in circulation' (1987, pp. 100–111).

17 The discourse of 'the field' is one that is undergoing an interesting deconstruction within anthropology. See, for example, D'Amico-Samuels (1991). A similar exercise is well overdue in geography and, not surprisingly, it is feminist geographers who have taken up the challenge. See, for example, Nast *et al.* (1994).

18 See Coleman (1986) for a summary of this literature.

19 There is only a very limited literature on the impact of these more unusual work schedules upon the partners, families and communities of shift workers. This is surprising in that, of all types of work, rotating shifts produce the most disruption of the family and household unit which functions around the more common day-oriented timetable of work, leisure and rest. See Banks (1956), and Hertz and Charlton (1989) for some examples of the limited literature on this topic.

20 In framing the research project I was conscious of the disciplinary power that men, unions, the companies and the social service gatekeepers exercised over miners' wives in these communities. Negotiating permission from husbands, the unions, service providers and the mining companies for these women to be involved was itself an interesting political exercise which is documented and discussed at length in a longer draft version of this paper.

21 The four mining towns, Collinsville, Moranbah, Moura and Tieri were selected as the sites for the study on the basis of their differing ages and company affiliation.

22 The interview schedule was straightforward and directed towards issues of concern to women (and, I would argue, men), rather than those of corporate interest. The questionnaire was designed to elicit information around the following topics: Who is providing all the unpaid labour which supports the physical and emotional needs of the shift worker? How much of the increased productivity gained by continuous production is being fuelled by an intensification of household labour? How does the increased tiredness and lack of weekends affect relations between workers and partners, workers and children and partners and children – and workers and workers? How might a better understanding of women's experience of their partner's shift work patterns help men and women alike? What general feelings did women have about their town?

23 In fact there are two (and probably multiple) communities in mining towns, one centred around mine management and professionals and the other around workers, each with their own disciplinary and homogenising practices.

24 Mies (1991) has discussed this process of partial identification in feminist research whereby

> . . . we proceed from our own contradictory state of being and consciousness. That is to say, not only do the 'other' women have a problem, but I do, too. . . . This enables recognition of that which binds me to the 'other women' as well as that which separates me from them.
>
> (1991, p. 79)

25 The Queensland branch of the United Mineworkers Federation has printed 5000 copies of *Different Merry-Go-Rounds: families, communities and the 7-day roster* for distribution to its members and to the communities of central Queensland.

References

Banks, O. (1956) Continuous shift work: the attitudes of wives, *Occupational Psychology*, 30, 69–84.

Barrett, M. (1991) *The Politics of Truth: from Marx to Foucault* (Cambridge, Polity Press).

Barrett, M. (1992) Words and things: materialism and method in contemporary feminist analysis, in: M. Barrett and A. Phillips (Eds) *Destabilizing Theory: contemporary feminist debates* (Stanford, CA, Stanford University Press).

Brown, W. (1991) Feminist hesitations, postmodern exposures, *Differences*, 3, 63–84.

Butler, J. (1990) *Gender Trouble: feminism and the subversion of identity* (New York, Routledge).

Coleman, R. (1986) *Wide Awake at 3am: by choice or by chance?* (New York, W. H. Freeman and Company).

Connell, R.W. (1987) *Gender and Power* (Cambridge, Polity Press).

Cotterell, J.L. (1984) Social networks of mining town women, *Australian Journal of Social Issues*, 19, 101–112.

D'Amico-Samuels, D. (1991) Undoing fieldwork: personal, political, theoretical and methodological implications, in: F. Y. Harrison (Ed.) *Decolonizing Anthropology: moving further toward an anthropology for liberation* (Washington DC, Association of Black Anthropologists and American Anthropological Association).

Dyck, I. (1993) Ethnography: a feminist method? *The Canadian Geographer* 37, 52–57.

Ferguson, K. (1993) *The Man Question: visions of subjectivity in feminist theory* (Berkeley, CA, University of California Press).

Gibson, K. (1991a) Company towns and class processes: a study of Queensland's new coalfields. *Environment and Planning D: Society and Space* 9, 285–308.

Gibson, K. (1991b) Hewers of cake and drawers of tea: women and restructuring on the coalfields of the Bowen Basin, *Economic and Regional Restructuring Research Unit Working Paper 2* (Department of Geography, University of Sydney, NSW 2006 Australia).

Gibson, K. (1992a) Hewers of cake and drawers of tea: women, industrial restructuring and class processes on the central Queensland coalfields, *Rethinking Marxism*, 5, (4) 29–56.

Gibson, K. (1992b) 'There is no normal weekend anymore . . .' A report to the coal mining communities of central Queensland on family life and continuous shift work, unpublished report available from author on request.

Gibson, K. (1993) *Different Merry-Go-Rounds: families, communities and the 7-day roster* (Brisbane, Queensland Colliery Employees Union).

Gibson-Graham, J.K. (1994) Beyond capitalism and patriarchy: reflections on political subjectivity, in: B. Caine and R. Pringle (Eds) *Contemporary Australian Feminism* (Sydney, Allen & Unwin).

Grosz, E. (1984) Interview with Gayatri Spivak, *Thesis Eleven*, 10/11, 175–187.

Grosz, E. (1990) A note on essentialism and difference, in: S. Gunew (Ed.) *Feminist Knowledge as Critique and Construct* (London, Routledge).

Harding, S. (1986) *The Science Question in Feminism* (Ithaca, NY, Cornell University Press).

Hartsock, N. (1990) Foucault on power: a theory for women?, in: L. Nicholson (Ed.) *Feminism/Postmodernism* (New York, Routledge).

Hekman, S. (1992) *Gender and Knowledge: elements of a postmodern feminism* (Cambridge, Polity Press).

Hertz, R. and Charlton, J. (1989) Making family under a shiftwork schedule: air force security guards and their wives, *Social Problems*, 36, 491–507.

Kingsolver, B. (1989) *Holding the line: women in the Great Arizona Mine Strike of 1983* (Ithaca, NY, ILR Press).

Kondo, D.K. (1990) *Crafting Selves: power gender, and discourses of identity in a Japanese workplace* (Chicago, IL, University of Chicago Press).

McDowell, L. (1992) Doing gender: feminism, feminists and research methods in human geography, *Transactions of the Institute of British Geographers*, 17, 399–416.

Metcalfe, A. (1987) Manning the mines: organizing women out of class struggle, *Australian Feminist Studies*, 4, Autumn, 73–96.

Mies, M. (1991) Women's research or feminist research? The debate surrounding feminist science and methodology, in: M. M. Fonow and J.A. Cook (Eds) *Beyond Methodology: feminist scholarship as lived research*, trans. by A. Spencer (Indianapolis, IN, Indiana University Press).

Nast, H., Katz, C., Kobayashi, A., England, K.V.L., Gilbert, M. Staeheli, L.A. and Lawson, V.A. (1994) Women in the field: critical feminist methodologies and theoretical perspectives, *Professional Geographer*, 46, 54–102.

Neil, C.C., Brealey, T.B. and Jones, J.A. (1983) Delegitimisation of mental health myths of remote mining communities, *Community Health Studies*, 7, 42–53.

Nicholson, L. (1990) (Ed.) *Feminism/Postmodernism* (New York, Routledge).

Pratt, G. (1993) Reflections on poststructuralism and feminist empirics, theory and practice, *Antipode*, 25, 51–63.

Reinharz, S. (1992) *Feminist Methods in Social Research* (New York, Oxford University Press).

Rose, D. (1993) Of feminism, method and methods in human geography: an idiosyncratic overview, *The Canadian Geographer* 37, 57–61.

Spivak, G. (1989) A response to 'The Difference Within: Feminism and Ethical Theory', in: E. Meese and A. Parker (Eds), *The Difference Within Feminism and Critical Theory* (Amsterdam, John Benjamins).

Stanley, L. and Wise, S. (1983) *Breaking Out: feminist consciousness and feminist research* (London, Routledge and Kegan Paul).

Sturmey, R. (1989) *Women and Services in Remote Company Dominated Mining Towns* (Armidale, The Rural Development Centre, University of New England, NSW 2351).

Watson, S. and Gibson, K. (Eds) (1994) *Postmodern Cities and Spaces* (Oxford, Blackwell).

Weedon, C. (1987) *Feminist Practice and Poststructuralist Theory* (Oxford, Blackwell).

Williams, C. (1981) *Open Cut: the working class in an Australian mining town* (Sydney, Allen & Unwin).

Young, I.M. (1990) The ideal of community and the politics of difference, in: L. Nicholson (Ed.) *Feminism/Postmodernism* (New York, Routledge).

10 Vera Chouinard and Ali Grant

'On not Being Anywhere Near the "Project": Revolutionary Ways of Putting Ourselves in the Picture'

Excerpt from: *Antipode* **27**(2), 137–66 (1995)

Living exclusion: two geographers' tales

This section discusses what it is like to live in an ableist and heterosexist society. Ableism is defined here as any social relations, practices, and ideas which presume that all people are able-bodied. Examples include: evaluating disabled workers by the same criteria used to evaluate able-bodied employees, holding events in physically inaccessible locations, and treating not being able-bodied as defining a disabled person. Heterosexism refers to social relations, practices and ideas which work to construct heterosexuality as the only true, "natural" sexuality whilst negating all other sexualities as deviant and "un-natural." Examples include: legal definitions of "family" which do not include same-sex couples, assumptions that peoples' partners must be of the opposite sex, hostility toward lesbians and gays who make themselves visible in territory dominated by heterosexual relations and norms (e.g., public places and workplaces), and failing to recognize and appreciate lesbian and gay cultures.

We begin with personal experiences to help the reader "see" society and our discipline through others' eyes. The intention is not to evoke sympathy or pity but to encourage the reader to understand more fully the environments we both negotiate daily. While we present experiences of ableism and heterosexism separately, this should by no means suggest that these are separate from oppressions based on class, gender, race/ethnicity and age. Rather, there are multiple locations of oppression in patriarchal, capitalist societies including those based on both ableism and heterosexism.

The disabled woman

It is hard to think of any facet of my life which has been untouched by ableism and by struggles to occupy able-bodied spaces on my own terms.

For example, as a professor, my workplace, the University, has been a very significant site of my oppression. This has taken multiple forms, from physical barriers to access to the use of ableist standards to evaluate my contributions. One form of exclusion is very visible: after four years, I still lack physical access to my office. Two entrances which appeared as scooter and wheelchair accessible on our official map for disabled staff and students turned out to be nothing of the sort: there are no automatic doors, no working lift, and heavy internal fire doors block access to corridors. So, although I have acquired a scooter, I still cannot get into my official workplace independently. This situation sends out strong signals that my presence, and the presence of disabled colleagues and students, is not important; that we are not valued in the academic setting.

Ableist relations and practices are manifested in a number of other ways as well. For instance, there is no procedure to adjust the workload required of disabled professors, although a full workload for many disabled people should be defined not in terms of work expected from the able-bodied professors, but in terms of the capacity of the disabled academic. Other social barriers include a reluctance to be flexible in terms of how classes are taught (e.g., in the disabled professor's home or through interactive computer technology).

To these social manifestations of "ableism" one must add the little everyday practices of academic life which exacerbate the challenges of being disabled. Recently, in a feminist geography conference session, I was forced to stand because the room was filled. This was arguably my own "fault" as I arrived late (having had to walk a long distance from another session), but after about half an hour the pain in my feet, legs and hips was so intense that I was forced to ask a young woman if she and her companions could shift one chair over so that I could sit down (someone had left their seat, so there was an empty chair at the far end of the row). I apologized for asking, but explained that I was ill, very tired and in a lot of pain. She turned, looked very coldly at me and simply said "no, the seats are being used." She may well have been right, but I suddenly no longer felt part of a feminist geography session: I was invisible . . . and I was angry. Fighting a juvenile urge to bop her on the head with my cane, I began to see feminist geography through new eyes; eyes which recognized that the pain of being "the other" was far deeper and more complete than I ever imagined, and that words of inclusion were simply not enough.

Other negative reactions to my being disabled include direct challenges to my right to occupy able-bodied territory. Recently an older woman burst into a shop where I was sitting chatting with the owner (a friend of mine). Sticking her face uncomfortably close to mine (invading my territory!) she looked at my arm braces and walker and blurted out "My god, what the hell happened to you?" I explained that I was disabled by rheumatoid arthritis. "Oh." she replied, and walked away. She had no interest in my well-being; she was simply asserting her "right" as an able-bodied woman to demand explanations for the presence of the disabled in "her" space.

My sense of myself, as a disabled, academic woman has also been shaped by more subtle aspects of daily life. Walking on the university campus and in other public places, I am constantly conscious of frequent looks (often double and even triple-takes). I realize it is unusual to see a relatively young woman walking slowly with a cane or using a scooter and the looks reflect curiosity, but they are a constant reminder that I am different, that I don't "belong." It is painful for me to acknowledge this. I guess that is why I have learned to look away: to the ground, to the side . . . anywhere that lets me avoid facing up to being the "other."

It is remarkable how thoroughly ableist assumptions and practices permeate every facet of our lives, even though we often remain relatively sheltered from and insensitive to these forms of oppression. Yet disability in some form will come to each and every one of us someday, and when it does, and ableism rears its ugly head, one finds a topsy-turvy world in which none of the old rules apply and many "new" ones don't make sense. People develop new ways of relating to you often without recognizing it. For instance some of my students will not call me at home, despite instructions to do so, because I am "sick." Other students shy away from working with a disabled professor: some assume that the best, most "successful" supervisors must be able-bodied; others are unwilling to accommodate illness by, for example, occasionally substituting phone calls for face-to-face meetings or meeting at my home rather than the office. Of course this is not true of all students, but these practices are pervasive enough to hurt every day and to make it just a little harder to struggle to change relations, policies, practices and attitudes.

Some manifestations of ableism would be hilarious if they weren't so hurtful and damaging. An administrator heading a university disability program invited me to sit on a committee planning events for our annual disability awareness week. I explained that I would be happy to contribute but, as I was quite ill and immobile at the moment, we might at times have to settle for a phone conversation rather than my actually attending a meeting. The response was absolute silence, even though I was told those responsible would get in touch. It was a shock to realize that even those in charge of disability issues could act in such ways – a bit like being droppped in the middle of the Mad Hatter's tea party without being told what story you were in.

The dyke

Like disabled women, lesbians must struggle over the right to space, in a culture that constructs "woman" as both able-bodied and heterosexual. And as a lesbian who tries to be always "out" in everyday spaces, I spend most of my time fighting for some space, whether in a geography department, in a restaurant or on the street. Heterosexism pervades all environments and operates initially to presume that I am heterosexual. Thus when I socialize outside of lesbian and/or feminist company I am often asked about a male partner; when I try to access social or health services with my lover, no spaces exist on the paperwork for us; when we walk into restaurants or stores arm in arm, people often feel compelled to ask if we are sisters or if she is my mother, although we do not look at all alike. Environments of heterosexism permit women certain identities, but deny them others: women together can be mothers, daughters, sisters, roommates, friends – but not lovers. They cannot be lovers because this is profoundly political. It is profoundly political because it both resists and threatens the oppressive system of male dominance which requires women to be heterosexual for its very existence. Compulsory heterosexuality ensures that each member of the oppressed group – women – is individually coupled with a member of the dominant group – men. This assures male rights of access to women on an economic, emotional and physical level. Lesbian existence attacks this right. It challenges heterosexual hegemony.

Thus, when I walk down the street, day or night, outwardly expressing myself as a dyke, announcing my identity as a woman who other women may have access to, but men may not, I often experience open hostility and/or violence from both heterosexual men and heterosexual women. This is intensified if I am with a group of dykes, if we are taking up space on our own terms. If women refuse/resist the heterosexual identity that is the only one available to them in heterosexist society, they will be denied any others.

It is not as simple as "being out." Heterosexism works to either deny me/make me invisible, or force on me identities which, a) do not threaten the system of compulsory heterosexuality, and b) fit into heterosexist ideology. A sample of the most popular stereotypes about lesbians illustrates this point: all I need is a good man; I can't get a man; I am like a man; I was sexually traumatized by a man; I want to be a man; I am a man-hater. There is an obvious common denominator here – men. Yet what defines my existence as a lesbian is loving other women: men are totally irrelevant to this basic definition. All the myths and stereotypes are related to men because, in a heterosexist society, women together are not allowed the self-defined identity. Compulsory heterosexuality demands that women direct their energies towards men, and be accessible to men. Women who direct all energies to other women cannot be accessed by men. But men must be in the picture and one way to do this is through myths and stereotypes.

While lesbian existence is very different from heterosexual existence, it would be a mistake for heterosexuals to always make it "other." When you think of sexuality, don't automatically think lesbian, think self. Hetero-sexuality is as much a social construction as lesbian sexuality is. However, as Valentine (1993c: 396) notes, "such is the strength of the assumption of the 'naturalness' of heterosexual hegemony, that most people are oblivious to the way that it operates as a process of power relations in all spaces." Thus, for example, talking about the weekend with heterosexual colleagues, I might explain that on Friday I went to a lesbian dance, Saturday I relaxed with a great new lesbian novel, and Sunday I went to see a lesbian movie. This is often interpreted as "flaunting" or obsessing on my sexuality. My colleague can tell me that Friday night she went to a [————] nightclub with her husband, they spent Saturday morning in bed reading the [————] newspaper, and on Sunday they took the kids to a matinee to see a [————] movie. The point is that I will fill the empty spaces with "heterosexual," not she; that her (hetero) sexuality is always upfront and center – her wedding band announces it, walking down the street hand in hand with her husband announces it, her discussions of everyday life announce it – is not recognized. The spaces above are the spaces of privilege.

What's wrong with this picture? Missing sisters in geography

The silencing and exclusion of disabled and lesbian women is, not surprisingly, manifest in the geographic literature. In this section, we comment on the phenomena of missing sisters in our discipline.

Invisible sisters: disabled women

From the perspective of a disabled woman, the geographic literature is in many ways a wasteland. Few studies speak to the lives of the disabled, and even fewer grapple with the social processes through which mental and physical disabilities become bases for discrimination, marginalization and oppression. A computerized search of 5000 geographic journals found no references at all to disabled women. Issues such as access for the disabled were addressed in journals of related disciplines, in particular planning and engineering, but often from a physical rather than social planning perspective. A few geographers have studied facets of the lives of the disabled, such as service provision and coping amongst the psychiatrically disabled (e.g. Dear, 1981; Elliott, 1992; Taylor, 1989), work on the visually impaired (Golledge, 1993), the US disabled persons movement (Dorn, 1994) and Dear and Wolch's (1987) important work on homelessness. Some geographers report that there is significant ongoing research on such topics as multiple sclerosis but, due to lack of interest, this work has not been very visible in geographic journals (personal communications). More encouraging, graduate research on disability issues is increasing as I discovered while trying to organize a special journal issue on geography and disability research.

It is important to recognize that some work by non-geographers speaks to geographic aspects of disability and oppression. For example, Hahn (1986, 1989), a political scientist, has discussed the challenges of creating more inclusive urban built environments and the role of ableist "body images" in marginalizing the disabled.

Despite these encouraging signs, in radical geography disabled women have been rendered almost completely invisible and silent. Browse through the index of a major collection (like Peet and Thrift's *New Models* (1989)) and try to find a reference to the disabled or to ableism (I could not). Or turn to discussions of ways forward in feminist geography (e.g. Bowlby, *et al.* 1989) and note major research priorities. Sexuality is there. So too is "race, class and gender." Then why not ableism or ageism? We know from women's lives that these things matter. Why aren't we saying so? More importantly, why aren't we doing something about it?

To be fair, the "invisibility" of disabled women and men in radical geography undoubtedly reflects a real absence in academic and student ranks (in fact, geographers could use a good study of this). After all, performance and evaluation standards in academia are extremely ableist; allowing little or no room for differences in abilities to read, write, teach or do research, take on speaking engagements, and, most importantly, "produce" in general. Those who cannot perform to ableist standards are likely to find themselves pressured to give up their positions, even though they may be able to make important contributions, and despite laws requiring accommodation of the disabled in the workplace. If they resist, they face an often lonely battle to convey the need for non-ableist standards and practices. To modify an old adage, it is hard to understand what it means to live as a disabled person in an ableist society until you have walked (or wheeled) some miles in her shoes.

It is very likely, therefore, that many disabled women and men are quickly "pushed out" of the system. As women are under-represented in positions of power within the discipline, it is likely that disabled women are especially vulnerable to such pressures. They lack, for instance, the chance to be mentored by women, especially disabled female faculty, and mentoring is essential to coping with the gendered power relations of the "old boys' network." Class oppression also comes into play, as disabled women are especially vulnerable to poverty (National Council of Welfare, 1990). For disabled women, limited economic means translates into concrete difficulties in acquiring mobility aids needed to negotiate campuses as well as in raising the tuition needed to pursue higher education.

In a way, it is puzzling that radical geographers have had very little to say about these processes, especially considering the Marxist origins of this part of the discipline. For the exclusion and marginalization of the disabled is deeply intertwined with the "commodification" of human life; with valuing people for their capacity to produce commodities, services, and profit rather than for diverse talents, abilities and ways of being and becoming. This is one of the more damaging and insidious facets of patriarchal, capitalist societies for it encourages us to reduce human worth to "what we can get out of each other" and in the process helps marginalize those who, for various reasons, cannot "compete." There is abundant evidence of this economic and social devaluation. In Ontario, Canada, for instance, 80 percent of disabled persons live in poverty, a result of discrimination and exclusion in the job market, and relatively meager public and private support programs (Disabled Persons for Employment Equity, 1992).

Of course, the silences in the literature are just one sign of academic practices which marginalize the disabled. Traditional research methods, which construct the disabled as an "object" population rather than as experts in living as part of the disabled community, are another form of silencing and exclusion. So too are conferences which scatter sessions between buildings and floors, forcing mobility-impaired people to cover long distances and endure fatigue and often pain. In fact, at most conferences there is no sign of any accommodations for the disabled: no special information booth, no questions about special needs on registration forms, no aides available to assist people – with a wheelchair, or knowledge of sign language, or just indicating accessible elevators. Sadly, much the same can be said of our campuses, where many administrations refuse to spend the money needed to make access a reality.

Invisible sisters: lesbians

The geographic literature cannot be fairly called a wasteland when it comes to lesbians; however, analyses of the spatial expression of sexuality, have appeared only very recently in the margins of geographical research. Feminist geography is just beginning to recognize sexuality as an important part of that great abyss of "otherness" (England, 1994; McDowell, 1993a, 1993b; Peake, 1993). The emerging literature on lesbian and gay geographies to date, speaks much more to the experiences of gay men than to those of lesbians (Valentine, 1992, 1993a, 1993b, & 1993c is an exception). Nonetheless, it is very exciting to see a growing body of work on the impact of lesbians and gay men in the socio-spatial restructuring of the city (Adler and Brenner, 1991; Bell, 1991; Bell *et al.*, 1994; Knopp, 1987, 1990a, 1990b; Lauria and Knopp, 1985; and Winchester and White, 1988). However, there is an urgent need for a much more critical approach to the research issues addressed in this literature.

The dangers inherent in discussing lesbians and gay men in the same breath should be clear given the wealth of feminist work in geography illustrating the critical difference that gender makes (McDowell, 1993a, 1993b; Pratt, 1993; Rose, 1993). That it is impossible to ignore the fact that human experience is gendered is surely well established. It is clear, for example, that to discuss the "working class" is to ignore (amongst other things) the all-important differences between what it is to be the "woman on the street" as opposed to the "man on the street." This applies equally in the realm of sexuality. To state the obvious: lesbians are women, gay men are men and thus common experiences cannot be presumed. Although lesbians are not completely ignored, many authors – rather than recognizing that the socio-spatial experiences of lesbians and gay men may in fact constitute two separate and discrete research problems – are at great pains to explain the bases for the differences observed. For example, Johnston (in Lauria and Knopp, 1985) argues that these differences reflect the fact that gay men may perceive a greater need for territory. Discussing this and other explanations, Peake (1993: 425) argues that:

> Such empirical and conceptual generalizations smack of an inability to rise above the level of the patriarchal mire, of being unable to unpack the heterogeneity of class, "race", and other relations that characterize the lesbian community.

I would agree, but go further to suggest that the question of lesbians does not always have to be addressed in research focused on gay men. The obligation to attempt to explain differences in the socio-spatial experiences of lesbians and gay men only arises if the premise is that there should be any similarity. Both lesbians and gay men engage in same sex relationships and experience oppressive marginalization, but there is no reason to assume they must have any more in common than that.

Recognizing that the question of difference/similarity need not always be a question may avoid such dangerously misleading arguments as those made by Lauria and Knopp (1985) in discussing the reasons for differences in the impact lesbians and gays have had on the city's socio-spatial structure. Part of their explanation is based on their belief that lesbian sexuality has always been more accepted than gay male sexuality. They illustrate by stating, "lesbian sexuality has been accepted under certain conditions, as when it is 'performed' for men by women who conform to societal standards of beauty" (1985: 158). They fail to realize that this is not "lesbian sexuality."[1] The most superficial exploration of pornography will show that the scenario they describe has nothing to do with "lesbian sexuality" and everything to do with two or more women preparing each other for "the main act" (i.e., heterosexual intercourse). And "the main act" always ensures that there is no threat whatsoever to heterosexual hegemony.

Thus, given the critical importance of gender in structuring our experiences of everyday spaces, commonalities cannot be presumed between lesbian women and gay men. The frequent use of the terms "gay" and "homosexual" to mean all lesbians and gay men both affects and denies the realities of lesbian existence. It makes invisible the huge power differential based on gender between lesbians and gay men (and doesn't even begin to touch on the differences within each community). It fails to reflect the myriad ways in which gay men, as men, oppress lesbians, as women. And, as women's experiences are more often than not subsumed under the "norm" of men's in androcentric geographical research, so too are lesbians' experiences often subsumed under gay men's.

This problem of ignoring and/or trivializing the importance of differences based on gender – not to mention ability, class, "race," culture, and so on – points to another dangerous path that "lesbian and gay geographies" could easily follow: one which sees a dichotomy with heterosexuality as dominant and all other sexuality thrown together into one big oppositional construct. All of us who fall into this latter category are then defined by the fact that we do not engage exclusively in "normal" heterosexual sex. Those familiar with the politics of lesbian and of gay communities in many North American cities will be aware of the recent trend toward "queering" everyone who is not heterosexual. For example, it seems of late that Lesbian and Gay Pride Day has become Lesbian and Gay and Bisexual and Transsexual and Transgendered Peoples' Pride Day. While this alliance may be politically expedient, it muddies the distinctions between these groups and is likely to depoliticize lesbian existence in the process. For example, the unique threat which lesbian existence poses to heteropatriarchy gets lost:[2] in this crowd of lesbians, gay men, bisexuals, transsexuals, transgendered people and straight supporters, only one group denies men access to women – lesbians. Further, as lesbian autonomy is made invisible so too is the revolutionary message to other women that there can be life without men. Many feminist lesbians, especially those who came out through the so-called "second wave" of the modern women's movement, are wondering what ever happened to the consciously political struggle for a collective resistance/challenge to heteropatriarchy, and to our dream of the Lesbian Nation (Johnston, 1973).

Blurred distinctions pose another problem: just as feminists in Anglo-American goegraphy have painfully realized that there is no single category Woman (McDowell, 1993a), it must be recognized that there is no single category Lesbian, never mind Queer.[3] Discussing lesbian organizing in Toronto, Ross points out that:

> In large urban centers across Canada and other Western countries, the 1980s have heralded the subdivision of activist lesbians into specialized groupings: lesbians of color, Jewish lesbians, working class lesbians, leather dykes, lesbians against sado-masochism, older lesbians, lesbian youth, disabled lesbians and so on.
>
> (Ross, 1990: 88)

As a radical feminist Dyke who experiences the privileges of being white, formally educated and able-bodied, my resistance to and experiences of heteropatriarchal oppression, are as different from those of "homosexual ladies to really watch out for" (Bechdel, 1994), as they are from S/M Lesbians. And they are worlds apart from those of gay men, bisexuals, transsexuals and transgendered people.

Part of this conflation of very different experiences stems from approaching this topic as a question of sex – of something supposed to take place between individuals in private space – rather than as a question of power and oppression. For example, in discussing negative reactions to the appearance of Knopp's 1990 article in the *Geographical Magazine*, Bell (1991: 327–28) states:

> members of the academy do feel uneasy researching this topic. This *squeamishness* regarding sexual issues is partly homophobic and partly a "justifiable fear of never being cited, except in a list of interesting, albeit peripheral work" (Christopherson 1989, 88); while the study of the geographies of homosexuality remains marginalized and obscure, it will not attract career- and status-minded academics [my emphasis]

This quote raises two sets of issues. First, "homophobia" is a very small part of the explanation of why this subject is often treated with a certain squeamishness. "Homophobia" is to heterosexism what "prejudice" is to racism: neither comes close to describing the systems of oppression from which those who are "homophobic" and/or "prejudiced" greatly benefit. Heterosexism – which privileges heterosexuality as the only true, pure and natural sexuality and discredits and makes deviant any other expression of sexuality for women – is necessary to heteropatriarchy. Heterosexism works effectively to control women, all women, and should not be conceptualized as simply irrational fear of "different lifestyles." It ensures that women have little choice other than to enter into intimate relationships with members of the very group that oppresses them – men. Heterosexism bestows a whole array of privileges on heterosexuals whilst encouraging hatred of lesbians through harassment, discrimination and violence.[4] In other words, squeamishness about sex simply does not come close to explaining why this subject has rarely been subjected to a critical analysis in geography.

Second, the "justifiable fear of never being cited," can be described as part of the power relations at work in geographical research. Deciding to "play it safe" for the sake of career and status is thoroughly understandable in a patriarchal institution; however it has very real implications for those of us who are marginalized within the discipline.[5] As McDowell (1992b: 59) argues, "we cannot ignore our own positions as part of the conventional structures of power within the academy, nor, although it is often painful, can we afford to ignore the structures of power between women." Feminists doing work in this area must remain vigilant so that the question of power and oppression – who loses, who gains, in whose interests is our oppression – is not lost in efforts to make lesbians visible in geography.

Confronting ableism and heterosexism in geography

We want to underscore, again, the point that manifestations of these oppressions in workplaces, homes and communities often make the difference between disabled women and lesbians who can be active political advocates of change, and those who are too exhausted and discouraged to even pick up the phone. It is sobering, for example, to read that 50 percent of all rheumatoid arthritis patients suffer from clinically-defined depression. Pain and mobility limitations contribute to this, but are probably less significant than underfunding of medical research and services, and unsupportive medical practices based on some physicians' aversion to treating incurable and chronic diseases because they are more troublesome to deal with. For lesbians the penalties for coming out in the workplace and/or wider community (such as job loss and violence) are so severe that many women live a splintered existence: out in their "private" lives, but trying to "pass" as heterosexual in other facets of life. The damage involved in denying one's identity and culture in order to survive should not be underestimated. The personal costs are devastating; so, too, are the cost to all of us as members of society: the knowledges and experiences of disabled and lesbian women, and their capacities to understand, care, and contribute become lost to teaching, research, planning and policy-making. Our stories are hidden, our voices silenced; a lost heritage for us, our children and their children.

The social construction of disabled and lesbian women as oppressed others is a pervasive and complex process; it permeates many facets of academic life and so much of daily life that, for women like ourselves, it is inescapable. In this section we consider ways of challenging ableism and heterosexism in geography, an endeavor that requires fundamental re-thinking of our social theories, research methods and politics as academics.

Are geographers up to it? Facing up to ableism

As ableism is a pervasive set of social relations, practices and ideas affecting both our discipline and society, it follows that to attack ableism within the social sciences is to touch but the tip of the iceberg. These efforts are very important in their own right, but are unlikely to challenge ableism unless they are part of a comprehensive offensive against attitudes, practices, services, policies and power relations that imprison and maim disabled people because they are "different." This robs society of precious sources of knowledge and hope.

If ableism is so much a part of radical geography and daily life, what can we do to challenge it? Must disabled women remain excluded and silenced "others"?

One important first step is to insist that research by and for – as opposed to "on" – the disabled deserves greater priority in radical geography. We must also agree that it is important to fight for inclusion of more disabled people in the radical research community, not only by trying to increase representation within academia, but with a whole array of measures ensuring the disabled physical, economic and social access to workplaces and communities. Without solid support for rethinking how we work and live, for finding more inclusionary ways of producing and living, all the affirmative action initiatives in the world won't allow disabled women to define their own revolutionary terms and conditions of participating in the production of knowledge and daily life in general. Nor will it support disabled women's struggles against sexist ableism in its various guises – including efforts to reduce disabled women's comparatively greater vulnerability to male violence (Masuda and Ridington, 1992).[6]

Thinking through strategies for challenging ableism, it quickly becomes apparent that even the initial steps require us, in a very profound way, to relearn how we value differences in ourselves and in others. Performance standards which value quantitative output (such as grants and papers) and frequent conference travel, for example, not only devalue "ordinary" academic activities such as teaching a class, they fail to recognize just how amazing it may be for someone who is mentally or physically disabled to produce even a single paper. Somehow, we need to learn how to think and act through "other eyes" if we hope to challenge these oppressions and exclusions. This means, amongst other things, learning to "see," through our theories, research, and lives, the relations and processes through which disabled women are socially constructed as marginal and excluded others. It means, as well, learning to respect the pain and anger of the disabled at being cast as less important and less capable; it means supporting struggles to challenge the processes of exclusion and marginalization which sustain this "other" status.

A related task is to search for creative ways of giving voice to lived experiences of ableist relations and practices in our research designs. This means, among other things, relinquishing privileged academic viewpoints in favor of more inclusive modes of description and analysis, not simply giving "voice" and validity to "subjugated" knowledges (although this is important) but also developing research designs in which participants have a say in the conduct, interpretation and use of research, and where both researcher and participants "live the research process" in a very direct way (Chouinard, 1994).

It also means struggling to conceptualize disablement processes as part of the political economy of patriarchal, capitalist societies – treating disablement not just as an unrelated set of oppressive relations and practices added on to existing research agendas, but as processes rooted in significant ways in classist, sexist, heterosexist, ageist (as well as ableist) relations and practices which are part and parcel of the development of capitalist societies today. The growing body of radical and/or feminist literature can help geographers to conceptualize disablement processes in this comprehensive way. Work on the political economy of disability (Oliver, 1990) and feminist critiques of the social construction of the disabled and of disabled women in both the women's and disability rights movements (e.g., Findlay and Randall, 1988; Morris, 1991) has highlighted, amongst other things, the importance of cultural images in distorting peoples' understanding of disabled women's lives and how these distortions affect disabled women's sense of their own realities and struggles. Thompson writes:

> Anger felt by women because of our disabilities is rarely accepted in women's communities, or anywhere else for that matter. Disabled or not, most of us grew up with media images depicting pathetic little "cripple" children on various telethons or blind beggars with caps in hand ("handicap") or "brave" war heroes limping back to a home where they were promptly forgotten. Such individuals' anger was never seen, and still rarely is. Instead of acknowledging the basic humanity of our often-powerful emotions, able-bodied persons tend to view us either as helpless things to be pitied or as Super-Crips, gallantly fighting to overcome insurmountable odds. Such attitudes display a bizarre two tiered mindset: it is horrible beyond imagination to be disabled, but disabled people with guts can, if they only try hard enough, make themselves almost "normal." The absurdity of such all-or-nothing images is obvious. So, too, is the damage these images do to disabled people by robbing us of our sense of reality.
>
> (cited in Morris, 1991: 100)

Unless geographers manage to build on such sophisticated insights by, for example, considering the social construction of exclusionary territories, we risk perpetuating representations of the disabled as "special cases" rather than as people living through some of the most destructive manifestations of societies driven by profit, greed, intolerance and superficial types of individual success – qualities which translate into excluding those of us who are "different" from the spaces of the powerful and advantaged.

To further efforts to develop such theories (e.g. Oliver, 1990), geographers need input and guidance from those living with disabilities and struggling to challenge the multiple discriminations that go along with this type of "difference." We need to learn to open our conceptual and empirical debates to those living disablement. This means including disabled activists in the social construction of academic knowledges about ableism and in debates about its connections to broader lived relations such as class, and letting them bring their lived experiences of discrimination and struggle into the research process. In this way, we can build political challenges to ableism, within and outside of the research process, including alliances between researchers and disabled activists.

These reconstructions, focusing as they do on "empowering" the oppressed, will not be easy. Indeed, as McDowell (1992a) points out in her discussion of feminist methods, even the most progressive research designs raise very difficult ethical, practical and political questions. The researcher is never really "outside" the dilemmas of radical research but constantly struggles to handle them a little bit better, a little more fully.

Sensitivity to issues like exploitation of the "researched" is likely to be especially important in the case of geographic research focusing on disabled women. These women are (at least) "doubly disadvantaged" by gender and by mental or physical challenges. In most cases, limited finances and marginalization combine to limit their capacity to participate in society and in the research community. This means two things. First, researchers have a responsibility to further struggles to open the research process to disabled women and make their voices heard in the conduct and use of research. Second, politically, the research must be aligned with struggles against ableism and the relations, institutions and practices that support it. The method will vary from project to project, however it should be recognized that a research project which does not centrally contribute to the research and political priorities of the disabled women involved is exploitative and oppressive.

Challenging ableism in geographic research will be as revolutionary for our understanding of processes of urban and regional change as it will be for the politics and practices of research. For disabled women, it will help us better understand how exclusion, silencing and oppression are reproduced within the predominantly white, middle class ranks of the women's movement, as well as through patriarchal institutions. It will make it as important to understand the positioning of people within the "micro-relations" of power in daily life and life spaces, as it is to understand the role of major social divisions in empowering some groups at the expense of vulnerable "others." More importantly, challenging ableism will force us to grapple with time- and place-specific manifestations of ableism, and with how the living of these oppressive and exclusionary relations translates into resistance and rebellion.

Are geographers up to it? Facing up to heterosexism

Much of the geographic work to date on lesbians and on gays has concentrated on increasing visibility rather than critical analyses of heterosexual hegemony. That is, most authors concentrate on lesbian space and gay male space (e.g. meeting places, gentrified neighborhoods) rather than on everyday spaces – the heterosexual and hostile environments in which lesbians and gay men spend most of their time. Of course, given the dominance of heterosexuality in space, it is important to document and understand lesbian space and gay male space, but if this is all we do then it is surely a case of adding "queers" and stirring. It is striking how little critical political analysis there is in the literature, and how few connections are made to the wealth of feminist work in geography. Valentine's work is an important and refreshing exception in that it moves beyond the "impact on the city" approach to a critical examination of how lesbians create, transform and negotiate not only lesbian environments but the more day to day environments of heterosexism (see especially Valentine 1993b and 1993c). She argues (1993a: 114) that more research is needed "to gain a better understanding of how heterosexual hegemony which is so often taken for granted, is reproduced in space."

As public discourse in North America on lesbians changes, it is more critical than ever to illustrate the ways in which our material realities have not; that is, the ways in which heterosexual hegemony prevails. As Westenhoefer wryly points out in the infamous *Newsweek* article that put lesbians on almost every newsstand across North America, "We're like the Evian water of the '90s. Everybody wants to know a lesbian or to be with a lesbian or just to dress like one" (June 21, 1993). Knowing a lesbian may indeed be "trendy" in parts of white western culture but does this manifest itself in political solidarity? Does it lead to the thousands of heterosexuals who "know a lesbian" marching for lesbian rights? Does it lead to these same heterosexuals insisting on placing lesbian literature in the school system? Does it lead to mass heterosexual mobilization around the heterosexual bias of immigration, taxation and adoption laws? A change in discourse alone will not challenge structures of oppression.

Further, a change in discourse is often a double-edged sword. Although changing discourse undoubtedly has political implications, it can also make lesbians' struggles against heterosexism more difficult, as the ideology of heteropatriarchy works to suggest that we have less to struggle against. An analogy can be easily drawn: consider the feminist struggle against male violence against women, everyone is talking about it, it is squarely on the public agenda, but women are no less likely to experience male violence today than they were 20 years ago (Bart and Moran, 1993; Dobash and Dobash, 1992; Statistics Canada, 1993; Walker, 1990). Yet, despite ample hard evidence that this violence against women is endemic to heteropatriarchal culture, we are still likely to hear (and prefer to believe), that this violence is a product of poor anger management, women's low self-esteem, dysfunctional families and/or learned behavior. Similarly, everybody may be talking about (even to) lesbians in the "gay nineties," but heterosexism is still as powerful as, if not more powerful than, it was twenty years ago. Popular culture may tell us that being lesbian is fairly acceptable today, but there is no evidence that lesbian teachers are coming out in the thousands, that lesbians are holding hands in the street, that federal laws have been changed, or that violence against lesbians is decreasing. It is important to make lesbians visible in geography, but the concentration on lesbian and/or gay space has so far come at the expense of critical analyses of environments of heterosexism. It is time for more of the latter, and less of the former.

Notes

1 In fairness, the authors' approaches to the subject have changed considerably in the last ten years (personal communication with Larry Knopp).

2 The term heteropatriarchy is taken from Valentine (1993c: 396):

> To be gay [*sic*], therefore is not only to violate norms about sexual behavior and family structure – but also to deviate from the norms of 'natural' masculine and feminine behavior. These norms change over space and time, and hence sexuality is not merely defined by sexual acts but exists as a process of power relations. Heterosexuality in modern Western society can therefore be defined as a heteropatriarchy, that is, as a process of sociosexual power relations which reflect and reproduce male dominance.

3 Valentine (1993b) includes an interesting discussion of essentialism versus constructionism in understanding the roots of lesbianism.

4 For example, Valentine (1993c: 408) cites a San Francisco study of 400 lesbians: "84% had experienced antilesbian verbal harassment, 57% had been threatened with physical violence, and 12% had been punched, kicked or beaten . . ."

5 I want to be very clear that in using Bell's quotation from Christopherson's "On

Being Outside 'the Project' " article, and using it somewhat out of context, I am not criticizing Christopherson per se.

6 As Masuda and Ridington found in the DAWN study (1992: vii):

> Of the 245 women who participated in this survey, 40% had been raped, abused or assaulted, 64% had been verbally abused; girls with disabilities have less than an equal chance of escaping abuse than their non-disabled sisters; women with multiple disabilities experienced multiple-incidents of abuse, and only 10% of women who were abused sought help from transition houses, of which only half were accommodated.

References

Adler, S. and J. Brenner (1992) Gender and space: lesbians and gay men in the city. *International Journal of Urban and Regional Research* 16: 24–34.

Bart, P.B. and E.G.Moran (1993) *Violence Against Women: The Bloody Footprints*. London: Sage.

Bechdel, Alison (1994) *Dykes to Watch Out For: 1994 Calendar*. Ithaca, NY: Firebrand Books.

Bell, D. (1991) Insignificant others: lesbian and gay geographies. *Area* 23: 323–29.

Bell, D., J. Binnie, J. Cream and G. Valentine (1994) All hyped up and no place to go. *Gender, Place and Culture* 1: 31–48.

Bowlby, S., J. Lewis, L. McDowell and J. Foord (1989) The geography of gender. In R. Peet and N. Thrift (Eds.) *New Models in Geography*. Boston: Unwin Hyman, pp. 157–75.

Chouinard, V. (1994) Geography, law and legal struggles: Which ways ahead? *Progress in Human Geography* 18: 415–40.

Dear, M. (1981) Social and spatial reproduction of the mentally ill. In M. Dear and A.J. Scott (Eds.) *Urbanization and Planning in Capitalist Societies*. London: Methuen, pp. 481–97.

Dear M. and J. Wolch (1987) *Landscapes of Despair: From Deinstitutionalization to Homelessness*. Princeton, N.J.: Princeton University Press.

Disabled Persons for Employment Equity (1992), unpublished flyer, Toronto, Canada.

Dobash, R.E. and R.P. Dobash (1992) *Women, Violence and Social Change*. New York: Routledge.

Dorn, M. (1994) "Disability as Spatial Dissidence: a Cultural Geography of the Stigmatized Body." Unpublished MA thesis, Department of Geography, Pennsylvania State University.

Elliott S. (1992) "Psychosocial impacts in populations exposed to solid waste facilities." Unpublished Ph.D. dissertation, Geography Department, McMaster University, Hamilton, Canada.

England, K.V.L. (1994) Getting personal: Reflexivity, positionality, and feminist research. *Professional Geographer* 46: 80–89.

Findlay, S. and M. Randall (Eds.) (1988) Feminist perspectives on the Canadian state. Special issue on women and the state. *Resources for Feminist Research*, 17(3).

Golledge, R. (1993) Geography and the disabled: a survey with special reference to vision impaired and blind populations. *Transactions of the Institute of British Geographers*. 18: 63–85.

Hahn, H. (1986) Disability and the urban environment: a perspective on Los Angeles. *Environment and Planning D: Society and Space* 4: 273–88.

Hahn, H. (1989) Disability and the reproduction of bodily images: the dynamics of human appearances. In J. Wolch and M. Dear (Eds.) *The Power of Geography: How Territory Shapes Social Life*. Boston: Unwin Hyman, pp. 370–88.

Johnston, J. (1973) *Lesbian Nation: The Feminist Solution.* New York: Simon and Schuster.

Knopp, L. (1987) Social theory, social movements and public policy: recent accomplishments of the gay and lesbian movements in Minneapolis, Minnesota. *International Journal of Urban and Regional Research* 11: 43–261.

Knopp, L. (1990a) Social consequences of homosexuality. *Geographical Magazine* LXII: 20–25.

Knopp, L. (1990b) Some theoretical implications of gay involvement in an urban land market. *Political Geography Quarterly* 9: 337–52.

Lauria, M. and L. Knopp (1985) Toward an analysis of the role of gay communities in the urban renaissance. *Urban Geography* 6: 152–69.

Masuda, S. and J. Ridington (1992) *Meeting Our Needs: An Access Manual for Transition Houses*. Toronto: DAWN Canada.

McDowell, L. (1992a) Doing gender: feminism, feminists, and research methods in human geography. *Transactions: Institute of British Geographers* 17: 399–416.

McDowell, L. (1992b) Multiple voices: speaking from inside and outside 'The Project'. *Antipode* 24: 56–72.

McDowell, L. (1993a) Space, place, and gender relations: Part I. Feminist empiricism and the geography of social relations. *Progress in Human Geography* 17: 157–79.

McDowell, L. (1993b) Space, place and gender relations: Part II. Identity difference, feminist geometries and geographies. *Progress in Human Geography* 17: 305–18.

Morris, J. (1991) *Pride Against Prejudice*. London: The Women's Press Ltd.

National Council of Welfare (1990) *Women and Poverty Revisited* (Summer). Ottawa: Minister of Supply and Services.

Oliver, M. (1990) *The Politics of Disablement*. New York: MacMillan.

Peake, L. (1993) 'Race' and sexuality: challenging the patriarchal structuring of urban social space. *Environment and Planning D: Society and Space* 11: 415–32.

Peet R. and N. Thrift (Eds.) (1989) *New Models in Geography* Vols 1 and 2. Boston: Unwin Hyman.

Pratt, G. (1993) Feminist geography. *Progress Report in Urban Geography* 13: 385–91.

Rose, G. (1993) *Feminism & Geography: The Limits of Geographical Knowledge.* Minneapolis: University of Minnesota Press.

Ross, B. (1990) The house that Jill built: lesbian feminist organizing in Toronto, 1976–1980. *Feminist Review* 35: 75–91.

Statistics Canada (1993) *The Violence Against Women Survey: Highlights*. Ottawa: Minister Responsible for Statistics Canada.

Taylor, S.M. (1989) Community exclusion of the mentally ill. In J. Wolch and M. Dear (Eds.) *The Power of Geography: How Territory Shapes Social Life*. Unwin Hyman, 316–30.

Valentine, G. (1992) "Towards a geography of the lesbian community" paper presented at the Women in Cities conference, Hamburg, 10 April.

Valentine, G. (1993a) "Desperately seeking Susan": a geography of lesbian friendships. *Area* 25: 109–16.

Valentine, G. (1993b) Negotiating and managing multiple sexual identities: lesbian time-space strategies. *Transactions, Institute of British Geographers* 18: 237–48.

Valentine, G. (1993c) (Hetero) sexing space: lesbian perceptions and experiences of everyday space. *Society and Space* 11: 395–413.

Walker, G. (1990) *Family Violence and the Women's Movement: The Conceptual Politics of Struggle*. Toronto: University of Toronto Press.
Winchester, H.P.M. and P.E. White (1988) The location of marginalized groups in the inner city. *Environment and Planning: Society and Space* 6: 37–54.

SECTION THREE

THE NATURE OF GENDER

Editors' introduction

Western notions of gender binaries have been so pervasive, and have adapted to change so effectively because of their subtle reinforcement by other dualisms. Perhaps the most significant of these is the binary of culture/nature. Nesmith and Radcliffe (1993) have suggested that although theorising of nature by geographers has lagged behind the works on space, it is a vitally significant concept: the culture/nature dichotomy parallels and reinforces the dichotomy of man/woman, leading some feminists to suggest that there are important connections between the oppression and domination of women, and the exploitation of nature. As a result, it is argued that women have a particular interest in ending human domination of nature.

The gendering of nature in the west is not a recent phenomenon. Caroline Merchant has suggested for example that many traditional cultures saw mining as 'abortion' of their metal's natural growth cycle before its time. Many Roman philosophers including the Stoics openly opposed mining as abuse of their mother, the earth (Merchant, 1990 p. 3). Somewhere around the sixteenth to the seventeenth century, this living image of the cosmos with the female earth at its centre was replaced with a mechanistic world view within which nature was reconstructed as dead and passive, to be dominated and controlled by humans (Merchant, 1990 p. xvi). Runyan (1992) suggests that the rise of capitalist modes of production, modern scientific practice and enlightenment thought, together changed conceptions of nature as being (divine) order to being disordered and in need of the controlling influence of man's intervention. By this time, man was no longer seen as part of nature, as Donna Haraway has argued in her genealogy of primatology, 'it is the white man who has excluded himself from "nature" by both history and a Greek-Judeo Christian myth system' (Haraway, 1989 p. 159). Although not characterising nature as female (for she argues that gender and nature are historically and socially constructed and so offer no timeless essence) Merchant (1990 p. xxi) insists that 'we must re-examine the formation of a worldview and a science that, by reconceptualising reality as a machine rather than a living organism, sanctioned the domination of both nature and women'.

The binary of man/culture–woman/nature has both been enforced by hegemonic culture and also in some cases appropriated by women as they position themselves as guardians of nature. Many feminists have suggested that in hegemonic western culture, women have been placed alongside nature in the parallel binaries of man–woman, culture–nature. They insist that women have been understood to be more natural than men, tied to the rhythms of their bodies in menstruation and pregnancy. As a result of the privileging of culture – especially in its scientific manifestations – the patriarchal power structures of western culture and scientific control over the environment are understood by some feminists to be linked, even self-enforcing. There have been two major reactions to this. On the one hand, there have been figures such as Simone de Beauvoir who claim that the link between women and nature is not essential or natural, but instead the effect of a patriarchal culture that continually weakens women's agency. As a result, feminists should reject their links with nature and force their way into dominant masculinist culture, highlighting the hidden constructedness of its gendering as they do so.

On the other hand, feminists such as Mary Daly have insisted that women re-evaluate and reclaim their connections to nature, to use them in such a way as to oppose and undermine masculinist culture. Further, there has evolved a feminist ecology that insists that women have an essential ability to empathise with nature given the joint subjugation of both women and nature in patriarchal systems. As Costa Rican feminist Margarita Arias has claimed:

> Only those who have fought for the right to protect their own bodies from abuse can truly understand the rape and plunder of our forests.
>
> (Quoted in Seager, 1993 p. 7)

Many women have criticised development projects from ecofeminist perspectives. Although development as both a concept and a political practise is most commonly accepted as a 'liberal' post-colonial project, its roots can be found deep within colonial values of social evolution, 'the white man's burden' and so on. Development apparently allows a western-style lifestyle for all peoples. However, Rosa Luxemburg insisted that the west developed as it did *only* because of its exploitation of colonies. As a result, she argues, development elsewhere could only follow this path if internal colonies could be established. Following from this, development initiatives across the globe have necessarily involved the exploitation of marginalised groups (women's labour for instance), subjugated cultures such as Indian tribal peoples, and nature. In many cultures, women were traditionally 'managers' of the environment and its resources – this control has been appropriated and even destroyed by the development processes which redirects human–nature relations through the mediation of capitalistic principles. As women's agency is destroyed as a direct result of the exploitation of the environment, it has been argued that women's interests and those of nature are understood to be closely linked.

Writing from a post-colonial position, Vandana Shiva suggests that 'Feminists' interests in the environment can lead to an appreciation of the perspective of aboriginal people, whose cultural attitudes to the land tend to have been nonexploitative and partnership oriented' (Nesmith and Radcliffe, 1993, p. 386). Feminism would be enriched as a political movement with the inclusion of these non-western voices helping to define a political project in harmony with the needs of the environment.

Others have adopted a middle path, seeing a strategic essentialism between the position of women and nature in patriarchal societies. Joni Seager (1993) argues that environmental issues have too often been regarded as purely scientific concerns and thus beyond the purview of feminist analysis (although Donna Haraway and Sandra Harding have offered feminist analyses of scientific practice itself). Seager wants to look behind this scientific front to the interests that lie behind the destruction or preservation of the global environment. When the scientific narrative is replaced by one which entails stories of governments, militaries, transnational corporations and 'eco-establishments', Seager insists, power relations and agency come to the fore: now there is plenty of material for a feminist analysis. Although rejecting any essentialist links between women and nature, Seager sees that both are positioned so as to limit their agency by the institutions that control the environment, institutions that privilege men and male culture, and marginalise women and the environment.

The first paper in this section is from Joni **Seager**'s provocative book *Earth Follies*. This book challenges traditional explanations of the interconnections between international relations and ecological issues by openly challenging what she considers masculinist discourses and practices of statecraft, multi-national management and science. She challenges academic discourses of international relations in both the content and the form of her book: the 'usual suspects' in international affairs – statesmen, wars, international commerce, government bureaucracy – are critiqued from a feminist perspective. Seager's style is not to refer to the 'great thinkers' in this area but instead presents an angry and personal engagement with these issues. In the selection here Seager engages with the uses of images of the 'Earth as our Mother' to illustrate the ways in which the uncritical use of gender stereotyping of nature as female does not increase female agency, but instead relies upon a hidden masculinism that disguises power relations. In turn, this abrogates responsibility for global environmental destruction at the hands of capitalist accumulation or state territorial aquisition as it suggests that regardless of the damage, mother will always be there to clear up after us.

Vandana **Shiva**'s work on gender and nature both introduces non-western conceptualisations of the relationship between people and nature, and offers a key role to women as caretakers of the Earth. She provides an alternative to western binaries of woman–man, nature–culture from South Asian, especially Hindu, thought, replacing these dichotomies with 'dialectical unity' in which a trace of each side of the duality is always present. This challenges oppositional arguments and seeks to undermine essential connections between gender and nature. However, Shiva argues that as a result of their management of the environment and their treatment by patriarchal culture, women do share significant interests with nature.

However, some feminists have reservations about ecofeminist arguments. It has been argued that there is a danger in romanticising the struggles of 'Other' feminists and incorporating them into western feminist projects as an unproblematic category, without addressing the complex differences between women. The classic case of western romanticisation of Other feminist movements is the 'tree-hugging' Chipko movement in Northern India. Chipko has become symbolic of women's environmental politics but, as Nesmith and Radcliffe (1993 p. 387) have suggested, 'are not Indian women in Chipko playing the role for ecological feminists of suffering Third World women in struggle?' Although the inclusion of Other voices and perspectives into western feminist projects is only to be welcomed, it is misleading to suggest that Other feminist movements are *necessarily* living in a sustainable ecologically-conscious way, rather than the management of their environment being guided by survival issues. This would be a romantic essentialisation of Third World women that ignores the realities of their everyday lives and the struggles that they are engaged in. Furthermore, Nesmith and Radcliffe (1993 p. 388) suggest that 'it is generally true that only white, middle-class women can afford the time and the financial and emotional commitments to promote environmental feminism' either within their own or other societies. From the higher cost of 'environmentally friendly' alternative consumer goods, to time needed for political organisation, the ability to participate in ecofeminism is influenced by wealth.

It is not only femininity, however, that has been essentially connected to the environment. Robert Bly's 'Iron John' spiritual 'ecomasculinism' has insisted that masculinity has an intimate association with 'primal spaces' and that men have to return to nature in order to cut through their cultural affectations, and to release the wild man within. Bly and others within the mythopoetic men's movement (Bonnet, 1996) reverse feminist arguments about the impact of capitalistic industrialisation. Rather than women being unnaturally confined to the home as ecofeminists might argue, they suggest that it was men that were ripped away from their association with nature and home, and forced into the 'unnatural' environment and social relations of the workplace. Thus it is men who need to reclaim their lost identification with a more 'natural' manliness.

There is evidently a politics to the use of nature in the construction of identity and community. A feminist analysis must be aware of the power in women's (and men's) appropriation of nature.

In our next excerpt, Caroline **New** offers a critique of the essentialism inherent in Seager and Shiva's works, as she claims that they reduce women to guardians of nature. This constrains both male and female gendered identities, tying men and women to timeless subjectivities. Instead she wishes to approach women's and men's diverse interests and needs.

Finally Gillian **Rose**'s piece demonstrates how pervasive, and how silenced, the western gender and nature binaries are. Rose explains the role of these binaries throughout various traditions of geographic thought in her book *Feminism and Geography*. In the chosen section she uncovers the nature–culture binary in the very subdiscipline that seeks to reveal such hidden cultural values: the 'new cultural geography'. Rose seeks to demonstrate the inherent masculinity of the landscape tradition (see for example, her critique of Denis Cosgrove [Rose, 1993 pp. 90–91]). In the extract included here, Rose highlights silences in John Berger's reading of Gainsborough's painting 'Mr and Mrs Andrews'. Although Berger presents the class codings of the image, including the meaning of the countryside for the bourgeoisie, Rose claims that both Berger and Cosgrove's silence on the issue of gender renders them complicit with the sexism of the painting.

References and further reading

Agarwal, B 1992: The gender and environment debate: lessons from India. *Feminist Studies* **18**: 119-58.

Berger, J 1972: *Ways of Seeing*, BBC: London.

Bly, R 1990: *Iron John: a Book About Men*. Random House: New York.

Bonnett, A 1996: The New Primitives: Identity, Landscape and Cultural Appropriation in the mythopoetic men's movement. *Antipode* **28**, 273–92.

Cosgrove, D 1985: Prospect, perspective and the evolution of the landscape idea. *Transactions of the Institute of British Geographers* 10, 45–62.

Daly, M 1979: *Gyn/Ecology*. Women's Press: London.

Fitzsimmons, M 1989: The matter of nature. *Antipode* **21**(2), 106–20.

Fuss, D 1989: *Essentially Speaking: Feminism, Nature and Difference*. Routledge: New York.

Haraway, D 1989: *Primate Visions: Gender, Race, and Nature in the World of Modern Science*. Routledge: London.

Haraway, D 1991: *Simians, Cyborgs and Women: the Reinvention of Nature*. Free Association Books: London.

Harding, S 1986: *The Science Question in Feminism*. Open University Press: Milton Keynes.

Jackson, C 1995: Radical environmental myths: a gender perspective. *New Left Review* **210.**

MacCormack, C and Strathern M (eds) 1972: *Nature, Culture and Gender*. Cambridge University Press: Cambridge.

Merchant, C 1990: *The Death of Nature: Women, Ecology, and the Scientific Revolution*. Harper and Row: San Francisco.

Monk, J 1984: Approaches to the study of women and landscape. *Environmental Review* **8**(1), 23–33.

Nesmith, C and Radcliffe, S 1993: (Re)mapping Mother Earth: a geographical perspective on environmental feminism. *Environment and Planning D: Society and Space* 11, 379–94.

Norwood, V and Monk, J 1987: *The Desert is No Lady*. Yale University Press: New Haven, CT.

Porteous, J 1986: Bodyscape: the body-landscape metaphor. *The Canadian Geographer* **30**(1), 2–12.

Radcliffe, S and Westwood, S (eds) 1993: *Viva! Women and Popular Protest in Latin America*. Routledge: London.

Runyan, A 1992: The state of nature: a garden unfit for women and other living things. In V S Peterson (ed.) *Gendered States*. Lynne Rienner: London, 123–40.

Seager, J 1993: *Earth Follies*. Routledge: London.

Strathern, M 1992: *After Nature: English Kinship in the Late Twentieth Century*. Cambridge University Press: Cambridge.

Tickner, J A 1992: Man over nature: gendered perspectives on ecological security. In her *Gender and International Relations*. Columbia University Press: New York.

11 Joni Seager
'The Earth is Not Your Mother'

Excerpt from: *Earth Follies*, pp. 219–21. London: Routledge (1994)

This earth is not your mother

As the control of "motherhood" moves up on the environmental agenda, the reification of "Mother Earth" stands in ironic contrast. The most ubiquitous icon of modern environmentalism is the image of the earth, floating in black space, with the caption "Love Your Mother." The conceptualization of the earth as a mother has a long and honorable history: Earth as Mother, as a sacred and honored female life force, is a powerful icon in non-Christian, non-EuroAmerican, mostly agricultural, cosmography,[1] it rejuvenates a contemporary women-centered spirituality movement; it inspired a generation of Earth Day activists. But it is disingenuous for a spiritually hollow, urban, technical, male, ecobureaucracy (and one that is consciously becoming *more* invested in these characteristics) to adopt the mother imagery. Not only is this a terrible irony, but "Earth as Mother" is a deceptive paradigm for environmental politics.

The earth is *not* our mother. There is no warm, nurturing, anthropomorphized earth that will take care of us if only we treat her nicely. The complex, emotion-laden, quasi-sexualized, quasi-dependent mother relationship (and especially the relationship between *men* and their mothers) is not an effective metaphor for environmental action. It suggests a benign distribution of power and responsibility, one that establishes an erroneous and dangerous assumption of the relations between us and the environment. It obfuscates the power relations that are really involved when we try to sort out who's controlling what, and who's responsible for what, in the environmental crisis. It is not an effective political organizing tool: if the earth is really our mother, then we are children, and cannot be held fully responsible for our actions.

Beyond this, in a patriarchal culture in which female status is cast as subservient status, there are inherent pitfalls in sex-typing an inherently gender-free entity. A number of alternative environmental groups, deep ecologists and ecofeminists among them, are reclaiming the sex-typing of the planet as part of a radical environmental agenda. But sex-typing of a non-gendered entity invokes a male/female, greater than/lesser than cultural dualism; the limitations of female identification in a male-dominated culture undercut the claim that sex-typing the planet can be "radical."

Further, the sex-typing of the planet, in imposing on the earth a human imagery, also reinscribes an odd anthropomorphism. To describe the earth in human terms in order to understand it implies that, without this human veneer, we and it are separate and *other*.[2] Ascribing human archetypes to nature also

Fig. 11.1 The ubiquitous icon, Mother Earth

suggests design: that we are nature's favored progeny. The term "mother" is a human invention, and evokes uniquely human characteristics. To propose that this forms the essence of our relationship with nature exalts our place within nature, and reiterates claims that "man is the measure of all things."[3] To base this dualism on male and female identities only further reinforces cultural hierarchies – it certainly will do nothing to subvert the patriarchy, and will do little to further environmentalism.

Many people might argue that the "Mother Earth" metaphor is a harmless device, not worth dwelling on. And yet, the use of this metaphor is not always benign – "hiding behind Mom's skirts" is a convenient device to deflect accountability. In 1989, a vice-president of Exxon invoked Mother Earth imagery in defending his company's cleanup operations in Alaska after the Valdez oil spill. His words suggest the cynical and facile use of the Mother Earth "defense":

> I want to point out that water in the [Prince William] Sound replaces itself every 20 days. The Sound flushes itself out every 20 days. Mother Nature cleans up and does *quite* a cleaning job.
>
> – Charles Sitter, senior vice-president of Exxon, May 19, 1989[4]

There may be a broader subtext to the "Mom-will-pick-up-after-us" school of environmental philosophy. In 1989, a Boston journalist gave voice to the suspicions many women increasingly harbor:

> Sometimes I think the problem boils down to this: . . . most men have had women to clean up after them. In fact, it wouldn't surprise me one bit to find out that science has been covertly operating on the Mom-Will-Pick-Up-After-Me Assumption. . . . Men are the ones who imagine that clean laundry gets into their drawers as if by magic, that muddy footprints evaporate into thin air, that toilet bowls are self-cleaning. It's these over-indulged and over-aged boys who operate on the assumption that disorder – spilled oil, radioactive wastes, plastic debris – is someone else's worry, whether that someone else is their mother, their wife, or Mother Earth herself.
>
> – Linda Weltner, *Boston Globe* columnist, April 28, 1989[5]

Notes

1 See, for example, the description of the feminine life-force in Indian cosmology in Vandana Shiva, *Staying Alive*.
2 Patrick Murphy, "Sex-Typing the Planet," *Environmental Ethics*, Vol. 10, 1988.
3 W.J. Lines, "Is 'Deep Ecology' Deep Enough?" *Earth First*, May 1987.
4 An interview on National Public Radio, "All Things Considered," May 19, 1989.
5 Linda Weltner, "Even Mother Earth has Limits," *Boston Globe*, April 28, 1989.

12 Vandana Shiva
'Women in Nature'

Excerpt from: *Staying Alive: Women, Ecology and Development*, pp. 38–42. London: Zed Books (1989)

Nature as the feminine principle

Women in India are an intimate part of nature, both in imagination and in practise. At one level nature is symbolised as the embodiment of the feminine principle, and at another, she is nurtured by the feminine to produce life and provide sustenance.

From the point of view of Indian cosmology, in both the exoteric and esoteric traditions, the world is produced and renewed by the dialectical play of creation and destruction, cohesion and disintegration. The tension between the opposites from which motion and movement arises is depicted as the first appearance of dynamic energy (Shakti). All existence arises from this primordial energy which is the substance of everything, pervading everything. The manifestation of this power, this energy, is called nature (Prakriti).[1] Nature, both animate and inanimate, is thus an expression of Shakti, the feminine and creative principle of the cosmos; in conjunction with the masculine principle (Purusha), Prakriti creates the world.

Nature as Prakriti is inherently active, a powerful, productive force in the dialectic of the creation, renewal and sustenance of *all* life. In *Kulacudamim Nigama*, Prakriti says:

> There is none but Myself
> Who is the Mother to create.[2]

Without Shakti, Shiva, the symbol for the force of creation and destruction, is as powerless as a corpse. 'The quiescent aspect of Shiva is, by definition, inert . . . Activity is the nature of Nature (Prakriti).[3]

Prakriti is worshipped as Aditi, the primordial vastness, the inexhaustible, the source of abundance. She is worshipped as Adi Shakti, the primordial power. All the forms of nature and life in nature are the forms, the children, of the Mother of Nature who is nature itself born of the creative play of her thought.[4] Hence Prakriti is also called Lalitha,[5] the Player because *lila* or play, as free spontaneous activity, is her nature. The will to become many (Bahu-Syam-Prajayera) is her creative impulse and through this impulse, she creates the diversity of living forms in nature. The common yet multiple life of mountains, trees, rivers, animals is an expression of the diversity that Prakriti gives rise to. The creative force and the created world are not separate and distinct, nor is the created world uniform, static and fragmented. It is diverse, dynamic and inter-related.

The nature of Nature as Prakriti is activity *and* diversity. Nature symbols from every realm of nature are in a sense signed with the image of Nature. Prakriti lives in stone or tree, pool, fruit or animal, and is identified with them. According to the *Kalika Purana*:

> Rivers and mountains have a dual nature. A river is but a form of water, yet it has a distinct body. Mountains appear a motionless mass, yet their true form is not such. We cannot know, when looking at a lifeless shell, that it contains a living being. Similarly, within the apparently inanimate rivers and mountains there dwells a hidden consciousness. Rivers and mountains take the forms they wish.[6]

The living, nurturing relationship between man and nature here differs dramatically from the notion of man as separate from and dominating over nature. A good illustration of this difference is the daily worship of the sacred tulsi within Indian culture and outside it. Tulsi (*Ocimum sanctum*) is a little herb planted in every home, and worshipped daily. It has been used in Ayurveda for more than 3000 years, and is now also being legitimised as a source of diverse healing powers by western medicine. However, all this is incidental to its worship. The tulsi is sacred not merely as a plant with beneficial properties but as Brindavan, the symbol of the cosmos. In their daily watering and worship women renew the relationship of the home with the cosmos and with the world process. Nature as a creative expression of the feminine principle is both in ontological continuity with humans as well as above them. Ontologically, there is no divide between man and nature, or between man and woman, because life in all its forms arises from the feminine principle.

Contemporary western views of nature are fraught with the dichotomy or duality between man and woman, and person and nature. In Indian cosmology, by contrast, person and nature (Purusha-Prakriti) are a duality in unity. They are inseparable complements of one another in nature, in woman, in man. Every form of creation bears the sign of this dialectical unity, of diversity within a unifying principle, and this dialectical harmony between the male and female principles and between nature and man, becomes the basis of ecological thought and action in India. Since, ontologically, there is no dualism between man and nature and because nature as Prakriti sustains life, nature has been treated as integral and inviolable. Prakriti, far from being an esoteric abstraction, is an everyday concept which organises daily life. There is no separation here between the popular and elite imagery or between the sacred and secular traditions. As an embodiment and manifestation of the feminine principle it is characterised by (a) creativity, activity, productivity; (b) diversity in form and aspect; (c) connectedness and inter-relationship of all beings, including man; (d) continuity between the human and natural; and (e) sanctity of life in nature.

Conceptually, this differs radically from the Cartesian concept of nature as 'environment' or a 'resource'. In it, the environment is seen as separate from man: it is his surrounding, not his substance. The dualism between man and nature has allowed the subjugation of the latter by man and given rise to a new world-view in which nature is (a) inert and passive; (b) uniform and mechanistic; (c) separable and fragmented within itself; (d) separate from man; and (e) inferior, to be dominated and exploited by man.

The rupture within nature and between man and nature, and its associated transformation from a life-force that sustains to an exploitable resource characterises the Cartesian view which has displaced more ecological world-views and created a development paradigm which cripples nature and woman simultaneously.

The ontological shift for an ecologically sustainable future has much to gain from the world-views of ancient civilisations and diverse cultures which survived sustainably over centuries. These were based on an ontology of the feminine as the living principle, and on an ontological continuity between society and nature – the humanisation of nature and the naturalisation of society. Not merely did this result in an ethical context which excluded possibilities of exploitation and domination, it allowed the creation of an earth family.

The dichotomised ontology of man dominating woman and nature generates maldevelopment because it makes the colonising male the agent and model of 'development'. Women, the Third World and nature become underdeveloped, first by definition, and then, through the process of colonisation, in reality.

The ontology of dichotomisation generates an ontology of domination, over nature and people. Epistemologically, it leads to reductionism and fragmentation, thus violating women as subjects and nature as an object of knowledge. This violation becomes a source of epistemic and real violence – I would like to interpret ecological crises at both levels – as a disruption of ecological perceptions of nature.

Ecological ways of knowing nature are necessarily participatory. Nature herself is the experiment and women, as sylviculturalists, agriculturists and water resource managers, the traditional natural scientists. Their knowledge is ecological and plural, reflecting both the diversity of natural ecosystems and the diversity in cultures that nature-based living gives rise to. Throughout the world, the colonisation of diverse peoples was, at its root, a forced subjugation of ecological concepts of nature and of the Earth as the repository of all forms, latencies and powers of creation, the ground and cause of the world. The symbolism of Terra Mater, the earth in the form of the Great Mother, creative and protective, has been a shared but diverse symbol across space and time, and ecology movements in the West today are inspired in large part by the recovery of the concept of Gaia, the earth goddess.[7]

The shift from Prakriti to 'natural resources', from Mater to 'matter' was considered (and in many quarters is still considered) a progressive shift from superstition to rationality. Yet, viewed from the perspective of nature, or women embedded in nature, in the production and preservation of sustenance, the shift is regressive and violent. It entails the disruption of nature's processes and cycles, and her inter-connectedness. For women, whose productivity in the sustaining of life is based on nature's productivity, the death of Prakriti is simultaneously a beginning of their marginalisation, devaluation, displacement and ultimate dispensability. The ecological crisis is, at its root, the death of the feminine principle, symbolically as well as in contexts such as rural India, not merely in form and symbol, but also in the everyday processes of survival and sustenance.

Notes

1 'Prakriti' is a popular category, and one through which ordinary women in rural India relate to nature. It is also a highly evolved philosophical category in Indian cosmology. Even those philosophical streams of Indian thought which were patriarchal and did not give the supreme place to divinity as a woman, a mother, were permeated by the prehistoric cults and the living 'little' traditions of nature as the primordial mother goddess.
2 For an elaboration of the concept of the feminine principle in Indian thought see Alain Danielon, *The Gods of India*, New York: Inner Traditions International Ltd., 1985; Sir John Woodroffe, *The Serpent Power*, Madras: Ganesh and Co., 1931; and Sir John Woodroffe, *Shakti and Shakta*, London: Luzaz and Co., 1929.
3 Woodroffe, *op. cit.*, (1931), p. 27.
4 W.C. Beane, *Myth, Cult and Symbols in Sakta Hinduism: A Study of the Indian Mother Goddess*, Leiden: E.J. Brill, 1977.
5 *Lalitha Sahasranama*, (Reprint), Delhi: Giani Publishing House, 1986.
6 *Kalika Purana*, 22.10.13, Bombay: Venkateshwara Press, 1927.
7 Erich Neumann, *The Great Mother*, New York: Pantheon Books, 1955.

13 Caroline New

'Man Bad, Woman Good? Essentialisms and Ecofeminisms'

From: *New Left Review* **216**, 79–93 (1996)

Can socialists, radical environmentalists and feminists from other traditions safely dismiss ecofeminism? In this paper I offer both a critique of ecofeminism and a modified defence. On the one hand, I argue, ecofeminism is riddled with essentialism, and open to all the philosophical critiques levelled at any position which attributes timeless natures to women and men. I shall show that even 'social' ecofeminists, in Mellor's terminology,[1] who steadfastly denounce essentialism and dualism, frequently fall back on their own versions of these. Yet I shall also argue that ecofeminism must be taken seriously, both theoretically and strategically. I begin with that embodiment of dualism, Greenham Common.

At Greenham Common armed men guarded nuclear missiles behind three high perimeter fences topped with barbed wire. While men drove round in camouflaged vehicles and radioed to each other across the bare and muddy ground of the camp, peace-camp women slept on the ground, under canvas or plastic, attached images of their children to the wire, picnicked among the trees, built fires to sit round, holes to crap in, sat in front of the lorries carrying the weapons, cut the wires to enter the base and danced on the silos. Global splits between men and women, between militarized states and the homes women make, between North and South, between authoritarian–hierarchical and cooperative ways of living, all seemed condensed in this powerful symbol.

For me, Greenham Common was my first meeting with ecofeminism. I was shocked, both by the place and the arguments. I used to visit with a friend and stay up all night on guard to relieve the women who lived there. I was surprised to notice that policemen and soldiers were now young enough to be my sons. One night some of the young soldiers threw live coals and a dead rabbit at the sleeping women. I ran to the wire and told them off, and realized in their shame-faced response that I now had a mother's authority. This was also the first time I really listened to the argument that women, as a sex, by virtue of our actual or potential motherhood, have a particular interest in saving the planet and are particularly well-equipped to do it. It was the first time that I heard, and pondered on, Frankie Armstrong singing 'Will there be womanly times, or must we die?'

The reason for taking ecofeminism seriously, even away from the visual spell of Greenham Common, is well expressed by Joni Seager in her book *Earth Follies*:

> Militaries, multinationals, governments, the eco-establishment. When I write down this list of institutions on a piece of paper, the first thing that I notice, as a feminist, is that these are all . . . controlled by men (and a mere smattering of women). The culture of these institutions is shaped by power relations between men and women, and between groups of men in cooperation or in conflict. Institutional behaviour is informed by presumptions of appropriate and necessary behaviour for men and for women. Their actions, their interactions and the often catastrophic results of their policies cannot be separated from the social context that frames them.[2]

I shall argue that Seager falls into essentialism in her dependence on the concept of 'male culture'. But she is right, I think, to insist that the social reproduction of male domination and of ecologically destructive social practices are inseparable. We cannot explain environmental destruction simply by referring to capitalism's institutionalized greed, true as this is: such a schematic approach only begins the search for mechanisms producing this systematically blinkered agency. Ecofeminism points to a deep source of such mechanisms in gendered subjectivity. In other words, our ideas of ourselves as male or female, gender differences in feelings and habitual responses to the world, our gendered conceptions of our interests can all be seen as central to our participation in environmentally destructive practices.

In Plumwood's important book, *Feminism and the Mastery of Nature*, she asserts the need for 'a cultural ecological feminism' which involves 'a great cultural revaluation of the status of women, the feminine and the natural', without falling into the trap of seeking in women the salvation-bringing 'angel in the ecosystem'.[3] If Seager and Plumwood are right to see these concerns of ecofeminism as strategically central to environmentalism in *general*, a 'critical ecological feminism' would have to give up the claim to represent only gendered interests, and I come back to this issue at the end of the paper. I begin, though, by explaining what I mean by essentialism and why, if environmentalists are to be ecofeminists, we should seek a non-essentialist, non-dualistic variant.

Essentialisms and ecofeminisms

The term 'essentialism' used to be relatively straightforward. As applied in feminism, it referred to the naturalizing reduction of gender to sex; as if the meaning attributed to sexual difference in a particular socio-historical context had universal validity. Social constructionism has extended the meaning of essentialism. Diane Fuss describes it as 'the idea that any essential or natural givens precede the processes of social determination'.[4] For Fuss, I would be an essentialist, along with all realists who hold that extra-discursive reality exists and has a shape of its own, which affects the efficacy of our descriptions of it.

Social constructionist and deconstructionist feminists have trouble with the categories of 'women' and 'female', as evidenced by Wittig's claim that 'lesbians are not women'.[5] For Haraway, 'there is nothing about being "female" that naturally binds women. There is not even such a state as "being" female, itself a highly complex category constructed in contested sexual scientific discourse and other social practices'.[6] For Butler, sexed bodies are constructed as such retrospectively, from the standpoint of already dichotomized gender – that is, intersexed bodies are constructed, linguistically and medically, as male or female.[7] By contrast, I maintain that sexual difference is real, though it is not merely dichotomous. However, it is no accident that human beings have been able to dichotomize it more or less successfully: sexual differences do have a bipolar distribution, explicable by evolutionary accounts of sexuality. There *are* real male and female capacities and liabilities, although whether and how these are instantiated in particular cases depends on the entire causal context.

The trouble with essentialism is not its realism, but its lack of depth and its simplification of causal processes. Essentialism understands some appearances and events as expressions of the *essence* of the things involved, which are understood as spawning simulacra whatever else is happening. Women (in one version) are essentially nurturing and caring, and in their doings produce nurture and care – of the environment, for instance. Of course, there are always so many exceptions. To retain the plausibility of this view, it becomes necessary to privilege certain outcomes as 'of the essence', while others are contingent, 'beside the point'. Thus nineteenth-century biological essentialism maintained that women were 'naturally' weak, nurturing and ornamental, ignoring or pathologizing exceptions.[8] Essentialism does not recognize that while sexual difference, for instance, permits some things and forbids others, its mechanisms work in conjuction with many others, which combine to codetermine what actually happens in specific cases and situations. No cause *always* produces the same effect, whatever is happening around it – that is, whatever other causal factors are operating, sometimes in a contrary direction. Women may or may not be nurturing, but even where they are, environmental destruction may result from their care.

Essentialism is key in maintaining the power of the linked dualisms that so many theorists have identified in Western thought. It is commonplace in ecofeminism to list these oppositional couples: man–woman, culture–nature, reason–emotion, male–female, and so on. As Plumwood points out, what is wrong with these couples is not the distinction itself, nor even the fact that they are dichotomous terms. We need to be able to *speak* of culture and nature, male and female, mind and body, humans and animals, even if we deconstruct these oppositions in our next breath. What makes these linked couples into dualisms is not that they posit differences, but that these differences are maintained through what Plumwood calls 'radical exclusion', a denial of continuity and mutual dependency, an essentialist polarization which justifies and reproduces certain social practices. These couples are construed as fixed hierarchies which are 'closely associated with domination and accumulation'.[9] As 'the process of domination forms culture and constructs identity, the inferiorised group (unless it can marshal cultural resources for resistance) must internalize this inferiorisation in its identity and collude in this low valuation, honouring the values of the centre . . .'[10]

Linking postulates map the pairs onto each other – thus various cultural assumptions link the poor and/or the working class to animals (incapable of deferred gratification, interested only in instinctual satisfaction) and children (ignorant and unable to govern themselves), and thus to emotion, nature and body.[11] In this argument, Plumwood recognizes the political power of conceptualization, without embracing the absolute equation of power and knowledge that makes all conceptualization a hierarchical and restrictive act.

About these hierarchical dualisms there are two distinct positions, between which particular writers often vacillate. The first position is an ecofeminist view of these very dualisms with their value reversed – let us call it 'dualistic ecofeminism'. The second position views the dualisms as constructed and as having their own deleterious effects on the planet via the way they organize and reproduce social practices. This second position has a tendency to revert to dualistic ecofeminism, as we shall see. What then is ecofeminism's magnetic appeal, the dream of Greenham?

Dualistic ecofeminism

Dualistic ecofeminism is most straightforwardly held by those whom Mellor calls 'affinity' ecofeminists, who believe that women and nature are similarly treated by men, and perceived by all as linked, because of a *real* resemblance; and that there are real, inherent connections between women and the devalued side of each of the notorious couples.[12] On this view, men's opposition to nature, their instrumental use of it, is rooted in their lack of, and opposition to, such a connection. Thus Collard sees a *real* timeless resemblance between women and nature:

> Nothing links the human animal and nature so profoundly as women's reproductive system which enables her to share the experience of bringing forth and nourishing life with the rest of the living world. *Whether or not she personally experiences biological mothering*, it is in this that woman is most truly a child of nature and in this . . . lies the well-spring of her strength.[13]

Similarly, Spretnak describes men as 'not feeling intrinsically involved in the processes of birthing and nurture, nor strongly predisposed toward empathetic communion', and as therefore turning their attention to death rather than life.[14] This strong form of essentialist ecofeminism actually suggests a necessary causal relationshiop between sexual difference, its psychic correlates, and their expression in the ecologically destructive culture of patriarchy. Politically speaking, this is a pessimistic position, since if patriarchal culture is really an expression of the timeless essence of men, 'female culture' cannot change male nature, although by 'taking the toys from the boys' it may prevent the male principle destroying the planet and humankind with it. This position may permit us to imagine a good – female-dominated or even entirely female – society, but does not readily yield any strategy for getting there, other than the 'spinning' and 'sparking' and female bonding Mary Daly suggests,[15] or the gradual spread of feminist spirituality.

One version of dualistic ecofeminism claims that – unless prevented by poverty – women's greater affinity with nature makes them tend to act in eco-friendly ways. This view is criticized by Cecile Jackson in a detailed paper which rejects the discourse of feminism in favour of 'gender analysis'. She writes: 'The false idea of the positive synergism of women's gender interests and environmental interests seems strongly related to an essentialist denial of the . . . historical construction of gender and nature.'[16] Gender interests cannot be universalized, she (rightly) insists – though side-stepping the difficult task of theorizing 'interests'.[17] The Chipko 'tree-hugging' movement, extolled by Mellor as 'inspirational' and showing women's 'clear affinity with the ecological needs of their region',[18] is for Jackson one of the 'familiar icons' of ecofeminism,[19] and, in reality, 'part of a broader current of peasant protest'. Chipko women can be seen as defending a conservative 'moral economy . . . rather than trees as such . . .'[20]

Brandth, in her study of women's farming practice in Norway, similarly concludes that in their attitude to modern agriculture, despite some differences in actual practice, 'The way nature is socially conceived does not vary very much between groups of women and between women and men'. Even the differences in farming practice between groups of women, which do reflect different ideas of womanhood, do not differ in terms of the ideas of nature they express, nor in terms of their consequences for the environment.[21]

'Social' ecofeminism

Dualistic ecofeminism has other less open proponents who would not fall into the category of 'affinity' ecofeminism, but rather into Mellor's second category of 'social' ecofeminism. These 'see the relationship between women and nature as socially created and therefore capable of being socially resolved . . . [rather than as] transcending particular societies and eras.'[22] For 'social' ecofeminists, it is women's role in society which has given them a distinctive culture or a distinct perspective. This difference is historicized and seen as a temporary phenomenon. Its eventual transcendence, though, depends on its current strategic use. If 'male culture' is the source of danger, then 'female culture' may be a source of salvation. If both are socially constructed, male culture, and with it masculinity, may actually be transformed by women's insistence on bringing into the public sphere the nurturing values which women's culture usually practices in private, as a step towards abolishing this distinction. This is a common ecofeminist position. Thus Seager suggests that women's role in society has given them a distinctive culture which 'may suggest alternative 'ways of being' in the world'.[23] Ynestra King describes it as 'embodying what is best in women's life-oriented socialization'.[24] Some argue for replacing 'masculine' ways of being with 'feminine' ones, some for combining them.[25]

'Social' ecofeminism can be understood as a variant of 'feminist standpoint theory', first elaborated by Hartsock, and widely discussed in feminist literature. Hartsock's original article offered a theoretical justification of feminist empiricism: 'like the lives of proletarians according to Marxian theory, women's lives make available a particular and privileged vantage point on male supremacy . . . which can ground a powerful critique of . . . the capitalist form of patriarchy.'[26] She falls into essentialism herself at one point, reading off women's experience from the female body as firmly as any 'affinity' ecofeminist: 'There are a series of boundary challenges inherent in the female physiology . . . challenges which make it impossible to maintain rigid separation from the object world. Menstruation, coitus, pregnancy, childbirth, lactation – all represent challenges to bodily boundaries.'[27] But for the most part she sees women's experience, and the knowledge it gives rise to, as a function of social positioning: 'The unity of mental and manual labour, and the directly sensuous nature of much of women's work leads to a more profound unity of . . . social and natural worlds than is experienced by the male worker in capitalism.'[28] With their hands covered with shit and their breasts leaking in response to babies' cries, women are supposedly more down to earth, more respecting of all forms of life; too sensible not to see that biodiversity and a future for our descendants must be valued above capital accumulation.

This putative epistemological vantage point has important strategic value, according to many 'social' ecofeminists. The woman–nature connection, constructed as it is, represents 'a vantage point for creating a different kind of culture and politics that would integrate intuitive, spiritual and rational forms of knowledge embracing both science and magic in as far as they enable us to transform the nature–culture distinction.'[29] Yet the claim that women's position on the 'soft' side of the dualisms equips her to transcend them is itself an essentialist claim, a version of Plumwood's 'angel in the ecosystem'.[30] Greenham Common – and other women's camps – showed that the social positon of women and its symbolic values can be used as a mobilizing focus for a section of women, and that this has useful tactical force. But as the widespread opposition from women also showed, there was nothing automatic about this response.

'Social' ecofeminists have no intention of remaining within dualistic modes of thinking, which they denounce. They would reject the term 'dualistic ecofeminism' as applied to their own views – and indeed, few are *consistently* dualistic. However, I shall argue that even in their treatment of current, socially constructed gender difference, they often practice the conceptual techniques of radical exclusion, identified by Plumwood. And in their advocacy of the bringing together of the split halves of humanity it is noticeable that the male half is usually treated as distorted and in need of modification, while the female half is perfect as it is.[31] Although they recognize the dualistic couples as *constructed* – both conceptually and practically – these thinkers collude with dualism by driving a wedge between the two sides of the couple and attributing all the power and agency to the 'male' side, rather than seeing them as two sides of one coin. They want to have their cake and eat it – to reject dualism on the one hand, but on the other to explain its iniquitous domination in terms of the male principle: 'Men have not made a 'mistake' in their creation of nature and women as the dominated Other. The feminine is not the missing half of the masculine; the feminine is what men need to create the masculine in a patriarchal culture.'[32]

Seager's work offers a good example of such essentialist thinking, just because she 'firmly reject[s] "biology is destiny" arguments in whatever guise they take'.[33] In *Earth Follies*, she examines the environmental impact of the military, of governments, and of science, pointing out that in all these institutions – and in the eco-establishment which opposes them – men are in charge and numerically overwhelming.[34] She notes the gender-specific moral terms, such as 'patriot' or 'honourable', which men in the military-industrial complex tend to apply to other men they approve of, and points to the rituals of 'male bonding' and their institutional effects. But then – and this is the essentialist move – she tries to sum up and link these gendered phenomena by attributing them to *male culture, male consciousness*, or by calling them *male constructs*. She asks, for example, 'Are sovereignty, nationalism, territoriality and wars particularly male constructs?', and hesitantly implies that they are.[35] I think Seager's hesitation probably springs from the realization that to answer 'yes' to such questions will commit her to an essentialism she wants to avoid.

Essentially male?

What work does the word 'male' do here? What does it tell us? It might mean that certain aspects of culture, certain jobs, certain ways of behaving, certain modes of speech and ways of thought, are considered appropriate for men and partially restricted to them, and that these gender-specific attitudes and practices are crucially involved in the social structures of militarism, industry and government. This is certainly true. It is also true, as Seager says, that militarism and warfare operate as enforcement systems for everyday patriarchy, and, further, that 'there is a synchronicity between "hegemonic masculinity" and ordinary manhood . . . gender hierarchies privilege all men . . .'[36] More powerful *in general* in the public sphere, men have also had an overwhelmingly greater say in developing the ideology and the practice of national sovereignty. At one level, it seems obvious that men bear the burden of agency in military-industrial systems, and must therefore also bear the burden of blame for the environmental destruction resulting from their actions. Theirs, after all, are the hands that made the bomb, that are even now tossing spent nuclear waste into the Sea of Japan. Easlea describes how the Los Alamos physicists started their work as a 'lesser evil' in order to pre-empt the Nazis but actually intensified it after the German surrender, also how mesmerized they were by the 'technically sweet', irresistible discoveries they were making, and how they celebrated the vindication of their efforts – the destruction of Hiroshima – with a champagne party.[37]

However, to slip, as he does, into deploring men's unfortunate nature – their womb envy, their substitution of death for life, their unattractive and destructive 'male culture' – is to naturalize the very dualism 'social' ecofeminists denounce. No litany of men's crimes can *explain* environmental destruction *unless* we take the essentialist leap to seeing men as inherently more likely to develop disastrous social forms, and as transhistorically sufficiently powerful to enforce them on women. Such destructive acts have to be understood in historical context, in the late capitalist mid-twentieth century world of warring imperial powers, a world in which institutionalized social practices allocated public positions and scientific research to men. While the men made the bomb, their women were cooking, cleaning and bringing up their children. They were not slaves, but motivated by the desire to fit in, to be respected and approved of, to have a home and a place of their own, economic security and the love of children. Ordinary motives, which resulted in non-accidental ignorance and consequent collusion. The women's motives complement the equally ordinary motives of the men: to help defeat Hitler, to be successful, perhaps famous, to be part of a team doing something big, to be accepted, approved of, admired, to have a place in history, economic security, and the love and admiration of women and children. The women's silence was not an 'efficient cause', but it definitely contributed to the total causal picture. To see women as not acting, and therefore as orally neutral, is actually to accept the oppressive assumption that they are but powerless victims who can do no other. To theorize the actual relationship between structure and agency in this case we need to analyze the mechanisms which produced these motivations. We need, in other words, to understand the creation of gendered subjects who actively accept their places in the division of labour, a process which is still a crucial element of the social reproduction of late capitalism and therefore of environmental destruction.

Seager describes how, on one occasion, the women of the Los Alamos community did assert themselves. Soldiers had been detailed to fell some trees that shaded the children's play area. The women sat under the trees, preventing this action, until they won their case. This is a sad but familiar story: women are often heroic where their children's lives, health or safety are at stake, but lack the confidence to know their power and to use it in the transformation of the public sphere. Women's knowledge that wasteful and destructive social practices are wrong tends to be

> powerlessly ventilated in a running critique, a subordinates' critique, sealed off from the flow of formal historic events to which it refers. This societal safety valve . . . has channelled off potentially subversive female energy . . . letting the male-steered stream of public events move undeflected – and with substantial female consent, we must remember – toward what by now looks like all-but-inevitable . . . ecological hell.[38]

Only a dualistic – and thus essentialist – theory can unequivocally attribute sole agency to the hand that did the deed, ignoring the role of the hands that were washing-up at the time, and the mechanisms which allocate sexed humans to these social positions. 'Social' ecofeminism is still in the realm of essentialism, still clinging to teleology to make sense of the senseless. The constant return to men's agency is reminiscent of Freud's vain attempt to separate 'active–passive' from 'male–female'.[39] Try as he might, the 'linking postulates' crept back in and mapped the couples on top of each other. But here we have man–woman constantly returning, as if magnetized, to its place on top of subject–object. Braidotti's description of the 'double trap which threatens feminism' sums up the dilemma of ecofeminism: 'on the one hand a sociologizing reductivism which, on the binary model of the class struggle, sets the female individual in opposition to the male patriarchal system . . .' – as if an oppositional feminist standpoint were the simple result of women's oppression – 'on the other, the utopian model which makes "women" an entity (on the) outside, foreign to the dominant system and not contaminated by it',[40] but offering female culture and 'the authentic female mind'[41] as unguent for male-inflicted planetary wounds.

Gendered experience, gendered interests

Feminist standpoint theory, widely drawn on by 'social' ecofeminists, has two strands. As we have seen, it claims that women are epistemologically privileged: their experiences allow them to see more clearly, and their oppression makes them less motivated to disguise the truth. Second, from this social positioning and communality of experience arise common interests. Before examining this second claim in its ecofeminist versions, I will sum up the argument for common, knowledge-yielding experiences.

The capacity to bear and suckle children, whether or not these powers are ever used, and their bodily signs – even though they may be misleading – are the criteria for allocation of humans to the category of woman. These capacities have tremendous salience for all societies. All have their dualisms, and almost all involve hierarchy. Although its forms vary tremendously, women's subordination is arguably universal, at least in societies with political structures. In the abstract, the capitalist mode of production needs no gender system, but all real historical forms of capitalism have forged their own version of male domination from the materials they inherited. I do not propose to discuss the roots of male domination, but it is signficant that the most hopeful candidates for gender equality are pre-industrial gathering and hunting societies, on the one hand, and highly technological socialist societies, on the other.

If there were nothing more linking women than wombs and words, it would be a powerful connection. But there is more – subordination. Even though women are constructed as subjects on other dimensions, as black or white, as Jew or Gentile, as lesbian or heterosexual, as disabled or able-bodied, as workers, managers and so on, even at times and in places where class or ethnic group are the major determinants of life chances and solidarity, to be a black or working-class woman is always different from being a black or working-class man. We cannot talk of 'women's experience' in the singular. We are talking about the effect of gendered social practices – or what the French more accurately call 'social relations of sex' – on human subjects positioned in terms of sexual difference. 'Experience' is used to mean two things, related but non-identical: people's life events – what happens to them and what they do – and how they understand and respond to these. Being a woman in any one social context makes certain things more likely to happen to you and other things less likely or impossible; it offers you, as agent, a range of options and closes off others. It determines the sorts of interpretations that will be offered of your actions, and so on. And between social contexts this will still be true to varying extents, as many constituencies in the NGO Forum in Beijing demonstrated.

We can cautiously go along with standpoint theory this far: being positioned as a woman tends to lead to a recognizable range of life events, though this tendency is mediated through the effects of other forms of social positioning. It is not that the vantage point of the subjugated is true, unitary, or trustworthy, but that similar social positions both offer and produce certain similar happenings, similar sets of responses and ways of understanding, including ways of conceptualizing self and group identity. At any one time and place women's perspectives may take the form of a number of clusters of various or even conflicting ways of being a woman, that have inner coherence and intelligibility. Standpoint theorists usually end up idealizing one such perspective – a variant of feminism – as the right or true one, as an achievement rather than a given.[42] This is, of course, what 'social' ecofeminists do when they privilege felt closeness with nature. As soon as you privilege one standpoint above others in this way, you can no longer claim it as the necessary effect of women's experience. It now needs its own separate legitimization, through the development of a theory explaining why it is true or appropriate. For the theory to have any strategic value, it also needs to specify conditions in which women are likely to take that standpoint. Ecofeminist versions of standpoint theory need to show, if they can, that women's experiences can enable them to know their connection to non-human nature, and therefore make them less liable to behave in ecologically destructive ways. They also need to specify the conditions in which this may happen. I do not believe they have done this.

The second strand of feminist standpoint theory is the claim that the dualistic modes of thinking which justify, express and reproduce practices of male domination were created by men in their own interests. Thus Hartsock argues: 'Men's power to structure social relations in their own image means that women too must participate in social relations which manifest and express abstract masculinity.'[43] This 'abstract masculinity' results from 'the reversal of the proper order of things characteristic of the male experience, the substitution of death for life'.[44] And Mellor, while realizing that 'very few people can realize their full potential' in the 'ME-world' of capitalist advanced technology, insists that 'it still represents male priorities and male interests', and, like Hartsock, speaks of 'a public world created in the one-sided, distorted and damaging image of male experience'.[45] But if gendered experience, including the split between public and private, is *constructed*, how can 'male experience' precede this construction? And if men's orientation to death and disconnection from nature is a tragedy for men as well as for women, how can it be in their interest?

Objective and subjective interests

The logic of interests is complex, and here I can only put forward one or two schematic points with the shameless intention of complicating the issue. Callinicos remarks that in trying to understand how agents use the capacities they have by virtue of social positioning, and how structures relate to agents' conscious experience, 'a great virtue of the notion of interests is that, properly understood, it allows us to connect the two without reducing either to the other'.[46] Concepts of 'objective interests' can only play such an explanatory role if they can theorize the conditions under which such interests may become 'subjective' and thus motivating. On the other hand, *purely* subjective concepts of interests have no explanatory potential at all. To say simply that men act in their own interests without offering any separate path to characterizing these would be as vacuous as the assertion that 'people do what they want'. Surprisingly, Callinicos' own treatment is only one step away from subjectivism. Following Giddens, he argues that someone has an interest in a given course of action if it would facilitate the possibility of achieving her or his wants. To act in one's interests is, therefore, to act as effectively as information and resources permit to realize one's wants.[47] We *can* use such an approach to speak of gendered interests, but it is inadequate. It allows us to describe men's actions to defend the destructive status quo in which they are dominant as in their interests, but it does not allow women to have any emancipatory interests in ending their oppression, or in stopping environmental destruction, until and unless they themselves *want* to do so.[48]

Any interesting account which addresses what *could* happen, as well as what has happened and does happen, must problematize 'wants'. It is only in my interests to act to achieve my wants, if they are what I need – they will lead to my thriving. The logic of interests is nearer to the logic of needs than to that of wants, and similarly requires grounding in a theory of human nature, which sees humans as having species-characteristic needs, capacities and liabilities.[49] These only take concrete form in particular historical societies, and these forms will be correspondingly various. Even basic needs, like health, give rise to interests which are relative to social resources. And needs, wants and interests emerge from different social situations – academics need to be on the Internet. Projects, such as rock-climbing, produce their own needs and interests. Where people – such as women – are similarly disadvantaged on important dimensions by virtue of their membership of a group, we can certainly say, at a high level of abstraction, that it is in their interests to bring about social change which gets more of their needs met. But as soon as we try to pin this down, complications appear.

Women's diverse interests

What are women's needs, and women's interests?

1. Women do have needs consequent upon their reproductive role: their need for health, common to all humans, gives them an interest in measures to prevent breast and cervical cancer, for instance, in societies where such prevention is possible. The essentialist claim that women – constantly reminded of their earthiness by menstruation and birth – identify with nature, and therefore have special interests in environmental protection, would fit in here.

2. Not only men, but also women have interests in the status quo of male domination. In fact, in all societies almost everyone has *some* interest in preserving the status quo. The known and understood almost always has *some* advantages, some way in which it does satisfy needs. Much of the female opposition to the Equal Rights Amendment to the US constitution took the tack that women were advantaged by dependence on men and should not be robbed of it.[50] We are creatures of our societies, necessarily constructed to find satisfaction in some aspect of them. For us to reject their values and set ourselves up in opposition to them can be painful and dangerous.

3. Relatedly, the status quo within which women have to seek satisfaction of their needs itself gives rise to gendered interests. If women are required to be ornamental, they have an interest in plenty of choice in make-up and fashion products. While women are the primary carers of children, they have a particular interest in good public child care – or in generous, reliable and loving husbands, or in incomes of their own . . . Mothers have an interest in good 'mother and baby' facilities in cafes, shops, stations and so on, and in technological advances which assist their task – and these, like disposable nappies, may well be environmentally disastrous. They have an interest in part-time and flexible working hours which is arguably in conflict with their emancipatory interest in equality in the workplace and labour market.

4. Women also have an emancipatory interest in social change which ends their relative disadvantage, although exactly what form such change would take is the subject of another discussion. Unless we take a subjectivist view, women's liberation must involve more scope for women to use their capacities and to satisfy their needs and wants in ways that help them to thrive.

Jackson's evaluation of the effect of technology, or the market, on rural women's lives makes implicit use of a notion of emancipatory interests which includes self-determination.[51] For much ecofeminism, women's emancipatory interests are in an ecologically sustainable society. A society which respected women and met their human and special needs would also be one which recognized the fragility and value of the ecosystem around us.

These various sorts of interests may and do conflict. There is a conflict between short-term interests – my interest in using the car to ferry my children around – and long-term and emancipatory interests – in replacing private cars with good public transport and better local facilities. There can be a conflict between my interests as an individual who happens to be socially positioned as a woman, but is also positioned on other, equally salient, dimensions; and the collective interests of women everywhere. I would argue that the work of the Fourth World Conference on Women and its accompanying Forum were in the interests of women in general (in sense 4), but it might be in the interests of particular women (in sense 3), or particular groups of women to sabotage that work. Indeed, the 'Boycott Beijing' campaign did try to prevent the Forum taking place in China. (Of course, the women in that campaign may have had sectional needs – as Tibetan exiles, for instance – which led to a real conflict of interests, or they may have had the same interests as the Forum supporters but a different strategic understanding.) A depth realist account can acknowledge such layered complexity, which does leave everything contested, but which also shows us how we can conduct the argument. However, such an account certainly forbids any simple assertion of the congruent interests of women and nature.

Men's interests

What then of men's interests?

1. Obviously men, too, have special interests arising from reproductive difference. There is no reason to believe that men's sexual pleasure requires the control or abuse of women, or that men genuinely thrive when 'conquering' or destroying the natural world.

2. Men have an interest in preserving the status quo, because – like women – they have constructed their personal identities, values and ideas of themselves in terms of the options socially available. For men to become supporters of women's liberation, or to become committed to stopping environmental degradation, means becoming aware of the scale of both sexism and gratuitous destruction – both of which can be painful. The future is uncertain so, unless the present becomes unbearable, conservatism will always have points in its favour.

3. Men benefit materially from the oppression of women, from their privileges in the labour market, and in public life, and from women's greater share of unpaid labour.[52] But it would be strange to argue that the marriage of capitalism and patriarchy, in which most of the powerful positions in capitalist societies are distributed to men, gives *all men* an unequivocal interest in

the capitalist system – and thus in the ecological destruction it produces. What about the workers? In any case, material benefits are not the only source of interests. True, in money economies almost everyone needs money – as the universal equivalent it represents access to an enormous range of use values. That does not mean we are all economic maximizers, as some versions of rational-choice theory would have it, or that the other things we value are similarly calculable. In the chapter on the working day in Volume 1 of *Capital*, Marx describes the coercive process of the social construction of workers as economic maximizers, in terms comparable to Weber's in *The Protestant Ethic*.[53] Similarly, men's interests in using women's unpaid labour, in 'possessing' women and in controlling their sexuality, are rooted in needs and wants that are produced, not given. Women's and men's studies have begun to trace and describe the particular processes in particular historical contexts that give rise to these grim effects. These relatively contingent interests are in conflict with other, deeper ones.

4. Do men have any emancipatory interests as men? If women need social change in order to thrive, are men already thriving? Or are they thriving for their particular interests, given their attraction to death and domination? I find the ecofeminist idea that the military-industrial complex is in men's interests extraordinary, given the brutalization, abuse, wounding and killing of men in war. This idea rests on a deficit model of human health and thriving, which assumes that, since to be oppressed is bad for women, to be the agents of oppression must be good for men.[54] I suggest that men, as humans, as people in close relationships with women, as fathers and as sons, have themselves an emancipatory interest in ending the oppression of women and in putting a stop to the destruction of the environment. Undoubtedly the motivations in question may be hard to activate, and may require processes of some ferocity. The vested male interest in present structures under sense 3 is greater than that of women: this argument for women's leadership can be salvaged from standpoint theory.

As for 'social' ecofeminism, what we should save is a causal claim. In advanced capitalist societies, the social reproduction of the multiple forms of male domination is carried out through the *same mechanisms* as the simultaneous reproduction of environmentally destructive social practices. The processes that make boys and men reject their own weakness and hide their vulnerability – or entrust it to women – also inhibit clear awareness of the vulnerability of life on earth. I have been arguing that it is not just men's position on the top of the hierarchical dualisms that constructs them as the agents of destruction, but equally women's position on the lower side, the institutionalized splits themselves, and their cognitive and emotional hold. The enemy of women and nature is not men, but dichotomized gendered subjectivity itself. These causal claims, for there are several, need considerable teasing out, and the specification of socio-psychological mechanisms – though candidates for these already exist. Wherever this theoretical work was judged satisfactory, it would forbid the ghettoizing of gender issues in environmental politics. It would encourage the creative critique of current political forms and methods. It would require the extension of the practical critique of dualism to the overlapping dimensions of class, ethnicity and so on. It would offer, beyond the device of ecofeminism, alongside irreducibly real sectional interests, the ethical vision of grounded human solidarity, of human interests in reversing the current destructive trend.

Notes

1 M. Mellor, *Breaking the Boundaries: Towards a Feminist Green Socialism*, London, 1990, p. 51.
2 J. Seager, *Earth Follies: Feminism, Politics and the Environment*, London 1992, p. 5.
3 V. Plumwood, *Feminism and the Mastery of Nature*, London 1993, p. 9.
4 D. Fuss, *Essentially Speaking: Feminism, Nature and Difference*, London 1989, p. 3.
5 M. Wittig, 'One is not Born a Woman', *Feminist Issues*, no. 2, 1981, pp. 47–54.
6 D. Haraway, *Simians, Cyborgs and Women*, London 1991, p. 155.
7 J. Butler, *Gender Trouble: Feminism and the Subversion of Identity*, London 1990, p. 24.
8 J. Sayers, *Biological Politics: Feminist and Anti-Feminist Perspectives*, London 1982, p. 35.
9 Plumwood, *Feminism and the Mastery of Nature*, p. 42.
10 Plumwood, *Feminism and the Mastery of Nature*, p. 47.
11 Plumwood, *Feminism and the Mastery of Nature*, p. 45.
12 Mellor, *Breaking the Boundaries*.
13 A. Collard with J. Contrucci, *The Rape of the Wild*, London 1988, p. 106.
14 C. Spretnak, 'Towards an Ecofeminist Spirituality', in J. Plant (ed.), *Healing the Wounds: the Promise of Ecofeminism*, Philadelphia 1989, p. 129.
15 M. Daly, *Gyn/ecology*, London 1979.
16 C. Jackson, 'Gender Analysis and Environmentalisms', in T. Benton and M. Redclift (eds), *Social Theory and the Global Environment*, London 1994, p. 127.
17 C. Jackson, 'Gender Analysis and Environmentalisms', in T. Benton and M. Redclift (eds), *Social Theory and the Global Environment*, London 1994, p. 116.
18 M. Mellor, *Breaking the Boundaries: Towards a Feminist Green Socialism*, London 1992, p. 80.
19 C. Jackson, 'Radical Environmental Myths: A Gender Perspective', *NLR* 210, p. 134.
20 Jackson, 'Gender Analysis and Environmentalisms', p. 139.
21 B. Brandth, 'The Social Construction of the Woman–Nature Linkage', in *Feminist Perspectives on Technology, Work and Ecology*, Conference Proceedings of the Second European Feminist Research Conference, 1994, p. 280.
22 Mellor, *Breaking the Boundaries*, p. 51.
23 Seager, *Earth Follies*, p. 12.
24 Y. King, 'The Ecology of Feminism and the Feminism of Ecology', in Plant, *Healing the Wounds*, p. 26.
25 M. French, *Beyond Power: on Women, Men and Morals*, New York 1985, p. 443.
26 N. Hartsock, 'The Feminist Standpoint: Developing the Ground for a Specifically Feminist Historical Materialism', in S. Harding and M. Hintikka (eds), *Discovering Reality: Feminist Perspectives on Epistemology, Metaphysics, Methodology and Philosophy of Science*, Dordrecht 1983, p. 284.
27 N. Hartsock, 'The Feminist Standpoint: Developing the Ground for a Specifically

Feminist Historical Materialism', in S. Harding and M. Hintikka (eds), *Discovering Reality: Feminist Perspectives on Epistemology, Metaphysics, Methodology and Philosophy of Science*, Dordrecht 1983, p. 294.

28 N. Hartsock, 'The Feminist Standpoint: Developing the Ground for a Specifically Feminist Historical Materialism', in S. Harding and M. Hintikka (eds), *Discovering Reality: Feminist Perspectives on Epistemology, Metaphysics, Methodology and Philosphy of Science*, Dordrecht 1983, p. 299.

29 King, 'The Ecology of Feminism', p. 23.

30 Plumwood, *Feminism and the Mastery of Nature*, p. 10.

31 R. Tong, *Feminist Thought: A Comprehensive Introduction*, London 1992, p. 100.

32 Mellor, *Breaking the Boundaries*, p. 81.

33 Seager, *Earth Follies*, p. 6.

34 For instance, Seager, *Earth Follies*, p. 163.

35 Seager, *Earth Follies*, pp. 5, 42, 43.

36 Seager, *Earth Follies*, pp. 43, 8.

37 B. Easlea, *Fathering the Unthinkable: Masculinity Science and the Arms Race*, London 1983, ch. 3.

38 D. Dinnerstein, 'Survival on Earth', in Plant, *Healing the Wounds*, p. 197.

39 S. Freud, *Three Essays on Sexuality*, London 1962, p. 85.

40 R. Braidotti, *Patterns of Dissonance*, Cambridge 1991, p. 89.

41 Spretnak, 'Towards an Ecofeminist Spirituality', p. 573.

42 Haraway, *Simians, Cyborgs and Women*, p. 190.

43 Hartsock, 'The Feminist Standpoint', p. 302.

44 Hartsock, 'The Feminist Standpoint', p. 299.

45 Mellor, *Breaking the Boundaries*, pp. 259, 251.

46 A. Callinicos, *Making History*, Cambridge 1987, p. 129.

47 A. Callinicos, *Making History*, Cambridge 1987, p. 129.

48 See E. Laclau and C. Mouffe, (1985) *Hegemony and Socialist Strategy: Towards a Radical Democratic Politics*, Verso, London 1985, p. 153.

49 L. Doyal and I. Gough, *A Theory of Human Need*, London 1991.

50 B. Ehrenreich, *The Hearts of Men*, London 1983.

51 Jackson, 'Radical Environmental Myths', p. 133.

52 C. Delphy and D. Leonard, *Familiar Exploitation*, Cambridge 1992.

53 K. Marx, *Capital Volume 1*, London 1938; M. Weber, *The Protestant Ethic and the Spirit of Capitalism*, Hemel Hempstead 1985, p. 62.

54 C. New, *Agency, Health and Social Survival: the Eco-politics of Rival Psychologies*, London, 1996, ch. 8.

14 Gillian Rose

'Looking at Landscape: the Uneasy Pleasures of Power'

Excerpt from: *Feminism and Geography*, pp. 86–93.
Cambridge: Polity (1993)

Landscape is a central term in geographical studies because it refers to one of the discipline's most enduring interests: the relation between the natural environment and human society, or, to rephrase, between Nature and Culture. Landscape is a term especially associated with cultural geography, and although 'literally [the landscape] is the scene within the range of the observer's vision',[1] its conceptualization has changed through history. By the interwar period, for its leading exponents, such as Otto Schlüter in Germany, Jean Brunhes in France and Carl Sauer in the USA, the term 'landscape' was increasingly interpreted as a formulation of the dynamic relations between a society or culture and its environment: '*the process of human activity in time and area*'.[2] The interpretation of these processes depended in particular on fieldwork, and fieldwork is all about looking: 'the good geographers have first been to see, then they have stopped to think and to study the conclusions of others before finally recording their findings for us in maps and print'.[3] Just as fieldwork is central not only to cultural geography but also to the discipline as a whole, however, so too the visual is central to claims to geographical knowledge:[4] a president of the Association of American Geographers has argued that 'good regional geography, and I suspect most good geography of any stripe, begins by looking'.[5] The absence of knowledge, which is the condition for continuing to seek to know, is often metaphorically indicated in geographical discourse by an absence of insight, by mystery or by myopia; conversely, the desire for full knowledge is indicated by transparency, visibility and perception. Seeing and knowing are often conflated.

More recent work on landscape has begun to question the visuality of traditional cultural geography, however, as part of a wider critique of the latter's neglect of the power relations within which landscapes are embedded.[6] Some cultural geographers suggest that the discipline's visuality is not simple observation but, rather, is a sophisticated ideological device that enacts systematic erasures. They have begun to problematize the term 'landscape' as a reference to relations between society and the environment through contextual studies of the concept as it emerged and developed historically, and they have argued that it refers not only to the relationships between different objects caught in the fieldworker's gaze, but that it also implies a specific way of looking. They interpret landscape not as a material consequence of interactions between a society and an environment, observable in the field by the more-or-less objective gaze of the geographer, but rather as a gaze which itself helps to make sense of a particular relationship between society and land. They have stressed the importance of the look to the idea of landscape and have argued that landscape is a way of seeing which we learn; as a consequence, they argue that the gaze of the fieldworker is part of the problematic, not a tool of analysis. Indeed, they name this gaze at landscape a 'visual ideology', because it uncritically shows only the relationship of the powerful to their environment. This is an important critique of the unequal social relations implicit in one element of geographical epistemology, and the first section of this chapter examines these arguments.

Questions of gender and sexuality have not been raised by this newer work, however. This seems an important omission: the previous chapter cited Fitzsimmons's comment that cultural geography retained an interest in Nature, and also noted the feminization of Nature in geographical discourse. A consequence has been that, historically, in geographical discourse, landscapes are often seen in terms of the female body and the beauty of Nature. Here, for example, is one of the quotations from the previous chapter expanded to highiight the parallels that it makes between 'live, supple, sensitive, and active' Nature and a female body:

> It is [in] the face and features of Mother-Earth that we geographers are mainly interested. We must know something of the general principles of geology, as painters have to know something of the anatomy of the human or animal body . . . the characteristic of the face and features of the Earth most worth learning about, knowing and understanding is their beauty.[7]

Stoddart's celebration of geography's exploration and fieldwork tradition similarly conflates the exploration of Nature with the body of Woman; for example, his frontispiece is an eighteenth-century engraving representing Europe, Africa and America as three naked women.[8] This feminization of what is looked at does matter, because it is one half of what Berger characterizes as the dominant visual regime of white heterosexual masculinism: 'women appear', he says, but 'men act'.[9] This particular masculine position is to look actively, possessively, sexually and pleasurably, at women as objects. Now, Berger's comments refer to the female nude in Western art; but I will suggest in this chapter that the feminization of landscape in geography allows many of the arguments made about the masculinity of the gaze at the nude to work in the context of geography's landscape too, particularly in the context of geography's pleasure in landscape. The second section of this chapter suggests that geography's look at landscape draws on not only a complex discursive transcoding between Woman and Nature, as the previous chapter argued, but also on a specific masculine way of seeing: the men acting in the context of geography are the fieldworkers, and the Woman appearing is the landscape. This compelling figure of Woman both haunts a masculinist spectator of landscape and constitutes him.

The pleasures that geographers feel when they look at landscape are not innocent, then, but nor are they simple. The pleasure of the masculine gaze at beautiful Nature is tempered by geography's scientism, as the last chapter suggested. The gaze of the scientist has been described by Keller and Grontkowski as part of masculinist rationality,[10] and to admit an emotional response to Nature would destroy the anonymity on which that kind of scientific objectivity depends. Keller and Grontkowski trace the tradition of associating knowledge with vision back to Plato, and they argue that by the seventeenth century the equivalence of knowing with seeing was a commonplace of scientific discourse. It remains so today. But when Descartes discovered that the eye was a passive lens, in order to retain an understanding of the accession to knowledge as active he was forced to separate the seeing intellect from the seeing eye. This was one aspect of the split between the mind and the body so much associated with his work, and it rendered the objects of the gaze separate from the looking subject: 'Having made the eye purely passive, all intellectual activity is reserved to the "I", which, however, is radically separate from the body which houses it'.[11] Such disembodiment separated knowing from desire, and protected men's scientific neutrality from Woman's wild nature. For Keller, the scientific gaze is another aspect of the distanced, disembodied objectivity of science. However, geographers are constituted as sensitive artists as well as objective scientists in their approach to Nature and landscape. This contradiction produces a conflict between desire and fear in visual forms. It creates a tension between distance from the object of the gaze and merger with it, which is at work both in the conflict between knowledge and pleasure – a conflict between 'a highly individual response' and 'a distinterested search for evidence'[12] – and also within the pleasured gaze. These complex contradictions between and within (social-) scientific objectivity and aesthetic sensitivity disrupt cultural geography's claim to know landscape, as the second section argues. These disruptions are elaborated there through the work of psychoanalytic feminists who suggest that 'the specificity of visual performance and address has . . . a privileged relation to issues of sexuality'.[13] This second section is adopting one of the tactics outlined in the previous chapter, then – finding contradictions in the Same. I argue that the structure of aesthetic masculinity which studies landscape is inherently unstable, subverted by its own desire for the pleasures that it fears.

The third section uses another tactic of critique, and looks at various attempts to re-present a different relation between subject and environment from other spectating positions. None draws on the structure which posits Woman as Nature in order to establish Man as Culture, and all stress differences between women. They begin to imagine different kinds of landscape.

Landscape as visual ideology

Recent critiques of the landscape idea in geography insist that landscape is a form of representation and not an empirical object. As Daniels and Cosgrove remark, 'a landscape is a cultural image, a pictorial way of representing, structuring or symbolising surroundings'.[14] Whether written or painted, grown or built, a landscape's meanings draw on the cultural codes of the society for which it was made. These codes are embedded in social power structures, and theorization of the relationship between culture and society by these new cultural geographers has so far drawn on the humanist marxist tradition of Antonio Gramsci, Raymond Williams, E. P. Thompson and John

Berger. All of these authors see the material and symbolic dimensions of the production and reproduction of society as inextricably intertwined.[15] Cosgrove, one of the most prominent theorists of the new critique of the landscape idea, defines culture as:

> . . . symbolisation, grounded in the material world as symbolically appropriated and produced. In class societies, where surplus production is appropriated by the dominant group, symbolic production is likewise seized as hegemonic class culture to be imposed on all classes.[16]

In his work, landscape becomes a part of that hegemonic culture, a concept which helps to order society into hierarchical class relations.

Cosgrove points out that landscape first emerged as a term in fifteenth- and early sixteenth-century Italy, and he argues that it was bound up with both Renaissance theories of space and with the practical appropriation of space. Euclidean geometry was 'the guarantor of certainty in spatial conception, organisation and representation',[17] and its recovery paved the way of Alberti's explication of the technique of three-dimensional perspective in 1435. Other geometrical skills were being developed contemporaneously, especially by the urban merchant class, and these too involved the accurate representation of space: calculating the volume and thus the value of packaged commodities; map-making to guide the search for goods and markets; and surveying techniques to plot the estates that the bourgeoisie were buying in the countryside. All of these spatial techniques were implicated in relations of power and ownership. Cosgrove is particularly interested in Alberti because, using his manual, artists could render depth realistically, and so establish a particular viewpoint for the spectator in their painting – a single, fixed point of the bourgeois individual. (Cosgrove does remark that this individual was male, but does not develop the point.[18]) From this position, the spectator controlled the spatial organization of a composition, and Cosgrove argues that this was central to landscape images. Merchants often commissioned paintings of their newly acquired properties, and in these canvases, through perspective, they enjoyed perspectival as well as material control over their land. Cosgrove concludes that the idea of landscape is patrician because it is seen and understood from the social and visual position of the landowner. Other writers agree and emphasize the erasure of the waged labour relation in landscape painting. In the context of eighteenth-century English landscape painting, for example, Barrell notes that the labourers in these images are denied full humanity, and Bryson argues that the fine brushwork technique favoured in Western art until the late nineteenth century effaces the mark of the artist as waged worker.[19] It is argued then that landscape is meaningful as a 'way of seeing' bound into class relations, and Cosgrove describes landscape as a 'visual ideology' in the sense that it represents only a partial world view.[20]

This is an extremely important critique of the ideologies implicit in geographical discourse. Its strengths are evident in the interpretation shared by cultural geographers of the mid-eighteenth-century double portrait of Mr and Mrs Andrews, by the English artist Thomas Gainsborough (Fig. 14.1).[21] In

Fig. 14.1 *Mr and Mrs Andrews*, by Thomas Gainsborough. Reproduced by courtesy of the Trustees, The National Gallery, London

their discussions of this image, geographers concur that pleasure in the right-hand side of the canvas – those intense green fields, the heaviness of the sheaves of corn, the English sky threatening rain – is made problematic by the two figures on the left, Mr and Mrs Andrews. Berger, whose discussion of this painting geographers follow, insists that the fact that this couple owned the fields and trees about them is central to its creation and therefore to its meaning: 'they are landowners and their proprietary attitude towards what surrounds them is visible in their stance and their expressions'.[22] Their ownership of land is celebrated in the substantiality of the oil paints used to represent it, and in the vista opening up beyond them, which echoes in visual form the freedom to move over property which only landowners could enjoy. The absence in the painting's content of the people who work the fields, and the absence in its form of the signs of its production by an artist working for a fee on a commission, can be used to support Cosgrove's claim that landscape painting is a form of visual ideology: it denies the social relations of waged labour under capitalism. *Mr and Mrs Andrews*, then, is an image on which geographers are agreed: it is a symptom of the capitalist property relations that legitimate and are sanctioned by the visual sweep of a landscape prospect.

However, the painting of Mr and Mrs Andrews can also be read in other ways. In particular, it is possible to prise the couple – 'the landowners' – apart, and to differentiate between them. Although both figures are relaxed and share the sense of partnership so often found in eighteenth-century portraits of husband and wife, their unity is not entire: they are given rather different relationships to the land around them. Mr Andrews stands, gun on arm, ready to leave his pose and go shooting again; his hunting dog is at his feet, already urging him away. Meanwhile, Mrs Andrews sits impassively, rooted to her seat with its wrought iron branches and tendrils, her upright stance echoing that of the tree directly behind her. If Mr Andrews seems at any moment able to stride off into the vista, Mrs Andrews looks planted to the spot. This helps me to remember that, *contra* Berger, these two people are *not* both landowners – only Mr Andrews owns the land. His potential for activity, his free movement over his property, is in stark contrast not only to the harsh penalties awaiting poachers daring the same freedom of movement over his land (as Berger notes), but also to the frozen stillness of Mrs Andrews. Moreover, the shadow of the oak tree over her refers to the family tree she was expected to propagate and nurture; like the fields she sits beside, her role was to reproduce, and this role is itself naturalized by the references to trees and fields.[23] . . . This period saw the consolidation of an argument that women were more 'natural' than men. Medical, scientific, legal and political discourses concurred, and contextualize the image of Mr and Mrs Andrews in terms of a gendered difference in which the relationship to the land is a key signifier. Landscape painting then involves not only class relations, but also gender relations. Mr Andrews is represented as the owner of the land, while Mrs Andrews is painted almost as a part of that still and exquisite landscape: the tree and its roots bracketing her on one side, and the metal branches of her seat on the other.

Many feminist art historians have argued that heterosexual masculinism structures images of femininity: following that claim, my interpretation of the figure of Mrs Andrews stresses her representation as a natural mother. Obviously, her representation also draws on discourses of class and even nation. I emphasize her femininity, however, because there are feminist arguments which offer a critique not just of the discourses that pin Mrs Andrews to her seat, but also of the gaze that renders her as immobile, as natural, as productive and as decorative as the land. Such arguments consider the dynamics of a masculine gaze and its pleasures. More is involved in looking at landscape than property relations.

Notes

1 R. E. Dickinson, 'Landscape and society', *Scottish Geographical Magazine*, 55 (1939), pp. 1–14, p. 1.
2 Dickinson, 'Landscape and society', p. 6.
3 P. A. Jones, *Field Work in Geography* (Longmans, Green, London, 1968), p. 1.
4 On the importance of the visual to the contemporary discipline, see D. Cosgrove, 'Prospect, perspective and the evolution of the landscape idea', *Transactions of the Institute of British Geographers*, 10 (1985), pp. 45–62, p. 46, p. 58. For a baroque example of the visual as a metaphor of knowledge, see E. W. Soja, 'The spatiality of social life', in *Social Relations and Spatial Structures*, eds D. Gregory and J. Urry (Macmillan, London, 1985), pp. 90–127.
5 J. Fraser Hart, 'The highest form of the geographer's art', *Annals of the Association of American Geographers*, 72 (1982), pp. 1–29, p. 24.
6 See, for example, J. S. Duncan, 'The superorganic in American cultural geography', *Annals of the Association of American Geographers*, 70 (1980), pp. 181–98; P. Jackson, *Maps of Meaning: an Introduction to Cultural Geography* (Unwin Hyman, London, 1989). For a discussion of the exclusion of women from landscape studies, see J. Monk, 'Approaches to the study of women and landscape', *Environmental Review*, 8 (1984), pp. 23–33.
7 F. Younghusband, 'Natural beauty and geographical science', *The Geographical Journal*, 56 (1920), pp. 1–13, p. 3.
8 D. R. Stoddart, *On Geography and its History* (Blackwell, Oxford, 1986).
9 J. Berger, *Ways of Seeing* (British Broadcasting Corporation, London 1972), p. 47.
10 E. F. Keller and C. R. Grontkowski, 'The mind's eye', in *Discovering Reality: Feminist Perspectives on Epistemology, Metaphysics, Methodology and Philosophy of Science*, eds S. Harding and M. B. Hintikka (D. Reidel, Dordrecht, 1983), pp. 207–24.
11 Keller and Grontkowski, 'The mind's eye', p. 215.
12 D. Cosgrove, 'Geography is everywhere: culture and symbolism in human landscapes', in *Horizons in Human Geography*, eds D. Gregory and R. Walford (Macmillan, London, 1989), pp. 118–35, pp. 126 and 127.
13 G. Pollock, *Vision and Difference: Femininity, Feminism and Histories of Art* (Routledge, London, 1988), p. 123.
14 S. Daniels and D. Cosgrove, 'Introduction: the iconography of landscape', in *The Iconography of Landscape: Essays on the Symbolic Representation, Design and Use of Past Environments*, eds D. Cosgrove and S. Daniels (Cambridge University Press, Cambridge, 1988), pp. 1–10, p. 1.
15 For an excellent discussion of this tradition, see S. Daniels, 'Marxism, culture and the duplicity of landscape', in *New Models in Geography, Volume 2: The Political*

Economy Perspective, eds R. Peet and N. Thrift (Unwin Hyman, London, 1989), pp. 196–220.

16 D. Cosgrove, 'Towards a radical cultural geography: problems of theory', *Antipode*, 15 (1983), pp. 1–11, p. 5.

17 Cosgrove, 'Prospect, perspective and the evolution of the landscape idea', p. 46.

18 D. Cosgrove, 'Historical considerations on humanism, historical materialism and geography', in *Remaking Human Geography*, eds A. Kobayashi and S. Mackenzie (Unwin Hyman, London, 1989), pp. 189–226, p. 190.

19 J. Barrell, *The Dark Side of the Landscape: the Rural Poor in English Painting*, 1730–1840 (Cambridge University Press, Cambridge, 1980); N. Bryson, *Vision and Painting: The Logic of the Gaze* (Macmillan, London, 1983), esp. pp. 89–92.

20 Cosgrove, 'Prospect, perspective and the evolution of the landscape idea', p. 47. The term 'ways of seeing' is after Berger, *Ways of Seeing*.

21 Daniels, 'Marxism, culture and the duplicity of landscape', p. 213; Daniels and Cosgrove, 'Introduction: the iconography of landscape'; M. Gold, 'A history of nature', in *Geography Matters! A Reader*, eds D. Massey and J. Allen (Cambridge University Press, Cambridge, 1984), pp. 12–33, pp. 20–3; J. R. Short, *Imagined Country: Society, Culture and Environment* (Routledge, London, 1991), p. 170.

22 Berger, *Ways of Seeing*, p. 107.

23 A. Bermingham, *Landscape and Ideology: the English Rustic Tradition 1740–1860* (Thames & Hudson, London, 1987), pp. 14–16; S. Daniels, 'The political iconography of woodland in later Georgian England', in *The Iconography of Landscape: Essays on the Symbolic Representation, Design and Use of Past Environments*, eds D. Cosgrove and S. Daniels (Cambridge University Press, Cambridge, 1988), pp. 43–82.

SECTION FOUR

BODY MAPS

Editors' introduction

Although the academic literature about the body has expanded exponentially only in the last few years, the body has always been a key issue for feminism and for feminists. Of all the dualisms that have structured women's inferiority, it is perhaps the mind–body distinction that has posed the greatest challenge. Central in western enlightenment thought, this distinction – with its concomitant associations of lofty rational and disembodied thought (that is the mind) with masculinity, and the down to earth body with femininity – traps women in their bodies and excludes them from intellectual effort. Women's bodies: their fluid, changeable, powerful and life-giving properties have long been constructed as threatening by men, seen as a fleshy temptation or as pure embodied delight to worship. And, as we demonstrated in the previous section, the association of woman with nature is common not only in western thought: the nature–culture distinction has been traced by feminist anthropologists in many societies, in their myths, religions and social institutions (McCormack and Strathern, 1980).

Physical embodiment posed serious difficulties for second wave feminists. While not wanting to deny the significance of the female body and what then seemed its clear distinction from the male form, feminists were eager to escape from its confines. The distinction between sex and gender seemed one way to do it. Sex was defined as the biological differences between men and women whereas gender was theorised as the sets of cultural attributes mapped onto male and female forms: socially constructed and historically and spatially specific.

In the political arena, the apparent differences between male and female bodies and their capacities led to what has been termed the same–difference dilemma. Liberal feminists, anxious to claim equal rights for women in the political arena, the family and in the workplace, appealed to the values of the democratic state – that all individuals are equal before the law – to claim equal pay and freedom from sex discrimination. Whereas radical feminists argued that women's essential differences from men, especially the ability to give birth, required the recognition of women's particularity (see Phillips, 1987). Adrienne Rich is a key exponent of the difference argument. In her classic paper about compulsory heterosexuality published in 1980, she argued that women have a fundamental attachment to each other that is destroyed by the institutions of heterosexist society, as women are coerced into relationships with men. But Rich was no essentialist who denied differences between women. In a later paper, 'Notes toward a politics of location' (1984) (with its geographic allusion in the title) she struggled with the implications for feminist politics of the ways in which social institutions construct us as particular women and men; distinguished and differentially valued by a range of physical attributes – skin, hair colour, the possession or not of different organs, classed, raced and gendered by the circumstances in which we are born and raised. It is commonplace to argue that many of the earlier key thinkers of the post-war women's movement ignored the connections between class, ethnicity and gender in their analyses in their desire to establish the common bonds that united women as women, but Rich was keenly aware of the differences between women, well before difference became an essential term in feminist discourse.

It is undeniable, however, that many white women saw 'race' or colour as something that was specific to 'minority' women and ignored the social construction of whiteness and the privileges it brings. In the first extract in Section Four, Ruth **Frankenberg** specifically addresses the questions about the meaning of whiteness that had been raised by Rich a decade earlier, in her work on the social goegraphy of white women's childhoods. The piece we have chosen focuses on Frankenberg's own understanding of being white and the need for white women to become involved in anti-racist work, both in the academy and in the wider political arena. Frankenberg is not a geographer by discipline and it is interesting to think through the similarities and differences between her work and that of Cindi Katz (1993) who is one of the few geographers who has looked at the landscapes of childhood. A fuller version of Frankenberg's work is available in her book: *White Women: Race Matters* (1993).

Frankenburg's work, like that of Vron Ware (1992) and growing numbers of other white feminists, is an important corrective to the notion that only 'black' bodies are marked with a colour. Like the normal/abnormal, invisible/visible distinctions mapped onto men's and women's bodies, the white/black dualism is another extremely important way of creating an inferior/superior dualistic distinction. Iris Marion **Young** takes this distinction as the starting point for her discussion of the operation of the mechanisms of what she terms 'cultural imperialism'. In her work on the construction of a multiple and group-based definition of inequality, she draws specific attention to the significance of bodily distinctions in differentiating between dominant and inferior groups. Dominated groups are defined as nothing but their bodies, imprisoned in an undesirable body whereas the dominant groups occupy an unmarked, neutral, universal and disembodied position: masculine and white by default.

For women embodiment is contradictory. As **Young** suggests 'while a certain cultural space is reserved for revering feminine beauty and desirability, in part that very cameo idea renders most women drab, ugly, loathsome or fearful bodies'. Indeed Tseelon (1995) has argued that five paradoxes construct feminine embodiment. These are first, the modesty paradox – the woman is constructed as seduction to be punished for it; second, the duplicity paradox – the woman is constructed as artifice, and marginalised for lacking essence and authenticity; third, the visibility paradox – the woman is constructed as a spectacle while being culturally invisible; fourth the beauty paradox – the woman embodies ugliness while signifying beauty; and fifth the death paradox – the woman signifies death as well as the defence against it. (p. 5–6.) These five paradoxes are helpful ways in understanding the representations of woman and the social practices that result in women's exclusion from positions of power in the public arena. As Connell has suggested, in recent work,

> approaches that treat women's bodies as the object of social symbolism have flourished at the meeting point of cultural studies and feminism. Studies of the imagery of bodies and the production of femininity in film, photography and the other visual arts now number in the hundreds.
>
> (Connell, 1995 p. 49)

In geography too, as we suggested in the last section through the extract from Rose's work, there is a new emphasis on the visual and on representation among feminist scholars.

In more recent theorising about the body in feminist scholarship more generally, the sex–gender distinction has been dissolved and new ways of thinking about the body as what Grosz terms 'an inscriptive surface' have become common. The body, its size, shape, gestures, the very space it takes up, those masculine and feminine norms which mean that men sprawl and women don't; the differences in physicality that construct and reflect gender norms create ways of being in space. The body is an object over which we labour – dieting, working out, picking, pruning, squeezing and decorating to conform to some idealised view of an appropriate femininity or masculinity. Here the work of Michel Foucault has been extremely influential. He argued that the body is a field or surface on which the play of power, knowledge and resistance is worked out. Through social norms, self-surveillance and disciplinary practices, the body's materiality, and its desires and pleasures are produced and restricted to a narrow range of acceptable attitudes. In the extract from **Bordo**, however, we are reminded of the significance of explicit feminist theorising and action in constructing a sexual politics of the body. It is important to challenge masculine claims to dominate understandings of the body and to hold on to the materiality of the body. One of the key aims of feminist politics has been to challenge the medicalisation of 'female complaints' from menstruation and childbirth to the menopause. *Our Bodies, Ourselves* (Boston Collective, 1976), for example, was one of the most significant texts of the 1970s and early 1980s. In the excitement of new theoretical work on the body by theorists such as Butler and Grosz, whose work we include in the final two extracts, the same anxieties about the subject that we discussed in the introduction have begun to emerge. Zita's comment nicely sums up these reactions:

> Theory as a new discursive body snatcher relocated the body in its signs and lapsed into a crisis of referentiality, as feminism lost its subject. The body that had been the site of material struggle in 1970s feminism was disappearing from academic thought.
>
> (Zita, 1966 p. 786)

Bordo's paper reflects some of this anxiety but she also recommends the work of Foucault to feminists theorists of the body. In the fourth extract, Elizabeth **Grosz** provides a critical analysis of the work of a wider range of key male analysts of corporeality, although Foucault is chief among them, in an assessment of the utility of their work for feminist accounts of embodiment. Grosz emphasises the need to understand the body not as a neutral surface or as a naturally differentiated, already sexed form, but as socially-located morphologies. She argues that biological differences themselves are constructed and translated by social practices, and that sex and gender are inextricably interlinked rather than the latter being a cultural imposition on the (natural/biological) former.

Both the emphasis of the 'early' feminist activists – remember 'fat is a feminist issue' as a powerful slogan – and the later feminists influenced by post-structuralism and psychoanalysis on the body as a (relatively) fluid or flexible object to be struggled over to produce a particular image, parallel the understanding of gender as a social construction. Judith **Butler** in her work has brought these two notions together in her analysis of gender as a performance. Indeed, in the paper included here, she helps us to bring together several of the ideas discussed in the four previous pieces by her critical assessment of the essentialism that lies behind the very concept 'woman'. Butler comments on the implications of challenges to male disembodiment, and on the destabilisation of the subject and the maternal body, arguing for an alternative concept of the subject, a subject who is defined relationally and contextually, constructed in specific spatial and temporal locations. Like Rich, Butler argues that the rules and regulations of social life require the performance of a heterosexual gender identity. She suggests that a gender identity is neither natural nor fixed, but rather that the lived body in space is the outcome of culturally sanctioned fantasies. The appearance of a coherent gendered identity must be constantly created through the performative repetition of bodily acts and gestures. Gender is therefore a fiction or a fabrication that requires constant regulation. Butler uses the example of drag to destabilise the notion of true gender identities. She has written about the necessity of gender performance, not to reflect some 'deeper' gendered or sexed identity but to create it so that a gender becomes the effect of the performance rather than the creater of it. In her focus upon the performative aspects of gender, Butler draws attention to the necessity of this act for both gay and straight identities, so that apparently fictional gender acts – such as drag – are not simply copies of a heterosexual original, but come to indicate the necessity of repetitive performance to ensure the continued illusion of *all* gendered and sexed identities. As we argue later, in Section Six, the idea of gendered performances has become an important way of theorising the significance of the body in the new service sector occupations that now dominate the economies of many nations, especially in the west. The social relations of gender, in the workplace and in other spaces and locations, are symbolised and maintained through bodily performances which are place-specific and require constant surveillance and regulation.

These perspectives have opened up a valuable space for women to escape from the fiction of a 'natural' femininity, as well as from the fiction that women are nothing but bodies and men are all mind. The dominance of this view in high tech and scientific work has been illustrated by Massey (1995). But, in theory at least, as well as in the field of leisure and sport, and increasingly in some workplaces, it is now recognised that masculinity is also constituted through bodily performances. As Grosz argues:

> women are no more subject to this system of corporeal production than men; they are no more cultural, no more natural, than men. Patriarchal power relations do not function to make women the objects of disciplinary control while men remain outside disciplinary surveillance. It is a question not of more or less but of differential production.
>
> (Grosz, 1994 p. 144)

In his work on masculinity, Connell argues that the clearest examples of the constitution of masculinity through bodily performance are to be seen in sport and in manual labour. Middle class men are defined differently, as bearers of skill or rationality: here again we see the mind not the body used to define a particular class-based masculinity, but as the economy changes Connell argues, new definitions of male power are now being constructed to reproduce masculine superiority. He takes his example from the world of information technology and computing.

> The new information technology requires much sedentary keyboard work, which was initially classified as women's work. The marketing of personal computers, however, has redefined some of this work as an arena of competition and power – masculine, technical but not working class. These revised meanings are promoted in the texts and graphics of computer magazines, in manufacturers' advertising that emphasizes 'power' (Apple

> Computer named its laptop the 'PowerBook') and in the booming industry of violent computer games. Middle class male bodies, separated by an old class division from physical force, now find their powers spectacularly amplified in the man/machine systems (the gendered language is entirely appropriate) of modern cybernetics.
>
> (p. 55–6)

Cynthia Cockburn's magnificent work on skill and gender also examines these associations in different workplaces and production processes (Cockburn, 1985).

The emphasis on gendered performances, on parody and masquerade and on the contextual nature of gendered social identities is not, however, to deny the materiality of the body. It may be a surface to be inscribed, a copperplate not a white sheet in Grosz's vivid image in the piece included here, but it is more than a position to speak from. Bodies change over time, spread, age and wither, give birth and physical pleasure or pain, get sick and get better, and are a crucial part of ourselves as gendered individuals. Chouinard in her moving account of being a disabled woman geographer (in Section Two) made this plain.

References and further reading

Bell, D and Valentine, G 1995: *Mapping Desire: Geographies of Sexualities*. Routledge: London.

Boston Women's Health Book Collective, 1976: *Our Bodies, Ourselves*. Simon and Schuster: New York (also published in England, see Phillips, A 1989: *The New Our Bodies, Ourselves*. Penguin: Harmondsworth, second edition).

Butler, J 1993: *Bodies that Matter*. Routledge: London.

Cockburn, C 1985: *Machinery of Dominance: Women, Men and Technical Know-How*. Pluto Press: London.

Colomina, B (ed.) 1992: *Sexuality and Space*. Princeton Architectural Press: Princeton.

Connell, R W 1995: *Masculinities*. Polity: Cambridge.

Diamond, I and Quinby, L (eds) 1988: *Feminism and Foucault: Reflections on Resistance*. Northeastern University Press: Boston.

Featherstone, M, Hepworth, M and Turner, B (eds) 1991: *The Body: Social Process and Cultural Theory*. Sage: London.

Frankenberg, R 1993: *White Women, Race Matters: the Social Construction of Whiteness*. Routledge: London.

Grosz, E 1994: *Volatile Bodies: Toward a Corporeal Feminism*. Indiana University Press: Bloomington.

Jaggar, A and Bordo, S 1989: *Gender/Body/Knowledge: Feminist Reconstructions of Being and Knowing*. Rutgers University Press: New Brunswick, NJ.

Katz, C 1993: Growing girls/closing circles: limits on the spaces of knowing in rural Sudan and US cities. In Katz, C and Monk, J (eds) *Full Circles: Geographies of Women Over the Life Course*. Routledge: London. 88–106.

Massey, D 1995: Masculinity, dualisms and high technology. *Transactions, Institute of British Geographers* ***20***, 487–99.

McCormack, C P and Strathern, M (eds) 1980: *Nature, Culture and Gender*. Cambridge University Press: Cambridge.
Phillips, A 1987: *Feminism and Equality*. Blackwell: Oxford.
Rich, A 1980: The lesbian continuum and compulsory heterosexuality. In Rich, A 1987: *Blood, Bread and Poetry: Selected Prose, 1979–1985*. Virago: London.
Rich, A 1984: Notes towards a politics of location. In *Blood, Bread and Poetry: Selected Prose, 1979–1985*. Virago, London.
Sennett, R 1994: *Flesh and Stone: the Body and the City in Western Civilisation*. Faber and Faber: London.
Tseelon, E 1995: *The Masque of Femininity*. Sage: London.
Ware, V 1992: *Beyond the Pale: White Women, Racism and History*. Verso: London.
Zita, J 1996: Review of Bordo, Grosz and Butler. *Signs: Journal of Women in Society and Culture* ***21***, 786–93.

15 Ruth Frankenberg

'Growing Up White: Feminism, Racism and the Social Geography of Childhood'

Excerpts from: *Feminist Review* **45**, 51–84 (1993)

Whiteness: a privilege enjoyed but not acknowledged, a reality lived in but unknown.[1]

Introduction: personal and contextual notes

This essay is about the ways racism shapes white women's lives, the impact of race privilege on white women's experience and consciousness. Just as both men's and women's lives are shaped by their gender, and both heterosexual and lesbian women's experiences in the world are marked by their sexuality, white people and people of colour live racially structured lives. In other words, any system of differentiation shapes those upon whom it bestows privilege as well as those it oppresses. At a time in the histories of both the US and UK when we are encouraged, as white people, to view ourselves as racially and culturally 'neutral' rather than as members of racially and culturally *privileged* or *dominant* groups, it is doubly important to look at the 'racialness' of white experience.[2]

For the last two decades and more, women of colour have worked to transform feminism, challenging white feminists' inattention to race and other differences between women, and the falsely universalizing claims of much 'second wave' feminist analysis (for the 1970s see, among others: Cade, 1970; Garcia, 1990; for the 1980s and 1990s, key works include Moraga and Anzaldúa, 1981; hooks, 1981, 1984; Sandoval, 1982, 1991). Women of colour in North America and Black women in Britain have mapped the ways racism and ethnocentrism limit feminist theory and strategy over issues such as family structures (Carby, 1981; Bhavnani and Coulson, 1986) and reproductive rights (A. Y. Davis, 1981; see also S. E. Davis, 1988, for a more recent analysis that takes seriously the intersections of gender and sexuality with race and class) as well as making feminist institutions exclusive, be they workplaces, journals or conferences (Zinn *et al.*, 1986; Sandoval, 1982).[3] Alongside this critique feminist women of colour and, more recently, white feminists also, have analyzed women's lives as marked by the simultaneous impact of gender, sexuality, race and class (the founding text here is, I believe, Combahee River Collective, 1979; Zavella, 1987; Alarcon, 1990; Haraway, 1991; Sandoval, 1990) and generated visions and concepts of multiracial coalition work (notably Moraga and Anzaldúa, 1981: 195–6; Reagon, 1983). In this context, white feminists like myself have learned a great deal about the meaning of race privilege.

White Privilege
Today I got permission to do it in graduate school,
That which you have been lynched for,
That which you have been shot for,
That which you have been jailed for,
Sterilized for,
Raped for,
Told you were mad for –
By which I mean
Challenging racism –
Can you believe
The enormity
Of that?
(Frankenberg, 1985)

I came to the United States from Britain in 1979, a Marxist Feminist. My antiracist activism had involved participating in the anti-Nazi League, Rock Against Racism, and the All-Cambridge Campaign against Racism and Fascism – organizations that emerged in the mid-1970s in reaction to a resurgence of far-right, organized racism in the UK. I marched in London, picketed in Cambridge, and declared myself ready to join in physically defending the boundaries of Black neighbourhoods from the incursions of racist gangs.[4] In that context, I saw racism as entirely external to me, a characteristic of extermists or of the British State, but not a part of what made *me*, or shaped my activism. Ironically, however, and exemplifying the extent to which racism constructed my outlook, I barely noticed, much less questioned, the reality that the All-Cambridge Campaign was almost entirely white in its membership. My 'externalizing' of racisms changed in the United States, where, initially through university, I met, learned from, and later wrote, co-taught and lived with lesbian women of colour and white working-class women, who pushed the limits of my perceptions of racism beyond the purely external, so that I increasingly saw its ever-present impact on daily lives – mine and theirs. In the context of a group called the Women's Work and the Capitalist State Collective, we struggled to teach, write and analyze in new ways – to develop modes of working that might join, but not falsely unify all of our divergent, yet linked, experiences as women.[5]

Through the early 1980s I and my white feminist sisters groped for a language with which to talk about racism and feminism, racism and ourselves. By the mid-1980s as white feminists continued acting out elements of racism and ethnocentrism, I felt the need for a more systematic analysis of the situation. I wanted to ask *why* white feminist thought and practice replicated the racism of the dominant culture, about the social processes through which white women take our places in a racially hierarchical society like the US, and what we might do to challenge racism from within those places. My presumption here was one I'd held to since I first came to political activism in the early 1970s – that knowledge about a situation is a critical tool in dismantling it.

Between 1984 and 1986, I undertook in-depth life history interviews with thirty white women, ranging in age from twenty-one to ninety, diverse in class, family and household situation, sexuality, political orientation and geographical region of origin, but all living in northern California. While interviewing women and analyzing their narratives, I looked not only for the stories white women told, but also thought critically about the languages available to us in the 1980s and 1990s, for talking about race, culture, selfhood, otherness, whiteness. I wanted to think about what those languages made visible and what they concealed, as well as the historical moments in which they had come into being (I have written about this work in Frankenberg, 1993).

Racial identity is complex and in no way reducible to biological terms. In fact, as the history of the USA amply shows, race, race difference and racial identity are politically determined categories, intimately tied to racial inequality and racism, and constantly transformed through political struggles. Thus the names that groups of people give themselves, and the names ruling groups give to others, change over time.[6] My study worked with an avowedly political, provisional, historically and geographically specific understanding of race difference. As such, I viewed groups in US society who are targets of racism – including Native, Latino, African and Asian Americans and other immigrants of colour – as racially different from white people and from each other.

As is the case for people of colour in the US (itself a chosen name with the specific purpose of coalition-building in the United States), 'white' is neither a fixed nor a homogeneous label. My study set out to explore some of its meanings for white women and to understand the political and social shifts which have given whiteness its present shape. I came to view whiteness as having at least three dimensions. First, it is a position of structural advantage, associated with 'privileges' of the most basic kind, including, for example, higher wages, reduced chances of being impoverished, longer life, better access to health care, better treatment by the legal system, and so on. (Of course, access is influenced by class, sexuality, gender, age, and in fact 'privilege' is a misnomer here since this list addresses basic social rights.) Second, whiteness is a 'standpoint' or place from which to look at oneself, others and society. Thirdly, it carries with it a set of ways of being in the world, a set of cultural practices, often not named as 'white' by white folks, but looked on instead as 'American' or 'normal'.

My analysis of white women's childhoods is organized around what I call 'social geographies of race', exploring the ways racism as a system helps shape our daily environments, trying to identify the historical, social and political processes that brought these environments into being. *Geography* refers to the physical landscape – the home, the street, the neighbourhood, school, parts of town visited or driven through on rare or regular occasions, places visited on vacation. My interest is in how physical space was divided, who inhabited it, and of course for my purposes that 'who' is a racially or ethnically identified being. *Social* geography suggests that the physical landscape is peopled – by whom? How did the women I interviewed conceptualize and relate to the people around them? What were they encouraged, forced, or taught by example to do with the variously racially identified people in their environments? And how is the white sense of self constructed with reference to notions of race or ethnicity? *Racial* social geography, then, refers to the racial and ethnic mapping of a landscape in physical terms, and enables also a beginning sense of the conceptual mapping of self and other with respect to race operating in white women's lives.

Ultimately, the concept of social geography came to represent for me a complex mix of material and conceptual ingredients for I saw increasingly that, as much as white women are located in racially marked *physical* environments, we also inhabit 'conceptual environments' or environments of ideas, which frame and limit what we see, what we remember and how we interpret the physical world. They tell us, for example, what race is, what culture is, and even what racism means. Just as material environments have histories which are political and the product of social change or of oppression, so too do the conceptual frameworks through which we view them. For example, women I interviewed at times saw people of colour through the filter of racist ideas generated in the context of West European colonial expansion, or were raised to live by the rules of segregation associated with an explicit white supremacism. Other white women did not 'see' racism, and failed to recognize cultural difference, in part because of the 'melting pot' and 'colourblind society' myths which have dominated thinking on race in the US for much of the post-war era. Or more positively, some women had begun valuing cultural diversity and recognizing structural racism, here drawing on the anti-imperialist, antiracist and feminist struggles of recent decades. All of these ways of being and seeing with respect to race were, in short, the product of historical and political process – not surprisingly, for racism is as unnatural as the concept of race itself.

The landscapes of childhood are important because, from the standpoint of children they are received rather than chosen (although of course from an adult standpoint they *are* chosen and crafted in complex, conscious and unconscious ways). None of the women I interviewed was passive in relation to her childhood environment. However, beyond a point, children do not define the terms in which the world greets them; they can only respond. And while throughout their lives people can and do make profound changes in the ways they see themselves and the world, it seems to me that the landscapes of childhood are crucially important in creating the backdrop against which later transformations must take place. Looking at the social geography of race in white women's childhoods may then provide information and tools useful to us in the project of comprehending and changing our places in the relations of racism.

As I compared narratives with one another, common threads and lines of differentiation emerged. Narratives typified certain experiences, separable into four modes of social geography of race. Of these, one seemed at first to be characterized by an absence of people of colour from the narrator's life, but turns out to be only *apparently* all-white. Second, there were contexts organized in terms of *explicit* race conflict, hierarchy and boundary-marking. Third, there were contexts in which race difference was present, but unremarked, that is, in which race difference organized white women's perceptions, feelings and behaviour, but was not, for the most part, consciously perceived. Finally, white women described experiences I have interpreted as 'quasi-integrated', that is, as integrated, but not fully so.

White women's lives as sites for the reproduction of racism – and for challenges to it

In all five narratives landscape and the experience of it were racially structured. This was true whether those narratives were marked predominantly by the seeming presence or absence of people of colour. This is of course not to say that race was the only principle by which the social context was organized. For example, class intersects with race. Again, controls on sexuality link up with racism in creating frequent hostility towards relationships between African-American men and white women.

In addition, once in a landscape structured by racism, a conceptual mapping of race, of self and others, takes shape, which follows from and feeds the physical context. Even the presence and absence of people of colour seem to be as much social-mental as they are social-physical constructs. Here one can cite the positioning and invisibility of African-American and Latina domestic workers in some apparently all-white homes.

These narratives have some implications for a white feminist analysis of racism. To begin with, they clarify some of the forms, obvious and subtle, that racism and race privilege may take in the lives of white women: including educational and economic privilege, verbally expressed assertions of white superiority, the maintenance of all-white neighbourhoods, the 'invisibility' of Black and Latina domestic workers, white people's fear of people of colour and the 'colonial' notion that the cultures of peoples of colour were great only in the past. Racism thus appears not only as an ideology or political orientation chosen or rejected at will; it is also a system and set of ideas embedded in social relations.

My analysis underscores the idea that there is no place for us to stand 'outside' racism, any more than we can stand 'outside' sexism. In this context, it seems foolish to imagine that as individuals we can escape complicity with racism as a social system. We cannot, for example, simply 'give up' race privilege. I suggest that as white feminists we need to take cognizance both of the embeddedness of racism in all aspects of society and the ways this has shaped our own lives, theories and actions. Concretely, this means work in at least three linked areas: work on re-examining personal history and changing consciousness; thorough-going theoretical transformation within feminism; and participation in practical political work towards structural change.

Re-examining personal history is necessary in part because it is possible that white feminists continue relating to people of colour, as well as doing feminist work, on the basis of patterns and assumptions learned early on. For example, there could be a connexion between white women's 'not noticing' people of colour in their childhood environments, and white feminists' capacity to continue 'forgetting' to include women of colour in the planning of conferences and events. This forgetting may, in other words, be a socially constructed one.

Reconceptualization of past experience in fact marked each of the narratives discussed here. Although this was not the case for all the women I interviewed, this group indicated as they told their stories that, with hindsight, they had become more aware of how race differences and racism had been a feature of their childhoods. Phrases like, 'Now that we're talking about this I remember . . .' and 'I was so unaware of cultural differences that . . .' signal both lack of awareness of racism *and* moments of recognition or realization of it.

In addition, each woman's experience is complex and contradictory. Thus the two women most explicitly raised to 'be racist', Beth and Pat, found contexts and moments, however fleeting, to question the racist status quo. Conversely, Sandy and Louise, raised to find spaces *not* to be racist, were none the less in no sense outside the reach of racism: racism as well as antiracism shaped their environments, and both women at times drew on white-centred logics in describing and living their lives.

It is from these places of contradiction that the work of revisioning begins. For white women in 'mostly racist' contexts, the moments of questioning are perhaps moments when the door opens on to other realms of possibility, other ways of being. Those moments should give us hope. For white women who grew up in situations of 'quasi-integration', the racism that still pervades reminds us that this is not a simple struggle, that it is all too easy for us, as white women, to be ethnocentric or patronizing, at the same time as being consciously and purposively antiracist.

For these five white women (as for most white feminists) access to information about the impact of racism on people of colour, and/or direct or indirect access to the intellectual and political work of people of colour, seem to be crucial to the process of rethinking racism. Thus for example, Louise's awareness of racism first came about as she saw her friends of colour 'tracked' into vocational and remedial classes in high school. And, as an adult, Beth Ellison described how reading African-American women authors such as Toni Morrison and Alice Walker, as well as meeting African Americans, had pushed her to question many of the assumptions about racism and about African Americans with which she had been raised. Pat Bowen described participating in discussions about racism in university Women's Studies classes. And although her cohort was mainly white, the impetus for these discussions had come from the challenges posed by activist women of colour. Pat Bowen's experience raises a further point: the call to accountability raised by women of colour must be met in part by white women learning from one another, teaching each other, and thinking together, for example, about race privilege and its effects on feminism, rather than expecting women of colour to do all of this work for us. This kind of work is going on in the published writings of white antiracist women (Bulkin, Pratt and Smith, 1988; Segrest, 1985 among others) as well as through (usually multiracial) 'unlearning racism' workshops and racism consciousness-raising groups (sometimes multiracial, sometimes not) developed in feminist communities around the country. None the less, the painful truth is that white feminists continue to 'forget', to 'not think', and this means that the bulk of antiracist work is being done by people of colour.

There is another link between the reconceptualization of experience and the making of feminism, given that white feminists have often relied on notions of 'women's experience' in order to develop theory and strategy for feminism. As mentioned earlier, the experience referred to by white feminists has almost always been white women's experience, over generalized. Rarely and only recently have white feminists begun to examine the intersection of their gender and class positions with race privilege. Much white feminist theory generated 'from experience' has thus been flawed on two counts, very often assessing *neither* differences between white women and women of colour *nor* adequately describing the race-privileged positions of white women ourselves. (The same points may in fact be made in relation to other groups marginalized within feminism, such as women with disabilities.)

The relationship between experience and the process of interpreting or describing it is by no means simple: as the narratives showed, there were multiple ways in which women named, forgot, remembered and reinterpreted their experience, through the lenses of racism and antiracism. Accounts of experience are partial, and we must review them as always being open to change. It is also critical, as white women examine and re-examine our complicity with racism, that we go beyond our immediate daily environments to learn more about the history of racist ideas. We need to do this in order to understand the contexts for the production of our 'racist lenses', including, for example, the ways white women 'fear' people of colour, or the ways we view whiteness as 'neutral'. Reaching cognitive understandings of the history of white racist consciousness may be a valuable step towards loosening its grip on our daily lives and practice.

As we formulate antiracist practices, focusing on issues of consciousness is, while necessary, not sufficient. As should be clear from what has been said here, challenging racism is not a project that can take place only on the level of ideas, but one which calls for major changes in the social, economic and legislative orders. These hold in place the unequal life chances of white people and people of colour, and indeed create what I have called the 'social geographies of race' in white women's experience. By itself, reformulating ideas will change none of this. For example, I work in the arena of higher education. I am active in developing multiracial Women's Studies curricula. However, it is clear to me that we cannot progress very far towards multiracial curricula unless we make strides towards more multiracial composition of faculty and student bodies. This in turn implies that we must work to strengthen affirmative action, challenge cutbacks in student financial aid, demand day-care facilities on campus for women with children, improve funding for public high schools, and so on.

In short, there is a delicate balance to be maintained in white feminist practice, for it is precisely racist ideas and lack of awareness that have often prevented white women from challenging racism structurally. Unlearning racism, however, is not the same thing as ending it. Nor can we wait for a moment when we feel we have finished changing our 'race consciousness' before becoming active in working against racism in the world at large. Examining whiteness is as urgent now as it ever was, because of the persistence of systemic racism. In the last decade, at the urging of women of colour, white feminists like myself have learnt a lot about the meaning of race privilege. I believe that we can turn the corner into the next century knowing more about what 'being white' means than we did two decades ago. Moreover, we will know it from a standpoint that is specifically antiracist, one that will at the very least challenge the apparent invisibility or neutrality of whiteness,[7] and at best will also see whiteness as a place from which to participate actively in struggles for racial equality. The story of what we will, in practical terms, do with that consciousness in our politics and daily lives continues to unfold, as we transform feminist demands and agendas, and build feminist organizations and institutions. For as always in feminist thinking, new ideas are not merely ends in themselves, but tools to assist in the larger project of social change.

Notes

Ruth Frankenberg grew up in Manchester, England and moved to the United States in 1979. She now teaches in the American Studies Programme at the University of California, Davis. Her book, *White Women, Race Matters: The Social Construction of Whiteness*, was published in 1993 by University of Minnesota Press and Routledge, UK. She has been involved in pedagogical and activist antiracist work since the seventies.

1 These are the words of Cathy Thomas, one of the white women I interviewed in the research described here. Cathy Thomas is a pseudonym – unfortunately, given the confidentiality of the research, I cannot give Cathy credit by name for this acute observation about the meaning of whiteness for many white people socialized in contexts where the ideology of a 'colourblind' society overlays systemic racial inequality.

2 Here and throughout this paper, I assume neither that white women occupy equal positions of advantage in US society, nor that whites' cultures are equally powerful, equally formative of dominant cultural practices.

3 Racism and ethnocentrism are key terms in this paper, so that their meanings unfold throughout the piece. However, a preliminary note on how I would distinguish between these two terms may be in order. I use racism very broadly to denote the structures, institutions, practices and patterns of thought implicit in a system of domination, of unequal relations of power, constructed around the notion of 'race'. I do not view racism as an unchanging, timeless or inevitable system, but rather as one that is changing and historically specific, reproduced in conjunction with other social relations. In parallel, I view the term 'race' itself as referring to a socially constructed and historically specific categorization system, rather than to a set of essentially real, unchanging differences. I use ethnocentrism more narrowly than racism, to refer to the holding of presumptions about the universality, normalcy, superiority or 'generic' status of attitudes, practices and forms of social organization that are in reality culturally and historically specific. Ethnocentrism can then

refer simply to attitudes, or to patterns of thought embedded in particular instances of theorization and analysis, as well as in actions, practices and institutions.

4 I am using 'Black' here in the 'British' sense, to refer to people of Asian and African-Caribbean descent. However, it should be noted that elsewhere in this paper, when I and the women I interviewed spoke of 'Black people', we used the term in the 'American' sense, to refer to Americans of African descent.

5 None of that writing was ever published, although we did present our perspective in workshops, both at the University of California, Santa Cruz, and at the University of Oregon, Eugene, in the early 1980s. Our collectively taught class, 'Women and Work', University of California, Santa Cruz, January to March 1983, also exemplified our method. Finally, I have written about the collective and its method in, 'Different perspectives: interweaving theory and practice in women's work', History of Consciousness, University of California at Santa Cruz, June 1988.

6 For example, Gould (1985) describes the history of the reduction of race to biological differences; Omi and Winant (1986: 3–4 and elsewhere) indicate the historical mutability of racial naming and the strong ties between naming and political struggle; Baldwin (1984) analyzes the consolidation of white racial identity in the United States, again in context of political struggle, and the consolidation of power and privilege.

7 Conscious articulations of whiteness are, however, not necessarily anti-racist. For most of US history, use of the term 'white' has been deployed in the context of biology-based racist discourses and hierarchical constructions of 'difference'. Moreover, in the present, the US white supremacist movement continues to use the term 'white' to articulate its sense of white superiority. A good source of information about white supremacist activity is the Center for Democratic Renewal's newsletter, The Monitor, PO Box 50469, Atlanta, GA 30302.

References

Alarcon, Norma (1990) 'The theoretical subjects of *This Bridge Called My Back* and Anglo-American feminism' in Anzaldúa (1990) 356–69.

Baldwin, James (1984) 'On being white and other lies' *Essence* April, 80–1.

Bhavnani, Kum-Kum and Coulson, Margaret (1986) 'Transforming socialist feminism: the challenge of racism' *Feminist Review* 23.

Bulkin, Elly, Pratt, Minnie Bruce and Smith, Barbara (1988) *Yours in Struggle: Three Feminist Perspectives on Racism and Antisemitism* Ithaca, New York: Firebrand Books (original publisher, Long Haul Press, 1984).

Cade, Toni (1970) *The Black Woman: An Anthology* New York: Mentor.

Carby, Hazel (1981) 'White woman listen! Black feminism and the boundaries of sisterhood' in Center for contemporary cultural studies (1981) *The Empire Strikes Back: Race and Racism in '70s Britain* London: Hutchinson, 212–35.

Combahee River Collective (1979) 'A black feminist statement' in Eisenstein, Zillah R. (1979) editor, *Capitalist Patriarchy and the Case for Socialist Feminism* New York and London: Monthly Review Press.

Davis, Angela Y. (1981) *Women, Race and Class* New York: Random House.

Davis, Susan E. (1988) editor, *Women Under Attack: Victories, Backlash and the Fight for Reproductive Freedom* Boston: South End Press.

Frankenberg, Ruth (1985) 'White Privilege', unpublished.

—— (1993) *White Women, Race Matters: the Social Construction of Whiteness* Minneapolis: University of Minnesota Press. London: Routledge.

Garcia, Alma (1990) 'The development of Chicana feminist discourse, 1970–1980' in

DuBois, Ellen Carol and Ruiz, Vicki L. (1990) editors, *Unequal Sisters: A Multicultural Reader in US Women's History* New York: Routledge, 418–31.
Haraway, Donna J. (1991) 'Situated knowledges: the science question and the privilege of partial perspective' *Simians, Cyborgs and Women: The Reinvention of Nature* New York: Routledge, 183–202.
hooks, bell (1981) *Ain't I a Woman? Black Women and Feminism* Boston: South End Press.
—— (1984) *Feminist Theory: From Margin to Center* Boston: South End Press.
Moraga, Cherrie and Anzaldúa, Gloria (1981) editors, *This Bridge Called My Back: Writings by Radical Women of Color*, Watertown, MA, Persephone Press (reprinted by Kitchen Table/Women of Color Press, 1983).
Omi, Michael and Winnant, Howard (1986) *Racial Formation in the United States: From the 1960s to the 1980s* New York and London: Routledge & Kegan Paul, 3–4.
Reagon, Bernice Johnson (1983) 'Coalition politics: turning the century' in Barbara Smith (1983) editor, *Home Girls: a Black Feminist Anthology* New York: Kitchen Table/Women of Color Press, 356–69.
Sandoval, Chela (1982) 'The struggle within: women respond to racism – report on the National Women's Studies Conference, Storrs, Connecticut' Oakland, California: Occasional Paper, Center for Third World Organizing (revised version in Anzaldúa (1990) 55–71.
—— (1991) 'US Third World feminism: the theory and method of oppositional consciousness in the postmodern world' *Genders* 10, 1–24.
Segrest, Mab (1985) *My Mama's Dead Squirrel: Lesbian Essays on Southern Culture* Ithaca NY: Firebrand Books.
Zavella, Patricia (1987) 'The problematic relationship of feminism and Chicana studies' paper delivered to conference on 'Women: Culture, Conflict and Consensus' University of California, Los Angeles.
Zinn, Maxine Baca *et al.* (1986) 'The cost of exclusionary practices in Women's Studies' *Signs* Winter.

16 Iris Marion Young

'The Scaling of Bodies and the Politics of Identity'

Excerpt from: *Justice and the Politics of Difference*, pp. 122–37. Princeton, NJ: Princeton University Press (1990)

Racism and homophobia are real conditions of all our lives in this place and time. I urge each one of us here to reach down into that deep place of knowledge inside herself and touch that terror and loathing of any difference that lives there. See whose face it wears. Then the personal as the political can begin to illuminate all our choices.

Audre Lorde

My body was given back to me sprawled out, distorted, recolored, clad in mourning in that white winter day. The Negro is ugly, the Negro is animal, the Negro is bad,

> the Negro is mean, the Negro is ugly; look, a nigger, it's cold, the nigger is shivering, because he is cold, the little boy is trembling because he is afriad of the nigger, the nigger is shivering with cold, that cold goes through your bones, the handsome little boy is trembling because he thinks that the nigger is quivering with rage, the little white boy throws himself into his mother's arms; Momma, the nigger's going to eat me up.
>
> All round me the white man, above the sky tears at its navel, the earth rasps under my feet, and there is a white song, a white song. All this whiteness that burns me. . . .
>
> I sit down at the fire and I become aware of my uniform. I had not seen it. It is indeed ugly. I stop there, for who can tell me what beauty is?
>
> (Fanon, 1967, p. 114)

Racism, as well as other group oppressions, should be thought of not as a single structure, but in terms of several forms of oppression that in the United States condition the lives of most or all Blacks, Latinos, Asians, American Indians, and Semitic peoples. The oppressions experienced by many members of these groups are certainly conditioned by the specific structures and imperatives of American capitalism – structures of exploitation, segregated division of labor, and marginalization. Racism, like sexism, is a convenient means of dividing workers from one another and legitimating the superexploitation and marginalization of some. Clearly experiences like that evoked by Fanon above, however, cannot be reduced to capitalist processes or encompassed within the structures of oppression just mentioned. They belong instead to the general forms of oppression I have called cultural imperialism and violence. Cultural imperialism consists in a group's being invisible at the same time that it is marked out and stereotyped. Culturally imperialist groups project their own values, experience, and perspective as normative and universal. Victims of cultural imperialism are thereby rendered invisible as subjects, as persons with their own perspective and group-specific experience and interests. At the same time they are marked out, frozen into a being marked as Other, deviant in relation to the dominant norm. The dominant groups need not notice their own group being at all; they occupy an unmarked, neutral, apparently universal position. But victims of cultural imperialism cannot forget their group identity because the behavior and reactions of others call them back to it.

The Fanon passage evokes a particular and crucially important aspect of the oppression of cultural imperialism: the group-connected experience of being regarded by others with aversion. In principle, cultural imperialism need not be structured by the interactive dynamics of aversion, but at least in supposedly liberal and tolerant contemporary societies, such reactions of aversion deeply structure the oppression of all culturally imperialized groups. Much of the oppressive experience of cultural imperialism occurs in mundane contexts of interaction – in the gestures, speech, tone of voice, movement, and reactions of others (Brittan and Maynard, 1984, pp. 6–13). Pulses of attraction and aversion modulate all interactions, with specific consequences for experience of the body. When the dominant culture defines some groups as different, as the Other, the members of those groups are imprisoned in their bodies.

Dominant discourse defines them in terms of bodily characteristics, and constructs those bodies as ugly, dirty, defiled, impure, contaminated, or sick. Those who experience such an epidermalizing of their world (Slaughter, 1982), moreover, discover their status by means of the embodied behavior of others: in their gestures, a certain nervousness that they exhibit, their avoidance of eye contact, the distance they keep.

The experience of racial oppression entails in part existing as a group defined as having ugly bodies, and being feared, avoided, or hated on that account. Racialized groups, moreover, are by no means the only ones defined as ugly or fearful bodies. Women's oppression is also clearly structured by the interactive dynamics of desire, the pulses of attraction and aversion, and people's experience of bodies and embodiment. While a certain cultural space is reserved for revering feminine beauty and desirability, in part that very cameo ideal renders most women drab, ugly, loathsome, or fearful bodies. Old people, gay men and lesbians, disabled people and fat people also occupy as groups the position of ugly, fearful, or loathsome bodies. The interactive dynamics and cultural stereotypes that define groups as the ugly other have much to do with the oppressive harassment and physical violence that endangers the peace and bodies of most members of most of these groups.

This chapter explores the construction of ugly bodies and the implications of unconscious fears and aversions for the oppression of despised groups. Racist and sexist exclusions from the public have a source in the structure of modern reason and its self-made opposition to desire, body, and affectivity. Modern philosophy and science established unifying, controlling reason in opposition to and mastery over the body, and then identified some groups with reason and others with the body.

The objectification and overt domination of despised bodies that obtained in the nineteenth century, however, has receded in our own time, and a discursive commitment to equality for all has emerged. Racism, sexism, homophobia, ageism, and ableism, I argue, have not disappeared with that commitment, but have gone underground, dwelling in everyday habits and cultural meanings of which people are for the most part unaware. Through Kristeva's category of the abject, I explore how the habitual and unconscious fears and aversions that continue to define some groups as despised and ugly bodies modulate with anxieties over loss of identity. Our society enacts the oppression of cultural imperialism to a large degree through feelings and reactions, and in that respect oppression is beyond the reach of law and policy to remedy.

The analysis in this chapter raises questions for moral theory, about whether and how moral judgments can be made about unintended behavior. If unconscious behavior and practices reproduce oppression, they must be morally condemnable. I argue that moral theory must in such cases distinguish between blaming and holding responsible the perpetrators.

The dissolution of cultural imperialism thus requires a cultural revolution which also entails a revolution in subjectivity. Rather than seeking a wholeness of the self, we who are the subjects of this plural and complex society should affirm the otherness within ourselves, acknowledging that as subjects we are heterogeneous and multiple in our affiliations and desires. Social movement practices of consciousness raising, I note, offer beginning models of methods of revolutionizing the subject.

The scaling of bodies in modern discourse

Modern racism, sexism, homophobia, ageism, and ableism are not superstitious carry-overs from the Dark Ages that clash with Enlightenment reason. On the contrary, modern scientific and philosophical discourse explicitly propound and legitimate formal theories of race, sex, age, and national superiority. Nineteenth- and early twentieth-century scientific, aesthetic, and moral culture explicitly constructed some groups as ugly or degenerate bodies, in contrast to the purity and respectability of neutral, rational subjects.

Critical theoretical accounts of instrumental reason, postmodernist critiques of humanism and of the Cartesian subject, and feminist critiques of the disembodied coldness of modern reason all converge on a similar project of puncturing the authority of modern scientific reason. Modern science and philosophy construct a specific account of the subject as knower, as a self-present origin standing outside of and opposed to objects of knowledge – autonomous, neutral, abstract, and purified of particularity. They construct this modern subjectivity by fleeing from material reality, from the body's sensuous continuity with flowing, living things, to create a purified abstract idea of formal reason, disembodied and transcendent. With all its animation removed and placed in that abstract transcendent subject, nature is frozen into discrete, inert, solid objects, each identifiable as one and the same thing, which can be counted, measured, possessed, accumulated, and traded (Merchant, 1978; Kovel, 1970, chap. 5; Irigaray, 1985, pp. 26–28, 41).

An important element of the discourse of modern reason is the revival of visual metaphors to describe knowledge. Rational thought is defined as infallible vision; only what is seen clearly is real, and to see it clearly makes it real. One sees not with the fallible senses, but with the mind's eye, a vision standing outside all, surveying like a proud and watchful lord. This subject seeks to know a Truth as pure signifier that completely and accurately mirrors reality. The knowing subject is a gazer, an observer who stands above, outside of, the object of knowledge. In the visual metaphor the subject stands in the immediate presence of reality without any involvement with it. The sense of touch, by comparison, involves the perceiver with the perceived; one cannot touch something without being touched. Sight, however, is distanced, and conceived as unidirectional; the gazer is pure originating focusing agency and the object is a passive being-seen (Irigaray, 1985, pp. 133–51).

The gaze of modern scientific reason, moreover, is a normalizing gaze (Foucault, 1977; West, 1982). It is a gaze that assesses its object according to some hierarchical standard. The rational subject does not merely observe, passing from one sight to another like a tourist. In accordance with the logic of identity the scientific subject measures objects according to scales that reduce the plurality of attributes to unity. Forced to line up on calibrations that measure degrees of some general attribute, some of the particulars are devalued, defined as deviant in relation to the norm.

Foucault summarizes five operations that this normalizing gaze brings into play: comparison, differentiation, hierarchization, homogenization, and exclusion. Normalizing reason

> refers individual actions to a whole that is at once a field of comparison, a space of differentiation and the principle of a rule to be followed. It differentiates individuals from one another, in terms of the following overall rule: that the rule be made to function as a minimal threshold, as an average to be respected or an optimum towards which one must move. It measures in quantitative terms and hierarchizes in terms of value the abilities, the level, the 'nature' of individuals. It introduces, through this 'value-giving' measure, the constraint of conformity that must be achieved. Lastly, it traces the limit that will define difference in relation to all other differences, the external frontier of the abnormal.
>
> (Foucault, 1977, pp. 182–83)

Much recent scholarship has revealed the white, bourgeois, male, European biases that have attached to the expression of the idea of the rational subject in modern discourse. In thinly veiled metaphors of rape, the founders of modern science construct nature as the female mastered and controlled by the (masculine) investigator. The virtues of the scientist become also the virtues of masculinity – disembodied detachment, careful measurement and the manipulation of instruments, comprehensive generalizing and reasoning, authoritative speech backed by evidence (Keller, 1985; Merchant, 1978).

The attributes of the knowing subject and normative gazer become attached just as closely to class and race. Class position arises not from tradition or family, but from superior intelligence, knowledge, and rationality. Reason itself shifts in meaning. Its mission is no longer, as it was with the ancients, to contemplate the eternity of the heavens and the subtlety of the soul, but rather to figure out the workings of nature in order to direct its processes to productive ends. 'Intelligence' and 'rationality' now mean primarily the activity of strategic and calculative thinking, abstraction from particulars to formulate general laws of operation, logical organization of systems, the development and mastery of formalized and technical language, and the designing of systems of surveillance and supervision. Nature and the body are objects of such manipulation and observation. Moreover, this reason/body dichotomy also structures the modern division between 'mental' and 'material' labor. From the dawn of modern instrumental reason the idea of whiteness has been associated with reason purified of any material body, while body has been identified with blackness (Kovel, 1970, chaps. 5–7). This identification enables people who claim whiteness for themselves to put themselves in the position of the subject, and to identify people of color with the object of knowledge (Said, 1978, pp. 31–49).

It is important not to construe accounts such as these as claiming either that class, race, gender, and other oppressions are grounded in or caused by scientific reason or that scientific reason simply reflects the social relations of domination. Scientific and philosophic reason express a view of subjectivity and objectivity that has come to have enormous influence and repercussions in modern Western culture. The association of this reason with a white, male bourgeoisie arises and persists in the context of a society structured by hierarchical relations of class, race, gender, and nationality which have an independent dynamic.

Without doubt an association of abstract reason with masculinity and whiteness did emerge, but quite possibly it came about through a set of fateful historical accidents. Those articulating and following the codes of modern reason were white bourgeois men. In articulating their visual metaphors of reason they spoke for themselves, unmindful that there might be other positions to articulate. As this modern detached and objectifying reason assumed the meaning of humanity and subjectivity, and acquired the authoritative position of truth-seeing, privileged groups assumed the privilege of that authoritative subject of knowledge. Groups they defined as different thereby slid into the position of objects correlated with the subject's distancing and mastering gaze.

The imposition of scientific reason's dichotomy between subject and object on hierarchical relations of race, gender, class, and nationality, however, has deep and abiding consequences for the structuring of privilege and oppression. The privileged groups lose their particularity; in assuming the position of the scientific subject they become disembodied, transcending particularity and materiality, agents of a universal view from nowhere. The oppressed groups, on the other hand, are locked in their objectified bodies, blind, dumb, and passive. The normalizing gaze of science focused on the objectified bodies of women, Blacks, Jews, homosexuals, old people, the mad and feebleminded. From its observations emerged theories of sexual, racial, age, and mental or moral superiority. These are by no means the first discourses to legitimate the rule of the rich, or men, or Europeans. As Foucault argues, however, late eighteenth- and early nineteenth-century discourse instituted an epistemological break that found a theoretical expression in the 'sciences of man' (Foucault, 1970). In this *episteme* bodies are both naturalized, that is, conceived as subject to deterministic scientific laws, and normalized, that is, subject to evaluation in relation to a teleological hierarchy of the good. The naturalizing theories were biological or physiological, and explicitly associated with aesthetic standards of beautiful bodies and moral standards of upstanding character.

In the developing sciences of natural history, phrenology, physiognomy, ethnography, and medicine, the gaze of the scientific observer was applied to bodies, weighing, measuring, and classifying them according to a normative hierarchy. Nineteenth-century theorists of race explicitly assumed white European body types and facial features as the norm, the perfection of human form, in relation to which other body types were either degenerate or less developed. Bringing these norms into the discourse of science, however, *naturalized* them, gave the assertions of superiority an additional authority as truths of nature. In nineteenth-century biological and medical schemes white male bourgeois European bodies are the 'best' body types by nature, and their natural superiority determines directly the intellectual, aesthetic, and moral superiority of persons in this group over all other types (West, 1982, chap. 2).

In the nineteenth century in Europe and the United States the normalizing gaze of science endowed the aesthetic scaling of bodies with the authoritativeness of object truth. All bodies can be located on a single scale whose apex is the strong and beautiful youth and whose nadir is the degenerate. The scale measured at least three crucial attributes: physical health, moral soundness, and mental balance. The degenerate is physically weak, frail, diseased. Or the degenerate is mentally imbalanced: raving, irrational or childlike in mental simplicity. But most important, moral impropriety is a sign of degeneracy, and a cause of physical or mental disease. Moral degeneracy usually means sexual indulgence or deviant sexual behavior, though it also refers to indulgence in other physical pleasures. Thus the homosexual and the prostitute are primary degenerates, whose sexual behavior produces physical and mental disease.

In scientific discourse about the normal and the deviant, the healthy and the degenerate, it was crucial that any form of degeneracy, whether physical, mental, or moral, make itself manifest in physical signs identifiable by the scientific gaze. Degeneracy was thought to appear on the surface of the body, whose beauty or ugliness was objectively measurable according to detailed characteristics of facial features, degree and kind of hair, skin color and complexion, shape of head, location of eyes, and structure of the genitals, buttocks, hips, chest, and breasts (Gilman, 1985, pp. 64–70, 156–58, 191–94). The prostitute, the homosexual, the criminal, are all easy to identify because of the physical symptoms of ugliness and degeneracy they exhibit.

The nineteenth-century ideal of beauty was primarily an ideal of manly virtue (Mosse, 1985, pp. 31, 76–80), of the strong, self-controlled rational man distanced from sexuality, emotion, and all else disorderly and disturbing. Even white bourgeois men are capable of disease and deviance, especially if they give themselves over to sexual impulse. Manly men must therefore vigilantly defend their health and beauty through discipline and chastity (Takaki, 1979, chap. 2). In much nineteenth-century scientific discourse, however, whole groups of people are essentially and irrevocably degenerate: Blacks, Jews, homosexuals, poor and working people, and women.

As a group women are physically delicate and weak due to the specific constitution of their bodies, the operation of their reproductive and sexual parts. Because of their ovaries and uterus women are subject to madness, irrationality, and childlike stupidity, and they have greater tendencies toward sexual licentiousness than men. Beauty in women, like beauty in men, is a disembodied, desexualized, unfleshy aesthetic: light-colored hair and skin, and slenderness. Women of a certain class who are maintained under the disciplined rule of respectable and rational men can be saved from the insanity, degeneracy, and vice to which all women are prone.

Women are essentially identified with sexuality, as are the other groups scientifically classified as essentially degenerate: Blacks, Jews, homosexuals, in some places and times working people, and 'criminal elements.' A striking aspect of nineteenth-century discourse and iconography is the interchangeability of these categories: Jews and homosexuals are called black and often depicted as black, and all degenerate males are said to be effeminate. Medical science occupies itself with classifying the bodily features of members of all these groups, dissecting their corpses, often with particular attention to their sexual parts. The sexualization of racism in particular associates both men and women of degenerate races with unbridled sexuality. But scientists show particular fascination with Black, Jewish, and Arab women (Gilman, 1985, chap. 3).

The medicalization of difference brings about a strange and fearful logic. On the one hand, the normal/abnormal distinction is a pure good/bad exclusive opposition. On the other hand, since these opposites are located on one and the same scale, it is easy to slide from one to the other, the border is permeable. The normal and the abnormal are distinct natures, men and women, white and black, but it is possible to get sick, to lose moral vigilance, and to degenerate. Nineteenth-century moral and medical texts are full of male fears of becoming effeminate (Mosse, 1985, chap. 2).

In this context a new discourse about aging develops. Only in the nineteenth century does there emerge a general cultural and medical association between old age and disease, degeneracy, and death. Traditional patriarchal society more often revered the old man, and sometimes even the old woman, as an emblem of strength, endurance, and wisdom. Now old age comes increasingly to be associated with frailty, incontinence, senility, and madness (Cole, 1986). While such associations do not originate in the nineteenth century (witness King Lear), once again the normalizing discourse of science and medicine endows such associations with the authority of objective truth. The degeneracy of age, like that of race, is supposedly apparent in the objective ugliness of old people, especially old women.

Modern scientific reason thus generated theories of human physical, moral, and aesthetic superiority, which presumed the young white bourgeois man as the norm. The unifying structure of that reason, which presumed a knowing subject purified of sensuous immersion in things, made possible the objectification of other groups, and their placement under a normalizing gaze.

Conscious acceptance, unconscious aversion

So far I have addressed the question of how some groups become ugly and fearful bodies by reviewing the construction of theories of racial, sexual, and mental superiority generated by nineteenth-century scientific reason. Many of the writers I have cited suggest that these nineteenth-century structures condition the ideology and psychology of group-based fears and prejudices in contemporary Western capitalist societies. Cornel West asserts, for example, that the racist consequences of Enlightenment conceptions of reason and science 'continue to haunt the modern West: on the non-discursive level, in ghetto streets, and on the discursive level, in methodological assumptions in the disciplines of the humanities' (West, 1982, p. 48).

But can we assume such an easy continuity between the racist, sexist, homophobic, and ageist ideologies of the past and the contemporary social situation of Europe and North America? Many would argue that conditions have so changed as to render these nineteenth- and early twentieth-century theories and ideologies mere historical curiosities, with no relationship to contemporary thoughts, feelings, and behavior. Rational discussion and social movements discredited these tracts of nineteenth-century scientific reason. After much bitter struggle and not a few setbacks, legal and social rules now express commitment to equality among groups, to the principle that all persons deserve equal respect and consideration, whatever their race, gender, religion, age, or ethnic identification.

Those of us who argue that racism, sexism, homophobia, ageism, and ableism are deep structures in contemporary social relations cannot dismiss as illusory the common conviction that ideologies of natural inferiority and group domination no longer exercise significant influence in our society. Nor can we plausibly regard the aversions and stereotypes we claim perpetuate oppression today as simple, though perhaps weakened, extensions of the grosser xenophobia of the past. Many people deny claims that ours in a racist, sexist, ageist, ableist, heterosexist society, precisely because they identify these 'isms' with scientifically legitimated theories of group inferiority and socially sanctioned exclusion, domination, and denigration. To be clear and persuasive in our claims about contemporoary group oppression and its reproduction, we must affirm that explicit and discursively focused racism and sexism have lost considerable legitimacy. We must identify a different social manifestation of these forms of group oppression corresponding to specific contemporary circumstances, new forms which have both continuities and discontinuities with past structures.

To formulate such an account of contemporary manifestations of group oppression I adopt the three-leveled theory of subjectivity that Anthony Giddens (1984) proposes for understanding social relations and their reproduction in action and social structures. Action and interaction, says Giddens, involve discursive consciousness, practical consciousness, and a basic security system. Discursive consciousness refers to those aspects of action and situation which are either verbalized, founded on explicit verbal formula, or easily verbalizable. Practical consciousness, on the other hand, refers to those aspects of action and situation which involve often complex reflexive monitoring of the relation of the subject's body to those of other subjects and the surrounding environment, but which are on the fringe of consciousness, rather than the focus of discursive attention (Bourdieu, 1977). Practical consciousness is the habitual, routinized background awareness that enables persons to accomplish focused, immediately purposive action. For example, the action of driving to the grocery store and buying goods on my shopping list involves a highly complex set of actions at the level of practical consciousness, such as driving the car itself and maneuvering the cart in the grocery store, where I have acquired a habitual sense of the space in relation to the items I seek.

'Basic security system' for Giddens designates the basic level of identity security and sense of autonomy required for any coherent action in social contexts; one might call it the subject's ontological integrity. Psychotics are those for whom a basic security system has broken down or never been formed. Gidden's theory of structuration assumes that social structures exist only in their enactment through reflexively monitored action, the aggregate effects of that action, and the unintended consequences of action. Action, in its turn, involves the socially situated *body* in a dynamic of trust and anxiety in relation to its environment, and especially in relation to other actors:

> The prevalence of tact, trust or ontological security is achieved and sustained by a bewildering range of skills which agents deploy in the production and reproduction of interaction. Such skills are founded first and foremost in the normatively regulated control of what might seem . . . to be the tiniest, most insignificant details of bodily movement and expression.
>
> (Giddens, 1984, p. 79)

What psychoanalysis refers to as unconscious experience and motivation occurs at the level of this basic security system. In the personality development of each individual, some experiences are repressed in the process of constructing a basic sense of competence and autonomy. An independent unconscious 'language' results from the splitting of this experiential material off from the self's identity: it emerges in bodily behavior and reactions, including gestures, tone of voice, and even, as Freud found, certain forms of speech or symbolization themselves. In everyday action and interaction, the subject reacts, introjects, and reorients itself in order to maintain or reinstate its basic security system.

Racism, sexism, homophobia, ageism, and ableism, I suggest, have receded from the level that Giddens refers to as discursive consciousness. Most people in our society do not consciously believe that some groups are better than others and for this reason deserve different social benefits (see Hochschild, 1988, pp. 75–76). Public law in Western capitalist societies, as well as the explicit policies of corporations and other large institutions, has become committed to formal equality and equal opportunity for all groups. Explicit discrimination and exclusion are forbidden by the formal rules of our society for most groups in most situations.

Commitment to formal equality for all persons tends also to support a public etiquette that disapproves of speech and behavior calling attention in public settings to a person's sex, race, sexual orientation, class status, religion, and the like. In a fine restaurant waiters are supposed to be deferential to all patrons – whether Black or white, trucker or surgeon – as though they were aristocrats; on the supermarket line, on the other hand, nobody gets special privileges. Public etiquette demands that we relate to people as individuals only, according everyone the same respect and courtesies. Calling attention to a person's being Black, or Jewish, or Arab, or old, or handicapped, or rich, or poor, in public settings is in distinctly poor taste, as is behaving toward some people with obvious condescension, while deferring to others. Contemporary social etiquette remains more ambiguous about calling attention to a woman's femininity, but the women's movement has helped create social trends that make it poor taste to behave in deferential or patronizing ways to women as well. The ideal promoted by current social etiquette is that these group differences should not matter in our everyday encounters with one another, that especially in formal and impersonal dealings, but more generally in all non-familiar settings and situations, we should ignore facts of sex, race, ethnicity, class, physical ability, and age. These personal facts are supposed to make no difference to how we treat one another.

I should not exaggerate the degree to which beliefs about the inferiority, degeneracy, or malignancy of some groups have receded from consciousness. There continue to be individuals and groups who are committed sexists and racists, though in the dominant liberal context they must often be careful about how they make their claims if they wish to be heard. Theories of racial and sexual inferiority, moreover, continue to surface in our intellectual culture, as in Jensen's theory of IQ differential. They too, however, are on the defensive and generally fail to achieve wide acceptance. But although public etiquette may forbid discursively conscious racism and sexism, in the privacy of the living room or locker room people are often more frank about their prejudices and preferences. Self-conscious racism, sexism, homophobia, ageism, and ableism are fueled by unconscious meanings and reactions that take place at the levels Giddens calls practical consciousness and the basic security system. In a society committed to formal equality for all groups, these unconscious reactions are more widespread than discursive prejudice and devaluation, and do not need the latter to reproduce relations of privilege and oppression. Judgments of beauty or ugliness, attraction or aversion, cleverness or stupidity, competence or ineptness, and so on are made unconsciously in interactive contexts and in generalized media culture, and these judgments often mark, stereotype, devalue, or degrade some groups.

Group differences are not 'natural' facts. They are made and constantly remade in social interactions in which people identify themselves and one another. As long as group differences matter for the identification of self and others – as they certainly do in our society – it is impossible to ignore those differences in everyday encounters. In my interactions a person's sex, race, and age affect my behavior toward that person, and when a person's class status, occupation, sexual orientation, or other forms of social status are known or suspected, these also affect behavior. White people tend to be nervous around Black people, men nervous around women, especially in public settings. In social interaction the socially superior group often avoids being close to the lower-status group, avoids eye contact, does not keep the body open.

A Black man walks into a large room at a business convention and finds that the noise level reduces, not to a hush, but definitely reduces. A woman at a real estate office with her husband finds the dealer persistently failing to address her or to look at her, even when she speaks to him directly. A woman executive is annoyed that her male boss usually touches her when they talk, putting his hand on her elbow, his arm around her shoulder, in gestures of power and fatherliness. An eighty-year-old man whose hearing is as good as a twenty-year-old's finds that many people shout at him when they speak, using babylike short sentences they might also use to speak to a preschooler (Vesperi, 1985, pp. 50–59).

Members of oppressed groups frequently experience such avoidance, aversion, expressions of nervousness, condescension, and stereotyping. For them such behavior, indeed the whole encounter, often painfully fills their discursive consciousness. Such behavior throws them back onto their group identity, making them feel noticed, marked, or conversely invisible, not taken seriously, or worse, demeaned.

Those exhibiting such behavior, however, are rarely conscious of their actions or how they make the others feel. Many people are quite consciously committed to equality and respect for women, people of color, gays and lesbians, and disabled people, and nevertheless in their bodies and feeling have reactions of aversion or avoidance toward members of those groups. People suppress such reactions from their discursive consciousness for several reasons. First, as I will discuss in a later section, these encounters and the reactions they provoke threaten to some degree the structure of their basic security system. Second, our culture continues to separate reason from the body and affectivity, and therefore to ignore and devalue the significance of bodily reactions and feelings. Finally, the liberal imperative that differences should make no difference puts a sanction of silence on those things which at the level of practical consciousness people 'know' about the significance of group differences.

Groups oppressed by structures of cultural imperialism that mark them as the Others, as different, thus not only suffer the humiliation of aversive, avoiding, or condescending behavior, but must usually experience that behavior in silence, unable to check their perceptions against those of others. The dominant social etiquette often finds it indecorous and tactless to point out racial, sexual, age, or ableist difference in public and impersonal settings and encounters. The discomfort and anger of the oppressed at this behavior of others toward them must therefore remain unspoken if they expect to be included in those public contexts, and not disturb the routines by calling attention to forms of interaction. When the more bold of us do complain of these mundane signs of systematic oppression, we are accused of being picky, overreacting, making something out of nothing, or of completely misperceiving the situation. The courage to bring to discursive consciousness behavior and reactions occurring at the level of practical consciousness is met with denial and powerful gestures of silencing, which can make oppressed people feel slightly crazy.

Unconscious racism, sexism, homophobia, ageism, and ableism occur not only in bodily reactions and feelings and their expression in behaviour, but also in judgments about people or policies. When public morality is committed to principles of equal treatment and the equal worth of all persons, public morality requires that judgments about the superiority or inferiority of persons be made on an individual basis according to individual competences. However, fears, aversions, and devaluations of groups marked as different often unconsciously enter these judgments of competence. Through a phenomenon that Adrian Piper (1988) calls higher-order discrimination, people frequently disparage attributes that in another person would be considered praiseworthy, because they are attached to members of certain groups. Assertiveness and independent thinking may be regarded as signs of good character, of someone you would want on your team, but when found in a woman they can become stridency or inability to cooperate. A woman may value gentleness and softspokenness in a man, but find these attributes in a gay man signs of secretiveness and a lack of integrity. Aversion to or devaluation of certain groups is displaced onto a judgment of character or competence supposedly unconnected with group attributes. Because the judger recognizes and sincerley believes that people should not be devalued or avoided simply because of group membership, the judger denies that these judgments of competence have a racist, sexist, or homophobic basis.

Similar processes of displacement often occur in public policy judgments and the reasons given for them. Since law and policy are formally committed to equality, assertions of race or gender privilege come coded and under rubrics other than the assertion of racial or sexual superiority (Omi and Winant, 1983). Affirmative action discussion is an important locus of covert or unconscious racism and sexism. Charles Lawrence (1987) argues that unconscious racism underlies many public policy decisions where race is not explicitly at stake and the policymakers have no racist intentions. In the late 1970s, for example, the city of Memphis erected a wall between the white and Black sections of the city; city officials' motives were to preserve order and protect property. In many cities there are struggles over the location and character of public housing, in which the white participants do not discuss race and may not think in terms of race. Lawrence argues that in cases such as these unconscious racism has a powerful effect, and that one tests the presence of racism by looking at the cultural meaning of issues and decisions: walls mean separation, public housing means poor Black ghettos, in the cultural vocabulary of the society. The cultural meaning of AIDS in contemporary America associates it with gay men and gay life style, despite vigorous efforts on the part of many people to break this association; consequently much discussion about AIDS policy may involve homophobia, even when the discussants do not mention gay men.

Unconscious racism, sexism, homophobia, ageism, and ableism are often at work, I have suggested, in social interactions and policymaking. A final area where these aversions, fears, and devaluations are at work is the mass entertainment media – movies, television, magazines and their advertisements, and so forth. How is it possible, for example, for a society to proclaim in its formal rules and public institutions that women are as competent as men and should be considered on their merits for professional jobs, when that same society mass-produces and distributes slick magazines and movies depicting the abuse and degradation of women in images intended to be sexually arousing? There is no contradiction here if reality and reason are boarded off from fantasy and desire. The function of mass entertainment media in our society appears to be to express unbridled fantasy; so feelings, desires, fears, aversions, and attractions are expressed in the products of mass culture when they appear nowhere else. Racist, sexist, homophobic, ageist, and ableist stereotypes proliferate in these media, often in stark categories of the glamourously beautiful and the grotesquely ugly, the comforting good guy and the threatening evil one. If politicizing agents call attention to such stereotypes and devaluations as evidence of deep and harmful oppression of the groups stereotyped and degraded, they are often met with the response that they should not take these images seriously, because their viewers do not; these are only harmless fantasies, and everyone knows they have no relationship to reality. Once more reason is separated from the body and desire, and rational selves deny attachment to their bodies and desires.

References

Bourdieu, Pierre. 1977. *Outline of a Theory of Practice.* Cambridge: Cambridge University Press.

Brittan, Arthur and Mary Maynard. 1984. *Sexism, Racism and Oppression.* Oxford: Blackwell.

Cole, Thomas R. 1986. 'Putting Off the Old: Middle Class Morality, Antebellum Protestantism, and the Origins of Ageism.' In David Van Tassel and Peter N. Stearns, eds., *Old Age in a Bureaucratic Society*. New York: Greenwood.

Fanon, Frantz. 1967. *Black Skin, White Masks.* New York: Grove.

Foucault, Michel. 1977. *Discipline and Punish.* New York: Pantheon.

Giddens, Anthony. 1984. *The Constitution of Society.* Berkeley and Los Angeles: University of California Press.

Gilman, Sander L. 1985. *Difference and Pathology: Stereotypes of Sexuality, Race and Madness.* Ithaca: Cornell University Press.

Hochschild, Jennifer. 1988. 'Race, Class, Power, and Equal Opportunity.' In Norman Bowie, ed., *Equal Opportunity*. Boulder: Westview.

Irigaray, Luce. 1985. *Speculum of the Other Woman.* Ithaca: Cornell University Press.

Keller, Evelyn Fox. 1986. *Reflections on Gender and Science.* New Haven: Yale University Press.

Kovel, Joel. 1984. *White Racism: A Psychohistory.* 2d ed. New York: Columbia University Press.

Lawrence, Charles R. 1987. 'The Id, the Ego, and Equal Protection: Reckoning with Unconscious Racism.' *Stanford Law Review* 39 (January): 317–88.

Merchant, Carolyn. 1978. *The Death of Nature.* New York: Harper and Row.

Mosse, George. 1985. *Nationalism and Sexuality.* New York: Fertig.

Omi, Michael and Howard Winant. 1983. 'By the Rivers of Babylon: Race in the United States, Part I and II.' *Socialist Review* 71 (September–October): 31–66; 72 (November–December): 35–70.

Piper, Adrian. 1988. 'Higher-Order Discrimination.' Paper presented at the Conference on Moral Character, Radcliffe College, April.

Said, Edward. 1978. *Orientalism.* New York: Pantheon.

Slaughter, Thomas F. 1982. 'Epidermalizing the World: A Basic Mode of Being Black.' In Leonard Harris, ed., *Philosophy Born of Struggle*. Dubuque, Iowa: Hunt.

Takaki, Ronald. 1979. *Iron Cages: Race and Culture in Nineteenth Century America.* New York: Knopf.

Vesperi, Maria D. 1985. *City of Green Benches: Growing Old in a New Downtown.* Ithaca: Cornell University Press.

West, Cornel. 1982. *Prophesy Deliverance! An Afro-American Revolutionary Christianity.* Philadelphia: Westminster.

17 Susan Bordo

'Anglo-American Feminism, "Women's Liberation" and the Politics of the Body'

Excerpts from: *Unbearable Weight: Feminism, Western Culture and the Body*, pp. 15–32. London: University of California Press (1993)

It is no surprise that feminist theorists turned to Western representations of the body with an analytic, deconstructive eye. From their efforts we have learned to read all the various texts of Western culture – literary works, philosophical works, artworks, medical texts, film, fashion, soap operas – less naively and more completely, educated and attuned to the historically pervasive presence of gender-, class-, and race-coded dualities, alert to their continued embeddedness in the most mundane, seemingly innocent representations. Since these dualities (although not these alone) mediate a good deal of our cultural reality, few representations – from high religious art to depictions of life at the cellular level – can claim innocence.[1]

Feminists first began to develop a critique of the "politics of the body," however, not in terms of the body as represented (in medical, religious, and philosophical discourse, artworks, and other cultural "texts"), but in terms of the material body as a site of political struggle. When I use the term *material*, I do not mean it in the Aristotelian sense of *brute* matter, nor do I mean it in the sense of "natural" or "unmediated" (for our bodies are necessarily cultural forms; whatever roles anatomy and biology play they always interact with culture). I mean what Marx and, later, Foucault had in mind in focusing on the "direct grip" (as opposed to representational influence) that culture has on our bodies, through the practices and bodily habits of everyday life. Through routine, habitual activity, our bodies learn what is "inner" and what is "outer," which gestures are forbidden and which required, how violable or inviolable are the boundaries of our bodies, how much space around the body may be claimed, and so on. These are often far more powerful lessons than those we learn consciously, through explicit instruction concerning the appropriate behavior for our gender, race, and social class.

The role of American feminism in developing a "political" understanding of body practice is rarely acknowledged. In describing the historical emergence of such an understanding, Don Hanlon Johnson leaps straight from Marx to Foucault, effacing the intellectual role played by the social movements of the sixties (both black power and women's liberation) in awakening consciousness of the body as "an instrument of power".[2] . . . Not a few feminists, too, appear to accept this view of things. While honoring French feminists Irigaray, Wittig, Cixous, and Kristeva for their work on the body "as the site of the production of new modes of subjectivity" and Beauvoir for the "understanding of the body as a situation," Linda Zirelli credits Foucault with having "showed us how the body has been historically disciplined"; to Anglo-American feminism is simply attributed the "essentialist" view of the body as an "archaic natural."[3]

Almost everyone who does the "new scholarship" on the body claims Foucault as its founding father and guiding light. And certainly Foucault did articulate and delineate some of the central theoretical categories that influenced that scholarship as it developed in the late 1980s and early 1990s. "Docile bodies," "biopower," "micropractices" – these are useful concepts, and Foucault's analyses, which employ them in exploring historical changes in the organization and deployment of power, are brilliant.[4] But neither Foucault nor any other poststructuralist thinker discovered or invented the idea, to refer again to Johnson's account, that the "definition and shaping" of the body is "the focal point for struggles over the shape of power." *That* was discovered by feminism, and long before it entered into its marriage with poststructuralist thought.

"There is no private domain of a person's life that is not political and there is no political issue that is not ultimately personal. The old barriers have fallen." Charlotte Bunch made this statement in 1968, and although much has been written about "personal politics" in the emergence of the second wave of feminism, not enough attention has been paid, I would argue, to its significance as an *intellectual* paradigm, and in particular to the new understanding of the *body* that "personal politics" ushered in. What, after all, is more personal than the life of the body? And for women, associated with the body and largely confined to a life centered *on* the body (both the beautification of one's own body and the reproduction, care, and maintenance of the bodies of others), culture's grip on the body is a constant, intimate fact of everyday life. As early as 1792, Mary Wollstonecraft had provided a classic statement of this theme. As a privileged woman, she focuses on the social construction of femininity as delicacy and domesticity, and it is as clear an example of the production of a socially trained, "docile body" as Foucault ever articulated:

> To preserve personal beauty, woman's glory! the limbs and faculties are cramped with worse than Chinese bands, and the sedentary life which they are condemned to live, whilst boys frolic in the open air, weakens the muscles and relaxes the nerves. . . . Genteel women are, literally speaking, slaves to their bodies, and glory in their subjection, . . . women are everywhere in this deplorable state. . . . Taught from their infancy that beauty is woman's scepter, the mind shapes itself to the body, and, roaming round its gilt cage, only seeks to adorn its prison.[5]

A more activist generation urged escape from the prison, and, long before poststructuralist thought declared the body a political site, recognized that the most mundane, "trivial" aspects of women's bodily existence were in fact significant elements in the social construction of an oppressive feminine norm. In 1914, the first Feminist Mass Meeting in America – whose subject was "Breaking into the Human Race" – poignantly listed, among the various social and political rights demanded, "The right to ignore fashion."[6] Here, already, the material "micropractices" of everyday life – which would be extended by later feminists to include not only what one wears but who cooks and cleans and even, more recently, what one eats or does not eat – have been brought out of the realm of the purely personal and into the domain of the political. Here, for example, is a trenchant 1971 analysis, presented by way of a set of "consciousness-raising" exercises for men, of how female subjectivity is trained and subordinated by the everyday bodily requirements and vulnerabilities of "femininity":

> Sit down in a straight chair. Cross your legs at the ankles and keep your knees pressed together. Try to do this while you're having a conversation with someone, but pay attention at all times to keeping your knees pressed tightly together.
>
> Run a short distance, keeping your knees together. You'll find you have to take short, high steps if you run this way. Women have been taught it is unfeminine to run like a man with long, free strides. See how far you get running this way for 30 seconds.
>
> Walk down a city street. Pay a lot of attention to your clothing: make sure your pants are zipped, shirt tucked in, buttons done. Look straight ahead. Every time a man walks past you, avert your eyes and make your face expressionless. Most women learn to go through this act each time we leave our houses. It's a way to avoid at least some of the encounters we've all had with strange men who decided we looked available.[7]

As Andrea Dworkin described it:

> Standards of beauty describe in precise terms the relationship that an individual will have to her own body. They prescribe her motility, spontaneity, posture, gait, the uses to which she can put her body. *They define precisely the dimensions of her physical freedom.* And of course, the relationship between physical freedom and psychological development, intellectual possibility, and creative potential is an umbilical one.
>
> In our culture, not one part of a woman's body is left untouched, unaltered. No feature or extremity is spared the art, or pain, of improvement. . . . From head to toe, every feature of a woman's face, every section of her body, is subject to modification, alteration. This alteration is an ongoing, repetitive process. It is vital to the economy, the major substance of male-female differentiation, the most immediate physical and psychological reality of being a woman. From the age of 11 or 12 until she dies, a woman will spend a large part of her time, money, and energy on binding, plucking, painting and deodorizing herself. It is commonly and wrongly said that male transvestites through the use of makeup and costuming caricature the women they would become, but any real knowledge of the romantic ethos makes clear that these men have penetrated to the core experience of being a woman, a romanticized construct.[8]

Here, feminism inverted and converted the old metaphor of the Body Politic, found in Plato, Aristotle, Cicero, Seneca, Machiavelli, Hobbes, and many others, to a new metaphor: the politics of the body. In the old metaphor of the Body Politic, the state or society was imagined as a human body, with different organs and parts symbolizing different functions, needs, social constituents, forces, and so forth – the head or soul for the sovereign, the blood for the will of the people, or the nerves for the system of rewards and punishments. Now, feminism imagined the human body as *itself* a politically inscribed entity, its physiology and morphology shaped by histories and practices of containment and control – from foot-binding and corseting to rape and battering to compulsory heterosexuality, forced sterilization, unwanted pregnancy, and (in the case of the African American slave woman) explicit commodification.[9]

One might rightly object that the body's literal bondage in slavery is not to be compared to the metaphorical bondage of privileged nineteenth-century women to the corset, much less to the twentieth-century "tyranny of slenderness." No feminist writers considered them equivalent. But at the heart of the developing feminist model, for many writers, *was* the extension of the concept of enslavement to include the voluntary behaviors of privileged women. Problematic as this extension has come to seem, I think it is crucial to recognize that a staple of the prevailing sexist ideology against which the feminist model protested was the notion that in matters of beauty and femininity, it is women alone who are responsible for their sufferings from the whims and bodily tyrannies of fashion. According to that ideology, men's desires bear no responsibility, nor does the culture that subordinates women's desires to those of men, sexualizes and commodifies women's bodies, and offers them little other opportunity for social or personal power. Rather, it is in Woman's essential feminine nature to be (delightfully if incomprehensibly) drawn to such trivialities and to be willing to endure whatever physical inconvenience is entailed. In such matters, whether having her feet broken and shaped into four-inch "lotuses," or her waist straitlaced to fourteen inches, or her breasts surgically stuffed with plastic, she is her "own worst enemy." Set in cultural relief against this thesis, the feminist "anti-thesis" – the insistence that women are the *done* to, not the *doers*, here; that *men* and *their* desires bear the responsibility; and that female obedience to the dictates of fashion is better conceptualized as bondage than choice – was a crucial historical moment in the developing articulation of a new understanding of the sexual politics of the body.

Notes

1 See, for example, Catharine Gallagher and Thomas Laqueur, eds., issue called "Sexuality and the Social Body in the Nineteenth Century," *Representations* 14 (Spring 1986); Sander Gilman, "AIDS and Syphilis: The Iconography of Disease," *October* 43 (Winter 1987): 87–108; Mary Jacobus, Evelyn Fox Keller, and Sally Shuttleworth, eds., *Body/Politics: Women and the Discourses of Science* (New York: Routledge, 1990); Kathleen Kete, "*La Rage* and the Bourgeoisie: The Cultural Context of Rabies in the French Nineteenth Century," *Representations* 22 (Spring 1988): 89–107; Thomas Laqueur, *Making Sex: Body and Gender from the Greeks to Freud* (Cambridge: Harvard University Press, 1990); Emily Martin, *The Woman in the Body: A Cultural Analysis of Reproduction* (Boston: Beacon

Press, 1987); Margaret Miles, *Carnal Knowing: Female Nakedness and Religious Meaning in the Christian West* (Boston: Beacon, 1989); Susan Suleiman, ed., *The Female Body in Western Culture* (Cambridge: Harvard University Press, 1986); Simon Watney, "Aids, 'Africa,' and Race," *Differences* (Winter 1989): 83–86.

2 Don Hanlon Johnson, "The Body: Which One? Whose?" *Whole Earth Review* (Summer 1989): 4–8.

3 Linda Zirelli, "Rememoration or War? French Feminist Narrative and the Politics of Self-Representation," *Differences* (Spring 1991): 2–3.

4 Michel Foucault, *Discipline and Punish* (New York: Vintage, 1979); *The History of Sexuality*. Vol. 1: *An Introduction* (New York: Vintage, 1980).

5 Mary Wollstonecraft, "A Vindication of the Rights of Woman," in Alice Rossi, ed., *The Feminist Papers* (Boston: Northeastern University Press, 1988), 55–57.

6 Reproduced in Nancy Cott, *The Grounding of Modern Feminism* (New Haven: Yale University Press, 1987), 12.

7 Williamette Bridge Liberation News Service, "Exercises for Men," *The Radical Therapist* (Dec.–Jan. 1971).

8 Andrea Dworkin, *Woman-Hating* (New York: Dutton, 1974), 113–14 (emphasis in original).

9 Among the "classics": Susan Brownmiller, *Against Our Will* (New York: Bantam, 1975); Mary Daly, *Gyn-Ecology* (Boston: Beacon, 1978); Davis, *Women, Race, and Class* (New York: Vintage, 1983); Dworkin, *Woman-Hating*; Germaine Greer, *The Female Eunuch* (New York: McGraw-Hill, 1970); Susan Griffin, *Rape: The Power of Consciousness* (New York: Harper and Row, 1979), and *Woman and Nature* (New York: Harper and Row, 1978); Adrienne Rich, "Compulsory Heterosexuality and Lesbian Existence," *Signs* 5, no. 4 (1980): 631–60. Also see the anthologies collected by Morgan, *Sisterhood Is Powerful* (New York: Vintage, 1970) and Vivian Gornick and Barbara Moran, *Woman in Sexist Society* (New York: Mentor, 1971).

---

18 Elizabeth Grosz

'Inscriptions and Body Maps: Representations and the Corporeal'

Excerpt from: T. Threadgold and A. Cranny-Francis (eds) *Feminine, Masculine and Representation*, pp. 62–74. London: Allen and Unwin (1990)

. . . books are only metaphors of the body . . .
Michael de Certeau 'Des outils pour ecrire le corps' *Traverses* 14–15, 1979, p. 3

The book has somehow to be adapted to the body.
Virginia Woolf *A Room of One's Own* Penguin, 1963, p. 78

The body has figured in many recent texts as a writing surface on which messages can be inscribed. The metaphorics of *body-writing* poses the body, its epidermic surface, muscular-skeletal frame, ligaments, joints, blood vessels and internal organs, as *corporeal surfaces* on which engraving inscription or 'graffiti' are etched. The metaphor of the *textualised body* affirms the body as a page or material surface on which messages may be inscribed. The analogy between bodies and texts is a close one: tools of body-engraving – social, surgical, epistemic or disciplinary – mark bodies in culturally specific ways; writing instruments – the pen, stylus, or laser beam – inscribe the blank page of the body. The 'messages' or 'texts' produced by such procedures construct bodies as networks of social signification, meaningful and functional 'subjects' within assemblages composed with other subjects. Each gains a (provisional) identity from its constitutive relations with others. Inscriptions of the corporeal differences between bodies can be seen to produce body-subjects as living significations, social texts capable of being read or interpreted.

I want to explore and evaluate this metaphor of corporeal inscription. The metaphor does not, of course, originate with those associated with it today – Foucault, Deleuze, Irigaray, Lyotard, Lingis *et al.*; it is anticipated in considerable detail in Nietzsche's writings, and is strikingly evoked in Franz Kafka's short story, 'The Penal Settlement'.[1] Both Nietzsche and Kafka conjecture about the ways in which social power, especially punitive and moral systems, mark bodies in more or less violent, brutal and socially sanctioned ways, through institutionalised cruelty and torture. It is not my aim to encourage sadomasochism (though an analysis of cultural sadomasochism would be well worth undertaking) but to re-examine various presumptions within current feminist and leftist theory about the ways in which power functions to construct subjectivity.

This chapter will explore the following theses:

1 that, as a *material* series of processes, power actively marks or brands bodies as social, inscribing them with the attributes of subjectivity. (This is intended to challenge a prevailing model of power conceived as a system of ideas, concepts, values and beliefs, *ideology*, that primarily effect consciousness);
2 that consciousness is an *effect* or result, rather than the cause of the inscription of flesh and its conversion into a (social) body; and
3 that while relying on the work of a number of male theorists of the body (Foucault, Nietzsche and Lingis), feminist assertions of sexual difference simultaneously problematise their work. Thus, although this chapter focuses on male theorists, my objectives remain feminist: to see what in their works may be of use for a feminist account of sexed bodies.

Writing bodies

Foucault is probably the most well-known theorist of the body today. His disparate works cluster around a thematics of carnality and its relations to subjectivity, around, that is, the intricate history of the link between pleasure/pain/sensation/knowledge and power. Foucault's account of the internal relations between power and knowledge relies on a belief that power functions *directly* on bodies by means of disciplinary practices, which, while relying on knowledges, operate without mediation of *conceptual* or intellectual processes – that is, without resort to a concept like 'ideology'. He bypasses a primarily marxist-psychoanalytic-semiotic understanding of social power, which sees ideology as a system of representations, signs received by subjects regarding the social world and their place within it. For him, power is not a set of signs, or texts, meanings or conceptual functions. Power is a material force that *does* and *makes* things, it is a substrate of forces in play within a given socio-personal constellation. The body is its primary object. In *Discipline and Punish* (1977) and the first two volumes of *The History of Sexuality* (1977; 1985) he argues that power is inscribed on and by bodies through modes of social supervision and discipline as well as self-regulation. The bodies and behaviours of individuals are targets for the deployment of power, and they are also the means by which power functions and proliferates.

Power-knowledge is invested in producing determinate types of bodies, with correlative psychical, economic and socio-moral attributes. Bodies are objects of knowledges, which then reinvest the body in increasing spirals of knowledge-pleasure and power: 'The body is moulded by a great many distinct regimes; it is broken down by the rhythms of work, rest, and holidays; it is poisoned by food or values, through eating habits or moral laws; it constructs resistances' (Foucault, 1977: 153). If power is primarily *ideological*, that is, a system of conceptual distortion, if ideas, beliefs, ideologies, values – some kind of soul – are to be attributed to the human subject, this is an effect of a certain mode of corporeal inscription. For this reason, Foucault is irresistably led in his accounts of the history of knowledges and of truth to accounts of punishment, torture, medicalised observations, sexuality and pleasure – all processes that mark the body in specific ways of specific rituals and practices. But if the body is the strategic target of systems of codification, supervision and constraint, it is also because the body and its energies and capacities exert an uncontrollable, unpredictable threat to a regular, systematic mode of social organisation. As well as being the site of knowledge-power, the body is thus also a site of *resistance*, for it exerts a recalcitrance, and always entails the possibility of a counterstrategic reinscription, for it is capable of being self-marked, self-represented in alternative ways.

Within our own culture, the inscription of bodies occurs both *violently* – in prisons, juvenile homes, hospitals, psychiatric institutions – keeping the body confined, constrained, supervised and regimented, marked by 'body-writing-implements', such as handcuffs, traversing neural pathways by charges of electricity in shock therapy, the straitjacket, the regimen of drug habituation, chronologically regulated time-and-labour divisions, cellular and solitary confinement, and deprivation of mobility, the bruising of bodies in police interrogations etc.; and by *less openly aggressive* but no less coercive means, through cultural and personal values, norms and commitments. The latter involve a psychic inscription of the body through its adornment, its rituals of exercise and diet, all more or less 'voluntary' inscriptions by lifestyle, habits, and behaviours. Makeup, stilettos, bras, hairstyles, clothing, underclothing, mark women's bodies in ways other than the ways in which hairdos, professional training, personal grooming, body-building etc. mark men's. There is nothing natural or a priori about these modes of corporeal inscriptions: through them, bodies are marked so as to make them amenable to the prevailing exigencies of power. They make the body into a particular *kind* of body – pagan, primitive, medieval capitalist, Italian, American, Australian. What is sometimes loosely called 'body-language' is a not inappropriate description of the ways in which culturally specific grids of power, regulation and force condition and provide techniques for the formation of particular bodies.

Body-writing relies on the one hand on extraneous instruments, tools for marking the body's surface – the stylus, or cutting edge, the needle, the tatoo, the razor; and on interior, psychical and physiological body-products or objects to remake the body – moisturising cremes, makeup, exercise, the sensations, pleasures, pains, sweat and tears of the body-subject. The subject is *named* by being tagged or branded on its surface, creating a particular kind of 'depth-body' or interiority, a psychic layer the subject identifies as its (disembodied) core. Subjects thus produced are not simply the imposed results of alien, coercive forces; the body is internally lived, experienced and acted upon by the subject and the social collectivity. Messages coded onto the body can be 'read' only within a social system of organisation and meaning. They mark the subject by, and as, a series of signs within the collectivity of other signs, signs which bear the marks of a particular social law and organisation, and through a particular constellation of desires and pleasures.

The subject is marked as a series of (potential) messages from/of the (social) Other, the symbolic order. Its flesh is transformed into a *body*, organised and hierarchised according to the requirements of a particular social and family nexus. The body becomes a 'text' and is fictionalised and positioned within those myths that form a culture's social narratives and self-representations. In some cultural myths, this means the body can be read as an *agent*, a contractual, exchanging being, a subject of the social contract; while in others, it becomes a body-shell capable of being overtaken by the Other's messages, (e.g. in shamanism, or epilepsy). Social narratives create their 'characters' and 'plots' through the tracing of the body's biological contours and organic outlines by writing tools. Writing instruments confine corporeal capacities and values, proliferating the body's reactions and capacities, stimulating and stifling social conformity (the acting out of these narrative roles as 'live theatre') *and* a corporeal resistance to the social. The consequences for the social are twofold: the 'intextuation of bodies', which transform the discursive apparatus or social fiction/knowledge regimes, 'correcting' or updating them, rendering them more truthful, and ensuring their increasingly microscopic focus on bodies; and the *incarnation* of social law in the movements, actions and desires of bodies.

Social inscription

For Nietzsche, civilisation instils its basic requirements only by branding the law on bodies through a *mnemonics of pain*. Morality, shame and guilt are not the causes but the consequences of the subject's incorporation into collective memory or history. Nietzsche conditions history and social life on the provision of a kind of corporeal memory for each subject, a memory fashioned out of the suffering and pain of the body. For example, economic and social law functions only if the relation between debtors and creditors is founded on some sort of contractual *guarantee*, ensuring the payment of debts. For Nietzsche, justice does not originate in economic equivalences or some kind of mathematical computation, but from a compensatory equivalence of the economic with the corporeal. This equivalence ensures that, even in the case of economic bankruptcy, the debt is repayable corporeally. Nietzsche cites examples from Roman law where

> The creditor . . . could inflict every kind of indignity and torture upon the body of the debtor; for example, cut from it as much as seemed commensurate with the size of the debt – and everywhere and from early times one had exact evaluations, *legal* evaluations, of the individual limbs and parts of the body from this point of view, some of them going into horrible and minute detail.
>
> (Nietzsche, 1969: 64)

Damages are not measured by equivalent values which are substitutable for each other, but by forces, organs or parts extractable from the debtor's body – a recompense by sanctioned cruelty. Contractual relations are thus the foundation of justice, and are themselves founded on *blood*, suffering and sacrifice. Within such a corporeal economy, the creditor gains both the benefit of a value equivalent to the debt, and the pleasure of extracting it from the debtor's body.

> It was in *this* sphere, the sphere of legal obligations, that the moral conceptual world of 'guilt', 'conscience', 'duty', 'sacredness of duty' had its origin: its beginnings were, like the beginnings of everything great on earth, soaked in blood thoroughly and for a long time . . . this world has never since lost a certain odor of blood and torture. (Not even in good old Kant: the categorical imperative smells of cruelty.)
>
> (p. 65)

If morality and justice share a common genealogy in barter and cruelty, social history and memory are also instilled in individuals by being branded on flesh. The law functions because it is tattooed indelibly on the subject:

> Man could never do without blood, torture and sacrifices when he felt the need to create a memory for himself; the most dreadful sacrifices and pledges . . . the most repulsive mutilations . . . the cruellest rites of all the religious cults (and all religions are at the deepest level systems of cruelty) – all thus has its origin in the instinct that realized that pain is the most powerful aid to mnemonics . . . The worse man's memory has been, the more fearful has been the appearance of his customs; the severity of the penal code provides an especially significant measure of the degree of effort needed to overcome forgetfulness and to impose a few primitive demands of social existence as *present realities* upon these slaves of momentary affect and desire.
>
> (p. 61)

These inscriptive processes may be more easily recognisable in those forms of body-engraving designated as 'savage' or 'primitive' rituals and practices. In his two books, *Excesses: Eros and Culture* (1984), and *Libido* (1985), Alphonso Lingis sketches an account of the body as a surface of erotogenic intensity, a product of and material to be further inscribed by social norms and ideals. The processes by which the 'primitive body' is scarred seem to us barbaric and painful. Lingis argues that the incision of or writing on the body surface functions to intensify, proliferate and extend the body's erotogenic sensitivity. Welts, cuts, scars, tattoos, perforations, incisions, inlays, function to increase the surface space of the body, creating locations, zones, hollows, ridges: places of special meaning (in some cases) and libidinal intensity (in all cases). What he describes is the creation of erotogenic orifices, rims or libidinal zones. These produce erotic zones potentially at all points on the surface of the skin and within the body's skeletal and muscular frame, a kind of weaving of inscriptive incisions with the sensations, sexual intensities and pleasures of the body. This *creates* erotogenic surfaces, not simply through the displacement of pregiven libidinal zones (as occurs in the 'civilised' neurosis, hysteria – where, say, in Dora's case, the meaning of the phallus is displaced from the genitals to the throat and oral cavity):

> The savage inscription is a working over the skin, all surface effects. This cutting in orifices and raising tumescences does not contrive new receptor organs for a depth body . . . it extends an erotogenic surface . . . It's a multiplication of mouths, of lips, labia, anuses, these sweating and bleeding perforations and puncturings . . . these warts raised all over the abdomen, around the eyes . . .
>
> (Lingis, 1984: 34)

Primitive initiation ceremonies and our own more 'civilised' forms of permanent and semi-permanent social body-markings, designate the body as a socio-cultural and sexual body, a body positioned in relation to the social body, an environment, mythic affiliations with animals, plants, locations, sites etc. They not only mark the kind of individual, the position he or she may occupy, but also *the ways* in which these positions may be occupied.

Our own cultural practices are no less barbaric and no more civilised than those operating in so-called 'savage' or 'primitive' cultures. Primitive inscriptions, it seems, can be differentiated from so-called 'civilised' body-inscriptions in two broad ways: the 'savage' body is marked on its naked surface by signifiers, patterns, arrangements or organisations of marks, welts, cuts, perforations and swellings; and the 'savage body' does not presuppose, as does the 'modern' body, a latent or secret 'private' *depth*, a depth beneath the body's superficial or manifest surface. The 'modern body' is a body read symptomatically, in terms of what it hides. The primitive body, by contrast, is all surface: it is a proliferation or profusion of zones, indefinitely extending libidinal intensity unevenly over the body's surface, using pleasure and pain to somato-psychically mark the body (like the Medusa's Head) through a multiplication of phalluses that are peculiarly non-phallic because of their profusion. The processes of initiation and tattoo designated as 'primitive' intensify and unevenly spread over the whole of the body, along the lines marked by incisions of social position, location, name and function. These lines are inscribed in the case of the 'civilised body' as the lines of incision of surgical and chemical intervention, sites of social and personal remaking.

Primitive body-marking does not merely spread out a surface of sexual intensity across the subject's body, creating orifices, hollows, plateaux, rims where previously there were smooth spaces and unbroken surfaces. It divides up, or maps, the body in regular ordered sequences carefully specified in ritual form. Cicatrisations and scarifications mark the body as a public, collective or social object – as a map of social needs, requirements and excesses; and as a legible, mean-receiving, interiority, a subjectivity experiencing itself in and as a determinate form. The body and its privileged zones of sensation, reception and projection are coded by objects, categories, affiliations, lineages, which engender or make real the subject's social, familial, marital or economic position and/or identity within a social hierarchy:

> 'It is the incision and tumescence of new intensive points, pain-pleasure points, that first extends the erotogenic extension. What we have then, is a spacing, a distributive system of marks. They form not representations and not signifying chains, but figures, figures of intensive points, whose law of systematic distribution is lateral and immanent, horizontal and not traverse.'
>
> (Lingis, 1984: 38)

The primitive body is distinguished from the civilised body not by degrees of barbarism or pain, nor in terms of the writing implements and tools used, but by its *sign-ladenness*. In the case of the civilised body, bodies are created as sign-systems, cohesively meaningful and integrated into patterns that can be read in terms of personality; and above all, by the construction of a *depth* body, a body within which resides an interiority, a psyche or self. Ours is not a *superficial* identity but an enigma to be explored by reduction of the body to a symptom of the self:

> 'All that is civilised is significant . . . We find the ugliness of tattooed nakedness puerile and shallow . . . The savage fixing his identity on his skin . . . Our identity is inward, it is our functional integrity as machines to produce a certain civilised, that is, coded, type of action.'
>
> (Lingis, 1984: 43)

Inscriptions on the subject's body coagulate corporeal signifiers into signs, producing the effects of meanings, depth, representations within or subtending our social order. The intensity and flux of sensations traversing the body become fixed into consumable, gratifiable form, become needs, requirements and desires which can now be attributed to an underlying psyche or consciousness. Corporeal fragmentation, the unity and disunity of the perceptual body, becomes organised into the structure of an ego or consciousness, which marks a secret or private depth. These mark the 'modern' or civilised body as use and exchange-value, the production and exchanges of messages.

Sexed bodies

Do differently sexed bodies require different inscriptive tools to etch their different surfaces? Is it power which inscribes bodies as sexually different? Or does sexual difference simply require sexually differentiated regimes of power? Is *sex* or *gender* the object of power? These remain central dilemmas for feminists working on the texts of the male theorists of the body I have been discussing. Their work remains problematic, even if highly suggestive for feminists, in describing the interventions of power on *women's* bodies.

Foucault, for example, in certain texts (especially *The History of Sexuality*) implies that the divisions between the sexes, and the different characteristics attributed to them by knowledges, institutions and practices, are effects of power. He implies that, outside the deployments of power, there is nothing other than 'bodies, organs, somatic localizations, functions, anatomophysiological systems, sensations and pleasures' (1978: 152–53). For him sex, not gender, is the object of power. Women's medical, moral, psychological and domestic position is specified only as part of a regime which creates the category of 'sex', bringing together a heterogeneity of hitherto non-specific, disparate and not always commensurable elements of bodies and pleasures:

> 'Sex – that agency which appears to dominate us and that secret which seems to underlie all that we are, that point which enthralls us through the power it manifests and the meaning it conceals, and which we ask to reveal what we are, and to free us from what defines us – is doubtless but an ideal point in the deployment of sexuality and its operations.'
>
> (1978: 155)

In spite of his brilliant evocation of corporeal inscriptions, Lingis also remains committed to a paradoxically sexed yet neutral (neutered?) body underlying or forming the surface to be incised. It is as if the body were a pure plenitude of undifferentiated processes and functions that becomes sexed only by social marks: 'circumcision castrates the male of the labia about his penis, as the clitoridectomy castrates the female of her penis. It is through *castration of the natural bisexual* that the social animal is produced' (Lingis, 1984: 40, emphasis added). If these male theorists of the body are still relevant to feminist theory – as I think they are – then it seems essential that feminists distinguish their own positions as feminists from this commitment to a neutralised, non-specified corporeality; and also that they make clear what the raw materials and basic units of inscription are, that is, sexed, carnal, specific bodies: male, female, black, white, etc.

To conclude, I would like to make some suggestions about how feminists may use these conceptions of the body to articulate women's lived experiences and their potential for autonomy. These suggestions are, I think, consistent with the insights of these (male) philosophers of corporeality, and also with feminist commitments to explore and question prevailing categories of sexual polarisation. I will outline these briefly.

1. Biological, anatomical, physiological and neuro-physiological processes cannot be automatically attributed a *natural* status. It is not clear that what is

biological is necessarily natural. Biological or organic functions are the raw materials of any processes of production of determinate forms of subjectivity and material, including corporeal, existence. If this is the case, universal or quasi-universal physiological givens, such as menstrual, anatomical and hormonal factors, need to be carefully considered as irreducible features of the writing surface, distinct from the script inscribed: a kind of 'texture' more than a designated content for the 'text' or the 'intextuated body' produced. The raw materials themselves are not 'pure' in so far as culture, social and psychological factors intervene to give them their manifest forms: it is well known, for example, that menstrual patterns can be severely disrupted or stopped according to diet, exercise patterns, anxiety etc. Biology provides a *bedrock* for social inscription but is not a fixed or static substratum: it interacts with and is overlaid by psychic, social and signifying relations (see Grosz, 1987). The body can thus be seen not as a blank, passive page, a neutral ground of meaning, but as an active, productive, 'whiteness' that constitutes the writing surface as resistant to the imposition of any or all patterned arrangements. It has a texture, a tonus, a materiality that is an active ingredient in the messages produced. It is less like a blank, smooth, frictionless surface, a page, and more like a copperplate to be etched.

2. The anatomical differences between the sexes must be distinguished from the ways in which sexed bodies are culturally classified. *Differences between bodies* can be represented on a vast continuum which could include bodies typical or representative of each sex, but also, all those who fit into both categories (for example, hermaphrodites) or neither. Conceived on the model of 'pure difference', corporeality is potentially infinite in form, no mode exhibiting a prevalence over others. However, within our social and signifying systems, this plenum is divided and categorised according to binary pairs – male/female, black/white, young/old, etc. – which reduce ambiguous terms not amenable to binary hierarchisation, back into this polarised structure (hence Foucault's analysis of the hermaphrodite Herculine Barbin, who is legally required to change 'her' sex from female to male; there is no possibility of adopting a sexual position that is neither male nor female). From *pure differences* of a biological type, *distinctions* and *oppositions*, binary categories and mutually exclusive oppositions are formed. It is thus *not* simply a matter of the socially variable issue of *gender* being imposed on a biological neutral body, but rather, a *social mapping* of the body tracing its anatomical and physiological details by social representations. The procedures which mark male and female bodies ensure that the biological capacities of bodies are always socially coded into sexually distinct categories. It is the *social inscription of sexed bodies*, not the imposition of an acculturised, sexually neutral *gender* that is significant for feminist purposes (see Gatens, 1983).

3. While the sexes are represented according to a binary structure that reduces *n-sexes* to two, the binary structure itself reduces one term within the pair to a position definitionally dependent on the other, being defined as its negation, absence or lack. This is a *phallocentric* representational system in the sense in which women's corporeal specificity is defined and understood only in some relation to men's – as men's opposites, their doubles or their

complements. This means that women's autonomously defined carnal and bodily existence is buried beneath both male-developed biological scientific paradigms, and a male-centred system of social inscription that marks female bodies as men's (castrated, inferior, weaker, less capable) counterparts. In other words, not only is the corporeal surface to be inscribed differently, the social regimes of body-tattoo, incision and marking, the tools of body writing, are oppositionally used to produce male bodies as virile, strong, phallic, hierarchised, structured, teleological, in relation to women's passive, weak, castrated, disorganised bodily structure.

4. Inscriptive procedures marking the body and producing it as sexually determinant and coded are active in transforming the anatomo-physiological structure of the body as socially located *morphology*. Body-morphologies are the results of the *social meaning* of the body. The morphological surface is a retracing of the anatomical and physiological foundation of the body by systems of social signifiers and signs traversing and even penetrating bodies. Morphological differences between sexed bodies imply both a traced, 'biological' difference which is transcribed or translated by discursive, textual representations, *and* corporeal significations. It implies a productive, *changeable*, non-fixed biological substratum mapped by a social, political and familial grid of practices and meanings. The morphological dimension is a function of socialisation and apprenticeship, and produces as its consequences a subject, soul, personality or inner depth. This has direct implications for the beloved feminist category of 'gender' and its relation to its counterpart, 'sex'. Masculinity and femininity are not simply social categories as it were externally or arbitrarily imposed on the subject's sex. Masculine and feminine are necessarily related to the structure of the lived experience and meaning of bodies. As Gatens argues in her critique of the sex/gender distinction (1983), masculinity and femininity *mean different things* according to whether they are lived out on and experienced by male or female bodies. Gender is an effect of the body's social morphology. What is mapped onto the body is not unaffected by the body onto which it is projected.

5. Whereas morphological *oppositions* between sexed (male or female) bodies are prevalent in contemporary Western cultures (at the least), their oppositional or phallocentric representations are now capable of being challenged so that we need no longer accept an unchangeable natural basis for their social status. Their morphological status, as an effect of the transcription and transmission of meanings and values for the body and its specific parts and processes, can now be addressed by the kinds of theoretical, literary and cultural representations of an autonomously conceived and defined femininity, or woman-centredness, that feminist work has provided. The oppositional form of this morphology, moreover, can be contested, even if not readily overthrown, by demonstrating the ways in which male self-definitions require and produce definitions of the female as their inverted or complementary counterparts. This implies, among other things, an analysis of the ways in which masculine or phallocentric discourses and knowledges rely on images, metaphors and figures of woman and femininity to support and justify their speculations. It also, and perhaps more importantly, implies an exploration of

the disavowed corporeal and psychic dependence of the masculine, with its necessary foundation in women's bodies, on the female corporeality it cannot claim as its own territory (the maternal body).

Notes

1 Franz Kafka's short story, 'The Penal Settlement', describes an 'exquisite' punishment machine that will serve as an emblem of the concerns of this chapter: the socio-material, representational inscription of bodies. Kafka's machine is an ingenious device made of three parts: first, a 'Bed', onto which the prisoner is tied; second, the 'Designer', which determines what messages will be inscribed; and, third, 'the Harrow'. It executes the sentence on the prisoner's body, using a moving layer of needles to print the Designer's message:

> As soon as the man is strapped down, the bed is set in motion. It quivers in minute, very rapid vibrations . . . You will have seen similar apparatus in hospitals; but in our Bed, the movements all correspond very exactly to the movements of the Harrow . . . Our sentence does not sound severe. Whatever commandment the condemned man has disobeyed is written on his body: 'Honour thy superiors' . . . (although he doesn't know the sentence) he'll learn it corporeally, on his person . . . An ignorant onlooker would see no difference between one punishment and another. The Harrow appears to do its work with uniform regularity . . . the actual progress of the sentence can be watched, [for] the Harrow is made of glass . . . When the Harrow . . . finishes its first draft of the inscription on the back, the layer of cotton wool (on the Bed) begins to roll and slowly turns the body over, to give the Harrow fresh space for writing. Meanwhile the raw part that has been written on lies on the cotton wool, which is especially prepared to staunch the bleeding and so makes all ready for a new deepening of the script . . . So it keeps on writing deeper and deeper for the whole 12 hours . . . But how quiet he grows at just about the 6th hour! Enlightenment comes to the most dull-witted. It begins around the eyes. From there, it grows, it radiates. A moment that might tempt one to get under the Harrow with him. Nothing more happens after that, the man only begins to understand the inscriptions . . . You have seen how difficult it is to decipher the script with one's eyes; but our man deciphers it with his wounds. To be sure, that is a hard task; he needs 6 hours to accomplish it. By that time, the Harrow has pierced him quite through and casts him into the grave . . . Then the judgement has been fulfilled and we bury him.
>
> (Kafka, 'The Penal Settlement', *Metamorphosis and other Short Stories*)

Three elements of Kafka's description are relevant to my chapter. He links punishment to the operations of knowledges. Penal punishment requires a certain epistemic backup for knowledges are both the preconditions of power's concrete operations, and are also amenable to revision and transformation by information gathered from the processes of supervision, observation and inscription of the prisoner's suffering body. Second, Kafka explicitly describes the machine as a discursive or *writing* instrument. Like the stylus, pen or typewriter, this writing-machine is an instrument of material inscription producing propositions, texts and discourses: the surface to be inscribed in this case is human flesh and skin rather than the blank page of the book, and the consequence of this inscriptive procedure is not only a text, but a particular type of human subject. Third, messages or inscriptions are etched on the body's surface, without the prisoner knowing the crime with which he is charged, or the punishment he is to be apportioned. These punitive practices create an 'enlightenment', a consciousness or psychic effect solely by materially marking the prisoner's body. Kafka allows us to focus on relations between bodies, textuality, and consciousness (or mind).

References

Foucault, M. (1977) *Discipline and Punish. The Birth of the Prison* London: Allen Lane.

—— (1978) *The History of Sexuality* vol. 1 London: Allen Lane.

Gatens, M. (1983) 'A Critique of the sex/gender distinction' in J. Allen and P. Patton (eds) *Beyond Marx? Interventions after Marx* Sydney: Intervention.

Grosz, E. (1987) 'Notes Towards a Corporeal Feminism' in *Feminism and the Body* special issue, eds J. Allen and E. Grosz *Australian Feminist Studies* 5, 1–16.

Lingis, A. (1984) *Excesses. Eros and Culture* New York: State University of New York.

—— (1985) *Libido* New York: State University of New York.

Nietzsche, F. (1969) *On The Genealogy of Morals* New York: Vintage Press.

19 Judith Butler

'Gender Trouble, Feminist Theory and Psychoanalytic Discourse'

Excerpt from: L. Nicholson (ed.) *Feminism/Postmodernism*, pp. 324–40. London: Routledge (1990)

Within the terms of feminist theory, it has been quite important to refer to the category of "women" and to know what it is we mean. We tend to agree that women have been written out of the histories of culture and literature that men have written, that women have been silenced or distorted in the texts of philosophy, biology, and physics, and that there is a group of embodied beings socially positioned as "women" who now, under the name of feminism, have something quite different to say. Yet, this question of being a woman is more difficult than it perhaps originally appeared, for we refer not only to women as a social category but also as a felt sense of self, a culturally conditioned or constructed subjective identity. The descriptions of women's oppression, their historical situation or cultural perspective has seemed, to some, to require that women themselves will not only recognize the rightness of feminist claims made in their behalf, but that, together, they will discover a common identity, whether in their relational attitudes, in their embodied resistance to abstract and objectifying modes of thought and experience, their felt sense of their bodies, their capacity for maternal identification or maternal thinking, the nonlinear directionality of their pleasures or the elliptical and plurivocal possibilities of their writing.

But does feminist theory need to rely on a notion of what it is fundamentally or distinctively to be a "woman"? The question becomes a crucial one when we try to answer what it is that characterizes the world of women that is marginalized, distorted, or negated within various masculinist practices. Is there a specific femininity or a specific set of values that have been written out of various histories and descriptions that can be associated with women as a group? Does the category of woman maintain a meaning separate from the conditions of oppression against which it has been formulated?

For the most part, feminist theory has taken the category of women to be foundational to any further political claims without realizing that the category effects a political closure on the kinds of experiences articulable as part of a feminist discourse. When the category is understood as representing a set of values or dispositions, it becomes normative in character and, hence, exclusionary in principle. This move has created a problem both theoretical and political, namely, that a variety of women from various cultural positions have refused to recognize themselves as "women" in the terms articulated by feminist theory with the result that these women fall outside the category and are left to conclude that (1) either they are not women as they have perhaps previously assumed or (2) the category reflects the restricted location of its theoreticians and, hence, fails to recognize the intersection of gender with race, class, ethnicity, age, sexuality, and other currents which contribute to the formation of cultural (non)identity. In response to the radical exclusion of the category of women from hegemonic cultural formations on the one hand and the internal critique of the exclusionary effects of the category from within feminist discourse on the other, feminist theorists are now confronted with the problem of either redefining and expanding the category of women itself to become more inclusive (which requires also the political matter of settling who gets to make the designation and in the name of whom) or to challenge the place of the category as a part of a feminist normative discourse. Gayatri Spivak has argued that feminists need to rely on an operational essentialism, a false ontology of women as a universal in order to advance a feminist political program.[1] She concedes that the category of women is not fully expressive, that the multiplicity and discontinuity of the signified rebels against the univocity of the sign, but she suggests that we need to use it for strategic purposes. Julia Kristeva suggests something similar, I think, when she recommends that feminists use the category of women as a political tool without attributing ontological integrity to the term, and she adds that, strictly speaking, women cannot be said to exist.[2]

But is it the presumption of ontological integrity that needs to be dispelled, or does the practical redeployment of the category without any ontological commitments also effect a political consolidation of its semantic integrity with serious exclusionary implications? Is there another normative point of departure for feminist theory that does not require the reconstruction or rendering visible of a female subject who fails to represent, much less emancipate, the array of embodied beings culturally positioned as women?

Psychoanalytic theory has occupied an ambiguous position in the feminist quandary over whether the category of women has a rightful place within feminist political discourse. On the one hand, psychoanalysis has sought to identify the developmental moments in which gendered identity is acquired. Yet, those feminist positions which take their departure from the work of Jacques Lacan have sought to underscore the unconscious as the tenuous ground of any and all claims to identity. A work that makes both arguments, Juliet Mitchell's *Psycho-analysis and Feminism* (1974), sought not only to show that gender is constructed rather than biologically necessitated but to identify the precise developmental moments of that construction in the history of gendered subjects. Mitchell further argues on structuralist grounds that the narrative of infantile development enjoyed relative universality and that psychoanalytic theory seemed, therefore, to offer feminists a way to describe a psychological and cultural ground of shared gender identification.[3] In a similar position, Jacqueline Rose asserts:

> "The force of psychoanalysis is therefore precisely that it gives an account of patriarchal culture as a trans-historical and cross-cultural force. It therefore conforms to the feminist demand for a theory which can explain women's subordination across specific cultures and different historical moments."[4]

As much as psychoanalytic theory provided feminist theory with a way to identify and fix gender difference through a metanarrative of shared infantile development, it also helped feminists show how the very notion of the subject is a masculine prerogative within the terms of culture. The paternal law which Lacanian psychoanalysis takes to be the ground of all kinship and all cultural relations not only sanctions male subjects but institutes their very possibility through the denial of the feminine. Hence, far from being subjects, women are variously, the Other, a mysterious and unknowable lack, a sign of the forbidden and irrecoverable maternal body, or some unsavory mixture of the above.

Elaborating on Lacanian theory, but making significant departures from its presumptions of universal patriarchy, Luce Irigaray maintains that the very construct of an autonomous subject is a masculine cultural prerogative from which women have been excluded. She further claims that the subject is always already masculine, that it bespeaks a refusal of dependency required of male acculturation, understood originally as dependency on the mother, and that its "autonomy" is founded on a repression of its early and true helplessness, need, sexual desire for the mother, even identification with the maternal body. The subject thus becomes a fantasy of autogenesis, the refusal of maternal foundations and, in generalized form, a repudiation of the feminine. For Irigaray, then, it would make no sense to refer to a female subject or to women as subjects, for it is precisely the construct of the subject that necessitates relations of hierarchy, exclusion, and domination. In a word, there can be no subject without an Other.[5]

Psychoanalytic criticism of the epistemological point of departure, beginning with Freud's criticism of Enlightenment views of "man" as a rational being and later echoed in Lacan's critique of Cartesianism, has offered feminist theorists a way of criticizing the disembodied pretensions of the masculine knower and exposing the strategy of domination implicit in that disingenuous epistemological gesture. The destabilization of the subject within feminist criticism becomes a tactic in the exposure of masculine power and, in some French feminist contexts, the death of the subject spells the release or emancipation of the suppressed feminine sphere, the specific libidinal economy of women, the condition of *écriture feminine*.

But clearly, this set of moves raises a political problem: If it is not a female subject who provides the normative model for a feminist emancipatory politics, then what does? If we fail to recuperate the subject in feminist terms, are we not depriving feminist theory of a notion of agency that casts doubt on the viability of feminism as a normative model? Without a unified concept of woman or, minimally, a family resemblance among gender-related terms, it appears that feminist politics has lost the categorial basis of its own normative claims. What constitutes the "who," the subject, for whom feminism seeks emancipation? If there is no subject, who is left to emancipate?

The feminist resistance to the critique of the subject shares some concerns with other critical and emancipatory discourses: If oppression is to be defined in terms of a loss of autonomy by the oppressed, as well as a fragmentation or alienation within the psyche of the oppressed, then a theory which insists upon the inevitable fragmentation of the subject appears to reproduce and valorize the very oppression that must be overcome. We need perhaps to think about a typology of fragmentations or, at least, answer the question of whether oppression ought to be defined in terms of the fragmentation of identity and whether fragmentation *per se* is oppressive. Clearly, the category of women is internally fragmented by class, color, age, and ethnic lines, to name but a few; in this sense, honoring the diversity of the category and insisting upon its definitional nonclosure appears to be a necessary safeguard against substituting a reification of women's experience for the diversity that exists.[6] But how do we know what exists prior to its discursive articulation? Further, the critique of the subject means more than the rehabilitation of a multiple subject whose various "parts" are interrelated within an overriding unity, a coalitional subject, or an internal polity of pluralistically related points of view. Indeed, the political critique of the subject questions whether making a conception of identity into the ground of politics, however internally complicated, prematurely forecloses the possible cultural articulations of the subject-position that a new politics might well generate.

This kind of political position is clearly not in line with the humanist presuppositions of either feminism or related theories on the Left. At least since Marx's *Early Manuscripts*, the normative model of an integrated and unified self has served emancipatory discourses. Socialist feminism has clearly reformulated the doctrine of the integrated subject in opposition to the split between public and private spheres which has concealed domestic exploitation and generally failed to acknowledge the value of women's work, as well as the specific moral and cultural values which originate or are sustained within the private sphere. In a further challenge to the public/private distinction in moral life, Carol Gilligan and others have called for a reintegration of conventional feminine virtues, such as care and other relational attitudes, into conventional moral postures of distance and abstraction, a kind of reintegration of the human personality, conceived as a lost unity in need of restoration. Feminist psychoanalytic theory based in object-relations has similarly called for a restructuring of child-rearing practices which would narrow the schism between gender differences produced by the predominating presence of the mother in the nurturing role. Again, the integration of nurturance and dependency into the masculine sphere and the concomitant assimilation of autonomy into the feminine sphere suggests a normative model of a unified self which tends toward the androgynous solution. Others insist on the deep-seated specificity of the feminine rooted in a primary maternal identification which grounds an alternative feminine subject, who defines herself relationally and contextually and who fails to exhibit the inculcated masculine fear of dependency at the core of the repudiation of the maternal and, subsequently, of the feminine. In this case, the unified self reappears not in the figure of the androgyne but as a specifically feminine subject organized by a founding maternal identification.

The differences between Lacanian and post-Lacanian feminist psychoanalytic theories on the one hand and those steeped in the tradition of object-relations and ego psychology on the other center on the conception of the subject or the ego and its ostensible integrity. Lacanian feminists such as Jacqueline Rose argue that object-relations theorists fail to account for the unconscious and for the radical discontinuities which characterize the psyche prior to the formation of the ego and a distinct and separate sense of self. By claiming certain kinds of identifications are primary, object-relations theorists make the relational life of the infant primary to psychic development itself, conflating the psyche with the ego and relegating the unconscious to a less significant role. Lacanian theorists insist upon the unconscious as a source of discontinuous and chaotic drives or significations, and they claim that the ego is a perpetually unstable phenomenon, resting upon a primary repression of unconscious drives which return perpetually to haunt and undermine the ostensible unity of the ego.[7]

Although these theories tend to destabilize the subject as a construct of coherence, they nevertheless institute gender coherence through the stabilizing metanarrative of infantile development. According to Rose and to Juliet Mitchell, the unconscious is an open libidinal/linguistic field of discontinuities which contest the rigid and hierarchizing codes of sexual difference encoded in language, regulating cultural life. Although the unconscious thus becomes a locus of subversion, it remains unclear what changes the unconscious can provide considering the rigid synchronicity of the structuralist frame. The rules constituting and regulating sexual difference within Lacanian terms evince an immutability which seriously challenges their usefulness for any theory of social and cultural transformation. The failure to historicize the account of the rules governing sexual difference inevitably institutes that difference as the reified foundation of an intelligible culture, with the result that the paternal law becomes the invariant condition of intelligibility, and the variety of contestations not only can never undo that law but, in fact, require the abiding efficacy of that law in order to maintain any meaning at all.

In both sets of psychoanalytic analyses, a narrative of infantile development is constructed which assumes the existence of a primary identification (object-relations) or a primary repression (the *Ürverdrangung* which founds the Lacanian male subject and marks off the feminine through exclusion) which instantiates gender specificity and subsequently informs, organizes, and unifies identity. We hear time and again about *the* boy and *the* girl, a tactical distancing from spatial and temporal locations which elevates the narrative to the mythic tense of a reified history. Although object-relations poses an alternative version of the subject based in relational attitudes characteristically feminine and Lacanian (or anti-Lacanian) theories maintain the instability of the subject based in the disruptive potential of the unconscious manifest at the tentative boundaries of the ego; they each offer story lines about gender acquisition which effect a narrative closure on gender experience and a false stabilization of the category of woman. Whether as a linguistic and cultural law which makes itself known as the inevitable organizing principle of sexual difference or as the identity forged through a primary identification that the Oedipal complex requires, gender meanings are circumscribed within a narrative frame which both unifies certain legitimate sexual subjects and excludes from intelligibility sexual identities and discontinuities which challenge the narrative beginnings and closures offered by these competing psychoanalytic explanations.

Whether one begins with Freud's postulation of primary bisexuality (Juliet Mitchell and Jacqueline Rose) or with the primacy of object-relations (Chodorow, Benjamin), one tells a story that constructs a discrete gender identity and discursive location which remains relatively fixed. Such theories do not need to be explicitly essentialist in their arguments in order to be effectively essentialist in their narrative strategies. Indeed, most psychoanalytic feminist theories maintain that gender is constructed, and they view themselves (and Freud) as debunking the claims of essential femininity or essential masculinity.[8]

At its most general level of narrative development, the object-relations and Lacanian versions of gender development offer (1) a utopian postulation of an originally predifferentiated state of the sexes which (2) also preexists the postulation of hierarchy, and (3) gets ruined either by the sudden and swift action of the paternal law (Lacanian) or the anthropologically less ambitious Oedipal injunction to repudiate and devalue the mother (object-relations). In both cases, an originally undifferentiated state of the sexes suffers the process of differentiation and hierarchization through the advent of a repressive law. "In the beginning" is sexuality without power, then power arrives to create both culturally relevant sexual distinction (gender) and, along with that, gender hierarchy and dominance.

The Lacanian position proves problematic when we consider that the state prior to the law is, by definition, prior to language and yet, within the confines of language, we are said somehow to have access to it. The circularity of the reasoning becomes all the more dizzying when we realize that prior to language we had a diffuse and full pleasure which, unfortunately, we cannot remember, but which disrupts our speech and haunts our dreams. The object-relations postulation of an original identification and subsequent repudiation constructs the terms of a coherent narrative of infantile development which works to exclude all kinds of developmental histories in which the nurturing presence of the nuclear family cannot be presupposed.

By grounding the metanarratives in a myth of the origin, the psychoanalytic description of gender identity confers a false sense of legitimacy and universality to a culturally specific and, in some contexts, culturally oppressive version of gender identity. By claiming that some identifications are more primary than others, the complexity of the latter set of identifications is effectively assimilated into the primary one, and the "unity" of the identifications is preserved. Hence, because within object-relations the girl-mother identification is "founding," the girl-brother and girl-father identifications are easily assimilated under the already firmly established gender identification with women. Without the assumption of an orderly temporal development of identifications in which the first identifications serve to unify the latter ones, we would not be able to explain which identifications get assimilated into which others; in other words, we would lose the unifying thread of the narrative. Indeed, it is important to note that primary identifications establish gender in a substantive mode, and secondary identifications thus serve as attributes. Hence, we witness the discursive emergence of "feminine men" or "masculine women," or the meaningful redundancy of a "masculine man." Without the temporal prioritization of primary identifications, it would be unclear which characterizations were to serve as substance and which as attributes, and in the case in which that temporal ordering were fully contested, we would have, I suppose, the gender equivalent of an interplay of attributes without an abiding or unifying substance. I will suggest what I take to be the subversive possibilities of such a gender arrangement toward the end of my remarks.

Even within the psychoanalytic frame, however, we might press the question of identification and desire to a further limit. The primary identification in which gender becomes "fixed" forms a history of identifications in which the secondary ones revise and reform the primary one but in no way contest its structural primacy. Gender identities emerge and sexual desires shift and vary so that different "identifications" come into play depending upon the availability of legitimating cultural norms and opportunities. It is not always possible to relate those shifts back to a primary identification which is suddenly manifest. Within the terms of a psychoanalytic theory, then, it is quite possible to understand gendered subjectivity as a history of identifications, parts of which can be brought into play in given contexts and which, precisely because they encode the contingencies of personal history, do not always point back to an internal coherence of any kind.

Of course, it is important to distinguish between two very different ways in which psychoanalysis and narrative theory work together. Within psychoanalytic literary criticism, and within feminist psychoanalytic criticism in particular, the operation of the unconscious makes all narrative coherence suspect; indeed, the defenders of that critical enterprise tend to argue that the narrative capacity is seriously undermined by that which is necessarily excluded or repressed in the manifest text and that a serious effort to admit the unconscious, whether conceived in terms of a repressed set of drives (Kristeva) or as an excluded field of metonymic associations (Rose), into the text disrupts and inverts the linear assumptions of coherent narrativity. In this sense, the text always exceeds the narrative; as the field of excluded meanings, it returns, invariably, to contest and subvert the explicitly attempted narrative coherence of the text.

The multiplication of narrative standpoints within the literary text corresponds to an internally fragmented psyche which can achieve no final, integrated understanding or "mastery" of its component parts. Hence, the literary work offers a textual means of dramatizing Freud's topographical model of mind in motion. The nonliterary use of psychoanalysis, however, as a psychological explanatory model for the acquisition and consolidation of gender identification and, hence, identity generally fails to take account of itself as a narrative. Subject to the feminist aim to delimit and define a shared femininity, these narratives attempt to construct a coherent female subject. As a result, psychoanalysis as feminist metatheory reproduces that false coherence in the form of a story line about infantile development where it ought to investigate genealogically the exclusionary practices which condition that particular narrative of identity formation. Although Rose, Mitchell, and other Lacanian feminists insist that identity is always a tenuous and unstable affair, they nevertheless fix the terms of that instability with respect to a paternal law which is culturally invariant. The result is a narrativized myth of origins in which primary bisexuality is arduously rendered into a melancholic heterosexuality through the inexorable force of the law.

Juliet Mitchell claims that it is only possible to be in one position or the other in a sexual relation and never in both at once. But the binary disjunction implicit to this gendered law of noncontradiction suggests that desire functions through a gender difference instituted at the level of the symbolic that necessarily represses whatever unconscious multiplications of positions might be at work. Kristeva argues similarly that the requirements of intelligible culture imply that female homosexuality is a contradiction in terms, with the consequence that this particular cultural manifestation is, even within culture, outside it, in the mode of psychosis. The only intelligible female homosexuality within Kristeva's frame is in the prohibited incestuous love between daughter and mother, one that can only be resolved through a maternal identification and the quite literal process of becoming a mother.[9]

Within these appropriations of psychoanalytic theory, gender identity and sexual orientation are accomplished at once. Although the story of sexual development is complicated and quite different for *the* girl than *the* boy, it appeals in both contexts to an operative disjunction that remains stable throughout: one identifies with one sex and, in so doing, desires the other, that desire being the elaboration of that identity, the mode by which it creates its opposite and defines itself in that opposition. But what about primary bisexuality, the source of disruption and discontinuity that Rose locates as the subversive potential of the unconscious? A close examination of what precisely Freud means by bisexuality, however, turns out to be a kind of bisexedness of libidinal dispositions. In other words, there are male and female libidinal dispositions in every psyche which are directed heterosexually toward opposite sexes. When bisexuality is relieved of its basis in the drive theory, it reduces, finally, to the coincidence of two heterosexual desires, each proceeding from oppositional identifications or dispositions, depending on the theory, so that desire, strictly speaking, can issue only from a male-identification to a female object or from a female-identification to a male object. Granted, it may well be a woman, male-identified, who desires another woman, or a man, female-identified, who desires another man, and it may also be a woman, male-identified, who desires a man, female-identified, or similarly, a man, female-identified, who desires a woman, male-identified. One either identifies with a sex or desires it, but only those two relations are possible.

But is identification always restricted within the binary disjunction in which it has been framed so far? Clearly, within psychoanalytic theory, another set of possibilities emerges whereby identifications work not to consolidate identity but to condition the interplay and the subversive recombination of gender meanings. Consider that in the previous sketch, identifications exist in a mutually exclusive binary matrix conditioned by the cultural necessity of occupying one position to the exclusion of the other. But in fantasy, a variety of positions can be entertained even though they may not constitute culturally intelligible possibilities. Hence, for Kristeva, for instance, the semiotic designates precisely those sets of unconscious fantasies and wishes that exceed the legitimating bounds of paternally organized culture; the semiotic domain, the body's subversive eruption into language, becomes the transcription of the unconscious from the topographical model into a structuralist discourse. The tenuousness of all identity is exposed through the proliferation of fantasies that exceed and contest the "identity" that forms the conscious sense of self. But are identity and fantasy as mutually exclusive as the previous explanation suggests? Consider the claim, integral to much psychoanalytic theory, that identifications and, hence, identity, are in fact *constituted* by fantasy.

Roy Schafer argues in *A New Language for Psychoanalysis* that when identifications are understood as internalizations, they imply a trope of inner psychic space that is ontologically insupportable. He further suggests that internalization is understood better not as a process but as a fantasy.[10] As a result, it is not possible to attribute some kind of ontological meaning to the spatial internality of internalizations, for they are only fantasied as internal. I would further argue that this very fantasy internal psychic space is essentially conditioned and mediated by a language that regularly figures interior psychic locations of various kinds, a language, in other words, that not only produces that fantasy but then redescribes that figuration within an uncritically accepted topographical discourse. Fantasies themselves are often imagined as mental contents somehow projected onto an interior screen, a conception conditioned by a cinematic metaphorics of the psyche. However, identifications are not merely fantasies of internally located objects or features, but they stand in a transfigurative relation to the very objects they purport to internalize. In other words, within psychoanalytic theory, to identify with a figure from the past is to figure that figure within the configuration of interior psychic space. Identification is never simply mimetic but involves a strategy of wish fulfillment; one identifies not with an empirical person but with a fantasy, the mother one wishes one had, the father one thought one had but didn't, with the posture of the parent or sibling which seems to ward off a perceived threat from some other, or with the posture of some imagined relation whom one also imagines to be the recipient of love. We take up identifications not only to receive love but also to deflect from it and its dangers; we also take up identifications in order to facilitate or prohibit our own desires. In each case of identification, there is an interpretation at work, a wish and/or a fear as well, the effect of a prohibition, and a strategy of resolution.

What is commonly called an introject is, thus, a fantasied figure within a fantasied locale, a double imagining that produces the effect of the empirical other fixed in an interior topos. As figurative productions, these identifications constitute impossible desires that figure the body, active principles of incorporation, modes of structuring and signifying the enactment of the lived body in social space. Hence, the gender fantasies constitutive of identifications are not part of the set of properties that a subject might be said to have, but they constitute the genealogy of that embodied/psychic identity, the mechanism of its construction. One does not have the fantasies, and neither is there a one who lives them, but the fantasies condition and construct the specificity of the gendered subject with the enormously important qualification that these fantasies are themselves disciplinary productions of grounding cultural sanctions and taboos – a theme to which I will momentarily turn. If gender is constituted by identification and identification is invariably a fantasy within a fantasy, a double figuration, then gender is precisely the fantasy enacted by and through the corporeal styles that constitute bodily significations.

In a separate context, Michel Foucault challenges the language of internalization as it operates in the service of the repressive hypothesis. In *Discipline and Punish*, Foucault rewrites the doctrine of internalization found in Nietzsche's *On the Genealogy of Morals* through the language of *inscription*. In the context of prisoners, Foucault writes, the strategy has not been to enforce a repression of their desires but to compel their bodies to signify the prohibitive law as their ownmost essence, style, necessity. That law is not internalized, but it is incorporated, with the consequence that bodies are produced which signify that law as the essence of their selves, the meaning of their soul, their conscience, the law of their desire. In effect, the law is at once fully manifest and fully latent, for it never appears as external to the bodies it subjects and subjectivates. "It would be wrong", Foucault writes, "to say that the soul is an illusion, or an ideological effect. On the contrary, it exists, it has a reality, it is produced permanently around, on, within, the body by the functioning of a power that is exercised on those that are punished. . . ."[11] The figure of the interior soul understood as "within" the body is signified through its inscription *on* the body, even though its primary mode of signification is through its very absence, its potent invisibility, for it is through that invisibility that the effect of a structuring inner space is produced. The soul is precisely what the body lacks; hence, that lack produces the body as its other and as its means of expression. In this sense, then, the soul is a surface signification that contests and displaces the inner/outer distinction itself, a figure of interior psychic space inscribed on the body as a social signification that perpetually renounces itself as such. In Foucault's terms, the soul is not imprisoned by the body, as some Christian imagery would suggest, but "the body becomes a prisoner of the soul".[12]

The redescription of intrapsychic processes in terms of the surface politics of the body implies a corollary redescription of gender as the disciplinary production of the figures of gender fantasy through the play of presence and absence in the body's surface, the construction of the gendered body through a series of exclusions and denials, signifying absences.

But what determines the manifest and latent text of the body politic? What is the prohibitive law that generates the corporeal stylization of gender, the fantasied and fantastic figuration of the gendered body? Clearly, Freud points to the incest taboo and the prior taboo against homosexuality as the generative moments of gender identity, the moments in which gender becomes fixed (meaning both immobilized and, in some sense, repaired). The acquisition of gender identity is thus simultaneous with the accomplishment of coherent heterosexuality. The taboo against incest, which presupposes and includes the taboo against homosexuality, works to sanction and produce identity at the same time that it is said to repress the very identity it produces. This disciplinary production of gender effects a false stabilization of gender in the interests of the heterosexual construction and regulation of sexuality. That the model seeks to produce and sustain coherent identities and that it requires a heterosexual construction of sexuality in no way implies that practicing heterosexuals embody or exemplify this model with any kind of regularity. Indeed, I would argue that in principle no one can embody this regulatory ideal at the same time that the compulsion to embody the fiction, to figure the body in accord with its requirements, is everywhere. This is a fiction that operates within discourse, and which, discursively and institutionally sustained, wields enormous power.

I noted earlier the kinds of coherences instituted through some feminist appropriations of psychoanalysis but would now suggest further that the localization of identity in an interior psychic space characteristic of these theories implies an expressive model of gender whereby identity is first fixed internally and only subsequently manifest in some exterior way. When gender identity is understood as causally or mimetically related to sex, then the order of appearance that governs gendered subjectivity is understood as one in which sex conditions gender, and gender determines sexuality and desire; although both psychoanalytic and feminist theory tend to disjoin sex from gender, the restriction of gender within a binary relation suggests a relation of residual mimeticism between sex, conceived as binary[13] and gender. Indeed, the view of sex, gender, and desire that presupposes a metaphysics of substance suggests that gender and desire are understood as attributes that refer back to the substance of sex and make sense only as its reflection.

I am not arguing that psychoanalytic theory is a form of such substantive theorizing, but I would suggest that the lines that establish coherence between sex, gender, and desire, where they exist, tend to reenforce that conceptualization and to constitute its contemporary legacy. The construction of coherence conceals the gender discontinuities that run rampant within heterosexual, bisexual, and gay and lesbian contexts in which gender does not necessarily follow from sex, and desire, or sexuality generally, does not seem to follow from gender; indeed, where none of these dimensions of significant corporeality "express" or reflect one another. When the disorganization and disaggregation of the field of bodies disrupts the regulatory fiction of heterosexual coherence, it seems that the expressive model loses its descriptive force, and that regulatory ideal is exposed as a norm and a fiction that disguises itself as a developmental law that regulates the sexual field that it purports to describe.

According to the understanding of identification as fantasy, however, it is clear that coherence is desired, wished for, idealized, and that this idealization is an effect of a corporeal signification. In other words, acts, gestures, and desire produce the effect of an internal core or substance, but produce this on the surface of the body, through the play of signifying absences that suggest, but never reveal, the organizing principle of identity as a cause. Such acts, gestures, enactments, generally construed, are performative in the sense that the essence of identity that they otherwise purport to express becomes a *fabrication* manufactured and sustained through corporeal signs and other discursive means. That the gendered body is performative suggests that it has no ontological status apart from the various acts which constitute its reality, and if that reality is fabricated as an interior essence, that very interiority is a function of a decidedly public and social discourse, the public regulation of fantasy through the surface politics of the body. In other words, acts and gestures articulate and enacted desires create the illusion of an interior and organizing gender core, an illusion discursively maintained for the purposes of the regulation of sexuality within the obligatory frame of reproductive heterosexuality. If the "cause" of desire, gesture, and act can be localized within the "self" of the actor, then the political regulations and disciplinary practices which produce that ostensibly coherent gender are effectively displaced from view. The displacement of a political and discursive origin of gender identity onto a psychological "core" precludes an analysis of the political constitution of the gendered subject and its fabricated notions about the ineffable interiority of its sex or of its true identity.

If the inner truth of gender is a fabrication and if a true gender is a fantasy instituted and inscribed on the surface of bodies, then it seems that genders can be neither true nor false but are only produced as the truth effects of a discourse of primary and stable identity.

In *Mother Camp: Female Impersonators in America*, anthropologist Esther Newton suggests that the structure of impersonation reveals one of the key fabricating mechanisms through which the social construction of gender takes place. I would suggest as well that drag fully subverts the distinction between inner and outer psychic space and effectively mocks both the expressive model of gender and of the notion of a true gender identity. "At its most complex," Newton writes, "[drag] is a double inversion that says, 'appearance is an illusion.' Drag says [Newton's curious personification], my 'outside' appearance is feminine, but my essence 'inside' {the body} is masculine." At the same time it symbolizes the opposite inversion: "my appearance 'outside' {my body, my gender} is masculine but my essence 'inside' myself is feminine."[14] Both claims to truth contradict one another and so displace the entire enactment of gender significations from the discourse of truth and falsity.

The notion of an original or primary gender identity is often parodied within the cultural practices of drag, cross-dressing, and the sexual stylization of butch/femme identities. Within feminist theory, such parodic identities have been understood to be either degrading to women, in the case of drag and cross-dressing, or an uncritical appropriation of sex-role stereotyping from within the practice of heterosexuality, especially in the case of butch/femme lesbian identities. But the relation between the "imitation" and the "original" is, I think, more complicated than that critique generally allows. Moreover, it gives us a clue to the way in which the relationship between primary identification, that is, the original meanings accorded to gender, and subsequent gender experience might be reframed.

The performance of drag plays upon the distinction between the anatomy of the performer and the gender that is being performed. But we are actually in the presence of three separate dimensions of significant corporeality: anatomical sex, gender identity and gender performance. If the anatomy of the performer is already distinct from the gender of the performer, and both of those distinct from the gender of the performance, then the performance suggests a dissonance not only between sex and performance but between sex and gender, and gender and performance. As much as drag creates a unified picture of "woman" (what its critics often oppose), it also reveals the distinctness of those aspects of gendered experience which are falsely naturalized as a unity through the regulatory fiction of heterosexual coherence. In imitating gender, drag implicitly reveals the imitative structure of gender itself – as well as its contingency. Indeed, part of the pleasure, the giddiness of the performance is in the recognition of a radical contingency in the relation between sex and gender in the face of cultural configurations of causal unities that are regularly assumed to be natural and necessary. In the place of the law of heterosexual coherence, we see sex and gender denaturalized by means of a performance which avows their distinctness and dramatizes the cultural mechanism of their fabricated unity.

The notion of gender parody defended here does not assume that there is an original which such parodic identities imitate. Indeed, the parody is *of* the very notion of an original; just as the psychoanalytic notion of gender identification is constituted by a fantasy of a fantasy, the transfiguration of an other who is always already a "figure" in that double sense, so gender parody reveals that the original identity after which gender fashions itself is itself an imitation without an origin. To be more precise, it is a production which, in effect, that is, in its effect, postures as an imitation. This perpetual displacement constitutes a fluidity of identities that suggests an openness to resignification and recontextualization, and it deprives hegemonic culture and its critics of the claim to essentialist accounts of gender identity. Although the gender meanings which are taken up in these parodic styles are clearly part of hegemonic, misogynist culture, they are nevertheless denaturalized and mobilized through their parodic recontextualization. As imitations which effectively displace the meaning of the original, they imitate the myth of originality itself. In the place of an original identification which serves as a determining cause, gender identity might be reconceived as a personal/cultural history of received meanings subject to a set of imitative practices which refer laterally to other imitations, and which, jointly, construct the illusion of a primary and interior gendered self or which parody the mechanism of that construction.

Inasmuch as the construct of women presupposes a specificity and coherence that differentiates it from that of men, the categories of gender appear as an unproblematic point of departure for feminist politics. But if we take the critique of Monique Wittig seriously, namely, that "sex' itself is a category produced in the interests of the heterosexual contract,[15] or if we consider Foucault's suggestion that "sex" designates an artificial unity that works to maintain and amplify the regulation of sexuality within the reproductive domain, then it seems that gender coherence operates in much the same way, not as a ground of politics but as its effect. The political task that emerges in the wake of this critique requires that we understand not only the "interests" that a given cultural identity has, but, more importantly, the interests and the power relations that establish that identity in its reified mode to begin with. The proliferation of gender style and identity, if that word still makes sense, implicitly contests the always already political binary distinction between genders that is often taken for granted. The loss of that reification of gender relations ought not to be lamented as the failure of a feminist political theory, but, rather, affirmed as the promise of the possibility of complex and generative subject-positions as well as coalitional strategies that neither presuppose nor fix their constitutive subjects in their place.

The fixity of gender identification, its presumed cultural invariance, its status as an interior and hidden cause may well serve the goals of the feminist project to establish a transhistorical commonality between us, but the "us" who gets joined through such a narration is a construction built upon the denial of a decidedly more complex cultural identity – or non-identity, as the case may be. The psychological language which purports to describe the interior fixity of our identities as men or women works to enforce a certain coherence and to foreclose convergences of gender identity and all manner of gender dissonance – or, where that exists, to relegate it to the early stages of a developmental and, hence, normative history. It may be that standards of narrative coherence must be radically revised and that narrative strategies for locating and articulating gender identity ought to admit to a greater complexity or it may be that performance may preempt narrative as the scene of gender production. In either case, it seems crucial to resist the myth of interior origins, understood either as naturalized or culturally fixed. Only then, gender coherence might be understood as the regulatory fiction it is – rather than the common point of our liberation.

Notes

1 Remarks, Center for the Humanities, Wesleyan University, Spring 1985.
2 Julia Kristeva, "Woman Can Never Be Defined," *New French Feminisms*, ed. Elaine Marks and Isabelle de Courtivron, (New York: Schocken, 1984).
3 Juliet Mitchell, *Psycho-analysis and Feminism*, (New York: Vintage, 1975), p. 377.
4 Jacqueline Rose, "Femininity and its Discontents," *Sexuality in the Field of Vision* (London: Verso, 1987), p. 90.
5 Luce Irigaray, "Any Theory of the Subject Has Already Been Appropriated by the Masculine," *Speculum of the Other Woman*, trans. Gillian Gill (Ithaca, NY: Cornell University Press, 1985), p. 140. See also "Is the Subject of Science Sexed?," *Cultural Critique*, Vol. 1, Fall 1985, p. 11.
6 For an interesting discussion of the political desirability of keeping the feminist subject incoherent, see Sandra Harding, "The Instability of the Analytical Categories of Feminist Theory," *Sex and Scientific Inquiry*, ed. Sandra Harding and Jean F. O'Barr (Chicago: University of Chicago Press, 1987).
7 See Jacqueline Rose's argument in "Femininity and its Discontents," *Sexuality in the Field of Vision*, pp. 90–94.
8 Sigmund Freud, *Three Essays on the Theory of Sexuality*, trans. James Strachey (New York: Basic Books, 1975), p. 1.
9 For a fuller exposition of Kristeva's positions, see my article "The Body Politics of Julia Kristeva" in the French Feminism issue of *Hypatia: A Journal of Feminist Philosophy*, Vol. 3, no. 3, pp. 104–108.
10 Roy Schafer, *A New Language for Psychoanalysis*, (New Haven, CT: Yale University Press, 1976), p. 177.
11 Michel Foucault, *Discipline and Punish* (New York: Panthenon, 1977), p. 29.
12 Foucault, *Discipline and Punish*, p. 30.
13 The assumption of binary sex is in no sense stable. For an interesting article on the complicated "sexes" of some female athletes and the medicolegal disputes about how and whether to render their sex decidable, see Jerold M. Loewenstein, "The Conundrum of Gender Identification, Two Sexes Are Not Enough," *Pacific Discovery*, Vol. 40, No. 2, 1987, pp. 38–39. See also Michel Foucault's *The History of Sexuality, Volume 1: An Introduction*, trans. Robert Hurley (New York: Vintage, 1980), pp. 154–155, and *Herculine Barbin, Being the Recently Discovered Memoirs of a Nineteenth-Century French Hermaphrodite*, trans. Richard McDougall (New York: Pantheon, 1986), pp. vii–xvii. For a feminist analysis of recent research into "the sex gene," a DNA sequence which is alleged to "decide" the sex of otherwise ambiguous bodies, see Anne Fausto-Sterling, "Recent Trends in Developmental Biology: A Feminist Perspective" (Departments of Biology and Medicine, Brown University).
14 Esther Newton, *Mother Camp: Female Impersonators in America* (Chicago: University of Chicago, 1972), p. 103.
15 Monique Wittig, "The Category of Sex," *Feminist Issues*, Vol. 2, p. 2.

SECTION FIVE

GENDERING EVERYDAY SPACES

Editors' introduction

Home is where the heart is; an Englishman's home is his castle; home sweet home; home town; homeland; born in the USA: all these common and sentimental sayings are bound up with ideas of space, place and gender. Less sentimental is the blunt British colloquial saying 'a woman's place is in the home' which must surely have its counterpart elsewhere in the world.

The social construction of the home as a place of familial pleasures, a place of leisure and rest – for men a sylvan and tranquil respite from the rigours of the city or the workplace and for women a supposedly safe haven – has a long history in the west. The dualism of home/workplace, the public sphere and the private arena, is mapped onto and constructs a gendered difference between male and female that has taken a particular spatial form in western nations since the industrial revolution. Although it has long been recognised that this division is an imperfect one, as millions of women have worked for wages of some sort outside and inside their homes, its significance rests in its normative function. Women who are 'out of place' – in the streets, in the workplace, in public – are seen in some senses as unnatural or unsexed. The early anxieties of the English factory inspectors that work would unfit young women for marriage was paralleled by Engels' anxiety that women's participation in the workforce, and ironically their prospect of involvement in radical politics, created a 'world turned upside down'. In contemporary Britain, the feminisation of the labour market has made this ideology hard to maintain, as we show in the next section, but yet women 'on the streets' still has a connotation of immorality. In stark contrast to the romanticisation of men who walked the streets of modern cities as flaneurs, women's unaccompanied appearance in public had different implications. 'Street walkers' is still the colloquial expression for prostitutes. The apparent belief of the British judiciary (out of touch as most judges are with the contemporary realities of women's lives) that women who venture out of doors in the late evening are tempting fate, and even 'asking' to be attacked, followed, harassed and raped led to feminist campaigns to 'reclaim the night'.

Perhaps the earliest feminist work by geographers and others, in Britain and the USA at least, was to overturn the idealistic notion that the home was either a haven or a place for leisure. Instead the domestic labour debate drew out the significance of all that reproductive labour done 'for love' in what the early feminist campaigner Margery Spring Rice termed in 1939, 'the small dark workshop of the home'. The myth that the home was a safe place for women and children was also challenged by analyses of the gendered power relations and male dominance that created home as a 'man's castle' – the space where his dependents, as tax and welfare services term women and children, were at their master's mercy. Studies of marital rape, domestic violence and child abuse revealed what police statistics had always shown: that the greatest risk of harm is from 'loved' ones.

This work had a salutory effect on many debates, both policy oriented and academic, but also lead to hysterical reactions in some cases – the 'Cleveland affair' in Britain in which accusations of widespread child abuse led to the removal of dozens of children from their homes is a 'good' example (see Campbell, 1988 and Cream, 1993). It also neglected the ways in which the home is also the arena of pleasure, desire, refuge and comfort, of socialisation as a gendered and sexed individual and the site of early childraising. This ambivalence in the home as a site of social relations was not apparent in the early critiques. They also neglected the differential significance of the home as a site of resistance and a longed for sphere for women in different class and racialised positions. For many working class women, their aim is to spend more time at home, out of reach of exploitative employers for example, and for many women in minority communities, their home and the local area is a network of resistance to the prying eyes and long arm of the state, and also to the dominant hegemonic social relations that construct them as 'outsiders' or the 'other'. The earlier characterisation of the home thus highlights western dominance of feminist theorising. The public–private divide is specific and does not naturally collapse into the home as the preceding work might suggest. For example, in Eastern European countries under state socialism, the home was not necessarily a place of domestication in the ways in which western feminism might suggest. In a system where public politics might be nothing less than a sham, the private spaces of home represented space of 'anti-politics' where discussion could occur. Whereas public space was constrained, private space offered the possibility of liberatory discussion (Einhorn, 1993; Goven, 1993).

Perhaps a greater sense of ambivalence to the idea of home and its gendered associations is evident among gay men and lesbians, as the very built form of the 'ideal home' embodies heterosexist distinctions and hegemonic ways of living. The master bedroom or the den, for example, are imbued with a particular way of being a man, and through the picture windows of modern homes in dormitory suburbs and new communities family life is open to surveillance by neighbours. Gloria Anzaldúa tells of a student who claimed that she had always thought that 'homophobia' refered to a lesbian fear of returning to the centre of patriarchial domination (Anzaldúa, 1987 p. 20).

In the articles that we include in this section we have attempted to make transparent and critique the naturalised assumptions about gender and gendered responsibilities, that are literally built into the form of western cities. In the short extract from **Hayden**'s magnificent feminist analysis of US landuse and planning policies in the postwar period, an early experiment in *Redesigning the American Dream* – the title of her book – is outlined. In the particular circumstances of the second world war when women's labour was essential to the war effort, an experimental new town was built on the coast of the Pacific northwest to facilitate the waged and domestic labour of (temporarily) single mothers. Sadly, this radical experiment in town planning did not last and the 'home, Mom and apple pie' principles of patriarchal design were re-established after the war. Many feminist urbanists and planners (Boys, 1990; Mackenzie, 1988; McDowell, 1983; Little, 1994; Little, Peake and Richardson, 1988; Roberts, 1991; Watson, 1988) have documented the ways in which women are restricted by the assumptions embedded in housing and land use policies – that their primary purpose in life is childrearing and dusting. Elizabeth **Wilson** reversed their arguments and suggested that it was the potential freedoms of the streets and the city that led male planners, architects and politicians, to attempt to pen women up in suburbs and new towns. Here we reproduce her introduction to the *Sphinx and the City* in which she recalls her personal pleasure in discovering London, as well as suggesting that cities are the site of a struggle between masculinised rigid order and feminised pleasurable anarchy. Her argument is provocative but overdrawn, reliant as it is on this single dichotomy. It also underplays the fear that many women have of public city spaces and the ambivalence and dangers for lesbian women and gay men of public displays of alternative sexualities in the predominantly heterosexist spaces of the city.

Gill **Valentine**'s paper explores this latter issue, drawing out the ways in which spatial arrangements reflect and reinforce gendered power relations. Except on Gay Pride days, where the gender identities and performances of homo- and bi-sexual communities become visible and overwhelm the heterosexist 'norm', 'alternative' ways of living have to remain private. Valentine perhaps underestimates the traditional significance of the city for all those seeking an escape from, or an alternative to, familial ways of living. Landuse divisions and the built form is above all based on the norm of a heterosexual family, rather than gender or sexuality *per se*. For single and group households alike the city may be a place of masquerade and freedom as well as a place to hide.

The somewhat romanticised rhetoric of Wilson and our comments so far neglect the sombre events of the 1990s when freedom and liberation are words far from the political agendas of both the right and the left. The last decade of the millenium has been marked a backlash against libertarian ideals (Faludi, 1992) by the re-establishment of what **Leslie**, in the extract from her paper, terms a 'new traditionalism' encouraging women to re-embrace the domestic ideals of the 1950s. Judith Stacey, author of the first extract in Section Two, has also documented, in her research on familial and gender relations in Silicon Valley in California (published as *Brave New Families*) the appeal of various forms of new traditionalism. She explores the appeal of evangelical movements to a number of women who in many ways we would have termed feminists in an earlier era, rejecting as they do traditional forms of domestic arrangements. But these women have not only rejected the 'romance' of heterosexual marriage and the nuclear family but have also turned against the rhetoric of feminism. Interestingly, Leslie comments on Wilson's work in her paper, suggesting that perhaps Wilson underplays the significance of the home in women's lives.

The 1990s have also seen the deepening of social divisions between the affluent and the poor, the increasing feminisation of poverty and a housing crisis that has left increasing numbers of households in severe debt and actually homeless. Growing numbers of women and children are joining the ranks of the dispossessed in British and US cities, living in appalling conditions in temporary accommodation, in non-residential spaces and on the streets. Attempts to analyse the extent and causes of homelessness, however, bring us right back to the definition of a home. Watson and Austerberry (1986), Venness (1992) and Somerville (1992), among others, have demonstrated a home is more than a house, more than a roof over one's head. If current housing conditions do not provide acceptable shelter in the sense that they are insecure, unsafe, not private, overcrowded, then they do not meet the idealised notion of a home. This is not to deny the absolute need of the growing numbers of women and children, as well as men who are completely without shelter, forced to sleep rough, in tunnels and on gratings, in boxes and temporary structures in too many cities in the world; nor to deny the sorts of emotional ties to people and places that may be forged by homeless people.

We are conscious, in this section above all, of the western biases in our own knowledge and in our selection of articles. The meaning of home in a range of other societies faced by famine or warfare, by the chillingly misnamed ethnic cleansing and by police violence in the ghettos and shanty towns of both First and Third World cities is an area of huge dispute, as people struggle to maintain even the barest minimum of anything approaching acceptable living standards. Comparative examinations of the meaning of space and of home in cross-cultural contexts owes much to the pioneering work of anthropologists in Oxford, especially Shirley Ardener (1981). More recently Daphne Spain (1992) has attempted to draw together work on the gendered division of space in a range of historical periods and different societies. There is also a rich and exciting literature about the implications and consequences of the huge variety of circumstances in which people grow up male and female (see for example Katz and Monk, 1993). Space alone precluded a selection of this work here.

The last decades have also been an era in which the relationships between place, nationality and identity have been transformed by the movement of people and capital and by technological developments. Seen from the west these changes have compressed space and time (Harvey, 1989), and facilitated greater and greater contacts between peoples who were previously divided by space and culture. The stretching out of social relations over bigger and bigger spaces – 'time–space distanciation', the phrase coined by the sociologist Giddens, is the obverse of the geographer's view of space–time compression. For the affluent westerner, the consequences may be largely positive, as 'we' enjoy travel and 'exotic' experiences, both at 'home' and abroad. But for millions of men, women and children, the struggle is to hold onto a sense of place, of belonging to a community or a nation state, against enforced movement. The internationalisation of capital and labour has 'displaced' innumerable people, forcing them to recreate their sense of home in a foreign place, challenging older associations between a people and a territory. As Stuart Hall has so eloquently rephrased Edward Said's distinction between the west and the rest: the rest are now in the west, forcing 'us' to rethink notions of home and homeland, and the relationships between identity and place. bell hooks has also talked about the ways in which moving, although painful, also enables migrants to push against 'oppressive boundaries set by race, sex and class domination'. The spatial language of marginality, of boundaries and the 'betweenness' of belonging to different locations is celebrated in a great deal of contemporary western post-enlightenment thought. The creative challenges of moving, evidenced in the remarkable flowering of art by migrant and dispossessed peoples, may be clear but neither should we underestimate the pain and costs of movement.

References and further reading

Anzaldúa, G 1987: *Borderlands/La Frontera: the New Mestiza*. Spinsters/Aunt Lute: San Francisco.

Ardener, S 1981: *Women and Space: Ground Rules and Social Maps*. Croom Helm: London.

Boys, J 1990: Women and the designed environment: dealing with difference. *Built Environment* **16**, 249–56.

Campbell, B 1988: *Unofficial Secrets*. Virago: London.

Cream, J 1993: Child sexual abuse and the symbolic geographies of Cleveland. *Environment and Planning A: Society and Space* **11**, 231–46.

Einhorn, B 1993: *Cinderella Goes to Market: Citizenship, Gender and Women's Movements in East Central Europe*. Verso: New York.

Faludi, S 1992: *Backlash: The Undeclared War Against Women*. Chatto and Windus: London.

Giddens, A 1990: *The Consequences of Modernity*. Polity: Cambridge.

Goven, J 1993: Gender politics in Hungary: autonomy and antifeminism. In Funk, N and Mueller, M (eds) *Gender Politics in Post-Communism*. Routledge: London. 224–40.

Hall, S 1991: The local and the global: globalization and ethnicity. In King, A (ed.) *Culture, Globalization and the World Systems*. Macmillan: London. 19–39.

Harvey, D 1989: *The Condition of Postmodernity*. Blackwell: Oxford.

Katz, C and Monk, J (eds) 1993: *Full Circles: Geographies of Women Over the Life Course*. Routledge: London.

Little, J 1994: *Gender, Planning and the Policy Process*. Pergamon: London.

Little, J, Peake, L and Richardson, P (eds) 1988: *Women in Cities: Gender and the Urban Environment*. Macmillan: Basingstoke.

McDowell, L 1983: Towards an understanding of the gender division of urban space. *Environment and Planning D: Society and Space* **1**, 59–72.

20 Delores Hayden

'Housing and American Life'

Excerpt from: *Redesigning the American Dream: the Future of Housing, Work and Family Life,* pp. 3–12. New York and London: W. W. Norton and Co. (1984)

Does our housekeeping raise and inspire us, or does it cripple us?

Ralph Waldo Emerson

Mired in spring mud, striped with the treads of bulldozers, Vanport City, Oregon, is a new town under construction. Concrete trucks pour foundations and give way to flatbed trucks that deliver cedar siding from the forests of the North-west. Carpenters, plumbers, and electricians try to stay out of each other's way as they work evenings, Saturdays, and Sundays. Architects from the firm of Wolff and Phillips confer on the site six, ten, a dozen times a day. "All my life I have wanted to build a new town," the project architect confides to a reporter, "but – *not this fast.* We hardly have time to print the working drawings before the buildings are out of the ground."

Near the town site, steel deliveries arrive at several shipyards on the Columbia River, where production is geared to an even more frenetic pace. Twenty-four hours a day the yards are open; cranes move against the sky, shifting materials. Tired workers pour out the gates at 8 AM, 4 PM, and midnight, each shift replaced by fresh arrivals – women and men in coveralls who carry protective goggles and headgear. The personnel office is recruiting as far away as New York and Los Angeles. They want welders, riveters, electricians. They offer on-the-job training, housing, child care, all fringe benefits. They also advertise for maintenance workers, nursery school teachers, elementary school teachers, and nurses. In ten months the personnel office does enough hiring to populate a new town of forty thousand people, white, black, Asian, and Hispanic workers and their families. This is the first time that an integrated, publically subsidized new town of this type has ever been built in the United States.

The chief engineer from the Federal Public Housing Authority is checking the last of the construction details as the residents' cars, pickup trucks, and moving vans start to arrive. It has been ten months from schematic designs to occupancy. The project architect is exhausted; never has he had a more demanding design program to meet, never a more impossible timetable. He has to rethink many basic questions in very little time, especially every idea he has ever had about normal family life, about men, women, and children. The program specified that he design affordable housing for all types and sizes of households, including single people, single-parent families, and non-family groups. He also had to design for low maintenance costs, and for energy efficiency, to make the maximum use of very scarce natural resources.

He was directed to emphasize public transportation by bus. His housing also had to be positioned in relation to several child care centers and job sites: "On a straight line," said James Hymes, the client in charge of child care,[1] because he didn't want parents to have to make long journeys to drop off or pick up their children.

"They certainly should become famous for that," the architect asserts, considering the large child care centers, open twenty-four hours a day, seven days a week (just like the shipyards), complete with infirmaries for sick children, child-sized bathtubs so that mothers don't need to bathe children at home, cooked food services so that mothers can pick up hot casseroles along with their children, and, most important of all, large windows with views of the river, so that children can watch the launchings at the yards. "There goes mommy's ship!" said one excited five-year-old. It all seems to work very well. And it costs seventy-five cents per day for each child.

It is March 1943. This new town, a product of World War II, is nicknamed Kaiserville after the industrialist who owns the shipyards. Everywhere, at home and abroad, Americans are singing a song at the top of the wartime hit parade . . . The amazing American woman has been the client as much as Henry J. Kaiser, who has built this town for her: Rosie the Riveter.

Six years later, another new town for seventy-five thousand people is being built at the same frantic pace in Hicksville, Long Island. In Hicksville, nothing is on a straight line. Roads curve to lead the eye around the corner, but every road is lined with identical houses. There is no industry in Hicksville except the construction industry. Each new Cape Cod house is designed to be a self-contained world, with white picket fence, green lawn, living room

Fig. 20.1 Women working as riveters and welders, 1944

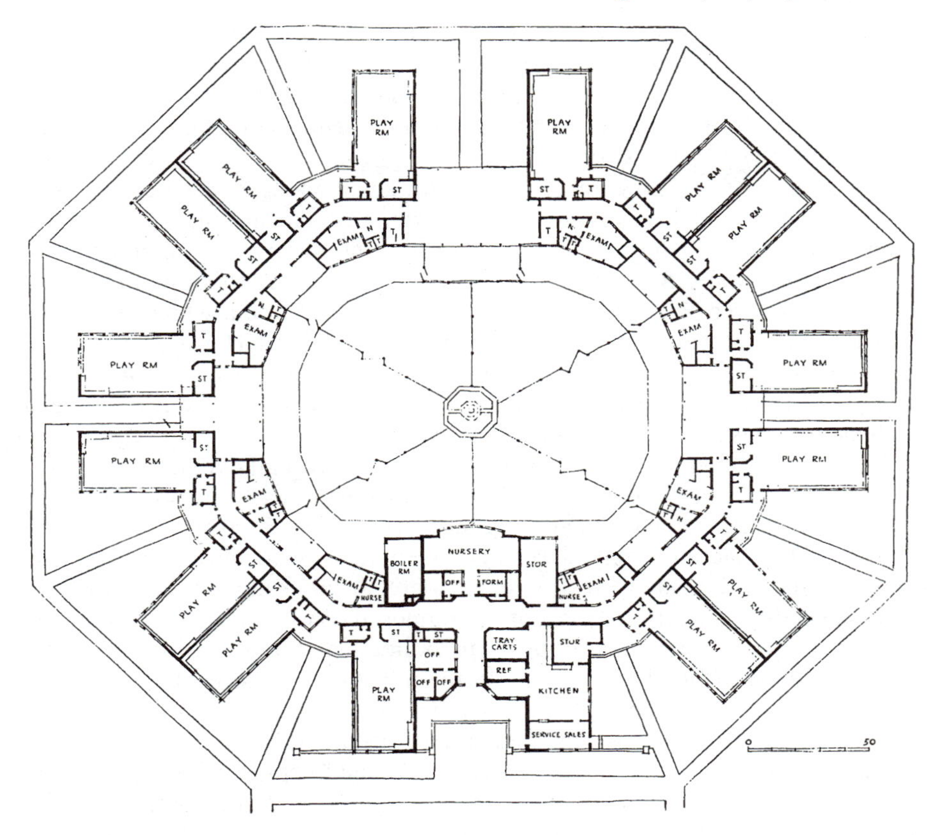

Fig. 20.2 Kaiser and Wolff, architects, plan of a day-care center at Vanport City, Oregon, 1944. This project represents the industrial strategy of building housing as a support for the female and male workers in the industrial labor force

with television set built into the wall, kitchen with Bendix washing machine built into the laundry alcove. Every family is expected to consist of male breadwinner, female housewife, and their children. Energy conservation is not a design issue, nor is low maintenance, nor is public transportation, nor is child care. A few parks and public swimming pools are planned to provide recreation.

In March 1949, the developer in Hicksville is ready to sell his houses. On a Wednesday the first prospective buyers appear to camp out in front of the sales office that will open the following Monday. It is the end of winter on Long Island: raw, wet, and cold. One of the women on the line of buyers camping out is pregnant; the developer's assistant rushes her to the hospital so she doesn't have her baby in the street. He returns and sets up a canteen for hot coffee and hot soup. News photographers come by and take pictures. On Monday night, in three and a half hours, the developer sells $11 million worth of identical houses. His company emerges as one of the great business successes of the postwar era. His Cape Cod house becomes the single most powerful symbol of the dream of upward mobility and homeownership for American families. Because of mortgage subsidies and tax deductions for homeowners, it is cheaper to buy a house in Hicksville than to rent an apartment in New York City.[2]

The creator of this new town, Bill Levitt, acknowledges that Levittown is not integrated, and he explains to a reporter that this is "not a matter of prejudice, but one of business. As a Jew I have no room in my mind or heart for racial prejudice. But, by various means, I have come to know that if we sell one house to a Negro family, then 90 to 95 percent of our white customers will not buy into the community."[3] In fact, the Federal Housing Authority does not, at this time, approve mortgage funds for integrated communities, or mortgages for female-headed families.[4] The prospective customers do not get a chance to make this choice for themselves.

This second new town – Levittown – becomes known all over the world as a model of American know-how just as the first new town – Vanport City – is being dismantled, some of its housing taken apart piece by piece. Yet both of these ventures had great appeal as solutions to the housing needs of American families, and both made their developers a great deal of money. Vanport City met the needs of a wartime labor force, composed of women and men of many diverse racial and economic groups. The builders of Vanport City responded to the need for affordable housing, on-the-job training and economic development for workers. They recognized that single parents and two-earner families required extensive child care services in order to give their best energies to production. The site design and landscaping of Vanport City were good, the economic organization was good, and the social services were superb (down to maintenance crews who would fix leaky faucets or repair broken windows), but the housing lacked charm. It looked like a "housing project," and the residents were renters not owners. Yet it was the most ambitious attempt ever made in the United States to shape space for employed women and their families. The US government supplied $26 million to build the housing. Kaiser made only a $2 profit on it, but he made a fortune on the ships the war workers built for him.[5]

Levittown met rather different needs from the ones provided for by Vanport City. Levitt's client was the returning veteran, the beribboned male war hero who wanted his wife to stay home. Women in Levittown were expected to be too busy tending their children to care about a paying job. The Cape Cod houses recalled traditional American colonial housing (although they were very awkwardly proportioned). They emphasized privacy. Large-scale plans for public space and social services were sacrificed to private acreage. Although they were small, a husband could convert his attic and then build an addition quite easily, since the houses covered only 15 percent of the lots. Levitt liked to think of the husband as a weekend do-it-yourself builder and gardener: "No man who owns his house and lot can be a Communist. He has too much to do," asserted Levitt in 1948.[6] His town was as ambitious as Vanport City, but Levitt aimed to shape private space for white working class males and their dependents. The pressures of war and the communal style of

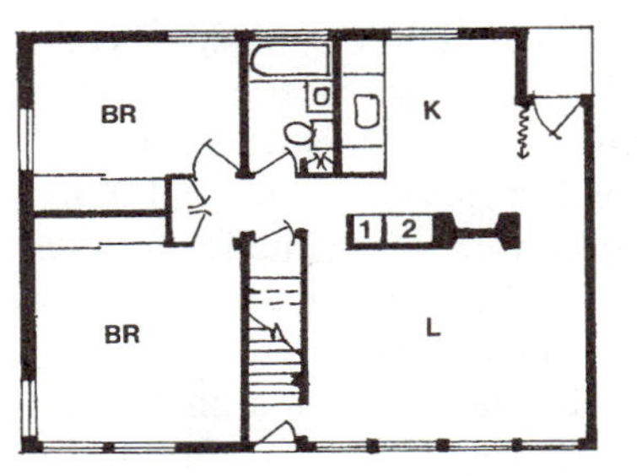

Figure 20.3 Plan of a Levitt house, 1952 model. (1) Bendix washing machine; (2) water heater

Figure 20.4 Levittown, 1955

military barracks living made suburban privacy attractive to many veterans, especially those with new cars to go with their new houses. Levitt made his fortune on the potato farms that he subdivided with the help of both federal financing programs for FHA and VA mortgages and federal highway programs to get people to remote suburbs. And as the landscaping matured, Levittown began to look better than the acres of little boxes some visitors perceived at the start.

Ironically, although Kaiser's highly praised wartime town lost the public relations battle to Levitt's postwar suburb, Kaiser himself was not a loser in this contest. He understood changing federal subsidy programs for housing, and after receiving wartime Lanham Act funds for Vanport City, which enabled him to expand his shipyards with new workers, Kaiser entered the post-World War II housing arena with new housing developments suited to

FHA and VA subsidies. On the West Coast, he built thousands of single-family houses in subdivisions much like Levitt's.[7] "Vets! No down!" read his signs. The losers were not the housing developers but the skilled white female and minority male and female workers, who lost their wartime jobs to returning white male veterans and found there were no postwar housing subsidies designed to help them find new jobs, new homes, and mortgages with easy terms.

In the same era a third new town was launched – Baldwin Hills Village, in Los Angeles, California. It did not make anyone a fortune: neither an industrialist like Kaiser nor a developer like Levitt. Funded by FHA and the Reconstruction Finance Corporation, its designers had sophisticated professional ambitions: to reinterpret the tradition of common land at the heart of New England's Puritan communities in a way that could be copied throughout the United States; to adapt the best low-cost European public housing designs of the previous decades to American housing programs and life styles; and to keep the car in its proper place for the sake of air quality, children's safety, and open space design.

Unlike the other two projects, the construction of the Baldwin Hills Village dragged on in the early 1940s. City engineers made complaints about the designers' refusal to cut roads through the site; the building department didn't like the great variety of apartment and townhouse layouts, and the plans had to be redrawn no less than ten times. Budget cuts removed three child care centers and a shopping center; land acquisition problems canceled the second phase of the project; Clarence Stein, the overall designer, discovered that his proposal for community kitchens had not been funded.[8]

Yet when the project finally opened as subsidized rental housing, several of the collaborating local architects moved to Baldwin Hills Village. As a statement of support for their values about good housing, they left elegant private homes in other parts of Los Angeles to be part of the new experiment and to make sure it worked. They felt extremely pleased that they had created low-rise, medium density housing with generous floor plans, sunlight, and lush landscaping. The cost was almost as low as that of other local public housing "projects." The residents enjoyed a belt of three parks running through the center of the site, as well as smaller landscaped courtyards, tot lots, and private fenced-in outdoor space for each family. There were common laundries and drying yards, common garages, and a community center with a swimming pool.

Baldwin Hills Village was integrated at the start, but within ten years many white tenants left and were replaced by nonwhite and female-headed households who were considered "problem families" in comparison with the homeowners living on suburban quarter acre plots around them.[9] Eventually a group that was formed to rescue the buildings turned the Village into condominiums, prohibited children under eighteen, tore out the tot lots, and installed a miniature golf course on the central green. Today the children for whom the village was designed are gone, and many of the elderly residents are still too afraid of crime to use its three magnificent parks. Yet the Baldwin Hills Village is not as much of a ghost town as Vanport City. Part of Vanport

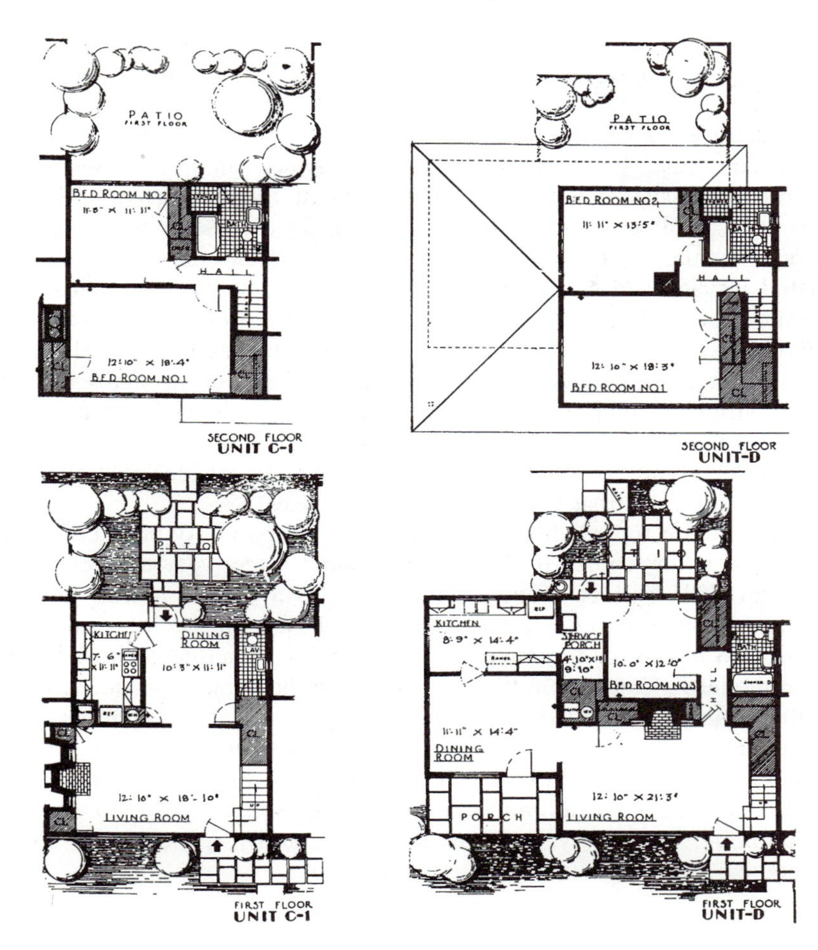

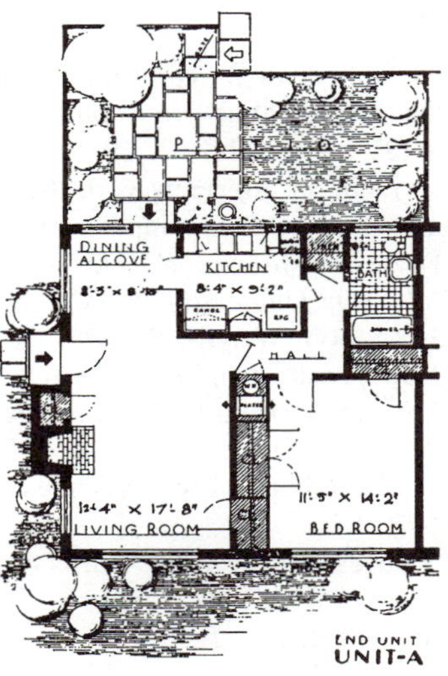

Figure 20.5 Clarence Stein, Robert Alexander, *et al.*, Baldwin Hills Village, Los Angeles, 1938–1942. Baldwin Hills represents the neighborhood strategy of building homes. The one-bedroom (A), two-bedroom (C-1), and three-bedroom (D) houses were subsidized, rental units organized around shared open spaces

City was dismantled after World War II. The rest was destroyed in a flood, and today the site of what was once the fifth-largest city in the Northwest is a park.[10]

Baldwin Hills Village and Vanport City whisper the stories of planned settlements based on complex visions of the American dream. Both sites raise the broadest issues in housing and urban design: the relationship of housing to jobs and social services, the need to design for diverse household types, the rights of female and minority workers to housing and jobs, the need for both spatial privacy and spatial community, the need for the regulation of automobiles, the problem of affordability, and the question of homeownership or tenancy as it concerns the stability of residential neighborhoods. Baldwin Hills Village and Vanport City are models of earlier struggles to come to terms with the social and economic programing of affordable housing. These projects, now largely forgotten, remind us that the need for affordable housing for all Americans is not a new problem, nor are the design problems and political questions that housing raises novel ones.

Very little of today's housing follows the Vanport City model of the home as a support for women in the industrial labor force; very little emulates the Baldwin Hills Village model of the home as a part of a well thought-out neighborhood. Most American housing is based on Levitt's model of the home as a haven for the male worker's family. Americans chose the Levittown model for housing in the late 1940s; we have mass-produced the home as haven and transformed our cities to fit this model and its particular social, economic, and environmental shortcomings. This choice is at the heart of the housing problem of the 1980s. Americans cannot solve their current housing problems without reexamining the ideal of the single-family house – that is, reexamining its history, and the ideals of family, gender, and society it embodies, as well as its design and financing.

Notes

1 "Designed for 24 Hour Child Care," *Architectural Record* (March 1944) 86.
2 "Nation's Biggest Housebuilder," *Life* (May 22, 1950) 75–76; "Up from the Potato Fields," *Time* (July 3, 1950) 68.
3 Levitt in an interview with the *Saturday Evening Post* (August 7, 1954) 72. For a full study of this company and its policies, see Herbert Gans, *The Levittowners: Ways of Life and Politics in a New Suburban Community* (1967: New York: Columbia University Press, 1982).
4 Gwendolyn Wright, *Building the Dream: A Social History of Housing in America* (New York: Pantheon, 1981) 247–48; Esther McCoy, "Gregory Ain's Social Housing," *Arts and Architecture* 1 (Winter 1981) 66–70.
5 "Vanport City," *Architectural Forum* (August 1943) 53.
6 Eric Larrabee, "The Six Thousand Houses that Levitt Built," *Harpers* (September 1948) 84.
7 "Houses Off the Line: Burns and Kaiser Throw the Switch," *Architectural Forum* (October 1946) 10.
8 Clarence Stein, *Toward New Towns for America* (1957: Cambridge, Mass: MIT Press, 1971) 188–216, 218.

9 Personal interview with a former administrator for the schools in that area, 1982.
10 Leverett Richards, "And So a City Died," *Oregontan* (December 3, 1975); Kate Huang, "Women, War, and Work," unpublished paper, 1982; the Northwest Women's History Project in Portland, Oregon, has done some oral history and a slide show on shipyard workers; Sherna Gluck of California State University, Long Beach, has also researched an extensive oral history "Rosie the Riveter Revisited."

21 Elizabeth Wilson

'Into the Labyrinth'

Excerpts from: *The Sphinx in the City: Urban Life, the Control of Disorder and Women*, pp. 1–11. London: Virago (1991)

> Autobiography has to do with time, with sequence and with what makes up the continuous flow of life. Here, I am talking of a space, of moments and discontinuities.
>
> Walter Benjamin, *One Way Street* (1932)

'Now let me call back those who introduced me to the city', wrote Walter Benjamin in the 1930s, remembering his childhood in old Berlin. In his case, it was nursemaids; in mine, my mother. She planted within me, never to be eradicated, a conviction of the fateful pleasures to be enjoyed and the enormous anxieties to be overcome in discovering the city.

Every excursion we made together was an immense labour, a strenuous and fraught journey to a treacherous destination: we waited for buses that never came, were marshalled into queues that never grew shorter, walked down endless streets in the hot sun. Our destinations also were terrible. The Tower of London, Hampton Court and Madame Tussaud's were theatres of cruelty: *here* was the exact spot upon which Anne Boleyn was beheaded; *this* was the gallery along which Catherine Howard ran desperately to beg Henry the Eighth for mercy; here was the Chamber of Horrors with its electric chair.

There were also the crowds of that first, weary, hot, London summer. I had never seen crowds like those. The insolence, the promiscuity of the crowd, jostling my mother and myself, seemed like a vast yawn of indifference. The stale suits and rayon dresses brushed against us, bodies against bodies. The air seemed yellow with a kind of blasé fatigue. My mother tried to keep her hat tipped forward, her little veil in place, her corsage of soft suede anemones – blue, rose-red and purple – crisply pinned against the navy crêpe of her dress, but I felt the vulnerability of her pretensions exposed, and together we seemed so insignificant and lost.

Our visits to the Zoo and to Kensington Gardens expressed some longing for what was so absent from the stony streets in which we lived and wandered:

a memory of the rural life we had left behind. Walter Benjamin recalled the park as a scene of bourgeois domestic harmony:

> There were serpentine paths near the lake and . . . benches . . . at the edge of the sand pit with its ditches, where toddlers dig or stand sunk in thought until bumped by a playmate or roused by the voice of a nursemaid from the bench of command; there she sits stern and studious, reading her novel and keeping the child in check while hardly raising an eyelid until, her labour done, she changes places with the nurse at the other end of the bench, who is holding the baby between her knees and knitting. Old, solitary men found their way here, paying due honour, amid these scatterbrained womenfolk, among the shrieking children, to the serious side of life: the newspaper.[1]

and perhaps my mother hoped to find a lost tranquillity in the green vista with its lines of trees in faultless perspective. The flowers and especially the spring blossoms, like all flowers in cities, appeared as a luxury item set against the urban fabric, rather than as an invasion of nature or a rural enclave; they symbolised some other, idealised world.

The Zoo was a very different experience, for there again were the crowds, jostling to stare at the infant gorilla and the apes. This was an old-time crowd, more of an eighteenth-century 'mob' come to stare at whatever exotic spectacle was on offer – a hanging, lunatics at Bedlam. Screams of laughter greeted the antics of the chimpanzees, those caricatures of humanity. Family groups approached the tiger's cage with a frisson of fear. Always for me the great question was whether to brave the reptile house, where huge snakes lay so creepily still. Their malevolent, horrible inertia gave me nightmares, yet I could never resist. 'I won't look' – but I always did.

The reptile house was for me that Minotaur's chamber cited by so many writers who liken the city to a labyrinth. Benjamin's Minotaur was 'three-headed', being the three prostitutes in a small Parisian brothel. In either case, fear mixed with an obscure or suspect pleasure lay at the heart of the city's secret courtyards and alleyways.

In Benjamin's adolescence the Berlin cafés played their part in introducing him to the world of pleasure that is one layer in the geology of the social city, and years later he remembered the names of those cafés like an incantation: the Romanisches Café, the Viktoria, the West End Café. Those salons were neither exactly public nor private space, and yet partook of both, and in them bohemia and the bourgeoisie mingled as part of the quintessential urban spectacle:

> For one of the most elementary and indispensable diversions of the citizen of a great metropolis, wedged, day in, day out, in the structure of his office and family amid an infinitely variegated social environment, is to plunge into another world, the more exotic the better. Hence the bars haunted by artists and criminals. The distinction between the two, from this point of view, is slight. The history of the Berlin coffeehouses is largely that of different strata of the public, those who first conquered the floor being obliged to make way for others gradually pressing forward, and thus to ascend the stage.[2]

There were, of course, no comparable cafés in London in the mid 1950s, when I was myself of an age to explore the city alone, coffee bars and jazz clubs offering a poor substitute. Soho drinking clubs were barred to me, in any case unknown. I nevertheless roamed London, solitary, engaged in that urban search for mysteries, extremes and revelations, a quest quite other than that of the wanderer through the natural landscape: a search less hallowed, yet no less spiritual.

Christine Mallet Joris's *Into the Labyrinth* was the title of the second lesbian novel I ever read (the first being, of course, Radclyffe Hall's *The Well of Loneliness*). *Into the Labyrinth* was French, and, unlike Radclyffe Hall's Edwardian romance, fitted precisely into my aimless, desperate walks and rides around London's streets, squares and inner suburbs. The heroine, a schoolgirl, discovered love in a house on a street called, romantically – and inappropriately – the *Rempart des Béguines* (the Rampart of Nuns). The adventures and sufferings attendant upon her sexual initiation took place in the bedrooms, hotels, the theatres and cafés of a great city – a city like a magic set of boxes, with, inside each box, a yet smaller and more secret one.

This recurring image, of the city as a maze, as having a secret centre, contradicts that other and equally common metaphor for the city as labyrinthine and centreless. Even if the labyrinth does have a centre, one image of the discovery of the city, or of exploring the city, is not so much finally reaching this centre, as of an endlessly circular journey, and of the retracing of the same pathways over time.

Yet one never retraces the same pathway twice, for the city is in a constant process of change, and thus becomes dreamlike and magical, yet also terrifying in the way a dream can be. Life and its certainties slither away from underfoot. This continual flux and change is one of the most disquieting aspects of the modern city. We expect permanence and stability from the city. Its monuments are solid stone and embody a history that goes back many generations. Rome was known as the 'Eternal City'. Yet, far from being eternal, in the sense of being outside time, Rome, like all cities, is deeply time-bound.[3] Although its history gives it its character, and a patina of durability, in modernity especially the city becomes ever more changing. That which we thought was most permanent dissolves as rapidly as the kaleidoscopic spectacle of the crowds and vehicles that pass through its streets.

The London of the 1990s, for all the destruction that has occurred, is a livelier place than gloomy 1950s London. Today I am nevertheless sometimes conscious of a nostalgia for that vanished city: for the hushed interior spaces of long-defunct department stores with their carpeted trying-on rooms; for the French provision stores of Soho, replaced first by stripshows, later by fashion boutiques; but most of all for the very gloom and shabbiness now banished by gentrification, redevelopment and the commercialisation of leisure. It felt safe, and as you wandered through the streets you sensed always that pervasive English privacy, of lives veiled by lace curtains, of a prim respectability hiding strange secrets behind those inexpressive Earls Court porticoes.

In my mid teens I was unfamiliar with the writings of Benjamin, but I intuitively identified with an urban consciousness of which his reminiscences are one of the most beautiful examples. This consciousness had been developed by the dandies and '*flâneurs*' (strollers, loiterers) of mid-nineteenth-century Paris. They had relished the kaleidoscope of urban public life and had created from it a new aesthetic, perceiving a novel kind of beauty in streets, factories and urban blight. In the 1930s the anthropologist Claude Lévi-Strauss discovered this beauty in an even more intense form in the Latin American cities he visited. He wrote that although 'São Paulo was said at the time to be an ugly town . . . I never thought São Paulo was ugly; it was a "wild" town, as are all American towns.' This quality of 'wildness' was, Lévi-Strauss felt, due to exaggerated and surreal contrasts. Extremes of wealth and poverty, of enjoyment and misery, made an essential contribution to this perception of the city. It was just those things that were shoddy and awful about city life that constituted its seduction, its peculiar beauty. What Lévi-Strauss found strange and evocative about the cities of the New World was their premature decrepitude, the incongruity of concrete skyscrapers alongside shanty towns, of Victorian Gothic churches jumbled up with bleak warehouses, creating a stone landscape as melancholy as it was striking.[4]

His perception, like that of the dandies, 'makes strange' the familiar and disregarded aspects of city life. It inverts our values: what was once seen as marginal becomes the essence of city life and that which makes it truly beautiful, even if its beauty is a beauty of ugliness. This new definition of beauty and meaning places the underside or 'Other' of city existence at the centre of consciousness.

Lévi-Strauss was a latterday *flâneur* who discovered in the streets of São Paulo and Chicago a heartrending nostalgia not for the past but for the future. Their street canyons and windswept vistas suggested a lost future that was never to be, and ached with the yearning of human aspirations destined ever to fall short of the grandiose hopes that inspired them.

This sophisticated urban consciousness, which, as we shall see, reached a high point in central Europe in the early twentieth-century, was an essentially male consciousness. Sexual unease and the pursuit of sexuality outside the constraints of the family were one of its major preoccupations.

This in itself made women's very presence in cities a problem. The city offers untrammelled sexual experience; in the city the forbidden – what is most feared and desired – becomes possible. Woman is present in cities as temptress, as whore, as fallen woman, as lesbian, but also as virtuous womanhood in danger, as heroic womanhood who triumphs over temptation and tribulation. Writers such as Benjamin concentrated upon their own experience of strangeness in the city, on their own longings and desires, but many writers more definitely and clearly posed the presence of women as a problem of order, partly *because* their presence symbolised the promise of sexual adventure. This promise was converted into a general moral and political threat.

Nineteenth-century planning reports, government papers and journalism created an interpretation of urban experience as a new version of Hell, and it would even be possible to describe the emergent town-planning movement – a movement that has changed our cities almost beyond recognition – as an organised campaign to exclude women and children, along with other disruptive elements – the working class, the poor, and minorities – from this infernal urban space altogether.

Sexuality, was only one source of threatening ambiguity and disorder in the city. The industrial city became a crucible of intense and unnerving contrasts. The hero, or less often the heroine, of urban literature was lured by the astonishing wealth and opportunity, threatened by the crushing poverty and despair of city life. Escape and entrapment, success and disaster offered heightened, exaggerated scenarios of personal triumph or loss of identity.

There was another contradictory aspect of city life. The sociologist Max Weber argued that the western city developed a typical form of political organisation: the democracy. Feudal lords found that they were unable to retain their hold over their vassals, bondsmen and serfs once these had settled in cities. It was in the western late medieval city that men and women for the first time came together as individuals rather than as members of a kin group, clan or feudal entourage. The western city evolved political organisations which displaced existing paternalistic and patriarchal forms, and so the way was opened both to individualism and to democracy during the transition from feudalism to capitalism.

By the nineteenth century this had become contradictory because commentators and reformers of that period claimed to value individualism and democracy, but as cities grew, the mob became a revolutionary threat.

There were women as well as men in the urban crowd. Indeed the crowd was increasingly invested with female characteristics, while retaining its association with criminals and minorities. The threatening masses were described in feminine terms: as hysterical, or, in images of feminine instability and sexuality, as a flood or swamp. Like women, crowds were liable to rush to extremes of emotion. As the rightwing theorist of the crowd, Le Bon, put it, 'Crowds are like the sphinx of ancient fable; it is necessary to arrive at a solution of the problems offered by their psychology or to resign ourselves to being devoured by them.' At the heart of the urban labyrinth lurked not the Minotaur, a bull-like male monster, but the female Sphinx, the 'strangling one', who was so called because she strangled all those who could not answer her riddle: female sexuality, womanhood out of control, lost nature, loss of identity.[5]

Yet the city, a place of growing threat and paranoia to men, might be a place of liberation for women. The city offers women freedom. After all, the city normalises the carnivalesque aspects of life. True, on the one hand it makes necessary routinised rituals of transportation and clock watching, factory discipline and timetables, but despite its crowds and the mass nature of its life, and despite its bureaucratic conformity, at every turn the city dweller is also offered the opposite – pleasure, deviation, disruption. In this sense it would be possible to say that the male and female 'principles' war with each other at the very heart of city life. The city is 'masculine' in its triumphal scale, its towers and vistas and arid industrial regions; it is 'feminine' in its enclosing embrace, in its indeterminacy and labyrinthine uncentredness. We might even go so far as to claim that urban life is actually based on this perpetual struggle between rigid, routinised order and pleasurable anarchy, the male–female dichotomy.

Perhaps the 'disorder' of urban life does not so much disturb women. If this is so, it may be because they have not internalised as rigidly as men a need for over-rationalistic control and authoritarian order. The socialisation of women renders them less dependent on duality and opposition; instead of setting nature against the city, they find nature *in* the city. For them, that invisible city, the 'second city', the underworld or secret labyrinth, instead of being sinister or diseased as in the works of Charles Dickens and many other writers is an Aladdin's cave of riches. Yet at the same time, it is a place of danger for women. Prostitutes and prostitution recur continually in the discussion of urban life, until it almost seems as though to be a woman – an individual, not part of a family or kin group – in the city, is to become a prostitute – a public woman.

The city – as experience, environment, concept – is constructed by means of multiple contrasts: natural, unnatural; monolithic, fragmented; secret, public; pitiless, enveloping; rich, poor; sublime, beautiful. Behind all these lies the ultimate and major contrast: male, female; culture, nature; city, country. In saying this I am not arguing (as do some feminists) that male–female difference creates the deepest and most fundamental of all political divisions. Nor am I arguing that the male/female stereotypes to which I refer accurately reflect the nature of actual, individual men and women. In the industrial period, nonetheless, that particular division became inscribed on urban life and determined the development and planning of cities to a surprising degree and in an extraordinarily unremarked way. It will be one purpose of this book to explore how underlying assumptions, based both on this unconscious division and on consciously spelt-out ideas about women's rightful place, have determined the shape of contemporary cities.

The contradictions and intensity of urban life have produced strong responses, one of which has been a corrosive anti-urbanism. For many years I took for granted the assumption that a great city was the best place to live, and Paris and New York seemed the only possible – and even more magical – alternatives to the shabbier but comfortable and accommodating ambience of sub-bohemian London. It was only my involvement in 'alternative' radical politics in the 1970s which alerted me to the hatred many 'progressive' people feel for cities, and to an alien point of view, which self-righteously attacked the ugliness and vulgarity of urban life while setting out some rural or small town idyll as the desired alternative. I had known that many rightwing writers feared the modern city as destructive of the traditional patriarchal order; but to me the anti-urbanism of the left seemed like a betrayal, and made me permanently disillusioned with utopianism. William Morris in particular – a writer who seems exempt from any criticism by socialists to this day – demonstrated in his utopian *News from Nowhere* a retreat from modernity and a nostalgia for patriarchalism that I found suffocating.

Anti-urbanism has a long history, partly related to industrialisation; developments in the 1980s and early 1990s have served to make such ideas even more threatening and more plausible. One development is our growing ecological consciousness; another the redevelopment of inner cities as uninhabited office or business districts; a third the parallel growth of inner-city ghettoes inhabited by a so-called 'underclass'; fourthly, the simultaneous suburbanisation of more and more of the countryside. The result is that today in many cities we have the worst of all worlds: danger without pleasure, safety without stimulation, consumerism without choice, monumentality without diversity. At the same time, larger and larger numbers of people inhabit zones that are no longer really either town or countryside.

We need a radically new approach to the city. We will never solve the problems of living in cities until we welcome and maximise the freedom and autonomy they offer and make these available to all classes and groups. We must cease to perceive the city as a dangerous and disorderly zone from which women – and others – must be largely excluded for their own protection. There are other issues, of course, equally important. Leisure and consumption must cease to be treated purely as commodities controlled by market forces, nor can adequate housing ever be provided so long as it is regarded as a mere byproduct of urban development and property speculation.

Yet at the 'commonsense' level of our deepest philosophical and emotional assumptions, the unconscious bedrock of western culture, it is the male–female dichotomy that has so damagingly translated itself into a conception of city culture as pertaining to men. Consequently, women have become an irruption in the city, a symptom of disorder, and a problem: the Sphinx in the city.

Women are placed at the centre of my argument for this reason. For a woman to make an argument in favour of urban life may come as a surprise. Many women and much feminist writing have been hostile to the city, and recent feminist contributions to the discussion of urban problems have tended to restrict themselves narrowly to issues of safety, welfare and protection.[6] This is a mistake, since it re-creates the traditional paternalism of most town planning. Women's experience of urban life is even more ambiguous than that of men, and safety is a crucial issue. Yet it is necessary also to emphasise the other side of city life and to insist on women's right to the carnival, intensity and even the risks of the city. Surely it is possible to be both pro-cities and pro-women, to hold in balance an awareness of both the pleasures and the dangers that the city offers women, and to judge that in the end, urban life, however fraught with difficulty, has emancipated women more than rural life or suburban domesticity.

Perhaps we should be happier in our cities were we to respond to them as to nature or dreams: as objects of exploration, investigation and interpretation, settings for voyages of discovery. The 'discourse' that has shaped our cities – the utilitarian plans of experts whose goal was social engineering – has limited our vision and almost destroyed our cities. It is time for a new vision, a new ideal of life in the city – and a new, 'feminine' voice in praise of cities.

Notes

1 Benjamin (1979), p. 296.
2 Benjamin (1979), p. 312.

3 See Pike (1981), passim, to which I am indebted for the ideas developed in this section. See also Sizemore (1984).
4 Lévi-Strauss (1976), pp. 120–21.
5 Huyssen (1986), pp. 52–3, quoting Le Bon, Gustave (1981) *The Crowd*, Harmondsworth: Penguin, pp. 39, 52.
6 See, particularly, Little, Peake and Richardson, eds. (1988).

References

Benjamin, Walter (1979) *One Way Street*, London: New Left Books
Huyssen, Andreas (1986) *After the Great Divide: Modernism, Mass Culture, Postmodernism*, London: Macmillan
Lévi-Strauss, Claude (1976) *Tristes Tropiques*, Harmondsworth: Penguin
Little, Jo, Peake, Linda and Richardson, Pat, eds (1988) *Women in Cities: Gender and the Urban Environment*, London: Macmillan
Pike, Burton (1981) *The Image of the City in Modern Literature*, Princeton, New Jersey: Princeton University Press
Sizemore, Christine (1984) 'Reading the City as Palimpsest: The Experiential Perception of the City in Doris Lessing's *The Four Gated City*', in Squier, ed. (1984)
Squier, Susan Merrill ed. (1984) *Women Writers and the City: Essays in Feminist Literary Criticism*, Knoxville: University of Tennessee Press

22 Gill Valentine

'(Hetero)Sexing Space: Lesbian Perceptions and Experiences of Everyday Spaces'

Excerpts from: *Environment and Planning D: Society and Space* **11, 395–413 (1993)**

> There's nothing like a Saturday morning in the town centre to make you feel unconventional
>
> (Lesbian, middle class, 20s).[1]

It is well established in the geographical literature that age and gender have a profound impact on individuals' perceptions and experiences of everyday spaces (Hart, 1978; Valentine, 1989). It is argued that, in particular, differences between the sexes stem from inequalities of power between men and women which are reflected in the way space is designed, occupied, and controlled. But, as the quote above suggests, the ability to appropriate and dominate places and hence influence the use of space by other groups is not only the product of gender; heterosexuality is also powerfully expressed in space.

The myth of a private–public dichotomy

The dominant form of sexuality in modern Western culture is heterosexuality, despite the fact that same-sex relationships have occurred throughout time and across different societies and cultures with varying degrees of acceptability and frequency (D'Emilio and Freedman, 1988). The term homosexuality was coined by the medical profession in the late 19th century, "it was primarily viewed through a medical framework as a pathology, its causes were located in biological degeneracy or family pathology, and treatments ranged from castration to psychoanalysis" (Plummer, 1988, p. 23). Although homosexuality is no longer treated as a mental illness, the stigma and negativity surrounding same-sex relationships prevail despite the fact that there has been a shift in social consensus about the role of sexuality: "from reproduction to intimacy and personal happiness, and from family and community to the individual" (Herek, 1992a, p. 93).

Ideologically, heterosexuality is also linked to the notion of gender identity, that is, the shared beliefs and meanings attributed to what it means to be a man or a woman (masculinity and femininity). This is because the notion of opposite-sex relationships presumes, first, that there is a binary distinction between being a man and being a woman, and, second, that these binary gender identities (masculinity–femininity) map neatly onto binary sexed bodies (man–woman) (Butler, 1990).

'Normal' masculinity and feminity are defined in relation to one another such that the construction and reproduction of gender identities both create and perpetuate male superiority, or patriarchy (Coveney *et al.*, 1984). The asymmetrical (opposite-sex parents) family is by definition a heterosexual concept and hence childrearing is also heterosexually identified (Herek, 1992a).

To be gay, therefore, is not only to violate norms about sexual behaviour and family structure but also to deviate from the norms of 'natural' masculine or feminine behaviour. These norms change over space and time, and hence sexuality is not defined merely by sexual acts but exists as a process of power relations (Foucault, 1988). Heterosexuality in modern Western society can therefore be described as a heteropatriarchy, that is, a process of sociosexual power relations which reflects and reproduces male dominance.

Ostensibly, sexuality would appear to belong in the private space of the home, not the public sphere of the office or the restaurant. This assumption is reflected in a US survey of heterosexual attitudes to homosexuals, which produced a common response from participants that they had no objection to homosexuals as long as they did not flaunt their sexuality in public (Herek, 1987), an assumption repeated in similar UK surveys. But this cultural dichotomy (sic) locating sexuality in private rather than public space, is based on the *false* premise that heterosexuality is also defined by private sexual acts and is not expressed in the public arena. Yet, heterosexuality is institutionalised in marriage and in the law, tax, and welfare systems, and is celebrated in public rituals such as weddings and christenings. This therefore highlights the error of drawing a simple polar distinction between public and private activities, for heterosexuality is clearly the dominant sexuality in most everyday environments, not just private spaces, with all interactions taking place between sexed actors. However, such is the strength of the assumption of the 'naturalness' of heterosexual hegemony, that most people are oblivious to the way it operates as a process of power relations in *all* spaces. However, to be lesbian or gay[2] is both to perceive and to experience the heterosexuality of the majority of environments.

This paper will therefore use research carried out in a town in England[3] to explore lesbian perceptions and experiences of everyday spaces (home, workplace, social spaces, service environments, and public open spaces). The findings are based on forty in-depth interviews (which were taped and transcribed) with women aged between 18 and 60 years who currently identify themselves as lesbian and are either in a lesbian relationship or are seeking a female partner. Some of these women previously identified themselves as heterosexual. Of these, some made a distinct break between their heterosexual and gay lives; others made a more gradual transition to a lesbian identity, living with a male partner whilst coming to identify themselves as lesbian. Thirteen have been married, and eight have children from previous heterosexual relationships.

By only concentrating on the perceptions and experiences of women who currently identify themselves as lesbians, this paper appears to dichotomise sexuality into 'gay' or 'straight'. However, I recognise that sexual identities can be fluid; and that there are multiple sexual identities within and outside the dominant heterosexual–homosexual discourses. For example, bisexuals are commonly 'outsiders' in environments appropriated and controlled by heterosexuals and lesbians and/or gay men (Eadie, 1992).

Heterosexualised spaces

House and home

Housing in 19th and 20th century Britain has been and is "primarily desgined, built, financed and intended for nuclear families – reinforcing a cultural norm of family life with heterosexuality and patriarchy high on the agenda" (Bell, 1991, p. 325). For example, common features such as 'master' bedroom and smaller bedrooms for children physically represent and reinforce the cultural norm of the reproductive monogamous family unit. Although the significance and use of different rooms have changed over time with changing class and gender relations (for example, the decline in domestic labour and the mechanisation of domestic tasks have made the kitchen a more 'respectable' room), housing design continues to express a privatised form of family life (Matrix, 1984) in which all tasks such as cooking, eating, and childcare are contained within the family.

However, lesbians are less likely to have children than heterosexual couples; the most common estimate is that only 25% of lesbians are parenting (Adler and Brenner, 1992), and those influenced by feminist politics are more likely to be nonmonogamous and to want to organise childcare and domestic chores on a collective basis (Ettorre, 1978). Yet there is no housing stock designed and built for nonheteropatriarchial life-styles. In the 1970s, therefore, a significant trend was evident amongst lesbian feminists of creating their own housing forums through squatting and communal living (Egerton, 1990).

Housing provision is also orientated towards the asymmetrical family. Many lesbians share the economic marginalisation of heterosexual women, but public-housing providers and managers often do not recognise same-sex 'family units', and those without children are rarely eligible for the declining stock of public-authority housing (Anlin, 1989). In addition, gay partners do not have the same legal rights to succeed to a tenancy on the death of a partner. Although women with sufficient income to buy their own homes can overcome barriers of access to the housing market, some lesbians interviewed also claimed that their house purchases were influenced by their perceptions of the sexuality of space. In particular, women claim that they have or would consciously avoid living in rural communities because they perceive towns as more likely to have a gay community. Also, urban areas are seen as more anonymous, and hence lesbians believe it is easier to manage and control others' images of their sexual identity in such an environment (Valentine, 1993a). Similarly, some towns, such as Brighton, are perceived to have a large and active gay community (Valentine, 1993b), whereas others have a heterosexual image because of their association with suburban family life.

Decisions about specific locations are also motivated by perceptions of the heterosexuality of space at a local level. A number of women said they had chosen to avoid modern middle-class housing estates because they were conscious that as two women they would stand out in neighbourhoods they perceived to contain predominantly asymmetrical families, and that this would make them feel 'out of place'. More insidiously, they were also aware that by 'standing out' as an 'abnormal' family unit their property could become a target for antigay violence. This is reflected in the fact that five out of the forty women have experienced violence or other forms of harassment from neighbours because of their sexuality, two know of friends whose property has been attacked, and others have overheard neighbours' aggressive comments about their sexuality, such as 'bloody lezzies'.

It is because of such incidents or their possibility that a number of the women interviewed have consciously chosen to live in neighbourhoods of mixed age and race where they perceive it is easier to blend in. In particular, one housing area has developed a reputation as a lesbian residential (though not as an institutional) ghetto (Valentine, 1993b) and consequently a snowball effect appears to be in operation, with other lesbians being drawn to the area to be near friends and because the neighbourhood is perceived to be tolerant.

The older women interviewed (aged 30–60 years) who were living with a female partner were most conscious of this dual risk of feeling out of place or being harassed as an outsider, because they were conscious that the absence of a male partner highlighted the fact that they were obviously fulfilling neither the gender role nor the expectations of the majority of their peer group. The younger women were less concerned about the sexuality of residential areas, they reasoned that because of their youth they were not expected to have a husband or male cohabitee and hence landlords and/or neighbours would assume two women living together to be students or friends sharing for financial reasons rather than lesbian partners.

But it is not only housing which reflects heterosexual life-styles, the ideology of the home also derives much of its meaning from this identification with the asymmetrical family. The home is "the spatial location of family identity and the place within which family relations are played out" (Bowlby *et al.*, 1985, p. 8). Therefore, because of its association with the family and child-rearing, and hence with emotional and physical sustenance, the home is perceived as a haven or refuge from the stresses and anxieties of the public world of work and strangers. For some lesbians, the private space of their own 'home' is the only place where they feel safe and able to express their sexual identity without fear of exposure or violence, because they can control access to it and the behaviour of others and the expression of sexuality within it. Hence 'home' can be a haven where they can forget the habit of self-concealment and be themselves.

But for others who live or spend time in matrimonial or parental houses, the heterosexual family-based ideology of the 'home' makes them sites of alienation. For it is in the heteropatriarchal home, which is controlled by the extended family, that many lesbians (both those who are open and those who are secretive about their sexual orientation) become particularly conscious that they do not fit in with the asymmetrical 'family' identity because they do not conform to a particular form of heterosexual and gendered relations.

> I mean, as much as I love my family I always feel I don't fit in. The only place I feel at ease is with gay people . . . I feel I sit there in a room full of my family and feel I'm just not part of this, I don't fit in. I feel as if I'm stuck on a pedestal you know, not that I'm better than them but that everybody's looking at me, that I don't blend in (working class, 30s).

> I do sometimes find it hard when the normal straight world impinges on me. Like when I go home to my parents' for example and my sister's there with her husband or my cousins come over for the day and they all live, well as far as I know anyway, straight lives, I mean I don't feel it so much now but there was a time when I really felt that tug of wanting to be like everyone else. There's such a lot of pressure to conform, to be like everyone else even though we know what other people have doesn't necessarily make them happy, you know, the family, the man, the woman and the kids, all that stuff (middle class, 30s).

This perception of being out of place in the family home is made apparent not only through relatives' overtly heterosexual behaviour and rituals but also through the taken-for-granted way in which they assume all members of the family will share antigay sentiments or join in with antigay comments. Consequently, for lesbians the parental or matrimonial home is devoid of many of the shared meanings, experiences, and values which are simultaneously taken for granted by heterosexuals but which also serve to shape or reinforce the asymmetrical identity of the family. In this way, heterosexual power is invested in and expressed through so-called private spaces.

So, far from the heteropatriarchal home representing a great mixture of associations, actions, and emotions which contribute to a person's identity, for many lesbians 'the family home' symbolises everything they do not want or are unable to be.

For example, the home is perceived as one of the few places where you can impose something of your own identity on the environment. But for young lesbians living with parents or friends, or women who identify as lesbians but are living with male partners who are unaware of their sexuality, the lack of privacy or sanctity in the 'home' because it is a space controlled by others means that it becomes a site where identity is concealed or suppressed: for example, through hiding lesbian books or pictures of lovers; or there is deliberate misrepresentation by the display of heterosexual images, such as posters of male stars. Similarly, when the heterosexual world of cleaners, builders, meter readers, and visitors impinge on houses controlled by lesbians, some women attempt to maintain the sanctity of their home by hiding lesbian signifiers or by employing gay tradespeople to carry out work in the house.

Those who choose to disclose their sexuality to relatives or male partners risk exclusion from the family or marital home, rejection by relatives, and losing custody of children.

Far from being a haven, therefore, and an antidote to the pressure lesbians experience outside, the heteropatriarchal power which is invested in the matrimonial or parental home means it often becomes the site where gay women are put under most pressure to conform to a heterosexual identity of the family or to conceal their lesbianism. This desire to please relatives or to conform pressurises many gay women into heterosexual relationships which they then regret. Home for many lesbians is therefore not where the heart is but the place they need to escape from to express their heart's desire.

The workplace

A national survey of attitudes to lesbians and gay men in the USA revealed that 25% of respondents to the poll would strongly object to working around people who are gay. A further 27% said they would prefer not to do so (Herek, 1992b). Similarly, a survey of the heads of 640 sociology departments in the early 1980s showed that 63% held reservations about hiring a known homosexual (D'Emilio, 1989). British figures suggest that actual discrimination is commonplace. The Lesbian Employment Rights group found that 151 out of 171 gay women questioned in London in 1984 had experienced some form of antilesbianism in the workplace (Hall, 1989). Nine of the forty women interviewed have actually been discriminated against for being gay or have witnessed the negative way in which those who are 'out' are treated in the workplace. Lesbians are therefore very conscious that employers perceive gay sexuality as negative and inferior. One of three lesbians who came 'out' at work said:

> They've already got rid of one, hounded her out . . . Because the management committee, right-wing middle-class fogeys, that were there didn't like it [her open-

> ness about her sexuality] and harassed her so much she said that was it, she couldn't work with them, and our employer didn't back her up at all . . . They don't support us even though they have this equal opportunities policy, it's not worth the paper its written on. And the other one has definitely been stopped for promotion and is not chosen for lots of training events, so she's definitely being kept down. Me, well the head of the service has told me I'm too aggressive so I know that I'll probably be rotting here for the rest of my life (middle class, 30s).

But sexuality in the workplace is not confined to the attitudes of employers. Organisations themselves are not asexual but heterosexual. The whole organisation of production has evolved in parallel to the social organisation of reproduction. The heterosexual family, therefore, is seen to complement working organisations by "providing continuity and the rest and recreation workers need to be productive", whereas "the gay lifestyle is not perceived to be stable or to offer the same restoratives" (Hall, 1989, p. 126). Correspondingly, many organisations adopt a paternalistic approach to workers and their families which is reflected in the way companies provide, for example, life assurance, private health care, and other benefits for heterosexual family units only. Therefore, employers both organise and represent a particular form of power relations, heterosexual, in the workplace.

Similarly, expectations about gender roles and behaviour are also transferred to the workplace, a process described by Nieva and Gutek (1981, p. 59) as "sex role spillover". For example, women in modern Western culture are currently associated with characteristics related to their domestic role, such as being passive, caring, emotional, tidy, clean, whereas men are associated with dirt, danger, and assertiveness. As a result, workplaces commonly develop asymmetrical structures with complementary roles for men and women which reflect these constructions of masculinity and femininity. These constructions change over both time and space, but the binary distinction and the patriarchal power relationship between masculine and feminine is reproduced (Cockburn, 1983). The gendering of jobs in this way therefore establishes and effectively polices heteropatriarchy hegemony in the workplace, so that women who do 'masculine' jobs, such as engineering, run the risk of being labelled butch and therefore lesbian, whereas men in so-called 'feminine' roles, such as nursing, are perceived as effeminate and hence gay (Bowlby *et al.*, 1987). Lesbian and gay sexualities are therefore represented in the workplace as abnormal and inferior, or at best as a personal problem (Burrell and Hearn, 1989).

The (hetero)sexualisation of the workplace is not limited to the asymmetrical gendering of jobs. Gutek (1989) cites Schneider's (1982) research to support her claim that women at work are perceived to be inherently sexual in appearance, dress, and behaviour: "Because it is expected, people notice female sexuality, and believe it is normal, natural, an outgrowth of being female" (Gutek, 1989, p. 60). Therefore, women's behaviour and dress are often interpreted in a sexual way by men, even though they were not intended as such (Abbey *et al.*, 1987). Those women who do not conform to expectations of femininity, by, for example, not wearing makeup or by not flirting or responding to male overtures, risk being labelled lesbian and therefore as unsuccessful or inferior women.

Women interviewed who have not dislosed their sexual identity at work said that in order to operate successfully in a patriarchal workplace they feel pressurised into passing as heterosexual by conforming to a feminine identity, for example, by wearing makeup and skirts and feigning sexual interest in men. As a result of adopting a gender–sexual identity which is devoid of meaning for them, some lesbians feel out of place at work. Such subterfuge also means that it is less easy for lesbians to identify each other at work and therefore it perpetuates the isolation of gay individuals and the invisibility of the homosexual population.

This sociosexual behaviour in the workplace is not confined to asymmetrical interactions between sexually labelled employees; individuals' private lives and experiences are also used as common currency in exchanges between colleagues, particularly women, in the public arena of the workplace. For example, heterosexuals talk about what they have done in their leisure time with their partners, share marital difficulties or confidences, freely speak to lovers on the telephone, and display heterosexual signifiers such as photographs and wedding rings. Therefore most workplaces come to reflect physically and socially the ideology and social relations of the majority of the inhabitants, and so this reinforces the heterosexual identity of the employees as a group.

Whereas heterosexuals take for granted their freedom to express their sexuality publicly and therefore transcend the so-called public–private dichotomy, lesbians are alienated from colleagues by the need to keep their private lives out of the workplace. The most common coping strategy used to separate public and private identities is to avoid any mention of a partner or relationship at work (Valentine, 1993a). As a result, lesbians also have to avoid situations where the physical divide between home and work social relations is breached. For example, by not inviting colleagues home or attending social events organised by employers or amongst colleagues where there is pressure to produce a partner of the opposite sex. But, although this may maintain a neutral or asexual front, the women interviewed said that as a consequence they feel isolated because they are unable to share their personal problems and experiences with others. More significantly, this inability to join in makes them appear aloof, and so they are unable to develop authentic friendships with workmates, so tending to undermine their working relationships with colleagues and their ability to network.

For many lesbians, therefore, the workplace is not experienced as an asexual environment but as a *heterosexual* environment. This is because workplaces are physically and socially organised to reflect and reproduce asymmetrical sociosexual relations. As a result of this expression and representation of heteropatriarchal relations in space, heterosexual employees as a group appropriate the space through (hetero)sexualised signifiers, conversations, and behaviour.

Social spaces

Just as heterosexuality spills over from the home into the workplace, so it also imbues social spaces such as hotels and restaurants. In particular, hotels have a dual image; first, they represent a surrogate home for families on holiday and therefore are associated with heterosexual family units; second, they are effectively surrogate bedrooms having specific (hetero)sexual associations as a site for adultery and 'dirty weekends'.

Lesbian couples are therefore conscious that booking a double room implies that a sexual relationship is taking place between the women. In a survey by the *Pink Paper* (1991) it was found that hotels rejected bookings by single-sex couples, claiming there were no vacancies, but rooms were made available to subsequent heterosexual callers. Only one woman interviewed had been rejected in this way. But others said that they felt inhibited and embarrassed trying to make reservations because they anticipated they may be refused a room.

Although hotel and bed-and-breakfast (B&B) receptionists may be prepared to accept bookings from any paying customer, other guests and staff often appear to be less tolerant of difference. The women interviewed claimed to have been stared at, talked about, and verbally abused by fellow guests and intimidated by aggressive staff. They attributed this to the fact that they were identified as lesbians by the absence of male partners, an insufficiently feminine appearance, and intimate body language and behaviour. In other words, they failed to dress and behave according to their gender identity.

Even if they do not encounter any adverse reactions, women also claim to feel out of place in hotels and B&Bs because they are conscious of being the only single-sex couple present in an overwhelmingly and overtly heterosexual environment. A common response, therefore, is to avoid 'straight' places and to seek out accommodation run for or by homosexuals.

Lesbians also report similar experiences of hostility and discomfort in some restaurants, which, like hotels, are environments associated with intimacy and heterosexual courting rituals. They are also places where people commonly 'dress up' reflecting asymmetrical gender roles; for example, when going out women put on makeup and jewellery and men put on jackets, and so emphasise their heterosexuality. Consequently, women claim that when dining with a female partner they have been given poor tables 'out of sight' and hostile service by the staff; fellow diners have stared at them, and they have felt inhibited and unable to hold normal conversations or to touch and exchange intimacies. Women who have been married contrasted these reactions with the way they took for granted their ability to express their sexuality over dinner with a male partner without fear of incurring a hostile response or of feeling out of place.

Whereas hotels and restaurants are environments of intimacy, public houses, particularly at night, are traditionally identified as male-dominated environments. Women's access to pubs has historically been constrained by norms about morality and respectability. This meant that females entering public houses had to be accompanied by men and were restricted to set bars or times (Green *et al.*, 1987). Although it is more common now for women to go to pubs and clubs in the company of other women, women still avoid going into pubs and night clubs alone. This is not only because women's access to these places at night is limited by fear of travelling through public space alone at night (Valentine, 1992) but also because women alone in such venues are assumed to be available and receptive to sexual overtures from men and therefore encounter high levels of sexual harassment (Westwood, 1984). But sexual attention is not always unwanted. Women actively dress up and go out to pubs and clubs 'with the girls' with the intention or hope of finding a new partner (Burgoyne and Clark, 1984). Consequently, lesbians who make it clear through appearance or behaviour that they are not interested in men or that they are together as a couple stand out as different. Therefore, lesbians often feel conscious of being out of place or are actively made aware of this by hostile reactions from men.

Like housing and the workplace, therefore, most social spaces are organised to reflect and express heterosexual sociosexual relations. In particular, hotels and restaurants are environments of intimacy associated with heterosexual romance, dating, and sex; and pubs and clubs are environments where women receive and are expected to be receptive to male sexual advances. Lesbians can therefore feel out of place because of the orientation of these places towards heterosexual couples, or they are made to feel out of place by the hostility of others who identify them as outsiders through their dress, body language, and disinterest in men.

Public open spaces

> I feel very angry that wherever you go, that you're on the outside . . . I've always hated not being able to touch my partner in public. You know, you see everyone else walking hand-in-hand or arm-in-arm down the street on a Saturday, in all the shops. And that's never been a possibility for me.
>
> (middle class, 40s)

As this quote and the opening quote in this paper suggest, certain forms of overt displays of affection between men and women are commonplace in public places such as the high street. Such behaviour is particularly evident in open spaces such as the park and the beach during hot summer weather.

The taken-for-granted way in which asymmetrical couples and families take up public space serves to alienate lesbians who are rarely able to procure space in the same way. However, when lesbian, gay, and bisexual communities are mobilised and make their presence visible they can appropriate public space. By turning the tables on heterosexuals in this way, Gay Pride demonstrates that space is sexualised, and, more specifically, that it is 'usually' heterosexual.

Lesbians who do make the nature of their relationship apparent in public spaces risk a violent response. Like sexual abuse perpetrated against all women, antigay violence exists on a continuum from comments to threats, assault to murder. Kelly (1987) makes the point that the continuum of sexual violence does not refer to a linear line from least to most serious, despite the fact that some forms of violence are perceived as more common and therefore less serious than others, because individuals react differently to different experiences depending on their background and the way in which they perceive the incident as it happens. In particular, it has been argued that offences are subjectively linked, that is, one offence tends to accompany or follow from another (Warr, 1987). For example, verbal abuse is often a prelude to physical assault. 'Minor' incidents are therefore often very traumatic because of the implication that something 'more serious' could have followed.

Of the forty women interviewed 75% have been verbally abused at least once because of their sexuality, and three women have been chased and threatened and/or assaulted. Most also know of others who have been attacked, including one whose colleague on a helpline was murdered.

These figures are low compared with a study of 400 lesbians in San Francisco, CA, which found that 84% had experienced antilesbian verbal harassment, 57% had been threatened with physical violence, and 12% had been punched, kicked, or beaten (von Schulthess, 1992). Such violence also appears to be on the increase. In a survey of lesbians and gay men in Philadelphia, PA, in 1986–87, Gros *et al.* (1988) found that the number who had experienced criminal violence because of their sexuality had doubled since a previous survey in 1983–84 (Aurand *et al.*, 1985).

Of the sixty-one incidents recalled[4] by the women, 84% took place in 'ordinary' public spaces, whereas only 16% occurred in gay-identified places such as outside gay pubs, and all but one were exclusively perpetrated by men or boys. The women said that some of this public harassment was triggered because they had been seen expressing affection, such as holding hands, and therefore were known to be gay; others because they had not responded to male sexual overtures and therefore had been accused of being lesbians. This may, however, have been used as a sexual insult on a par with calling the women frigid, from men who felt their masculinity had been challenged, rather than being an intentionally accurate observation. But fourteen of the women said that the only explanation for incidents they had experienced was the fact that they had short hair and were wearing trousers and in most cases were in the company of another woman. By implication, therefore, they were not conforming to the dress and behaviour expected of a heterosexual woman in an 'ordinary' public space. This is in contrast to surveys about the victimisation of gay men which show that men are primarily attacked in gay spaces (again by men, not women) such as pubs or well-known cruising areas rather than in spaces that are not identified as gay (Berrill, 1992).

The difference between the geography of antilesbian attacks and the geography of assaults against gay men therefore implies that antilesbian violence is not only an attempt by heterosexuals to police the expression of gay sexual identities, but also reflects the fact that, although men are freely able to use and occupy public space alone or with other men without fear of sexual harassment, women who do so without male companions are open to comments about their appearance or to sexual overtures from men (Valentine, 1989; Westwood, 1984). Antilesbian abuse which is directed at women in public spaces reflects men's attempts to police independent women's behaviour, and hence reflects patriarchal power relations.

Conclusion – (hetero)sexing space

The evidence presented in this paper confirms that heterosexuality is the dominant sexuality in modern Western culture. This supremacy is attributed to the fact that opposite-sex sex is constructed as natural and therefore superior to homosexuality because of its association with procreation (Burrell and Hearn, 1989; Schneider and Gould, 1987). However, heterosexuality is not defined merely by sexual acts in private space. As the quote above implies, it is a *taken for granted* process of power relations which operates in most everyday environments, thus highlighting the inaccuracy of assuming a sexual public–private dichotomy.

Heterosexuality is expressed in the way spaces are physically and socially organised; from houses to the workplace, restaurants to insurance companies, spaces reflect and support asymmetrical family units. The lack of recognition of alternative sexual identities means that places and organisations exclude lesbian and gay life-styles and so unconsciously reproduce heterosexual hegemony. As a result of this expression and representation of heterosexual relations in space, heterosexuals as a group are allowed to appropriate and take up space, for example, with heterosexual signifiers such as pictures of partners or through constant (hetero)sexualised dialogue. Although the workplace and houses are perceived as asexual despite their heterosexual orientation, certain social spaces such as hotels have generally recognised (hetero)sexual associations which can directly inhibit and restrict their use by lesbians.

The dominance of heterosexuality is therefore perpetuated because lesbians feel out of place because space is organised for and appropriated by heterosexuals and so expresses and reproduces asymmetrical sociosexual relations. As a result, many lesbians practice self-censorship by avoiding or minimising the time spent in (hetero)sexualised space where they feel they do not belong, choosing, for example, where possible to socialise in gay spaces or self-created spaces where they feel at home. But more insidiously, heterosexual hegemony is maintained and policed through homophobia. Strictly, this means fear of homosexuals but it is commonly used to describe hatred and negative treatment of homosexuals. This includes the use of rejection, discrimination, and, ultimately, violence to oppress lesbians, gay men, and bisexuals. Many gay women therefore avoid publicly expressing their sexuality in environments where they perceive they will encounter such hostility. By concealing their identity in this way, lesbians become invisible in everyday environments. This fear of disclosure feeds the spatial supremacy of heterosexuality in three ways. First, it masks the number of lesbians present and so reinforces the heterosexual identity of environments. Second, it facilitates the perpetuation of negative stereotypes about what lesbians are like. Third, it ghettoises gay sexuality by making it difficult for lesbians to identify and meet other lesbians except in gay-defined spaces.

But lesbian identities are policed not only by homophobia but also by patriarchy. Heterosexuality is ideologically linked to the notion of gender identities (masculinity and femininity) because the notion of opposite-sex relationships presumes a binary distinction between what it means to be a man or a woman. Masculinity and femininity have been and are constructed and reconstructed in relation to one another to create and perpetuate male supremacy (Coveney *et al.*, 1984). In particular, women are perceived to be inherently sexual in appearance and behaviour and, in the last analysis, submissive to men, whereas male sexual behaviour is interpreted in terms of dominance and power. This asymmetry of gender identities is reflected in the behaviour and dress ascribed to and expected of each sex. Women are therefore expected to dress to be sexually attractive to men, to respond to male sexual overtures and dialogue, but to avoid public space alone at night or specific male-dominated environments, such as pubs, when unaccompanied by others. Heterosexuality in modern Western societies is therefore patriarchal, that is, it reflects male dominance.

This was recognised by feminists in the 1970s. Lesbianism was therefore identified by radical feminists as a political choice, under the slogan 'Feminism is the theory, lesbianism is the practice'. As Bunch (1991, p. 320) states: "lesbian feminist politics is a political critique of the institution and ideology of heterosexuality as a cornerstone of male supremacy. It is an extension of the analysis of sexual politics to an analysis of sexuality itself as an institution".

Lesbian feminists have therefore challenged notions of femininity and women as "the feminine (inferior) side of the masculine/feminine couple" (Young, 1990, p. 74). The media have seized on this notion of lesbianism as a challenge or threat to the hegemonic strength of patriarchy and the asymmetrical family. Consequently, lesbianism is constructed and reproduced in the media and popular culture as synonymous not only with masculinity and ugliness but also with 'man-hating' and aggression (Young, 1990).

Women who dress, behave, do jobs, or go to places associated with men run the risk of being labelled 'butch' and hence as 'man-hating' lesbians. The stigma and negativity associated with being a lesbian therefore means that accusations of being a 'dyke' are used by some men to keep independent women in their place, and, similarly, some women use the accusation to pressurise other women into complying in their own oppression (Bunch, 1991). In this way, the stigma of lesbianism is used to police patriarchal gender identities. Consequently, because gay women commonly have lifestyles which are relatively independent of men – for example, they go to pubs, or restaurants, or hotels without male partners – they are often abused as 'dykes'. However, the evidence I have presented in the sections on social and public space suggests that such hostile comments are not always intended to be accurate observations of lesbians' sexuality but can be meant as a term of abuse for independent women who, for example do not dress and behave according to men's expectations of femininity.

Lesbians therefore feel out of place and fearful of discrimination or violence in certain environments not only because of homophobia directed at them because they have been identified as the homosexual 'other', but also because of a patriarchal backlash, directed at them because they are women who are relatively independent of men and therefore are a threat to the hegemony of patriarchy. This pressurises some lesbians to dress and behave in a highly feminine or heterosexually identified way to avoid the accusation 'dyke'. The adoption of these fictional sexual identities in different spaces means that gay women are unable to develop authentic relations with others, so hindering their working, social, and business relationships and their ability to network. So patriarchy also perpetuates the invisibility of lesbians in everyday spaces and pushes the expression of lesbian identities into gay-identified or self-created spaces.

Thus, although lesbians, as the homosexual 'other', experience a different form of oppression from heterosexual women, expressed through homophobia, all women are also touched by antilesbianism. However, by ignoring antilesbianism or collaborating in perpetuating it, some heterosexual women comply in their own oppression, because such antilesbianism is also used to police heterosexual women's dress, behaviour, and activities. Hence, if 'dyke' were not a term of oppression, heterosexual women would also have more freedom to define their own identities. However, in practice, actual strategies to work together are made difficult by the apparently different interests of heterosexual and gay women.

I therefore suggest there is a need for more research to explore the complex and perhaps contradictory experiences of lesbians, heterosexual women, gay men, and bisexuals in a heteropatriarchy and hence to highlight the most appropriate ways in which to challenge its hegemony.

Notes

1 The quotes used in the text from interviews are verbatim. Ellipsis dots indicate that a word or phrase has been removed. Those quoted are identified only by an age and a class label. The author recognises that class, like sexual and gender identities, can be fluid and that individuals can maintain multiple class positions (Graham, 1992). In addition, because of the life-style changes women sometimes go through when they adopt a lesbian identity, the class position of many lesbians is complex. Consequently, the terms middle and working class are used to indicate only the current occupational status of the woman concerned. No further information can be supplied about the interviewees because of the need to maintain their anonymity.

2 Strictly, homosexual is a biological term, gay is used to describe homosexual men and women, and lesbian is used by women who wish to be distinguished from gay men. However, some women prefer to be identified as gay rather than lesbian, and vice versa. Therefore, in this paper the terms have been used interchangeably to describe all homosexual women. Other terms used are 'to come out' – to be open about sexuality; 'straight' – gay word for 'heterosexuals' 'dyke' – used as a term of antigay abuse by heterosexuals but as a positive label by lesbians.

3 Fear of prejudice, discrimination, and violence causes many lesbians to conceal their sexual identity from colleagues, friends, and relatives. In order to protect the anonymity of the participants, the identity of the town where the research was conducted and all the names of people and places mentioned in this paper have been changed.

4 The figure of sixty one is the number of incidents described in the interviews, but several women said they had experienced other episodes too numerous to recall. In addition, other researchers have found that respondents tend to underreport 'minor'

incidents because they are 'taken for granted' as common experiences. Therefore the actual levels of abuse may be much higher.

References

Abbey A, Cozzorelli C, McLaughlin K, Hamish R, 1987, "The effects of clothing and dyad sex composition on perceptions of sexual intent: do women and men evaluate these cues differently?" *Journal of Applied Social Psychology* **17**, 108–126.

Adler S, Brenner J, 1992, "Gender and space: lesbians and gay men in the city" *International Journal of Urban and Regional Research* **16**, 24–34.

Anlin S, 1989 *Out But Not Down! The Housing Needs of Lesbians* Homeless Action, 52 Featherstone Street, London EC1Y 8RT.

Aurand S, Adessa R, Bush C, 1985 *Violence and Discrimination Against Lesbian and Gay People* available from Philadelphia Lesbian and Gay Task Force, 1501 Cherry Street, Philadelphia, PA 19102.

Bell D, 1991, "Insignificant others: lesbian and gay geographies" *Area* **23**, 323–329.

Berrill K, 1992, "Anti-gay violence and victimisation in the United States: an overview", in *Hate Crimes: Confronting Violence Against Lesbians and Gay Men* Eds G Herek, K Berrill (Sage, London) pp. 19–45.

Bowlby S, Foord J, McDowell L, 1985, "Love not money: gender relations in local areas", paper presented at the 5th Urban Conflict and Change Conference; available from S Bowlby, Department of Geography, University of Reading, Reading.

Bowlby S, Foord J, Lewis J, 1987, "Gender and place", paper presented at the Association of American Geographers Annual Conference, Portland, OR: available from S Bowlby, Department of Geography, University of Reading, Reading.

Bunch C, 1991, "Not for lesbians only", in *A Reader in Feminist Knowledge* Ed. S Gunew (Routledge, Chapman and Hall, Andover, Hants) pp. 319–325.

Burgoyne J, Clark D, 1984 *Making a Go of It* (Routledge, Chapman and Hall, Andover, Hants).

Burrell G, Hearn J, 1989, "The sexuality of organizations", in *The Sexuality of Organization* Eds J Hearn, D Sheppard, P Tancred-Sheriff, G Burrell (Sage, London) pp. 1–27.

Butler J, 1990 *Gender Trouble: Feminism and the Subversion of Identity* (Routledge, Chapman and Hall, Andover, Hants).

Cockburn C, 1983 *Brothers: Male Dominance and Technological Change* (Pluto Press, London).

Coveney L, Jackson M, Jeffrey S, Kay L, Mahony P, 1984 *The Sexuality Papers* (Hutchinson Education, London).

D'Emilio J, 1989, "Not a simple matter: gay history and gay historians" *Journal of American History* **76**, 435–442.

D'Emilio J, Freedman E, 1988 *Intimate Matters: A History of Sexuality in America* (Harper and Row, New York).

Eadie J, 1992, "The motley crew: what's at stake in the production of bisexual identity", paper presented at the Sexuality and Space Conference, London; available from D Bell, School of Geography, University of Birmingham, Birmingham.

Egerton J, 1990, "Out but not down: lesbians' experiences of housing" *Feminist Review* **36**, 75–88.

Ettorre E, 1978, "Women, urban social movements and the lesbian ghetto" *International Journal of Urban and Regional Research* **2**, 301–336.

Foucault M, 1988, "Technologies of the self", in *Technologies of The Self: A Seminar with Michel Foucault* Eds L Martin, H Gutman, P Hutton (Tavistock Publications, Andover, Hants) pp. 16–49.

Graham J, 1992, "Post-Fordism as politics: the political consequences of narratives on the left" *Environment and Planning D: Society and Space* **10**, 393–410.

Green E, Hebron S, Woodward D, 1987, "Women, leisure and social control", in *Women, Violence and Social Control* Eds J Hanmer, M Maynard (Macmillan, London) pp. 75–92.

Gross L, Aurand S, Adessa R, 1988 *Violence and Discrimination Against Lesbian and Gay People in Philadelphia and the Commonwealth of Pennsylvania* available from the Philadelphia Lesbian and Gay Task Force, 1501 Cherry Street, Philadelphia, PA 19102.

Gutek B, 1985 *Sex and the Workplace: Impact of Sexual Behaviour and Harassment on Women, Men and Organizations* (Jossey-Bass, San Francisco, CA).

Gutek B, 1989, "Sexuality in the workplace: key issues in social research and organizational practice", in *The Sexuality of Organization* Eds J Hearn, D Sheppard, P Tancred-Sheriff, G Burrell (Sage, London), pp. 56–70.

Hall M, 1989, "Private experiences in the public domain: lesbians in organizations", in *The Sexuality of Organization* Eds J Hearn, D Sheppard, P Tancred-Sheriff, G Burrell (Sage, London) pp. 125–140.

Hart R, 1978 *Children's Experience of Place* (Irvington Press, New York).

Hearn J, 1985, "Men's sexuality at work", in *The Sexuality of Men* Eds A Metcalf, M Humphries (Pluto Press, London) pp. 110–128.

Herek G, 1987, "Can functions be measured? A new perspective on the functional approach to attitudes" *Social Psychology Quarterly* **50**, 285–303.

Herek G, 1992a, "The social context of hate crimes: notes on cultural heterosexism", in *Hate Crimes: Confronting Violence Against Lesbians and Gay Men* Eds G Herek, K Berrill (Sage, London) pp. 89–104.

Herek G, 1992b, "Physiological heterosexism and anti-gay violence: the social psychology of bigotry and bashing", in *Hate Crimes: Confronting Violence Against Lesbians and Gay Men* Eds G Herek, K Berrill (Sage, London) pp. 149–169.

Kelly L, 1987, "The continuum of sexual violence", in *Women, Violence and Social Control* Eds J Hanmer, M Maynard (Macmillan, London) pp. 46–60.

Matrix (Eds), 1984 *Making Space: Women and the Man Made Environment* (Pluto Press, London).

Nieva V, Guetek B, 1981 *Women and Work: A Psychological Perspective* (Praeger, New York).

Pink Paper 1991 *Pink Paper: The National Newspaper for Lesbians and Gay Men* September, page 3.

Plummer K, 1988, "Homophobia, homosexuality and gay youth", in *Family, Gender and Welfare* (Open University Press, Milton Keynes) pp. 23–42.

Schneider B, 1982, "Consciousness about sexual harassment among heterosexual and lesbian women workers" *Sociological Perspectives* **27**, 443–464.

Schneider B, Gould M, 1987, "Female sexuality: looking back into the future", in *Analysing Gender: A Handbook of Social Science Research* Eds B Hess, M Ferree (Sage, Newbury Park, CA) pp. 120–153.

Valentine G, 1989, "The geography of women's fear" *Area* **21**, 385–390.

Valentine G, 1992, "Coping with fear of male violence: women's use of precautionary behaviour in public space", paper presented at the Women in Cities Conference, Hamburg, Germany; available from G Valentine, School of Geography, University of Manchester, Manchester.

Valentine G, 1993a, "Negotiating and managing multiple sexual identities: lesbian time–space strategies" *Transactions of the Institute of British Geographers* in press.
Valentine G, 1993b, "Out and about: a geography of a lesbian landscape"; copy available from author.
von Schulthess B, 1992, "Violence in the streets: anti-lesbian assault and harassment in San Francisco", in *Hate Crimes: Confronting Violence Against Lesbians and Gay Men* Eds G Herek, K Berrill (Sage, London) pp. 65–75.
Warr M, 1987, "Fear of victimisation and sensitivity to risk" *Journal of Quantitative Criminology* **3**, 73–79.
Westwood S, 1984 *All Day, Every Day* (Pluto Press, London).
Young A, 1990 *Femininity in Dissent* (Routledge, Chapman and Hall, Andover, Hants).

23 D. A. Leslie

'Femininity, Post-Fordism and the "New Traditionalism"'

Excerpts from: *Environment and Planning D: Society and Space* **11**, 689–708 (1993)

Introduction

> A new kind of woman with deep-rooted values is changing the way we live . . . To us, it's a woman who has found her identity in herself, her home, her family. She is the contemporary woman whose values are rooted in tradition.
>
> (*Good Housekeeping* advertisement, 1990)

Femininity is in large part constructed through representations in the cultural arena, and space is fundamental to the circulation of images and the constitution of identity. Representation is profoundly political, and there is a need to bridge the divide between social sciences and cultural studies, between discourses and the material conditions of women's existence (Barrett, 1992; Bondi, 1990; Deutsche, 1991; Grossberg *et al.*, 1992; Massey, 1991; Morris, 1992; Rose, 1992). In this paper I examine the growing tension between materialism and poststructuralism in feminism and argue that it is important to take cognizance of the fact that the redefinition of images and the creation of new forms of identity are of pivotal importance in the recent round of restructuring (Burgess and Wood, 1988; Watson, 1991). An interest in representation need not signal an abandonment of materialism.

In the current period of economic, political, and cultural upheaval, a crisis with profoundly spatial dimensions, advertising serves as a crucial point of mediation between production and consumption, where the emergence of new meanings of masculinity and femininity and their shifting geographies can be examined. In periods when gender roles undergo marked shifts, such as in the 1920s and the 1950s, advertising takes on a particularly important role

(Doane, 1987; Ewen, 1976; Miller, 1991). In the postwar period, for instance, an intense ideological campaign was waged in the discourses of advertising, calling for women to abandon the workplace and return to 'traditional' family values. The present phase of late consumer capitalism constitutes yet another crisis in the category of 'woman', both in theory and in advertising. The renegotiation of gender and the crisis in family form has paralleled an economic transition to post-Fordism. At the same time, representations of women's roles have also shifted: for example, images of the 'new traditionalist woman' attempt to resituate women in the home, with the home constituting the primary location of women's identities. These images deploy meanings associated with small town America, rurality and nature, in an effort to solidify the increasingly tenuous distinction between public/private, male/female, culture/nature, urban/rural, and work/home. Donna Haraway refers to these attempts to fix boundaries as the 'border wars' of contemporary culture (1991). Although images of tradition in women's roles constitute only one among multiple identities offered to consumers, 'new traditionalism' represents a particularly revealing site of contestation in contemporary advertising.

In this paper I explore the many ways in which issues of identity, femininity, and space are related, through an empirical examination of this new traditionalism in advertisements for *Good Housekeeping* and *Family Circle*.

The redefinition of gender relations

Stuart Ewen argues that, with the emergence of Fordism, advertising assumed a central role in the growth of mass consumption by expanding national markets, both socially and geographically, and by mediating crises in identity. The delineation of the form the family should take, in advertising discourse, represented an attempt to redefine women's roles in accordance with the new industrial order (Ewen, 1976). Although I reject the overly functionalist and deterministic character of Ewen's account, it seems fair to say that advertising helped in the consolidation of the suburban nuclear family and the separation of the predominantly female sphere of consumption from the predominantly male sphere of production in the period after World War 2.

Although the literature on the nature of the changes associated with 'post-Fordism' has tended to focus on the growth of flexible production processes, new technologies, and industrial agglomerations, without regard to changes in gender, family relations, consumption, or new political movements, some have seen women's labor and 'feminine attributes' themselves as pivotal elements of the current restructuring (McDowell, 1991a, p. 401). In place of the nuclear family, which many see breaking down as a dominant form, a more unstable and diverse order is emerging, premised on a growing divergence between women. Linda McDowell argues that an ideology of femininity and domesticity dovetails with the formation of a flexible female labor force, made up of part-time and temporary workers (1991b, p. 101). Discourses of domesticity, tradition, and familialism continue to support and define a woman's place in the labor market and in the home. Images in advertising are a persuasive and effective component of this discursive construction.

At the same time, the 1980s were marked by the rise of a politics calling for people to take responsibility for their own lives, rejecting dependency on the so-called 'nanny' state. The attack on welfare state institutions such as public housing, health care, daycare, and education has been promulgated in the name of increased consumer choice, and this expression of consumer sovereignty strikes a chord in popular discontent (Clarke, 1991, p. 158).

The crisis of patriarchy since the women's movement in the 1960s, has been followed by passionate debates over the meaning of the family (Faludi, 1991; Klein, 1992; ten Tusscher, 1986). The contemporary or 'postmodern' family encompasses both experimental and nostalgic facets (Stacey, 1990, p. 18). With few social measures to replace the private functions of the 'modern' family, many women have come to mistrust both the postmodern family and feminism itself. This less stable familial order, which has left many women less secure and economically disadvantaged, often feeds nostalgia for the traditional family. The 'pro-family' movement and evangelical religious groups have been successful in co-opting this sentiment (Stacey, 1990). In the 1992 US presidential election, the Republicans and the Democrats were both advised by their advertising teams to pursue campaigns emphasizing family values. These familial discourses exhibit contradictory tendencies: nostalgia for a more traditional family form coexisting with the breakdown of the nuclear family.

Discerning these anxieties, advertising researchers have encouraged a proliferation of images of home and family in contemporary advertising, directed at particular segments of women. This is particularly evident in recent trade campaigns for *Good Housekeeping* and *Family Circle* which announce a return to family values and traditional women's roles. As marketers have noted, in a time of deepening economic crisis and 'posts', there is a heightened longing for roots and a growing distaste for the endless spiral of consumption (Free, 1989; Rubin, 1989). The fragmentation, intertextuality, and swift commodification of everyday life that began with modernity once served a function by engendering faith in a progressive future, but this narrative has lost its allure for middle-class consumers (Olaquiaga, 1992). As a traditional sense of place has been eroded by the instantaneity of electronic culture and the proliferation of homogenized landscapes of consumption, it has been replaced by idealized images of community and place, such as the concept of the 'home' as it was constructed in the 1950s. Marketers see this return to the past, to tradition, and the home, family and community, as a useful way to repackage consumerism: Levi, for example, ran a very successful campaign which managed to connect Levis' 501 jeans to a longing for a simpler time (Moore, 1991, p. 28).

An emphasis on tradition also relates to changes in the class structure of advanced capitalist societies. Many have discussed the rise of enterprise culture and a new middle or service class made up of managers and professionals (Featherstone, 1991; Lash, 1990; Savage *et al.*, 1992; Thrift, 1989). According to Pierre Bourdieu (1984), this new service class encompasses those occupations involving symbolic goods and services, and therefore exerts considerable cultural influence. As such, this class fraction includes those who work in marketing, advertising, design, public relations, radio, television, fashion, social work, marriage counselling, and journalism. The new middle class forms a body of cultural intermediaries who produce, transmit, and consume postmodern culture. As Bourdieu notes (1984), the cultural sphere takes on increased importance as a means by which various social classes differentiate themselves; this class is preoccupied with identity, presentation, life-style, and appearance (Betz, 1992).

The growth of the service class has been linked to the spread of its culture and a set of values firmly anchored in consumer culture. This class gains much of its collective identity from consumption and has adopted and distributed notions of competitive individualism and enterprise culture. As Nigel Thrift (1989) has argued, a dominant social group like the service class needs to provide interpretations of community that stress continuity and consensus in the face of discontinuity and conflict. This newly formed but not yet hegemonic class has utilized notions of tradition and heritage as a way of defining its relatively insecure class position, seeking dignity by surrounding itself in images of the past. Tradition strengthens the service class's social cohesion by emphasizing the similarity of tastes, and helps to lend the class legitimacy by representing it as a 'natural' possessor of power and privilege.

The current period of crisis in family and femininity, has also seen a resurgence of notions of enterprise culture, choice, and individualism. Some women have become increasingly discontented with both their place in the male dominated workplace and changes in the family (Faludi, 1991). Professional women confront a glass ceiling as well as the increased pressures of combining home and work roles. At the same time, although the breakdown of the nuclear family has been liberating for many women, for others it has generated a new set of anxieties. Advertisers are positioned between producers and female consumers and attempt to negotiate these tensions. In this period of instability, images are contested and conflicting, ranging from reverential and celebratory images of tradition, to parodic images, challenging dominant notions of family and femininity (Deveny, 1990, p. B1).

Representation, space, and femininity

Representations of women in new traditionalist campaigns show how images share in the constitution of the boundaries of femininity. I want to turn now to the issue of identity, which has emerged as the crucial issue for feminist theory in recent years, to explore the role of representation and space in the construction of female identity. Looking at advertising will show us how identities are generated out of relationships between power, knowledge, and spatiality. It is not just that there are variations in masculinity and femininity across space, but that notions and masculinity and femininity are actively constituted through distinctions of space and place, public and private, visible and invisible.

A tension has emerged within feminism between those who postulate a prior category of 'woman' and those who view 'woman' as an effect of discourse (see Barrett and Phillips, 1992; Butler and Scott, 1992; Coward, 1985; Deutsche, 1991; Haraway, 1991; Kolbowski, 1990; Nicholson, 1990; Pollock, 1990; Pratt, 1993). I would like to argue that femininity is the outcome of multiple discursive practices involved in the constant setting and shifting of the limits of identity. Donna Haraway refers to these practices as "boundary projects" insisting on a view of identity as an incomplete and contingent process, not a reified object:

> . . . bodies as objects of knowledge are material-semiotic generative nodes. Their boundaries materialize in social interaction. Boundaries are drawn by mapping practices; objects do not pre-exist as such. Objects are boundary projects. But boundaries shift from within; boundaries are very tricky.
>
> (1991, p. 201)

Although femininity is drawn through mapping practices which do not guarantee internal stability, this does not mean that femininity has no material foundation. Feminists must acknowledge the construction of 'woman' and the female body through the internalization of representations, but they must also avoid the tendency to conceive of a totally inscribed or passive body. The deployment of modern power and rationality, while it seeks an increasingly insidious form of bodily and spatial domination, is also resisted by women; its disciplinary regimes are never total (McNay, 1991). We need a feminist politics corresponding to contradictory forms of identity, family, and femininity, one which recognizes the need for a femininst identity, but immediately calls it into question.

Recent developments make the need for this type of resolution within feminist politics pressing. The recent growth of observational science, electronic data gathering, and consumer surveillance magnifies the network of disciplinary mechanisms running throughout the society that Michel Foucault (1979) described as enhancing the exercise of power and the reproduction of disciplined behavior. While crises in identity and family have necessitated the development of increasingly sophisticated monitoring methods, advances in advertising research have augmented the attempt to get 'inside the mind of the consumer'. Women have been the primary target of an invasive science of consumer research; to have one's private life probed is to be feminized (Fraser, 1992). Sandra Lee Bartky has argued that a 'microphysics of power' fragmenting and partitioning the body's time, space, and movements, constitutes an unparalleled modern discipline against the body (1990, p. 63). Following Foucault, she charts the increasing phenomenon of self-surveillance, a situation in which instrumental reason envelops the body. Referring to domestic space, Mark Wigley writes, 'the wife learns her "natural" place by learning the place of things. She is "domesticated" by internalizing the very spatial order that confines her' (1992, p. 340).

Feminine identity has tended to be more spatially confined than that of men (Bartky, 1990). Whereas male subjectivity is defined in terms of control over space, female subjectivity becomes that which must be controlled by being bounded: the house itself may be seen as a system of control and surveillance

(Wigley, 1992). Woman is not so much confined within space, however, as she is space itself; femininity comes to be associated with the house as a space. When a woman ventures outside the house into the city, she does not necessarily lose her femininity and become masculine, but becomes more dangerous and uncontrollable in her femininity. Contemporary representations of the 'new traditionalist woman' respond to the increasing presence of women in public space by attempting to reaffirm the blurring distinctions between public and private and to reinforce the association of woman with the home, and the house itself as essential to the definition of a passive consumer femininity.

Consumer research on women's families, consumption patterns, and spatial movements has highlighted lingering tensions in the women's movement and the family. Some advertisers have responded to this research by co-opting themes of feminist empowerment, depicting images of female independence and spatial mobility. Another response in the opposite direction has been the new traditionalist campaigns, which attempt to mediate tensions in femininity by linking women to the home and private sphere. In spite of the resistance offered by individuals to their inscription within power-knowledge relations, the heightened surveillance present in advertising research should cause us to be wary of celebrating consumer agency. As Wally Olins, a design firm executive, notes, 'Marketing is to the 1980's what sociology was to the 1960's' (quoted in Moore, 1991, p. 209). Market segmentation raises the question of how advertising strategies are linked to problems of identity stabilization and formation. Although contradictions in female subjectivity have always defined the feminine condition, producing identities more unstable than those of men, women face an intensified condition of fragmentation and destabilization which is reflected spatially. Gender identities are in large part constructed through the act of consumption and the discourses of advertising negotiate these meanings through the constitution of the space of consumer identity.

Situating the female consumer

A recent study indicates that women continue to be the principal consumers in capitalist society, still doing most of the buying in 95% of the 1000 universal product code categories (Neilsen Inc., 1992).[1] In spatial terms, consumption has been identified most closely with the home, and the house itself is often associated with women's bodies.[2] Historically, the private space of the home has offered an ideal place to reach women in the city. However with the waning of the mass consumption household of Fordism, female markets and traditional roles have become increasingly fragmented and contested. Peter Jackson (1991) emphasizes the fracturing effects of the urban landscape on identity, noting that individuals represent themselves differently in different spatial settings (Mort, 1988). The current blurring of identities opens up possibilities for challenging the traditional cultural alignment of femininity, consumption, and private space. The feminist critique of identity has ironically coincided with the increasing fragmentation of the female subject in advertising: concepts of fixed identity sit awkwardly with the fluidity of images found in the media (see Griggers, 1990). Advertisers recognize that advertising cannot reach back to its 'ideal' 1950s mass market housewife, instead it 'has to be anywhere, anytime, anyplace, because women are now anywhere, anytime, anyplace' (Sheth quoted in Golman, 1990). Here I examine the relationship between femininity, consumption, and the home, and the ways in which advertisers are grappling with spatial confusion.

Because women become consuming subjects through the consumption of goods, what they are offered by advertising are images of themselves.[3] Luce Irigaray's theory of woman as commodity and the historical positioning of woman as consumer is only apparently contradictory (Doane, 1987). This seeming paradox requires rethinking the dichotomy between subject and object and analyzing the ways in which women actively participate in their own oppression. The consumer is not necessarily an active appropriator of the commodity when the identity of the social subject itself is dependent on the acquisition of objects (Baudrillard quoted in Bowlby, 1985).[4]

Mary Ann Doane argues that, when women are spoken to 'as women' in advertising, the physical separation between subject and object, conducive to both flânerie and voyeurism, is annihilated. Woman, in her double role as consumer and image, maintains an intimacy with advertising which militates against the ability to see herself at a distance. Walter Benjamin argued that a detached mode of spectatorship collapsed with the final incarnation of the flaneur, the sandwichman. Unlike the bourgeois flaneur, the sandwichman, a human billboard advertising products and events, was a poor, sometimes female, casual laborer. The display of the sandwichboard was not far removed from the display of one's own body for sale (Benjamin quoted in Buck-Morss, 1986). A woman who loitered in the public sphere was therefore given the name of prostitute. Benjamin writes "the modern advertisement demonstrates . . . how much the attraction of woman and commodity can fuse together" (quoted in Buck-Morss, 1986, p. 120). Woman's position is analogous to the position of the sandwichman, unable to maintain a distance between subject and object, the advertisement affixed to her body. She further demonstrates her commodity status if she ventures into public space in the city. Feminist theorists have taken the gendering of public and private space to imply the impossibility of a female flaneur, and both femininity and consumption have come to be identified with the private space of the home.

Others are more skeptical about the impossibility of a female flaneur. According to Elizabeth Wilson (1992, p. 100), those who see an ideology of a woman's place in the domestic realm permeating the whoie of society ignore empirical fact to the extent that 'reality becomes but a pale shadow of ideology, or even bears no relation to it at all'. For Wilson, the private sphere is also a masculine domain, as well as the workplace of women. Furthermore, she argues that women were entering public places in the city, such as department stores and tearooms, by the late 19th century. Neither shopping nor women were invisible. In these half-private, half-public spaces, she asserts, women too could be flaneurs (p. 101). Although I agree that women were not 'turned to stone by the male gaze' (Wilson, 1992, p. 102) I would like to argue that the home as a place has been central to the ideological construction of female identity and that an ideology of women's place in the home followed women into the city. Although women contested and resisted dominant ideologies of femininity, given the passive nature of consumer subjectivity and the control of public sites of consumption, a female consumer could not be a flaneur.

The private space of the home serves as an ideal conduit – a controlled point for the dissemination of advertising and other public messages. Postwar suburban houses, for example, mediated the opposing ideals of separation and interaction, privatization and community through the television and other media (Spiegel, 1992). These communication spaces negotiated relations between public and private spheres, diffusing the threat of difference by defining the boundaries of active and passive space, production and reception, male and female. According to Lynn Spiegel, 'television's incorporation of the public sphere into the home did not bring "male" space into female space; instead it transposed one system of sexually organized space onto another' (1992, p. 209). The spaces in which gender identities were ideologically centered were not erased by women's appearance in the city or by men's presence in the home, but each carried over into the other space. Today, these tensions have heightened as public and private become even more intertwined, feeding nostalgia for older more strictly defined forms of community. Advertisers have responded to these 'border wars' in part with mapping practices that represent a return to traditionalism and older boundaries.

It is important to consider the methods by which advertisers create spaces of reception and femininity, locations for the deployment of power/knowledge in both production and reception. Jodi Berland (1992) asks why the literature on consumption is empty of cars, shopping malls, buses, houses, streets, offices, hotels, urban landscapes, rural settings, and, indeed, of any form of place. Advertising images circulate through space, producing not only texts, but a spatial reorganization of social life. As audiences become more spatially dispersed, they also become more spatially defined. Advertising centralized in space as viewers moved into the home, and today fragments as female viewers abandon the domestic scene (Berland, 1992). The role of advertising in constituting the mass market housewife of the 1950s parallels its contemporary role in resituating women in decentered or segmented markets, or recentering and repackaging the traditionalism of the 1950s. By highlighting the way in which national, familial, or feminine identities are being reconstituted within the discourses of advertising, and by delving into the social and cultural fabric of the crisis of Fordism, we can observe the role of representations in mediating economic crises.

Space, place, and the new traditionalism

If nostalgia is a central element of postmodern culture, then it is certainly made manifest in contemporary advertisements which announce the new traditionalism, postfeminism, return of the family and the renewed popularity of products from the 1950s. These advertisements are attempts to reify notions of femininity and to resituate women in the home, reinforcing gendered divisions between public and private. In an era of generalized homelessness and spatial displacement, where the notion of 'home' as a fixed location seems in doubt, constructions of the 'home' form a powerful unifying symbol (Gupta and Ferguson, 1992). The home as a place is central to the delineation of the new traditionalist woman. Two examples of the new traditionalism are found in trade advertisements for *Good Housekeeping* and *Family Circle*. Trade advertisements are directed primarily at the advertising community to encourage media planners to buy advertising space. The main target group for these magazines is middle-class white women over the age of 35. Both campaigns ran in newspapers, advertising trade journals, and in the streets of New York, Los Angeles, and Chicago, where advertising executives and the manufacturers were likely to be found. A residual aim of the campaigns is to reach women as potential readers.

The advertisements for *Good Housekeeping* announce the appearance in America of the new traditionalist woman; 'She started a revolution with some not-so revolutionary ideals. She was searching for something to believe in and look what she found. Her husband, her children, herself'. Women's identities are inescapably focused on the home. The return to family values and to tradition is tied to women's roles in the family, and the home represents a haven in the face of recession. A recent advertisement appearing in major US newspapers read,

> Good News. If you're tired of hearing doom and gloom about America, the New *Good Housekeeping*/Roper Study will get your attention. Traditional values are flourishing, the family is becoming 'much stronger', working mothers are more optimistic, more confident . . . And despite the pessimistic predictions of other studies, she [the new traditionalist] believes she can help provide her family with a higher standard of living than that of her parents.
>
> (*Wall Street Journal* 2 March 1992, p. B6A)

Many advertisers characterize the new traditionalism as the largest 'social movement' since the 1960s; noting that, 'America's new emphasis on home and family, fuelled by the aging baby boomers, has narrowed the gap between homemakers and "modern" women' (Fannin, 1989, p. 38).

The new traditionalist advertisements did not go uncontested. No other recent campaign has garnered as much attention among both advertisers and women's groups: a sign that it strikes a central issue in contemporary discourses on femininity. Marcy Darnovsky argues that the advertisements target tensions left unresolved by the feminist movement itself (Darnovsky, 1991; see also Faludi, 1991; Probyn, 1990). These advertisements respond to advertising research revealing women's anxieties about changing family forms and roles. Consumer research has been adept at isolating these anxieties in both family and femininity and at adapting advertising to address them. Whereas some advertisers explain the new traditionalism in terms of professional women's disillusionment with the workplace and the glass ceiling, others have suggested that it speaks to the concerns of pink-collar women who maintain that their jobs are not central to defining their identity (*Adweek*

1988).[5] Judith Stacey (1990) points to an atmosphere of postfeminism today which simultaneously incorporates, adapts, and depoliticizes aspects of the feminist movement.

The new traditionalist campaign has influenced the advertising of many product manufacturers, such as Ralph Lauren, and even *The New York Times*. As one reporter puts it, the *Good Housekeeping* advertisements 'almost singlehandedly made it more acceptable for advertisers to take women out of suits and spectacles and put them back into the kitchen, fretting about cleansers' (Lipman, 1992). One recent AT&T advertisement depicts a woman who gave up corporate life to work at home; 'As she faxes a document, she says she doesn't miss the power lunches, because she's eating with a more important person – her child' (Piirto, 1989). Alan Waxenberg, publisher of *Good Housekeeping*, labels the 1960s as 'protest', the 1970s as 'feminist', 1980s as 'yuppy' and the 1990s as the 'new traditionalists'; he claims that individuals in 'the 60s and 70s were idealists; the 80s, materialists, and the 90s, realists' (quoted in Dougherty, 1988, p. 18). A chart developed by Waxenberg and his director of marketing suggests that 'idealists' and 'yuppies' took women away from their traditional roles in the home and family. Now their slogan, 'American is *coming home* to Good Housekeeping' implies that women are returning to the home (Dougherty, 1988, p. 18, emphasis added).

Market fragmentation has had a pronounced impact on the advertising community. *Good Housekeeping* had argued that magazines offer a more specialized and defined audience; 'The vast exposure facilitated by TV is clearly the key to its wide impact. But with fragmentation of broadcast audiences, the cost of reaching these target audiences can be quite high' (CMA, 1991, fourth quarter, p. 4). In the past, *Good Housekeeping* advertised itself as 'Everywoman's Magazine', but now mass market magazines like the 'Seven Sisters' have been especially hard hit, forcing them to seek more individuality (Carmody, 1990).[6]

Like many advertisements today which use black and white and other techniques to connote realism, these trade advertisements, produced by a photojournalist, feature 'actual' women in an attempt to distinguish them from the multitude of mostly glamorous images found in magazines. The campaign seeks to mediate the crisis of Fordism, which is in part a crisis in family form and notions of the 'American Dream'. It represents one attempt by those in the advertising community to grapple with the instability in family and femininity in a way that is reconcilable with the product in question: a mass market magazine.[7] The trade advertisements suggest that change has been exaggerated, that female consumers still place faith in the claims both of advertisers and of manufacturers. A *Good Housekeeping* study reasserts the continued belief in the American Dream and traditional values of equality of opportunity, freedom of choice, owning a home, and having both a rewarding career and family life (CMA, 1991, second quarter, p. 7). Although women believe the American Dream to be further from reach than it was in the previous generation, the underlying affirmation of the consumerism as a fundamental aspect of American Dream is evident in this survey.

The *Good Housekeeping* campaign represents a structured dialogue among companies, advertising executives, and women. Criticisms of the advertisements by women's groups were accommodated through later revisions, illuminating processes of mediation of identity by advertising. This process of experimentation is evident in the fact that the advertisements were released in three stages: the first advertisements depicted women as housewives, whereas later advertisements responded to the outrage of feminist groups by recognizing women's 'secondary' roles in the workplace. When women's paid labor-force activity is acknowledged, their femininity must also be reasserted. Almost all women in the advertisements are employed in traditional female occupations such as teaching, interior design, or forms of self-employment, all of which are compatible with women's roles in the household. Careful attention is paid to the women's attire to ensure that they are feminine. The advertisements define the task of women to learn to manage femininity and their image, and to balance work with the traditionally feminine domains of family, domesticity, and romance (Newman, 1991, p. 245).

The advertisements articulate an ideology of choice, familialism, self-help, and responsibility. Sandra Lee Spaeth, the magazine's associate publisher employs an ideology of choice in her defense of the campaign: '"New traditionalism" is "more liberating than the feminist movement'. It allows you to work part time or full time or not at all' (quoted in Lipman, 1992, p. 10). In the British context, Janet Newman shows that

> femininity itself is inflected by the Thatcherite discourse of choice: here as elsewhere the individualism of the enterprise culture is linked to power through consumer choice in the marketplace. Feminine style can now be selected along with a choice of lifestyle, choice at work, choice as consumers.
>
> (1991, p. 246)

In the final analysis, this choice becomes one in which the needs of family must always be privileged. Although the growth of enterprise culture has opened new spaces for some women to own their own businesses or to pursue professional or managerial careers, home is still constructed as the defining space of femininity. It is the home as a place that consolidates a woman's identity.

Many of the advertisements, stress that these 'new' women are not that different from their mothers. 'Woman', like 'nature', stands for the eternal and unchanging. The *Good Housekeeping*-Roper report, 'Inside the mind of the new traditionalist', underlines this point: 'Family is job one for the new traditionalist. Family life on the decline? Not according to the New Traditionalist' (Roper, 1992, p. 8). The same report notes, 'Whatever her duties outside the home may be, in the home she is still wife, mother, teacher, counsellor, activities coordinator and nutritionist' (p. 13). Although home and family are defined as the most central aspects of a woman's life, it is the home as a place which is insisted upon: 'The home and hearth is where the meeting of her traditional values, self-confident spirit and contemporary lifestyle is most striking' (p. 20).[8]

The study also sets out to validate the work of homemakers and the standards of 'good housekeeping' that they maintain: 'the business of caring for the home is, itself, a full-time job – a job which provides stimulation and satisfaction' (p. 29). Commodities link public and private and transform the home into an expression of feminine identity. *Good Housekeeping* and *Family Circle* both present interiors of real homes as models depicting products that can be used to furnish identity. Similarly, the increased incidence of paid employment in the home is also cited as a positive trend; 'Blazing new trails, one in ten employed women are working from their home' (p. 29). Contradictions and tensions involved in homework are elided. The new traditionalist campaign constructs home and work as compatible by reinforcing unequal power relations in the household. The report suggests the new traditionalist values honesty, getting along with others, and *respects authority* (p. 15). Femininity is defined by discipline and respect for 'women's place'.

Suzanne Moore (1991) points out that the literature on 'cocooning' (returning to the home) has been limited to white middle-class contexts. The class-based nature of the images of femininity in the *Good Housekeeping* advertisements is very striking, with all advertisements depicting middle-class to upper-middle-class women. One later advertisement in the *Good Housekeeping* series attempts to compensate for the images of mostly white women by incorporating a photograph of an Asian woman with her son. The advertisement, which reads, 'Who says you have to discard your old values to be a modern American mother?', emphasizes women's role as mother and attempts to forge links between notions of Asian femininity and American traditionalism. This example parallels a newspaper series in Britain which described Muslims as 'the new traditionalists' and 'More British than the British' (Jackson, 1989, p. 10). This series positions Muslim identity in a way that collapses ethnic difference under a common ideology of 'tradition': 'Devout, hard-working, disciplined, they have all the qualities that made us a great nation' (quoted in Jackson, 1989, p. 10). Interestingly, it is their traditional attitudes towards gender and sexuality that are given most attention. Muslim culture is described as time-honoured and predominantly masculine. Like Victorian women in Britain, Muslim women are kept in their place, 'veiled' from the public sphere. This resembles the new traditionalist attempt to reinforce divisions between public and private. An erosion of the 'natural' links between gender and place has thus led to increased ideas about distinct places and 'imagined communities' (Anderson, 1983; Gupta and Ferguson, 1992).

In all of these advertisements, nature is not very far away. Fall leaves, large windows looking outdoors, flowers, forest views, apples, hay, and farmhouses signify the ties between nature and femininity. The houses shown are old or decorated in a traditional style, resonating with the trend toward rural gentrification (see Mills, 1992). Karen Till (1993) discusses the traditionalism deployed in constructing Rancho Santa Margarita as a historic small town, removed from the excesses and sprawling chaos characteristic of Los Angeles. The construction of this new urban form reinforces an ideology of conservatism and naturalness. Similarly, one *Good Housekeeping* advertisement questions, 'What keeps a woman down on the farm after she's seen the big city?'. The return to tradition, an anti-urban sentiment, clearly associates women with nature and rurality. This contradicts a view of women's progress linked to their movement from one gendered order (private space) into the gendered domain of public space and the city. Akhil Gupta and James Ferguson argue that a popular politics of place can be reactionary as well as revolutionary. They note that

> the association of place with memory, loss and nostalgia plays directly into the hands of reactionary popular movements. This is true not only of explicitly national images long associated with the Right, but also of imagined locales and nostalgic settings such as 'small-town America', or 'the frontier' which often play into and complement antifeminist idealizations of 'the home' and 'family'.
>
> (1992, p. 13)

Despite women's fear of public space in the city, they can occupy these spaces with a degree of ambiguity. The retreat to rural life signals an even more strictly controlled milieu for women; it also implies a return to paid or unpaid work in the home.

Conclusion

With the breakdown of mass markets, advertisers have found it difficult to respond to fragmentation. As female identities and spaces have become more decentralized and lost their stable location in the home, advertisers have attempted to resituate women. The reworking of public and private in the new traditionalist images constitutes one strategy among many, but it is a significant one. Because it deploys images of rurality, small town America, and nature, it reinforces an idea of women's identities centered on the private realm of the home. At a time when the boundaries between rural and urban, nature and culture, female and male, home and work, private and public seem most weakened, the new traditionalist images attempt to recreate old forms of gendered space and difference, reinforcing women's familial and domestic responsibilities.

The role of images in the current period of restructuring should not be overlooked. Dramatic shifts in the family and consumption have proceeded alongside changes in representations of women. Advertisers tap into women's frustrations with the workplace, family, recession, and environment, and in turn play a role in mediating these frustrations through the images they create to sell products. Images of traditionalism and a woman's place in the home in advertising, coincide and interact in revealing ways with the growth of a part-time and temporary, female labor force. Yet it is also apparent that the domain of advertising forms one site where the boundaries of gender are under construction, open to a degree of revision and dispute.

The crisis of identity in social theory is reflected in the discourses of advertising. Alongside the crisis in female identities, there has been a respatialization of identity. These nostalgic black and white images offer women overt representations of representations and therefore make possible a critical engagement with the processes by which femininity is constructed. Instabilities and slippages can, in some cases, provide women with a space to contest the repressive effects of regulatory practice. At the same time, the increasing theoretical emphasis on the practices by which consumers subvert and appropriate the meanings of commodities needs to be tempered with an awareness of the control of these meanings by cultural industries such as advertising. The spatial dynamics of these changes are complex and the creation of texts cannot be conceived outside of institutions and the production of spaces (Berland, 1992). In terms of feminist theory, dereification needs to be supplemented with a feminist project based on normative critique and construction recognizing the weight of shared history and collective affinities. Feminists should not fall victim to the same nostalgia for unified identities and spaces as advertisers. Women need to challenge and affirm identities in order to engage with the contradictory sides of the post-Fordist coin. After all, feminism is partly about leaving the home, and entering another place, risky and unknown.

Notes

1 Historically, the female consumer in advertising has been a middle-class white Western woman. The notion of a unitary female consumer or spectator obscures the very different relationship to consumption according to class, sexuality, race, or place, and needs to be challenged through further research.

2 The department store and shopping mall, although no longer exclusively female spaces, are also key sites of consumption and femininity (Dowling, 1991; Morris, 1988; Shields, 1992; Sorkin, 1992).

3 Although men are increasingly targeted in advertising, the solicitation of the male consumer has been accomplished with great difficulty, because of an inherent antagonism between a hegemonic masculinity and consumption (Fraser, 1989). Similarly, although representations of women have changed to show women in a variety of locations outside the home, including the workplace, this has been achieved only by reasserting that they have not lost their femininity. Maidenform came out with a US campaign in the 1980s in which a woman in lingerie is pictured in a stock market setting surrounded by the gaze of businessmen, illustrating the continuing opposition between femininity and work. The copy read: 'The Maidenform woman. You never know where she'll turn up'.

4 Although the representation of femininity in advertising reinforces sexual difference, in re-speaking this difference, there is an inevitable slippage of meaning. There are also limited possibilities for resistance in the process of reception: women do not simply absorb representations of femininity, but receive them differently and actively transform their meanings. Among representations, those which foreground their contingency and mediated dimensions invite a greater critical awareness on the part of viewers; for example, see a series of art works done by Barbara Kruger on the 'family values' political campaign which appeared in *Newsweek* on 8 June 1992 (alongside Klein, 1992).

5 The interpretation of these results as a return to the home was later denied by the research firm hired for the study. Yankelovich Clancy Shulman charged *Good Housekeeping* with misrepresenting its findings.

6 The Seven Sisters of women's service magazines include *Good Housekeeping*, *Family Circle*, *Redbook*, *Woman's Day*, *Ladies Home Journal*, *Better Homes and*

Gardens, and *McCalls*. With the proliferation of women's magazines in the fragmented market of the 1980s, *Good Housekeeping* lost advertising revenue.

7 *Good Housekeeping* attempted to cut its rates as a short-term fix, but later hired a new agency to initiate a trade campaign, representing an attempt by a mass market magazine to recapture advertising pages. It was immensely successful within six months of its release and by mid-1988, ad revenues had greatly increased. By 1991, however, advertising pages fell again by 7.1%, prompting a new series of advertisements and the latest study, "Inside the mind of the new traditionalist" (Roper, 1992), a comprehensive effort to get into the psyche of contemporary women (Lipman, 1992).

8 The study reports that, compared with 1981 when 56% of American women said they were optimistic about the institutions of marriage and family, by 1991 63% proclaimed their optimism (Roper, 1992, p. 20).

References

Adweek 1988, 'Beaver's mom', in *Adweek Sixth Annual Women's Survey* p. 5.

Anderson B, 1983 *Imagined Communities: Reflections on the Origin and Spread of Nationalism* (Verso, London).

Barrett M, 1992, 'Words and things: materialism and method in contemporary feminist analysis', in *Destablizing Theory: Contemporary Feminist Debates* Eds M Barrett, A Phillips (Polity Press, Cambridge), pp. 201–219.

Barrett M, Phillips A (Eds), 1992 *Destabilizing Theory: Contemporary Feminist Debates* (Polity Press, Cambridge).

Bartky S L, 1990, *Femininity and Domination: Studies in the Phenomenology of Oppression* (Routledge: New York).

Berland J, 1992, 'Angels dancing: cultural technologies and the production of space', in *Cultural Studies* Eds L Grossberg, C Nelson, P Treichler (Routledge, New York) pp. 38–55.

Betz H, 1992, 'Postmodernism and the new middle class' *Theory, Culture and Society* 9, 96–115.

Bondi L, 1990, 'Feminism, postmodernism and geography: space for women?' *Antipode* 22, 156–167.

Bondi L, Domosh M, 1992, 'Other figures in other places: on feminism, postmodernism and geography' *Environment and Planning D: Society and Space* 10, 199–213.

Bourdieu P, 1984 *Distinction: A Social Critique of the Judgement of Taste* translated by R. Nice (Harvard University Press, Cambridge, MA).

Bowlby R, 1985 *Just Looking: Consumer Culture in Dreiser, Gissing and Zola* (Methuen, New York).

Buck-Morss S, 1986, 'The flaneur, the sandwichman and the whore: the politics of loitering' *New German Critique*, 39, 99–140.

Burgess J, Wood P, 1988, 'Decoding Docklands: place advertising and decision-making strategies of the small firm', in *Qualitative Methods in Human Geography* Eds J Eyles, D M Smith (Polity Press, Cambridge) pp. 94–117.

Butler J, Scott J (Eds), 1992 *Feminists Theorize the Political* (Routledge, New York).

Carmody D, 1990, 'Identity crisis for "seven sisters"' *New York Times* 6 August, p. B1.

Clarke J, 1991 *New Times and Old Enemies: Essays on Cultural Studies and America* (Harper Collins Academic, London).

CMA, 1991 *Changing Mood of America* a quarterly magazine by Good Housekeeping and the Roper Organization, various issues; Good Housekeeping, 959 Eighth Avenue, New York, NY 10019.

Coward R, 1985 *Female Desires: How They are Sought, Bought and Packaged* (Grove Press, New York).

Darnovsky M, 1991, 'The new traditionalism: repackaging ms consumer' *Social Text* 29, 72–96.

Deutsche R, 1991, 'Boys town' *Environment and Planning D: Society and Space* 9, 5–30.

Deveny K, 1990, 'Grappling with women's evolving roles' *The Wall Street Journal* 5 September, p. B1.

Doane M A, 1987 *The Desire to Desire: The Woman's Film of the 1940s* (Indiana University Press, Bloomington, IN).

Doane M A, 1991 *Femmes Fatales: Feminism, Film Theory, Psychoanalysis* (Routledge, New York).

Dougherty P, 1988, 'Advertising social analysis from Good Housekeeping' *New York Times* 11 August, part 4, p. 18.

Dowling R, 1991, 'Shopping and femininity', MA thesis, Department of Geography, University of British Columbia, Vancouver, BC.

Ewen S, 1976 *Captains of Consciousness: Advertising and the Social Roots of Consumer Culture* (McGraw-Hill, New York).

Faludi S, 1991 *Backlash: The Undeclared War Against American Women* (Crown Publishers, New York).

Fannin R, 1989, 'The growing sisterhood' *Marketing and Media Decisions* October, pp. 38–44.

Featherstone M, 1991 *Consumer Culture and Postmodernism* (Sage, London).

Foucault M, 1979 *Discipline and Punish: The Birth of the Prison* (Vintage Books, New York).

Fraser N, 1989 *Unruly Practices. Power, Discourse and Gender in Contemporary Feminist Thought* (University of Minnesota Press, Minneapolis, MN).

Fraser N, 1992, 'Sex, lies, and the public sphere: some reflections on the confirmation of Clarence Thomas' *Critical Inquiry* 18, 595–612.

Free V, 1989, 'Present, past, perfect' *Marketing Insights* 1, 41–48.

Golman D, 1990, 'Adweek's marketing week' *Adweek* 25 June, pp. 39–40.

Griggers C, 1990, 'A certain tension in the visual/cultural field: Helmut Newton, Deborah Turbeville and the Vogue fashion layout' *differences* 2(2), 76–104.

Grossberg L, Nelson C, Treichler P (Eds), 1992 *Cultural Studies* (Routledge, New York).

Gupta A, Ferguson J, 1992, 'Beyond "culture": space, identity, and the politics of difference' *Cultural Anthropology* 7(1), 6–23.

Haraway D, 1991 *Simians, Cyborgs and Women: The Reinvention of Nature* (Routledge: New York).

Jackson P, 1989, 'The urban jungle? Images of "race" and place in British political discourse', paper presented to an ERASMUS workshop on City and State at University College London, 4–6 May; copy available from the author, Department of Geography, University of Sheffield, Sheffield.

Jackson P, 1991, 'The cultural politics of masculinity: towards a social geography' *Transactions of the Institute of British Geographers* 16, 199–213.

Klein J, 1992, 'Whose values?' *Newsweek* 8 June, pp. 18–22.

Kolbowski S, 1990, 'Playing with dolls', in *The Critical Image: Essays on Contemporary Photography* Ed. C Squiers (Bay Press, Seattle, WA) pp. 139–154.

Lash S, 1990 *The Sociology of Postmodernism* (Routledge, Chapman and Hall, Andover, Hants).

Lipman J, 1992, 'Magazine on defensive, to use ad to support "new traditionalist"' *Wall Street Journal* 28 February, section B, p. 10.

McDowell L, 1991a, 'Life without father and Ford: the new gender order of post-Fordism' *Transactions of the Institute of British Geographers* 16, 400–419.

McDowell L, 1991b, 'Restructuring production and reproduction: some theoretical and empirical issues relating to gender, or women in Britain', in *Urban Affairs Annual Reviews 39. Urban Life in Transition* Eds M Gottdiener, C Pickvance (Sage, London) pp. 77–105.

McNay L, 1991, 'The Foucauldian body and the exclusion of experience' *Hypatia* 6(3), 126–136.

Marketing to Women 5(3), December, 1991.

Massey D, 1991, 'Flexible sexism' *Environment and Planning D: Society and Space* 9, 31–57.

Miller R, 1991, 'Selling mrs. consumer: advertising and the creation of suburban socio-spatial relations, 1910–1930' *Antipode* 23, 263–301.

Mills C, 1992, 'For home and country: the cultural construction of femininity and rurality in England', paper presented at the Annual Meetings of the American Association of Geographers, San Diego; copy available from the author, Department of Geography and Geology, Cheltenham and Gloucester College of Higher Education, Cheltenham, Glos.

Moore S, 1991 *Looking for Trouble: On Shopping, Gender and the Cinema* (Serpents' Tail, London).

Morris M, 1988, 'Things to do with shopping centres', in *Grafts: Feminist Cultural Criticism* Ed. S Sheridan (Verso, London) pp. 193–225.

Morris M, 1992, 'The man in the mirror: David Harvey's "condition" of postmodernity' *Theory, Culture and Society* 9, 253–279.

Mort F, 1988, 'Boy's own? Masculinity, style and popular culture', in *Male Order: Unwrapping Maculinity* Eds R Chapman, J Rutherford (Lawrence and Wishart, London), pp. 193–224.

Neilsen Inc., 1992, 'Women still do most of the buying' *Marketing Pulse* 12, 7.

Newman J, 1991, 'Enterprising women: images of success', in *Off-centre: Feminism and Cultural Studies* Eds S Franklin, C Lury, J Stacey (Harper-Collins, London) pp. 241–259.

Nicholson L (Ed.), 1990 *Feminism/Postmodernism* (Routledge, Chapman and Hall, Andover, Hants).

Olaquiaga C, 1992 *Megalopolis: Contemporary Cultural Sensibilities* (University of Minnesota Press, Minneapolis, MN).

Piirto R, 1989, 'The romantic sell' *American Demographics* August, 38–41.

Pollock G, 1990, 'Missing women: rethinking early thoughts on images of women', in *The Critical Image: Essays on Contemporary Photography* Ed. C Squiers (Bay Press, Seattle, WA) pp. 202–219.

Pratt G, 1993, 'Reflections on poststructuralism and feminist empirics, theory and practice' *Antipode* 25, 51–63.

Probyn E, 1990, 'New traditionalism and post-feminism: TV does the home' *Screen* 31(2), 147–159.

Roper, 1992, 'Inside the mind of the new traditionalist: marketing strategies for the 90's', Roper Organization and Good Housekeeping; Good Housekeeping, 959 Eighth Avenue, New York, NY 10019.

Rose G, 1992, 'Feminist voices and geographical knowledge' *Antipode* 24, 218–237.

Rubin H, 1989, 'You can go home again' *Adweek* 30 October, pp. 4–5.
Savage M, Barlow J, Dickens P, Fielding T, 1992 *Property, Bureaucracy and Culture: Middle Class Formation in Contemporary Britain* (Routledge, Chapman and Hall, Andover, Hants).
Sheilds R (Ed.), 1992 *Lifestyle Shopping: The Subject of Consumption* (Routledge, Chapman and Hall, Andover, Hants).
Sorkin M (Ed.), 1992 *Variations on a Theme Park: The New American City and the End of Public Space* (Hill and Wang, New York).
Spiegel L, 1992, 'The suburban home companion: television and the neighbourhood ideal in postwar America', in *Sexuality and Space* Ed. B Colomina (Princeton Architectural Press, Princeton, NJ) pp. 187–217.
Stacey J, 1990 *Brave New Families: Stories of Domestic Upheaval in Late Twentieth Century America* (Basic Books, New York).
ten Tusscher T, 1986, "Patriarchy, capitalism, and the New Right', in *Feminism and Political Theory* Eds J Evans, J Hills, K Hunt, E Meehan, T ten Tusscher, K Vogel, G Waylen (Sage, London) pp. 67–80.
Thrift N, 1989, 'Images of social change', in *The Changing Social Structure* Eds C Hamnett, L McDowell, P Sarre (Sage, London) pp. 13–42.
Till K, 1993, 'Neotraditional towns and urban villages: the cultural production of a geography of "otherness"' *Environment and Planning D: Society and Space* 11, 709–732.
Watson S, 1991, 'Gilding the smokestacks: the new symbolic representations of deindustrialized regions' *Environment and Planning D: Society and Space* 9, 59–70.
Wigley M, 1992, 'Untitled: the housing of gender', in *Sexuality and Space* Ed. B Colomina (Princeton Architectural Press, Princeton, NJ) pp. 327–389.
Wilson E, 1992, 'The invisible flaneur' *New Left Review* number 191, 90–110.

SECTION SIX

GENDERING WORK

Editors' introduction

Work, whether waged or not, is central to women's and men's lives and in the social construction of masculine and feminine identities. As feminist scholars have long argued the very terms used to describe jobs and occupations and the places within which they are undertaken, are imbued with connotations of masculinity and femininity. To mention manual or mental labour conjures up masculine associations of brawn and brain respectively: the male solidarity, cameraderie and strength of miners versus the cool rationality of the scientist or the masculinist rhetorical power of the politician. As feminists have argued, that spatial distinction between public workplaces and the private sphere of the home – woman's place – that has structured capitalist societies from the industrial revolution, has been a central part of the construction of women's social inferiority. In other societies too the association of women's work with no pay or low pay and lack of status and power is a seemingly universal division between men and women. The work that women do is undervalued for the very reason that it is women that do it.

As a now well-known postcard based on a UN report so succinctly states:

Women constitute
half
the world's population,
perform nearly
two-thirds
of its work hours,
receive
one-tenth
of the world's income
and own less than
one-hundreth
of the world's property

(United Nations report, 1980)

This seeming universalism, however, disguises significant spatial and temporal variations in the position of women in the labour market, in gender divisions of labour and in associated rewards – property ownership, social status and the possibilities of maintaining independence.

Perhaps of all the issues that feminist geographers have investigated, it has been work – in the widest sense of domestic, community and waged labour – that, until recently at least has dominated feminist inspired scholarship by geographers. This reflects the early influence of socialist and marxist theories on feminist geography – in particular the centrality of production as the determinant of social relations between the classes and the sexes and an associated belief that participation in waged labour was a prerequisite for women's liberation. In marxist theory women were seen as a reserve army of labour able to be pulled into and expelled from the labour force in relation to economic cycles of expansion and recession, although somewhat paradoxically their domestic labour was also theorised as 'essential' for the maintenance of capital and the daily and generational reproduction of the labour force. Perhaps this essential domestic labour was really not so essential in boom periods after all? Women's extra wages may have resigned other household members to cold suppers and crumpled clothes!

Detailed empirical investigations of the ways in which women have been drawn into and excluded from waged labour, however, have revealed what has become known as the 'feminisation' of the labour force. Although there are enormous variations in the extent to which women are part of the waged labour force and in their status and positions between, and indeed within, nation states – related to factors such as the level of 'development', the nature of the economy and the role of religion – women seem to be being drawn into the social relations of capitalist economies in growing numbers across the world as a whole. Despite periods of boom and recession, women have entered the labour market in growing numbers and have become a permanent, rather than a temporary part of the economy. What remains constant, however, in virtually all economies, is women's inferior position. Based on an assumption that women have male partners on whom they are dependent – the male breadwinners of theory and state policy – women are able to be constructed as secondary workers. Their primary role is assumed to be domestic labour within the heterosexual nuclear family. Despite remarkable advances in, for example, women's educational level and their possession of skills, as well as a great variety of household and family forms, in all the countries of the world women are constructed as the bearers of natural, and so unvalued, attributes – of manual dexterity and caring for example. They are paid less than men and work under generally inferior terms and conditions. This is not to deny, however, that for many individual women and for the households of which they are part, women's participation in waged labour has in many instances radically improved their living standards and women's own sense of their self-worth and independence.

In feminist economics, development studies and geography there is now a large and fascinating literature exploring a range of explanations for women's entry into waged labour and disputes about its implications. The debates have been particularly fierce in the literature about 'developing' or Third World societies, where many of the women who enter the manufacturing sector in particular are young and unmarried. There is disagreement about the extent to which these women are part of an underpaid and undervalued global labour force, exploited by capital because of their social definition as bearers of inferior labour. Or on the other hand, whether their entry into the relations of waged labour is instrumental in their escape from patriarchal control in the family, or enables them to negotiate more equitable familial or household relations. As a range of empirical studies in both Third and First World societies have shown, for example in the Phillipines (Chant and McIllwaine, 1995) or Java (Wolf, 1991) and among minority groups in cities in the west (Pessar, 1994), actual circumstances are often complex. For younger women waged labour may bring some independence or freedom but for many older women with children, waged labour becomes a 'double burden' – an essential element of family survival without which millions of women and children in the First and Third Worlds would descend into greater poverty.

In the papers that constitute Section Six, we have ranged between and across societies, and have also attempted to give an idea of the post-war changes in the position of women and men in waged labour. We have tried to give some indication of the different approaches to the explanation of the persistence of gender segregation and occupational stereotyping, despite the enormous variety in forms of waged labour. In the first paper, Sharon **Stichter** provides a summary of the changes in women's position in the family and in employment between the 1960s and the end of the 1980s, when enormous changes were taking place in many countries in women's labour market participation rates. She clearly demonstrates feminist assertions that 'the economic' is inseparable from 'the social'. Women's labour market position is inexplicable without a consideration of the nature and form of household, and familial structures and gendered divisions of responsibilities and tasks. Stichter examines both the range of assumptions that construct women as inferior workers in different societies and the consequences for waged work and for domestic, reproductive and productive work undertaken within the household. In the second paper, Stichter's spatial comparisons are complemented by Bradley's historical approach and also by her focus on industrial societies. **Bradley** examines the ways in which sex-typing of occupations means that different jobs are constructed as suitable for one or other sex. She shows that despite appeals to essentialised attributes of femininity in arguments about why women do certain jobs, gender attributes are flexible and vary over time. What seems to remain constant is women's generally inferior position.

Stitcher and Bradley, feminist scholars working outside geography, set the scene historically and internationally and explain some of the ways in which gender segregation has been theorised. In the third and fourth papers which are by British feminist geographers who work predominantly on the changing nature of the British economy, some of the particularity of waged labour in manufacturing and in the service sector in different parts of Britain is explored. The third paper is by Doreen **Massey** whose work in economic geography has been highly influential in changing the shape of our understanding of the British space economy. In her classic book *Spatial Divisions of Labour* (1984, and reissued 1994), she argued that the ways in which the geography of economies change are related to spatial differences in the social relations at the local level. Gender relations are an important aspect of these local differences, and in an approach that parallels the reserve army of labour approach, Massey has shown how different gender divisions of labour in particular localities affect how women are drawn into, or excluded from waged labour, in part dependent on the historical nature of women's domestic responsibilities. Here we have chosen to include her paper 'Industrial restructuring as class restructuring' which was first published in 1983. Despite it not having a proclaimed feminist slant in its title, this paper was extremely important in geographer's understanding of the different ways in which women were constructed as a reserve of green labour in different parts of Great Britain. Massey examines the class and gender implications of the changes that have occurred in the structure of manufacturing in industrial societies. In western economies, industrial restructuring from the mid 1960s onwards led to a decline in traditional heavy male manufacturing employment and the expansion of women's employment in light industries, particularly in electronics. Paralleling the situation in the Third World, women were drawn into British manufacturing for exactly the same reasons as in the rest of the world: their construction as cheap labour with 'nimble fingers'. Thus Massey shows the parallels in both the theoretical understanding and empirical analyses of changes in global manufacturing.

One of the most significant changes in the period since Massey's writings in the early 1980s, however, has been the shift towards an economy dominated by the production and exchange of services in Britain, the USA and other western nations. In the fourth extract, Linda **McDowell** and Gill **Court**, argue that the nature of service work demands a different understanding of the social construction of gendered identities. They suggest the 'division of labour'/reserve army approaches, be they at the level of the nation or the locality, provide an incomplete explanation of the construction of gender at work, as too much about gender tends to be assumed. Instead of taking the social characteristics of femininity and masculinity for granted, as already fixed in place as their bearers become labour power, McDowell and Court suggest that women and men 'do gender' at work, as part of a performance that is a crucial part of selling a service. In the growing service sector occupations, the embodied characteristics of an individual are a crucial element in production. The product and the worker cannot be separated in the same way as in manufacturing industries. Following a theoretical elaboration of these ideas about subjectivity at work, the authors illustrate their argument with a brief examination of the gendered performance of a number of occupations in the City of London.

The paper challenges the home/workplace distinction that has been a crucial part of geographers' explanations of gender divisions of labour. For many women, especially those from minority communities, work is undertaken in the home as part of a patriarchal division of labour, in which the husband, father or other male relative is the boss as well as a family member. The contractual nature of wage labour is replaced by a personal connection, which may or may not be rewarded by financial remuneration, but in which control is bound up with relations of obligation, familial responsibility, love and desire. Many sweated occupations in the clothing industry exemplify this blurring of the boundaries between home and work, and Westwood and Bhachu (1989) in their collection of case studies of 'ethnic enterprises' examine this. We have chosen the case of Chinese restaurants, takeaways and 'chippies' that have become ubiquitous features of British cities. **Baxter** and **Raw** examine the particularities of Chinese women's labour as well as drawing out the similarities between other groups who were drawn into the imperial heartlands by demands for cheap, colonial labour power.

While we do not have the space to include a paper about domestic labour, commodified domestic labour, carried out in other women's homes, is the lot in life for millions of women around the world. In Britain at the beginning of the century for example, 2 000 000 women were domestic servants. The First World War saw the decline of domestic work on a large scale in Britain, as alternative opportunities were created to draw women into the war effort, on a temporary basis or so it was believed at the time. But women were reluctant to return to the penury and restrictions of domestic work and the Second World War saw its virtually total demise. However, as Nicky Gregson and Michelle Lowe (1994) have shown in the UK, the entry of growing numbers of highly educated women into professional occupations, has led to the resurrection of demand for domestic and childcare workers in Britain, and to the development of what they have termed a 'new servant class' – to service the growing numbers of middle class, dual income families.

Elsewhere in the world, in both First and Third World societies, large numbers of women have always done other women's domestic work. The same complex questions about the consequences that we raised in the context of factory work arise. For some women, domestic work is an escape: from rural to urban areas, as Radcliffe (1990) has demonstrated in Peru; from the segregated homelands of apartheid South Africa to the cities (Bozzoli, 1991); and from less developed nations to richer countries (Enloe, 1989; Bakan and Stasiulis, 1995). But as well as escape, this form of work is perhaps the lowest paid and most exploited of all work for women. Like the examples of sweated labour, domestic labour is distinguished by the form of personal relations between the contractor and the provider of the service. For many domestic workers, this relationship is made more complicated by the fact that they live in the home of their employers and so are doubly dependent upon them. For the women who employ other women to clean their homes, and especially to care for their children, the personal nature of the relationship is also complicated to negotiate. Their dependence on their employee may not be so stark but they have to feel secure in the nature of the service that is being provided.

Perhaps the starkest example of geographic movement to do domestic work is the migration of millions of women from Mexico and Central America to California. Many women travel daily across what must be the most extreme example of a border dividing the rich from the poor, as they cross the Rio Grande from Mexico to Southern California, to do domestic work in San Diego. Millions of other Chicanas have moved across the border on a more permanent basis, to work as domestics elsewhere in California and the rest of the USA. In her book *Maid in the USA*, Mary Romero (1992) examines both the theoretical analyses of domestic work and some of the consequences for women who not only clean up for their own households for love, but for other people for wages.

As Romero points out, domestic work is perhaps the starkest example of the social divisions between women. Changes in the nature of work in many societies are opening up new divisions between women, and indeed closing some of the divisions between men and women. For the women who benefit from a university education and move into high status occupations, their living standards and lifestyles may differ relatively little from their male colleagues. Similarly at the bottom end of the labour market, for the millions of men and women struggling to achieve a livelihood strategy that holds body and soul together, or for those most threatened by the new 'flexible' labour market conditions, class divisions are as important as gender divisions (McDowell, 1992; Phillips, 1987). But while women continue to do the majority of domestic work and childcare provision, the position of mothers in the labour market will remain distinctive. Single women may escape this particular location, but they still enter the labour market as embodied women. While the consequences of femininity and female embodiment are inferior positions and rewards in the labour market, women seem to be trapped.

References and further reading

Bakan, A B and Stasiulis, D K 1995: Making the match: domestic placement and the racialization of women's household work. *Signs: Journal of Women in Culture and Society* **20**, 303–35.

Bakker, I (ed) 1996: *Rethinking Restructuring: Gender and Change in Canada*. University of Toronto Press: Toronto and London.

Bozzoli, B with Nkotsoe, M 1991: *Women of Phokeng: Consciousness, Life Strategy and Migrancy in South Africa 1890–1983*. Raven Press: Johannesburg.

Carney, J and Watts, M 1991: Disciplining women? Rice, mechanization and the evolution of Mandinka gender relations in SeneGambia. *Signs: Journal of women in culture and society* **16**, 651–81.

Chant, S and McIllwaine, C 1995: Gender and export manufacturing in the Philippines: continuity or change in female employment? *Gender, Place and Culture* 2, 147–76.

Einhorn, B 1993: *Cinderella Goes to Market: Citizenship, Gender and Women's Movement in East Central Europe*. Verso: New York.

Enloe, C 1989: *Bananas, Beaches and Bases*. Pandora: London.

Gregson, J and Lowe, M 1994: *Servicing the Middle Classes: Class, Gender and Waged Domestic Labour in Contemporary Britain*. Routledge: London.

Hanson, S and Pratt, G 1995: *Gender Space and Work*. Routledge: London.

Kessler-Harris, A 1982: *Out to Work: A History of Wage-earning Women in the United States*. Oxford University Press: Oxford.

Kobayashi, A (ed) 1995: *Women Work and Place*. McGill-Queens University Press: London and Montreal.

Massey, D 1984/1994: *Spatial Divisions of Labour*. Macmillan: London.

McDowell, L 1992: Life without father and Ford: the new gender order of postfordism. *Transactions, Institute of British Geographers* **16**, 400–19.

McDowell, L 1997: *Capital Culture: Gender at Work in the City*. Blackwell: Oxford.

Mitter, S 1986: *Common Fate, Common Bond: Women in the Global Economy*. Pluto Press: London.

Pearson, R 1988: Female labour in the First and Third Worlds: the greening of women's labour. In Pahl, R (ed) *On Work: Historical, Comparative and Theoretical Approaches*. Blackwell: Oxford. 449–66 (see also many other interesting chapters about gender and work).

Pessar, P 1994: Sweatshop workers and domestic ideologies: Dominican women in New York's apparel industry. *International Journal of Urban and Regional Research* 18, 127–42.

Phillips, A 1987: *Divided Loyalties: Dilemmas of Sex and Class*. Virago: London.

Phizacklea, A and Walkowitz, C 1995: *Homeworking Women: Gender, Racism and Class at Work*. Sage: London.

Radcliffe, S 1990: Ethnicity, patriarchy and incorporation into the nation: female migrants as domestic servants in Peru. *Environment and Planning D: Society and Space* **8**, 379–93.

Romero, M 1992: *Maid in the USA*. Routledge: London.

Shaw, J and Perrons, D (eds) 1995: *Making Gender Work*. Open University Press: Buckingham.

Townsend, J G 1995: *Women's Voices from the Rainforest*. Routledge: London.

Westwood, S and Bhachu, P 1989: *Enterprising Women*. Routledge: London.

Wolf, D 1991: Female autonomy, the family and industrialization in Java. In Blumberg, R (ed.) *Gender, Family and the Economy: The Triple Overlap*. Sage: Newbury Park, CA. 128–48.

Women Working Worldwide (ed.) 1991: *Common Interests: Women Organising in Global Electronics*. Women Working Worldwide: London.

24 Sharon Stichter

'Women, Employment and the Family'

Excerpts from: S. Stichter and J. L. Parpart (eds) *Women, Employment and the Family in the International Division of Labour*, pp. 11–71. London: Macmillan (1988)

Throughout the 1960s, 1970s and 1980s, the participation of women in gainful employment has grown dramatically, not only in the industrialised nations but also in many parts of the Third World, especially those areas that have experienced increasing investment by multinational manufacturing firms with gender-specific hiring policies. In industrial societies, particularly the United States, the rise in women's labour force participation has been recognised as a major new social trend. In the developing world, the growth in women's employment has been uneven, being most marked in newly industrialising areas with strong export-oriented manufacturing sectors such as Mexico, Brazil, the Caribbean, and East and Southeast Asia. But even in many poorer nations of Africa and Central America, women's labour force participation is increasing, even while their formal, paid employment, and their incomes from such employment, may be stagnating or declining. Between 1985 and 2000, the female labour force is expected to increase faster than the male labour force in the more industrialised parts of Latin America, and to grow at the same rate as the male in East Asia and less industrial Latin America. In Africa, however, the male labour force is expected to increase faster than the female (UNESCO, 1986, p. 13).

National variations in female labour supply, employment levels and employment patterns are influenced to a very great extent by the overall level of world and national development, by the sectoral composition of employment growth or the international division of labour, by commercial cycles in the global economy, and by other national and international demand-side factors. To what extent, however, might these marked variations also be affected by the sex division of productive and reproductive labour in the family? In what ways might household-level factors such as fertility demands, sex role socialisation, and patriarchal controls affect women's relation to the labour market? Do they impact mainly on female labour supply, or do they also affect employers' demand for labour?

This chapter will explore this complex question, arguing that household factors are both essential and interesting, and that they directly influence age patterns of participation, time spent in the labour force, earnings, and even aggregate employment levels.

Market and production factors versus household factors in women's employment

Levels of employment

In virtually all societies, men have higher labour force participation rates than women, although in a few nations general female rates do approach those of men, notably in the USSR, Eastern Europe, and some Caribbean and African nations (Table 24.1). And, while variations in male labour force participation rates between nations, even between low and high income nations, are small, societal differences in female labour force participation rates are great (Table 24.1; Standing, 1982, 13ff). Similar observations may be made about rates of gainful employment (Table 24.2). Although both the male/female and the inter-country differences are doubtless affected by cultural practices as to the enumeration of women workers and the estimation of their work, it seems unlikely that these factors can account for all of the differentials. Nor can they account for the continuing gap between male and female average wages in most nations of the world (Table 24.3) and the marked segregation in occupations (Anker and Hein, 1986; House, 1986).

Despite the gender gap, female employment in the developing world has shown surprisingly rapid growth in recent years (Table 24.2), particularly in the industrial and service sectors. The overall proportion of women in the industrial labour force in developing countries rose from 21 per cent in 1960 to 26.5 per cent in 1980. (UNESCO, 1986, Table 2, p. 70; Joekes, 1987, p. 80). In particular areas growth was even more rapid: the female share of the paid labour force in Singapore, for example, rose from 17.5 per cent in 1957 to 33.6 per cent in 1979 (Wong, 1981). And despite the oil price shocks and the world economic recession of the 1970s and early 1980s, of the 32 major developing nations for which data were available, 28 showed increases in the female share of gainful employment between 1977 and the mid-1980s (Table 24.2).

Variations in female employment levels and patterns are affected by complex combinations of economic variables. Recent trends illustrate the effects of two broad categories of factors: firstly, changes in the organisation of production, that is, the growth of the industrial and service sectors and the impact of technological change in industry; and secondly changes in market conditions, particularly in product markets, but also in the availability of male labour. Examination of each of these factors uncovers points at which 'extra-economic' variables must be brought into the analysis.

Between 1960 and 1980, the decline in the number of workers in agriculture and the growth in importance of the industrial and service sectors was a broad sectoral change which had greater impact on women than on men. In these two decades, considering all developing countries as a whole, the total male labour force in agriculture dropped by 12.6 percentage points, but the total female labour force in agriculture dropped by 15.3 points. The percentage of the female labour force in industry doubled – from 8.2 per cent to 16.3 per cent – whereas that of the male labour force rose only from 15.1 per cent to

Table 24.1 Total economic activity rates

	Men	*Women*	*Year*	*Men*	*Women*	*Year**
Algeria	33.8	2.4	1983	42.4	4.4	1987
Argentina	55.3	19.9	1985	53.5	29.9	1995
Bangladesh	53.5	5.4	1983–84	88.3	62.6	1990–1
Barbados	52.2	39.7	1983	74.4	61.0	1994
Bolivia	48.7	14.6	1986	48.7	30.4	1992
Botswana	38.1	36.0	1984–85	42.8	24.5	1991
Brazil	56.6	27.9	1985	56.3	27.9	1989
Chile	50.2	20.6	1986	53.1	24.7	1994
China, People's Rep.	57.3	47.0	1982	57.3	47.0	1982
Costa Rica	53.2	18.5	1985	53.9	23.3	1994
Cuba	54.0	30.6	1986	55.4	31.7	1988
Dominican Republic	48.1	19.7	1981	48.1	19.7	1981
Egypt	49.8	12.5	1983	43.5	14.2	1992
El Salvador	47.9	23.7	1980	48.3	34.0	1991
Greece	51.4	27.3	1985	52.7	29.4	1994
Guatemala	48.1	8.0	1985	50.8	16.7	1989
Haiti	52.1	36.9	1983	51.4	33.6	1988
Hong Kong	61.9	39.5	1986	60.2	38.1	1994
Hungary	51.0	40.9	1987	46.2	35.0	1994
India	52.7	19.8	1981	51.6	22.3	1991
Indonesia	50.0	27.9	1985	82.6	49.1	1992
Iran	47.1	7.2	1982	45.6	5.5	1986
Ireland	51.7	21.7	1985	51.4	24.3	1991
Jamaica	49.6	41.1	1985	48.6	40.6	1990
Korea, South	46.8	30.6	1986	76.4	47.9	1992
Malawi	52.4	51.7	1983	43.9	42.8	1987
Malaysia	49.6	25.3	1980	47.3	22.6	1991
Mauritius	55.8	19.4	1984	81.9	43.2	1991
Mexico	48.2	18.2	1980	54.6	23.6	1993
Morocco	47.1	11.6	1982	48.6	16.8	1990
Nigeria	43.1	20.6	1983	41.1	20.9	1986
Peru	49.4	34.4	1986	53.7	33.6	1994
Senegal	55.3	39.1	1985	52.1	17.0	1988
Singapore	59.8	35.2	1986	79.6	50.9	1994
South Africa	46.2	22.8	1980	45.5	29.5	1991
Syria	40.3	6.8	1984	-	-	-
Tanzania	44.4	45.2	1978	-	-	-
Thailand	55.9	50.1	1984	58.9	46.4	1994
Tunisia	47.4	13.3	1984	46.5	12.7	1987
United Kingdom	59.5	35.8	1981	56.8	42.3	1993
United States	57.1	42.5	1986	57.0	45.8	1994
Venezuela	49.0	18.9	1986	49.5	22.8	1993
Zambia	46.0	17.4	1984	-	-	-
Zimbabwe	41.1	25.4	1980	42.8	26.7	1992

*Note: Additional figures from 1990 and 1995 Yearbooks supplied by editors.

Source: International Labour Organisation, *Yearbook of Labour Statistics* 1986, Table I, pp. 13ff; 1987, Table I, pp. 14ff; 1990, Table I, p. 12ff and 1995, p. 13ff.

Table 24.2 Female employment as a percentage of total employment, 1977–86

Country	*1977*	*1978*	*1979*	*1980*	*1981*	*1982*	*1983*	*1984*	*1985*	*1986*
Algeria	5.9	5.2	5.2	8.2	8.2	7.0	6.8	7.6	8.4	. . .
Botswana	21.9	21.6	22.0	23.3 |	21.8	22.9	23.3	23.9 |	29.1	. . .
Egypt	6.8	7.9	6.7	7.1	7.8	7.9	. . .	. . .	. . .	. . .
Kenya	17.1	17.4	17.0	17.6	18.3	18.4	17.8	18.7	19.7	20.4
Malawi	11.4	11.6	11.7	11.4	11.1	11.2	11.5	13.6	13.2	. . .
Mauritius	24.4	24.5	25.6	26.0	26.6	27.1	27.4	29.8	32.1	33.9
Zimbabwe	. . .	. . .	17.1	17.0	16.0	15.1	16.3	16.6	. . .	. . .
Barbados	40.7	42.1	42.4	44.2	42.8	43.3	43.3	42.9	43.5	44.3
Bolivia	22.9	23.2	23.4	23.4	24.0	25.1	25.3	25.6	25.6	25.9
Brazil	31.2	31.3	31.7	. . .	31.2	32.2	33.0	33.0	33.4	. . .
Colombia	37.8	37.6	38.1	38.2	. . .	. . .	. . .	. . .	38.2	38.9
Costa Rica	23.0	24.4	25.0	24.3	26.0	25.6	25.3	27.9	25.8	. . .
Cuba	29.0	29.5	31.2	31.5	33.1	34.5	35.6	36.4	37.2	37.5
Chile	26.1	28.0	28.2	29.5	29.2 |	30.4	30.8	. . . |	29.6	29.7
Haiti	47.0	47.0	47.0	48.9	49.3	40.3	40.3	. . .	. . .	. . .
Jamaica	39.6	39.5	38.1	39.2	38.9	38.3	38.5	39.9	39.2	40.0
Mexico	24.2	24.4	25.4	26.2	27.2	27.6	28.5	29.3	29.4	30.2
Puerto Rico	34.7	34.7	35.2	35.9	36.3	36.6	37.0	37.2	37.3	38.3
Venezuela	27.7	27.3	27.6	27.9	27.1	27.4	27.6	27.7	27.6	27.7
Hong Kong	. . .	35.2	34.8 |	34.8	35.9	36.2	36.4	37.0 |	36.6	36.6
India	11.9	12.0	11.9	12.1	12.2	12.3	12.5	12.6	12.9	. . .
Philippines	30.1	35.9	. . .	35.4	35.5	37.1	38.3	36.5	37.2	. . .
Singapore	31.8 |	33.1	33.6	35.0	35.5	35.6	35.5	36.3	36.4	37.4
South Korea	37.1	38.1	38.5 |	38.2	38.1	39.1	39.2	38.4	38.0	39.8
Sri Lanka	34.6	28.7	32.7	33.7	35.4	36.0	34.9	35.9	. . .	. . .
Syria	17.5	13.7	15.8	. . .	. . .	. . .	13.9	13.4	. . .	. . .
Thailand	44.4	44.3	43.9	47.3	45.8	. . . |	44.5	45.0	. . .	. . .
Ireland	28.0	28.3	28.4	29.2	29.8	30.5	31.2	31.1	31.3	. . .
Turkey	8.5	7.9	. . .	8.2	8.9	8.8	9.1	9.3	8.8	. . .
Trinidad/ Tobago	28.3	. . .	. . .	. . .	. . .	. . .	. . .	. . .	32.8	. . .
Panama	26.8	. . .	. . .	. . .	. . .	. . .	. . .	. . .	29.5	30.1
Indonesia	33.8	. . .	. . .	. . .	. . .	. . .	. . .	. . .	36.0	. . .

Note: Vertical line indicates break in comparability.

Source: ILO, *Yearbook of Labour Statistics*, 1986, Table III, pp. 325 ff.

21.6 per cent. The percentage of the female labour force in services also rose faster than did the male. Sectoral changes affected both men and women, but the change was greater for women. (UNESCO, 1986, pp. 72–73, Table 3).

An important reason for the increase in industrial employment in parts of the developing world has been the relocation of industrial production from developed to Third World countries, from whence the goods are re-exported back to developed nations. Relocation first took place in industries of high labour-intensity in which low wage costs were most critical (Safa, 1981). These were mainly textiles, clothing and food processing, in which the employment of large numbers of women was traditional. Often, these industries could utilise local raw materials. Improvements in world transport and communications facilitated the relocation process, as did the streamlining and standardisation of production processes and the increased

Table 24.3 Male–female wage differentials in manufacturing: 1975 and 1982

	Female wage as a percentage of male wage		
	1975	*1982*	
Cyprus	45.9	56.2	w
El Salvador	90.4	85.9	h
Greece	69.5	73.1	h
Ireland	60.9	68.5	
Japan	47.9	43.1	m
Kenya	66.1	75.8	m
South Korea	47.4	45.1	m
Tanzania	70.7	78.5*	m

* = 1980; w = weekly, h = hourly, m = monthly.

Source: International Labour Organisation, *World Labour Report 2*, Geneva 1985, Table 14.5, p. 224.

competitiveness of world markets. Gradually certain types of pharmaceutical production and much production in electronics, in which the full mechanisation of assembly work would have been more expensive than the cheap labour solution, were also transferred to 'offshore sources' in the Third World.

The transfer of low-skill manual assembly jobs to developing societies by transnational corporations is a fundamental restructuring of the world economy, and it has been closely associated with the increased role of women in the manufacturing labour force. From the point of view of strategies of Third World development, the older import-substitution approaches, which focused on the establishment of locally-owned industries catering mainly to the domestic market, tended to correlate with a lack of increase in female employment, whereas export-oriented approaches which welcome foreign-owned multinationals have been associated with increases in female employment (Safa, 1981; Chinchilla, 1977; Lim, 1978). Historically, industrial relocation first affected particularly the Caribbean and such Latin American countries as Mexico, Brazil and Colombia. The main relocation areas soon came to be in East and Southeast Asia: Korea, Hong Kong, Singapore, Malaysia, Philippines, Thailand. Some parts of South Asia were moderately affected (India, Pakistan, Sri Lanka), as were a few African nations (Tunisia, Egypt, Morocco, Mauritius and to a small extent, South Africa). For the most part, however, sub-Saharan Africa has not been a favoured site for export-oriented manufacturing. New locations are now being sought instead in China and low-wage areas of Europe. Numerically, the cases of large populous nations such as Brazil, India and China, all of which increased their international trade and the share of industrial goods exported in the 1960s and 1970s, account for a significant part of the total worldwide increase in female industrial employment (Joekes, 1987, p. 94).

It is important to note that the trend toward industrial relocation and employment of women has largely by-passed certain parts of the Third World, notably sub-Saharan Africa. Here, lack of any kind of industrial development, and stagnation in overall employment levels, have entrenched male/female inequities in industrial employment levels. Eleanor Fapohunda (1986) describes this situation in Nigeria, noting that in 1983 the female urban labour force participation rate was 41.9 per cent compared to the male rate of 75.4 per cent, and that the percentage of women in manufacturing declined between 1974 and 1983 (see also di Domenico, 1983). Economic factors play a key part in explaining this situation; the predominant types of industry do not convey any particular productivity advantage to women over men, and in addition there is an oversupply of low-skill, low-wage males in the urban labour market. But also, as described more fully below, African family patterns do not reinforce employer preferences for cheap, single, childless women.

Why have employers in the labour-intensive world market industries, such as garment and footwear assembly and electronics, discriminated positively in favour of women, in hiring if not in promotions or wages? Is it simply that stiff competition in the industry forces them to hire the cheapest labour possible? Elson and Pearson (1981) and others have argued that these hiring practices favouring women do not reflect simply the fact that women's wages are lower than men's, though that is so and is an important fact requiring explanation. It is also the case, they argue, that women are actually more productive than men in certain jobs because of their 'nimble fingers'. Manual dexterity, often cited by employers, represents not a natural advantage, as employers often believe, but a skill derived from prior training in the domestic tasks of sewing, weaving and chopping vegetables. Thus, they conclude, a household skill in which women have a trained productivity advantage has come to have a commercial market (also Lim, 1978, p. 15). This argument deserves further exploration.

Whether women's manual dexterity is learned or is a 'natural endowment', in either case it results in high productivity. Most economic analyses presume that at least in the long run high productivity will be reflected in higher wages. Yet in the world market factories the male/female wage gap runs in the opposite direction. As John Humphrey has noted, there seems in general to be a systematic over-valuation of male attributes and a corresponding under-valuation of female ones; for example, male physical strength commands a wage premium in metal-working industries, but female manual dexterity in the assembly-line industries does not (Humphrey, 1985, p. 223). How long can such a female productivity advantage coexist with lower wages, when from a neo-classical economic point of view it is an inherently unstable situation? Is the origin and persistence of this situation fundamentally dependent on the third factor that Elson and Pearson and many others adduce to explain employers' preference for women, that is, their docility, lack of labour mobility, and lesser likelihood of joining organised labour protests? (Fuentes and Ehrenreich, 1983, p. 12–15; Chapkis and Enloe, 1983). This docility, these writers suggest, is largely the result of subordination to patriarchal controls in the household which are extended into the work place. If this is correct, it is another important example of the effect of household patterns on employment behaviour, as we discuss more fully below.

Another aspect of the changing sectoral composition of the work force which serves to increase women's employment is the growth of the service sector. In Latin American and Caribbean economies today, for example, women constitute 39 per cent of the labour in the service sector, and fully 70 per cent of the entire female labour force is in that sector (Joekes, 1987, p. 107). Services include tourism, much of the informal sector, professional services such as nursing and teaching, domestic service, and transport and communications. As economic development proceeds, this sector is expected to increase and, with it, female employment. But why women? Employer prejudice, and perhaps the workers' own sex role socialisation, both perpetuated through childhood socialisation in the household, seem to be at work in defining such jobs as particularly appropriate for women.

In contrast to those production-related factors which favour women's employment, are those which work against it. A well-documented negative effect on women's employment comes about when more technically advanced, mechanised production methods are introduced in industries which had traditionally employed large numbers of women workers. The textile industry is the classic case in point. In Colombia, for example, as the industry was modernised between 1938 and 1979, more technically skilled labour was needed, and as a result the percentage of women employed fell from 74.3 per cent to 31.7 per cent. Even though the absolute number of women employed in the industry increased during these years, the number of men increased much faster, so the relative position of women suffered (UNESCO, 1986, p. 80; Joekes, 1987, p. 90). Why? It appears that training in the use of new machinery was not given to women at the same rate as to men. Similarly in Brazil: Saffioti has pointed out that the percentage of women in Brazil's textile industry dropped from 96.2 per cent in 1872 to 65.1 per cent in 1940 to 47.8 per cent in 1970 (1986, p. 110). Indeed, this trend, plus the expansion of other male-dominated industries in the twentieth century, meant that early industrialisation in Brazil led to an actual decline in the female share of total employment between 1872 and 1960 (Saffioti, 1978, p. 184–96). While women's share of employment in secondary industry continued to decline in Brazil between 1940 and 1970, their share in the tertiary sector rose (de Miranda, 1977). In the 1970s, however, the proportion of women employed in manufacturing in Brazil rose markedly, the rise continuing, if slowly, into the 1980s.

Here as elsewhere the rise in women's employment seems correlated with the introduction of the kind of automation which results in 'de-skilling', although, as Susan Joekes (1985) has pointed out, de-skilling does not completely explain employers' sex-selectivity. More particularly, women's employment is correlated with the rise of industries such as electronics in which the labour-intensive assembly-line techniques require high levels of managerial control and supervision to achieve high productivity. Ruth Milkman has suggested that it is actually this importance of discipline and control, arising from the technical nature of production, which is central in explaining why industries such as electrical manufacturing are predisposed toward hiring women (1983, p. 171–3).

The complex of technical and social factors which seems to favour high female employment may ultimately be seen as a particular phase of Third World industrialisation, one which could be undermined by new waves of capital-intensive production. For example, Aline Wong notes that Singapore is now moving to a 'second-stage' of export-oriented manufacturing, emphasising high technology, high value-added industries, in order to compete with neighbouring countries which offer even cheaper unskilled labour. She predicts that women will lose out relative to men in the needed upgrading of skills, and thus will bear the brunt of any layoffs (Wong, 1981, p. 443).

In addition to production factors, market factors also have a critical impact on women's employment, but here, too, the analysis cannot be divorced from considerations of gender discrimination deriving from women's position in the household. Prime among market factors in recent years has been world economic recession and declines or increased competitiveness in particular product markets. In the 1970s and early 1980s there was a slowdown in international trade, and in several countries production stagnated or actually decreased. In some, particularly in Latin America and the Caribbean, international indebtedness grew to enormous levels, forcing structural adjustments such as wage reductions, sometimes imposed by international agencies. The poor, both women and men, suffered most from price rises and drops in real income.

Recessions have had the effect not of reducing women's labour force participation, since poverty makes it increasingly difficult to withdraw from the labour force altogether, but of increasing women's unemployment, or of shifting them from full-time paid labour into casual labour, informal sector self-employment, sub-contracting or home-based outwork or piecework. In a little over half of the countries shown in Table 24.2, the percentage of women in total unemployment actually *increased* in the decade between 1976 and 1986. Some of this may be due to their increasing share in the labour force as a whole, but the rate of female jobs loss probably increased faster than that of men's. In 13 of the 21 countries for which ILO data were available, women's share of unemployment was greater than their share of employment in 1984–86; these included Barbados, Chile, Jamaica, India, Costa Rica, Argentina, Panama, Trinidad and Tobago, Uruguay, Philippines, Syria and Thailand. In times of economic downturn, employers often lay off women first. Joekes (1987, p. 96) points to the cases of Haiti, Venezuela, and Jamaica during the late 1970s, where, when their manufacturing sectors had difficulty in surviving international pressures, women had disproportionate job losses; this was also the case in industry in Sri Lanka and Taiwan. In so far as these reductions in the demand for female labour are due to gender discrimination on the part of employers, they would seem to reflect an ideology which assumes that male incomes are more important than female incomes, given women's subordinate position in the household.

In addition to general market factors, conditions in the labour market can affect female employment. For example, high female employment rates may come about simply as a result of shortage of males, due to wars, crises or out-migration. Shortages of male labour due to the war effort seem to be the main reason for the growth of women's labour force participation in Iran in recent years, a somewhat surprising development in view of the strength of the Islamic view that women's role in the family is primary (Moghadem, no date). Similarly, growth in female employment in Egypt is partly due to a shortage of skilled and unskilled male workers (Allam, 1986, p. 42). On the other hand, such an explanation does not appear to weigh heavily in the high employment of women in Mexican border industries, since there male unemployment is high (Fernandez-Kelly, 1983b). Still another pattern was found in Singapore in the 1970s, where high female employment was due to a high demand for labour generally coupled with full employment for males. Conversely, low female employment levels may be due in part to larger numbers of men available in the urban labour markets, a condition which is still found in many parts of sub-Saharan Africa.

Ultimately, however, a broader, more fundamental explanation for labour supply is needed. The availability of both men and women for employment and self-employment in the Third World is determined not simply by population levels but in addition by the degree of self-sufficiency of the rural household economy. Many peasant households and kin groups have traditionally subsisted largely outside the market economy, or only in localised market networks; today many pressures are forcing them to sell their labour or the products of their labour within the circuits of global capitalism. For female labour supply in particular, as argued more fully below, not only its level but also its duration and timing over the life cycle are profoundly influenced by the economic position of the household and by internal sex differences in the allocation of household productive and reproductive work.

This section has suggested that sectoral and technical changes in the organisation of production, as well as various market conditions, all fundamentally affect the levels of women's employment at any given time, and yet are still incomplete explanations. Factors such as women's skills, productivity and work attitudes, derived from sex role socialisation in the household, and also employer prejudices about women's proper productive and reproductive roles, have all been suggested as ways to extend and refine the analysis. These factors can affect not only the demand for women's waged labour but also the levels and patterns of its supply.

Age and marital patterns in women's employment

In export-oriented areas where female employment is high, the predominant characteristic of the female labour force is that it is young and single. Why? On the one hand, of course, the high turnover[1] and resultant low wages are to the advantage of, and imposed by, employers. Employers have an astonishing array of mechanisms which enforce or encourage women to drop out of employment after a number of years: (a) layoffs, either through temporary or permanent plant closings, are frequent, and when the plant reopens, hiring preference is given to young, unmarried women; (b) the lack of promotion or advancement incentives discourages women from long-term employment; (c) women's health often deteriorates from poor working conditions; (d) employment contracts are often temporary; (e) there are usually no maternity benefits; (f) employers can choose from an oversupply of female labour with the necessary skills. (Lim, 1978, p. 20; Safa, 1981).

Young women are not always cheaper because they are less skilled; sometimes they have higher levels of education and literacy than older women yet are still preferred for assembly jobs. One answer to this problem emerges when one considers the total social costs of reproduction of the labour force. That is, if the employer had to absorb any costs of child bearing and rearing such as maternity leaves, sick pay, frequent absences, higher health care bills, then the cheapness of women over men in the labour market would be undermined. A number of researchers have emphasised these potential costs of maternity and child care as an important factor in employer discrimination against women; Fapohunda (1986) for example, lays great stress on them as an explanation for employer resistance to hiring women in Nigeria (see also Safa, 1984). Considering women's family status, then, it is important to distinguish *daughters* and *childless single women*, on the one hand, from *single or married mothers*, on the other, since the former are the only ones who do not threaten to cost employers more than men.

But are the employer strategies and motivations the sum total of the picture? Are they a sufficient explanation for the prominence of young, single women in employment? Of the border industries in Mexico Maria Patricia Fernandez-Kelly writes:

> Other reasons given by women themselves for leaving their jobs are marriageable, unwed women's intention to have a child, and the desire of those who are married to give better attention to children and home. In both cases women, almost without exception, opt for this course of action to respond to the pressures of their male counterparts who urge them to leave their jobs in order to give full time to their homes, that is, to fulfill what is considered their 'normal' or 'proper' role. Such women rarely stay in a job long enough to acquire seniority benefits.
>
> (1983, p. 220)

The larger picture, Fernandez-Kelly believes, is that:

> At certain stages of its development, the domestic unit tends to produce and put into circulation young factory workers, that is, *daughters*, who after a few years of work in one or several factories tend to be reabsorbed by newly formed homes as wives, while a new wave of younger women take their place along the assembly lines.
>
> (1983, p. 220)

We should not reject the notion that women's preferences and/or the household constraints of reproduction figure into the analysis simply because this argument is also used by employers to justify their willingness to profit from family strategies. It is striking that the employment of young, single women fits in so well with the needs of families who have become dependent on a high fertility reproductive strategy and multiple wage earning by household members. This strategy comes about in the context of low prevailing wages for men and women, and low skill levels. In her comparison of women garment workers in the United States and Brazil, Helen Safa (1983) found that single women predominated in Brazil, whereas in New Jersey companies had mostly married women workers. In the advanced capitalist society, young women remained in school longer to try to qualify for white-collar jobs; they came from smaller families which could invest in the upward mobility of fewer offspring. The supply of single women for blue-collar work was therefore not sufficient, and married women constituted the cheaper labour pool. In Sao Paulo, in contrast, households were large and were forced into multiple wage earning strategies. The supply of less-educated single women was large enough to meet the demand, and their contributions were needed by their parents' households. Safa reports that today, however, with increasing fertility decline in Brazil, more and more women are continuing in the work force after marriage and pregnancy (personal communication).

We are suggesting that age and marital patterns in the female labour force dramatically reveal the direct (and not just indirect, through employer prejudices) impact of household factors such as fertility and multiple earning strategies. The fact that these patterns have shown marked cross-national variations indicates that differing household structures produce differing effects. One analysis based on data from the 1970s distinguished four main female age patterns: (a) the *central peak* or plateau (for example, Thailand), where there is no drop in labour force participation during child bearing years; (b) the *late peak* (for example, Ghana), in which women enter the labour force mainly after child bearing is completed, some perhaps forced to by widowhood; (c) the *early peak* (for example, Argentina) where the female labour force consists mainly of single or young married women and where participation occurs mainly before child bearing, dropping off steadily during and after it; and (d) the *double peak* (for example, Korea) in which labour force participation drops during child bearing but then rises again after it. (Figure 24.1; Durand, 1975, pp. 38–44).

The early peak pattern has been characteristic of Western Europe and Latin America and the central plateau pattern of the USSR, Eastern Europe and Southeast Asia. Salaff (1990) reports the early peak pattern for Hong Kong, Taiwan and Singapore. Most of Africa exhibits either the central plateau or the late peak pattern. The most characteristic feature of North Africa and the Middle East remains the lack of peaks and the overall low female labour force participation. The United States has exhibited an early peak pattern with the late peak a little higher. The years since 1945 have seen a conspicuous rise in overall female labour force participation and a rise in the numbers of women with small children entering the labour force, suggesting a move to the central plateau pattern.

The early age peak pattern suggests a situation in which marriage and fertility inhibit labour force participation throughout a woman's lifetime, whereas in the central peak there is obviously no such inhibition, and in the late and double peak patterns, the inhibition occurs for only a certain part of the life cycle. As one might expect from this line of reasoning, marital participation rates, like age rates, show much cross-national variation. It is usually but not always the case that single, widowed and divorced women

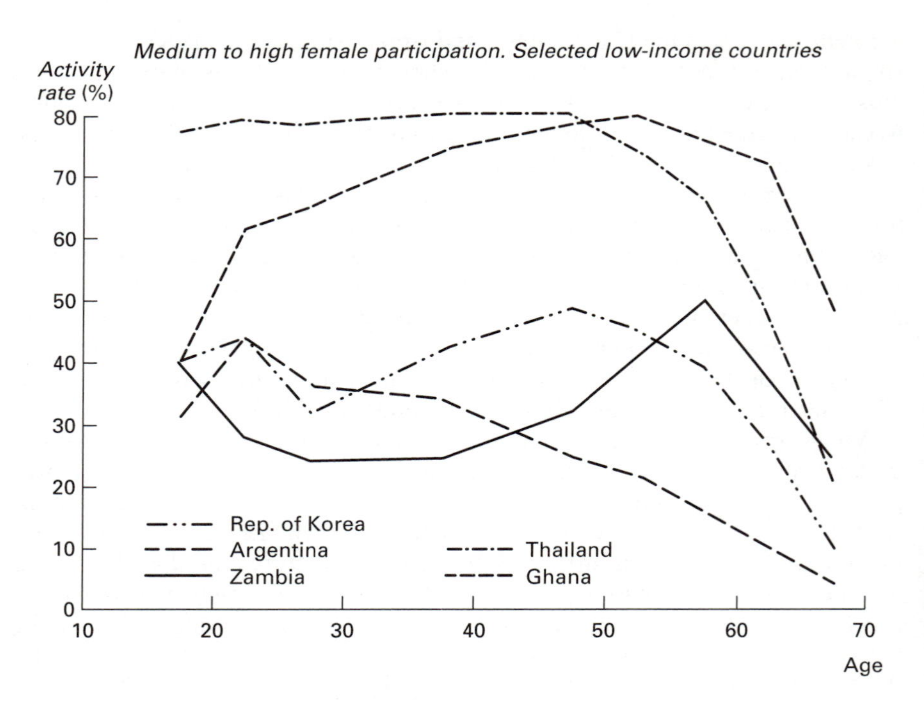

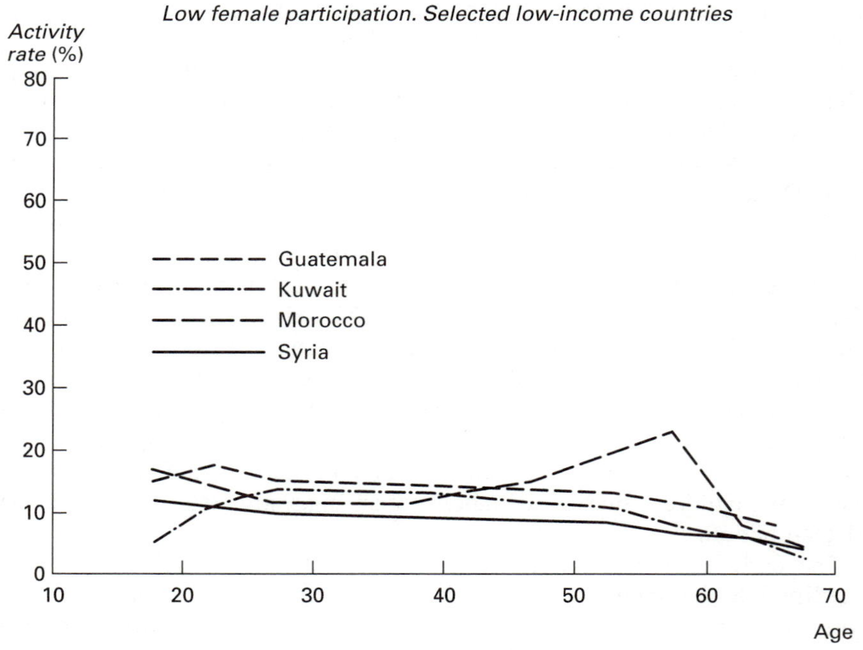

Figure 24.1 Age patterns in female labour force participation
Source: G. Standing, *Labour force participation and development*, 2nd edition (copyright, ILO Geneva, 1982)

have higher labour force participation rates than married women. In Brazil in 1970, single women had a rate three times that of married women (de Miranda, 1977, pp. 168, 170). In their 1985 sample of women industrial homeworkers in Mexico City, Lourdes Beneria and Martha Roldan (1987, p. 90) found that marriage or consensual union meant the interruption of paid work for 80 per cent of the women who had worked when single. On the other hand, Barbara Lewis (1977) found in the Ivory Coast that marriage affected female economic participation only slightly, and that age was a more important predictor; activity increases with age for all marital groups.

No adequate explanation has been offered in the literature as to why marriage and fertility should depress female labour force participation in some cases, but not in others, and only partially in still others. One study by Nadia Youssef (1974) did approach one aspect of this question. She compared female labour force participation rates in the Middle East and Latin America. She argued convincingly that level of economic development alone does not explain the extent of female non-agricultural participation. She linked the low rates of female employment in the Middle East to: (a) the near-universality of marriage; (b) the young age at marriage: and (c) the high fertility levels. By contrast, Latin America has high rates of marital instability especially among non-legalised relationships, postponement of marriage, and non-marriage. Youssef elaborated the argument by referring to the strong code of family honour in the Middle East, which is determined above all by the sexual conduct of the women, their premarital chastity and wifely fidelity. Strong social control thus deters much intermingling of the sexes, even in the job markets. In Latin America, on the other hand, values are similar but much less rigid, and there is a disjuncture between values and behaviour (Scott, 1988). Elsewhere, Youssef (1978) suggests that control over a woman by her agnatic group in the Middle East, together with the economic support the group provides her at all times, perpetuates women's position as economic dependents or home-based workers, and promotes a pro-natalist orientation. The crude birth rate and the child/woman ratio are in fact higher in Muslim societies than in non-Muslim Asia or Central or South America.

Youssef's study indicates that low levels of female employment can be linked with certain features of the marriage and fertility system. More detailed work is needed to explain the variability in age and marital participation patterns over the life cycle, and in general to elucidate the intricate connections between age, marriage, fertility, other household characteristics, and employment/self-employment in the various regions of the world. Some of these interrelations, especially that between fertility and employment, are explored more fully in Section III.

Household effects on employment

Having established the need to consider household impacts on the patterns of Third World women's employment, and reviewed theoretical approaches to the household from this perspective, we now turn to a more detailed examination of household/employment linkages.

I propose that the following aspects of the social relations of household production and reproduction can be expected to have an impact on women's employment patterns:

1 *Reproductive work*
(a) Amount of reproductive work: level of completed or expected fertility in the household
(b) Allocation of reproductive work: particularly the sex division of child-birth and child rearing, but also the distribution of such work among other household members
(c) Ability to transfer such work to others outside the household
2 *Productive work*
(a) Amount of productive work: varies according to how much the house-hold produces for itself and for the market economy, and how much it purchases from the market economy. Also includes shopping, entertaining, 'housework'
(b) Allocation of productive work: sex and age divisions of work; alloca-tions among all household members
(c) Ability to transfer such work to others outside the household
3 *Household structure*
Size, age/sex composition and persistence of the household unit. Important sub-types include single households and female-headed families.
4 *Income and resources*
(a) Total level of household income; total value of productive resources
(b) Distribution of income and resources among all household members. Includes immediate distribution as well as inheritance patterns
5 *Decision-making and power realations among members*
Particularly husband-wife relations but also parents' decisions about daughters, their schooling, training and employment.

From the vantage point of the labour market these are all supply-side factors, affecting which household members will offer for outside employment. I hypothesise that these factors will affect: (a) the aggregate supply of labour; (b) the age and sex composition of the labour force and patterns of participa-tion by age and sex; and (c) the supply price and thus wage levels for various categories of workers. Although these factors are expected to impact the labour market primarily through labour supply, other family variables may affect the market through the demand side, via employer discrimination. Ideological conceptions about normative family structure and division of labour is the aspect that is usually considered to affect employers' decisions (Scott, 1988). Other demand factors will also affect the final outcome, and there will be interactive or feedback effects. Contextual factors, such as the price and availability of non-familial child care, of domestic servants, and of other commoditised alternatives to domestic work, and the cost of schooling and health care for children, will also affect the results.

Conclusion

This overview has concentrated on fertility, domestic work, structure, income and decision-making as the household factors having the most important impacts on female employment. Inevitably, a range of other aspects of family relations has been neglected. In particular there are the bio-social-psychological dimensions, such as adult physical and mental health, stress, marital happiness, and children's self-esteem, social adjustment, and school performance. A particularly fertile new field of research lies in assessing the effects of maternal employment on these factors. Increasingly, work and family are coming to be recognised as closely interacting structures, perhaps the two social systems that most profoundly impact individuals' lives.

One aim of this review has been to suggest that research on the linkages between family/household and women's gainful employment is increasingly needed and relevant to the situation of Third World women. The changing international division of labour is resulting in the shift of more and more production jobs to certain areas of the Third World, and a large number of these jobs are going to women. What effect is this major global economic change having on the quality of women's lives? What changes are taking place in the relation between work and family systems, between women's productive and reproductive roles? To answer such questions concepts drawn from the study of nuclear families in industrial settings cannot be mechcanically applied. As this review has attempted to illustrate, world-wide variations in household and kinship structures pose challenges for the western-based economic theory of the household, yet the challenges are not insurmountable. In addition, we have attempted to show that household and kinship variations do affect female labour force participation in systematic and demonstrable ways. Existing research has provided a great deal of insight into these relations, yet in many ways it has barely scratched the surface. More research is needed, not only to understand women's lives, but to help them change them for the better.

Note

1 This refers to turnover between employment and non-employment over the life cycle, not to job turnover. Several writers, for example Joekes (1985) for Morocco and Humphrey (1985) for Brazil, point out that because of lack of promotion ladders and fewer job options, women actually have lower job turnover rates than men.

References

Allam, Etimad Mohammad (1986) 'Egypt: Islam, Women's Industrial Work Patterns and Male Labour Shortages', in United Nations Economic Commission for Africa, African Training and Research Centre for Women, *Women and the Industrial Development Decade in Africa*, pp. 23–58.

Anker, Richard and Catherine Hein (1986), 'Sex Inequalities in Third World Employment: Statistical Evidence', in R. Anker and C. Hein (eds), *Sex Inequalities in Urban Employment in the Third World* (London: Macmillan) pp. 63–115.

Beneria, Lourdes and Martha Roldan (1987) *The Crossroads of Class and Gender* (University of Chicago Press).

Chapkis, Wendy and Cynthia Enloe (1983) *Of Common Cloth* (Washington DC: Transnational Institute).

Chinchilla, Norma (1977) 'Industrialisation, Monopoly Capitalism and Women's Work in Guatemala', in Wellesley Editorial Committee, *Women and National Development* (University of Chicago Press) pp. 38–56.

di Domenico, Catherine (1983) 'Male and Female Factory Workers in Ibadan', in C. Oppong (ed.) *Female and Male in West Africa* (London: Allen and Unwin) pp. 256–65.

Durand, J.D. (1975) *The Labour Force in Economic Development: An International Comparison of Census Statistics* (Princeton University Press).

Elson, Diane and Pearson, Ruth (1981) '"Nimble Fingers make Cheap Workers": An Analysis of Women's Employment in Third World Export Manufacturing', *Feminist Review*, 7, pp. 87–107. Also appeared as 'The Subordination of Women and the Internationalisation of Factory Production', in K. Young, C. Wolkowitz and R. McCullagh, *Of Marriage and the Market* (London: CSE Books, 1981) pp. 144–66.

Fapohunda, Eleanor (1986) 'Nigeria: Women and Industrialization in Anglophone West Africa', in UN Economic Commission for Africa, African Training and Research Centre for Women, *Women and the Industrial Development Decade in Africa*, pp. 87–147.

Fernandez-Kelly, Maria Patricia (1983) 'Mexican Border Industrialization, Female Labor Force Participation and Migration', in J. Nash and M. Fernandez-Kelly (eds) *Women, Men and the International Division of Labor* (Albany, New York: SUNY Press) pp. 205–23.

Fuentes, Annette and Barbara Ehrenreich (1983) *Women in the Global Factory* (Boston: South End Press).

House, William J. (1986), 'The Status and Pay of Women in the Cyprus Labour Market', in R. Anker and C. Hein (eds), *Sex Inequalities in Urban Employment in the Third World* (London: Macmillan) pp. 117–69.

Humphrey, John (1985) 'Gender, pay, and skill: manual workers in Brazilian industry', in H. Afshar (ed.) *Women, Work and Ideology in the Third World* (London: Tavistock) pp. 214–31.

Joekes, Susan (1985) 'Working for lipstick? Male and female labour in the clothing industry in Morocco', in H. Afshar (ed.) *Women, Work and Ideology in the Third World* (London: Tavistock) pp. 183–213.

Joekes, Susan (1987) *Women in the World Economy: An INSTRAW Study* (New York: Oxford University Press).

Lewis, Barbara (1977) 'Economic Activity and Marriage among Ivoirian urban women', in A. Schlegel (ed.), *Sexual Stratification: A Cross-Cultural View* (New York: Columbia University Press) pp. 161–91.

Lim, Linda (1978) 'Women Workers in Multinational Corporations: The Case of the Electronics Industry in Malaysia and Singapore', Michigan Occasional Paper no. 9, Fall, University of Michigan.

Milkman, Ruth (1983) 'Female Factory Labor and Industrial Structure: Control and Conflict over "Women's Place" in Auto and Electrical Manufacturing', *Politics and Society*, vol. 12, no. 2.

de Miranda, Glaura Vasques (1977) 'Women's Labor Force Participation in a Developing Society: The Case of Brazil', in Wellesley Editorial Committee, *Women and*

National Development: The Complexities of Change (University of Chicago Press) pp. 261–274.
Moghadem, Val, (no date) 'Women, Work and Ideology in the Islamic Republic', unpublished paper.
Safa, Helen I. (1981) 'Runaway Shops and Female Employment: The Search for Cheap Labor', *Signs*, vol. 7, no. 2 (Winter), pp. 418–33.
Safa, Helen I. (1983) 'Women, Production, and Reproduction in Industrial Capitalism: A Comparison of Brazilian and US Factory Workers', in J. Nash and M. Fernandez-Kelly (eds) *Women, Men and the International Division of Labor* (Albany, New York: SUNY Press) pp. 95–116.
Safa, Helen I. (1984) 'Female Employment and the Social Reproduction of the Puerto Rican Working Class', *International Migration Review*, vol. 18, no. 4 (Winter), pp. 1168–87.
Saffioti, Heleieth I.B. (1978) *Women in Class Society* (New York: Monthly Review Press).
Saffioti, Heleieth I.B. (1986) 'Technological Change in Brazil: Its Effect on Men and Women in Two Firms', in J. Nash and H. Safa (eds) *Women and Change in Latin America* (South Hadley, Massachusetts, Bergin and Garvey) pp. 109–35.
Salaff, J. (1990) 'Women, the Family, and the State: Hong Kong, Taiwan, Singapore – Newly Industrialised Countries in Asia', in S. Stichter and J. Parpart (eds) *Women, Employment and the Family in the International Division of Labour* (London: Macmillan) pp. 98–136.
Scott, J. (1988) 'Deconstructing Equality Versus Difference: or the Uses of Post-structuralist Theory for Feminism', *Feminist Studies*, vol. 14, pp. 33–50.
Standing, Guy (1982) *Labour Force Participation and Development*, 2nd edition (Geneva: International Labour Office).
UNESCO (1986) *World Survey On The Role of Women in Development* (New York: United Nations).
Wong, Aline K. (1981) 'Planned Development, Social Stratification, and the Sexual Division of Labor in Singapore', *Signs*, vol. 7, no. 2, pp. 434–52.
Youssef, Nadia H. (1974) *Women and Work in Developing Societies* (Westport, Connecticut: Greenwood Press).
Youssef, Nadia H. (1978) 'The Status and Fertility Patterns of Muslim Women', in L. Beck and N. Keddie (eds), *Women in the Muslim World* (Cambridge, Massachusetts: Harvard University Press) pp. 70–85.

25 Harriet Bradley

'Gender Segregation and the Sex-typing of Jobs'

Excerpts from: *Men's Work, Women's Work*, pp. 7–26.
Cambridge: Polity (1989)

No more delightful wanderings . . . Henceforth it must be work, woman's work, dreary and monotonous sometimes, yet pleasant withal, as it rewarded me with the proud consciousness that I was not only able to eat my daily bread but earn it.

Wills, Lays of Lowly Life

In these words Ruth Wills, Leicester working woman and poet, looked back in 1861 at her transition from childhood to womanly status, when at the age of ten or eleven she gained a job in the warehouse at Corah's hosiery factory where she was to work for the rest of her life. Her comment reveals to us some of the ambiguous and contradictory attitudes common to most people who have to work under the prevailing arrangements and conditions of industrial capitalism. But it also tells us interesting things about social perceptions of 'women's work' in the middle of the nineteenth century. During the succeeding decades, ideas about 'men's work' and 'women's work' were to stabilise into the forms familiar to us today, after the period of tumultuous change in the nature of working arrangements and relationships and in working people's daily lives, which marked the long, slow, painful transition from one type of society to another that we now know by the shorthand label of the 'Industrial Revolution'.

Women's work in Victorian England was indeed often 'dreary and monotonous', both inside and outside the home. The focus of study here is employment outside the home, wage labour. Wage labour for Victorian women meant filling the less prestigious, more routinised, often less skilled tasks in both the new manufacturing and service industries and the traditional areas of agriculture and domestic service. While men's traditional trades and skills had often been challenged or destroyed by the new industrial system, they had succeeded in capturing for themselves the more responsible jobs and tasks which either were or could credibly be described as skilled. Women's work was low paid, often pitched at a level below subsistence needs, and few women, of any class, had expectations of anything better. Men were paid more and, while not yet attained, the 'breadwinner's' wage, sufficient to support a whole family, was their target, and claimed as their 'right' as head of a household. There was little sense that women's work was something that could be a source of pleasure and satisfaction in itself; it was undertaken solely out of necessity 'to earn one's daily bread'. The idea was gaining ground that all women, given the choice, would prefer to stay at home and devote themselves to things domestic. Few women had any real choice over the type of work they undertook, and the range of occupations open to them was limited. For men, however, a broader range of possibilities appeared, and for many middle-class and even some working-class men there was some prospect of a 'free' choice of trade and career. In any case men expected, if they did not always achieve it, some intrinsic satisfaction from their work, if only in the sense of the access it gave them to adult masculine status and the breadwinner role. Men's work offered an important source of social and personal identity, whereas for women the growing tendency was for their identity to be focused on their domestic roles as homemakers and mothers within the inturned, privatised family which was becoming the Victorian ideal. Finally, women's work was different in content from men's work: the characteristic features of women's and men's work as we know them today were becoming the norm in 1861.

It would be hard to find any single activity which has not been, at some time or place, 'women's work'. Yet over time and space we can discern some general trends, some gender allocations that are more common than others. Murdock and Provost's wide survey of the sexual division of labour as recorded in 185 societies shows, for example, that hunting large animals, fishing, smelting ores, metalwork, mining and quarrying and lumberwork are almost everywhere male tasks. Female tasks tend to be less sharply distinguished. However, Murdock and Provost found that dairy production, cooking, carrying water and gathering vegetables were very commonly performed by women (Murdock and Provost, 1973). Virginia Novarra (1980) has argued, more generally, that six key tasks are performed mainly by women in the majority of societies: provision of food, care of the home, child care, nursing the sick, teaching and manufacture of clothing. These tasks are frequently performed by women in the home, as subsistence labour or unpaid housework. They can be linked to the association of women with 'domestic' activity and, in the later stages of social development, the ascription of women to a home-centred existence. But when women move into the wage labour sphere they often perform commercialised forms of these same activities which centre on the care and servicing of other people, 'production of people' to borrow a phrase from Murgatroyd (1985). Around the world today women in vast numbers work as teachers, nurses, cleaners and garment-makers.

As well as this personal care motif, typical 'woman's work' displays certain common features, as Game and Pringle (1983) have argued, at least in the ideal visions we have of it. It is usually indoor work, considered to be 'lighter' than men's work; it is clean, safe, physically undemanding, often repetitive and considered boring, requires dexterity rather than 'skill', often has domestic associations; it tends to lack mobility, being tied to a particular work station; it may well have associations and requirements of beauty and glamour. By contrast, if we visualise typical men's work, we tend to evoke images of the outdoors, of strength and physicality; 'men's work' may be heavy, dirty, dangerous, it is often highly mobile (men have a curious monopoly of jobs in the transport sectors), it requires 'skill' and training. It is frequently highly technical, based on mechanical knowledge or scientific expertise; at the highest level, it requires characteristics of creativity, innovation, intelligence, responsibility, authority and power. Such qualities are rarely ones we would associate with women's work!

These typical ascriptions were already being elaborated in 1861, and jobs were allocated to men and women in terms of matching up to some or other of these criteria. During the remaining decades of the nineteenth century, the familiar modern sexual division of labour in factories and offices, schools and hospitals evolved, and the related notions of fit work for the sexes became a commonplace of our culture. By sex-typing, I mean the process by which jobs are 'gendered', ascribed to one sex or the other; while segregation refers to the way in which women and men are located in different types of jobs; in the words of Catherine Hakim: 'occupational segregation by gender exists when men and women do different kinds of work, so that one can speak of two separate labour forces, one male and one female, which are not in competition with each other for the same jobs' (Hakim, 1979, p. 1). This often involves the physical separation of the sexes in different spaces or locations. Sex-typing and segregation are thus analytically separable, although in practice they are almost always found in combination; in sociological terms, sex-typing can be seen as the ideological face of the structural process of segregation.

Since Ruth Wills wrote her autobiography in 1861 massive changes have occurred, which have affected women's social position and the relations between the sexes. Women have gained legal and political rights. The evolution of the welfare state has freed them from total dependence on fathers and husbands and the development of effective contraception has freed them from the burden of perpetual pregnancy and childbearing. Two important political waves of feminism have generated energetic campaigns for equal rights and opportunities and pushed women's view of their own disadvantages into public consciousness; partly as a result, higher education and many professional areas have opened up for women; two world wars have provided them with an opportunity to try their hand at almost every type of work deemed socially necessary; most recently, legislation has been passed outlawing discrimination in the field of employment. Yet, in spite of all this, the prevalence of sex-typing and segregation in paid work has diminished very little since 1861. In the words of one researcher:

> the most striking difference between men and women on the labor market is the work they do. Men and women are, to a great extent, found in different spheres of economic activity and occupations, and within most occupations they are employed at different levels and with different work tasks.
>
> (Jonung, 1984, p. 44)

Sex-typing and gender segregation

In her important and influential study, published in 1979, Catherine Hakim distinguished between two different aspects of segregation: horizontal segregation refers to the concentration of women and men in different types of work; vertical segregation to the concentration of men in higher grades, women in lower grades, both within and between occupations and industries. Taken together, these account for women's lower levels of pay and face us with a substantial picture of inequality.

Figures gathered by the Equal Opportunities Commission (EOC) for its Annual Reports reveal the strength of horizontal segregation. Many occupations and industries are heavily dominated by men. In 1981 men were 95 per cent of the work force in mining and quarrying, 91 per cent in shipbuilding and in construction, 88 per cent in vehicle construction and in metal manufacture, and 89 per cent in farming and fishing. Such 'heavy' industries maintain their traditional image as 'men's work'. Women, by contrast, are less markedly dominant in any given industry or occupation, as these are classified by the collectors of official statistics. However, in 1981 they constituted 76 per cent of clerical workers and of workers in catering, cleaning and personal services and 74 per cent of clothing and footwear workers (Huws, 1982). Hakim argues, on the basis of such figures, that men are more likely to find themselves working in a job where their own sex is numerically dominant.

However, within each industry and occupation there is a tendency to the sex-typing of particular jobs, which is not revealed by statistics based on such broad categories. Jonung argues, rightly, that 'segregation is higher the more detailed the occupational level studied' (Jonung, 1984, p. 53). Accordingly, over 90 per cent of secretaries, typists and receptionists, of cashiers, of nurses, of canteen assistants and of maids are women, for example, and over 75 per cent of chars, office cleaners, sewing machinists, laundry workers and kitchen maids (Lockwood and Knowles, 1984; Dex, 1985).

In another sense, women's employment can be seen as *more* concentrated than men's, in that the range of occupations and industries in which they are found is more restricted. Thus 85 per cent of female manual workers are found in three of the occupational categories used by government statisticians: catering, cleaning and other personal services; painting, repetitive assembly, inspecting, packaging and related activities; making and repairing (excluding electrical and metal goods). Among non-manual women concentration is even starker, as 91 per cent are to be found in three categories: selling; clerical work; professional work in education, welfare and health (Cockburn, 1985). Ursula Huws' useful profile of women's activity in one geographical area (West Yorkshire) reflects such a pattern. For every 100 women of all ages 32 are in employment; of these nine are in clerical work, eight in factories, seven in cleaning and catering, four in the teaching, nursing and social work professions, three in sales and only one in other occupations (Huws, 1982, p. 15).

The pattern of segregation has been largely undisturbed by the changes and upheavals in the economy which have resulted from the introduction of computers and information technology, from the restructuring of jobs and the shakeout of workers brought about by the recent recession, and the other changes which some have seen as a 'second industrial revolution'.

Indeed, Hakim's historical investigation revealed an extraordinary stability of patterns of segregation over the whole of this century. Examining data for the period between 1811 and 1871, she concluded that while there had been a decrease in the number of occupations which completely excluded one sex or the other, the likelihood of a man working in an occupation where his sex predominated has actually increased over the period. In 1971 50 per cent of men worked in occupations where they outnumbered women by nine to one, and two-thirds in occupations where the ratio was four to one. Where horizontal segregation was weakened over the period it was related to male inroads into 'female' occupations. There was little compensatory expansion of women into male areas. Moreover, she argues that vertical segregation also increased over the period: the exclusion of women from skilled manual jobs has cancelled out the slight increase of women in top managerial and professional posts. In a second article examining 1981 data, she argued that there had been no significant change in the intervening decade. Mallier and Rosser's later study (1987) slightly challenges this verdict, suggesting that in contrast to the earlier period women have recently been making some inroads into male-dominated occupations. This, however, they consider insufficient to change established notions of what is 'women's work', and since men are not entering female-dominated occupations in any numbers they conclude that the general pattern of segregation has been little disturbed. We can see that, by and large, the structure of segregation has persisted obstinately throughout the twentieth century.

Although this study is concerned with gender differences and not with ethnic disadvantage, it should be acknowledged that many have discerned a pattern of segregation within segregation when looking at the position of black women in Britain. Parmar (1982) argues that, while, as indicated above, all women face a restricted set of labour market opportunities, racism and discrimination ensure that the position of black women is even more restricted. West Indian women, for example, tend to be concentrated in unskilled jobs of very low status which white women are unwilling to take on, such as hotel and catering work, hospital domestic work and the lowest grades of nursing. Moreover, while many women's jobs are based on ideas of feminine attractiveness and charm in handling clients (secretaries, receptionists, bank clerks, shop assistants, hostesses), racial discrimination and stereotyping act to keep black women out of such 'visible' jobs and confine them to the backstage areas, such as kitchens and laundries. This social invisibility of black women means that their contribution to the economy and to the welfare of the community is obscured and unappreciated. West Indian women are also grossly under-represented in top jobs, only 2 per cent of them being in managerial and professional work as opposed to 17 per cent of white women and 20 per cent of Asian women. Asian women, however, are more likely than West Indian or white women not to enter the labour market at all (Beechey, 1986). Traditional family relations may compel many married women to remain at home. Language difficulties also make it hard for some of the older women to enter non-manual work, and in fact Asian women are quite heavily concentrated in unskilled factory work, especially in Yorkshire and the Midlands. They are also often found as homeworkers, which partly reflects the hostility of some Asian men to the idea of their wives leaving the home and exposing themselves to an alien culture with its threat of sexual temptation, which challenges traditional notions of *izzat* (family pride and male honour) (Wilson, 1978). But, as Allen and Wolkowitz (1987) point out, women tend to do homework only when they fail to find opportunities outside the home, and the discrimination which black women face in the labour market may also force them into this low-paid form of employment.

Homework is seen by many as epitomising the devaluation of women's work. Although it is hard to make an accurate assessment of the numbers currently employed as homeworkers, Allen and Wolkowitz suggest that this is not only a long-established form of female employment but also, in the context of the restructuring of the economy, a potentially growing trend. Often done for appalling pay (as little as 50 pence an hour), in unpleasant conditions, unprotected by union organisation and isolated from the companionship which for many women is a major attraction of work outside the home, it symbolises the subjugation of women to domestic obligations, which, as we shall see, has been such an important factor in promoting sex-typing and segregation.

Explanations of and 'orientations' to segregation and sex-typing

How, then, can we start to explain the origins of segregation ans sex-typing of work, and their remarkable persistence? Discussion has focused on a number of historical and sociological factors: the dynamics of capitalist industrial development and in particular the confrontations and negotiations between employers and male workers which have characterised it; the sexual division of labour within the household; women's childbearing role; authority relations in the family; prevailing ideologies and social attitudes of both men and women, and in particular the role of the 'domestic ideology' elaborated by the Victorian middle classes, as described, for example, in the work of Catherine Hall and Leonore Davidoff (Hall, 1979, 1980; Davidoff and Hall, 1987); the attitudes of employers to women workers; processes of socialisation in the home, at school and in later life; and the role of the state in fostering certain social definitions of male and female activities.

Attempts to explain sex-typing and segregation within various broader theoretical perspectives will also be considered. But I should say at this point that, in my opinion, there is currently no completely satisfactory sociological theory of gender divisions. It may be tempting to attribute this to the relative youth of feminist social analysis, the sexual division of labour having been a neglected topic within sociology (and especially within industrial sociology) until the late 1960s. Certainly this is implied in the work of many feminist sociologists, such as Michele Barrett and Sylvia Walby, who see themselves as laying the groundwork for a complete reconceptualisation of gender relations. My own belief is that sociologists should no longer aspire to *any* comprehensive and totalising theory of society; no single model which explains each and every aspect of social life can be constructed. The challenge posed by postmodern theory, like that of Lyotard (1984), to structural forms of explanation should, at the least, make us wary of over-ambitious claims in our attempts to understand and depict social reality. This is indeed one of the significant challenges that feminism has made to the more rigid forms of Marxist structuralism with which it first tended to engage. While exposing the limitations of an approach which wishes to subsume all social divisions into the category of class and the analysis of surplus and value, feminists should be alert to the danger of falling into the same kind of error themselves.

The belief that sociological analysis is necessarily incomplete and partial, that no single theory can explain everything, is not, of course, to reject theoretical work or the task of perpetual redefinitions of existing theories and concepts. One of the key features of sociological work is its 'essential contestability' (we shall *never* agree exactly what social reality *is*, let alone how to go about studying and analysing it). The quest for understanding proceeds through debates and it is through this very contestation that insights are gained. Feminism, in its debate with existing forms of sociological analysis, has opened up whole new areas of investigation: new topics and issues, new problems and agendas, new concepts and theories emerge from the process of argument.

Before I proceed to more specific issues I wish to look briefly at the attempts to develop a general analysis of gender differentiation in terms of existing sociological perspectives or, as I prefer to term them, orientations. We can discern three such orientations, or rough groupings or clusterings of ideas, within the existing debates on the sexual division of labour.

The first orientation, which we can call the 'production orientation', examines gender differences within existing frameworks for analysing other social divisions; it subsumes, as it were, the *sexual* division of labour into the *social* division of labour. This approach is particularly associated with Marxism, but has also been adopted by others, including Weberians or pluralists, who espouse a basically materialist approach. The focus of analysis here is firmly on the sphere of *work* or of *production* and it proceeds by *adding gender in*, adapting existing economic concepts to explain women's specific position. This has undoubtedly generated some of the most influential thinking about sex divisions at work. For example, Marxists have made use of the idea of the reserve army of labour or related women's position to the process of 'deskilling' (see Braverman, 1974; Beechey, 1977). An alternative conceptual frame, closer to the Weberian tradition, has drawn on the notion of dual or segmented labour markets (for example, Barron and Norris, 1976; Kreckel, 1980; Rubery, 1980). Economists have considered working women within the prevailing frameworks of the discipline of economics, those of calculative choice and consumer preference; for example, segregation is explained in terms of discrimination (that is, the preferences of employers for particular types of labour) (Becker, 1957). These theories fail to take account of the specific nature of gender divisions, which are seen as no different in kind from class, ethnic or other kinds of social division. Moreover, the almost exclusive focus on the work sphere leads to a neglect of family relationships, which are seen merely as a precondition for the exploitation of women by employers and/or male workers.

The second or 'reproduction orientation' reverses this order of priority. Here the causes of gender divisions are sought within the *domestic* sphere. The key concepts are culture, ideology and reproduction. Many different theoretical strands feed in to this orientation. Conventional mainstream sociology, especially American functionalism, with its stress on socialisation and sex roles, has been an influence, even where writers like Ann Oakley have rejected the slippage into biological determination which is often discernible in the work of sociologists such as Parsons and Murdock. Yet such a slippage may also be evident in the writings of those feminists often labelled as 'radical' who consider reproductive relations or constructions of sexuality to be the root of the sexual divisions and male power (for example Firestone, 1979; de Beauvoir, 1953). To avoid this problem, many feminist anthropologists have concentrated on the symbolic opposition of nature and culture seen as virtually universal in all societies, and the way that the linking of women with nature may serve to downgrade their economic contributions (Rosaldo, 1974; Ortner, 1974; Ardener, 1975). Marxism, too, can feed into this orientation: some Marxist feminists have argued for the existence of a mode of reproduction separate from any mode of production but interacting with it (Mitchell, 1975; Mackintosh, 1977). Alternatively others (Dalla Costa and James, 1972; Delphy, 1977) have seen domestic labour as itself constituting a distinct mode of production which generates sexual inequalities. The major problem with these approaches is exactly the reverse of those in the first orientation: inequalities in the economic sphere are assumed to be a simple reflection of inequalities produced in the family or the domestic sphere, or to arise from social attitudes learned through socialisation or generated by patriarchal ideology or culture. The way in which the economy itself breeds sexual divisions is ignored. As Mies (1986) has argued, these approaches are more useful in explaining why segregation at work persists than why it developed in the first place.

The third orientation (and the reader may perhaps have already guessed that it is one to which the writer is most sympathetic!) is based on the argument that an understanding of sexual divisions must embrace both work and home, production and reproduction. We might call it the 'joint orientation'. It involves an explicit claim that concepts drawn from Marxism, structuralism, functionalist sociology or cultural studies are insufficient to that understanding, and that new concepts, based on the specificity of gender divisions, must be elaborated. The key concepts here have been patriarchy and male dominance. These are seen to transcend the familiar sociological distinctions between production and reproduction, private and public, even perhaps, more controversially, structure and culture, material and ideological. Thus, for example, Walby has defined patriarchy as a set of interlinking institutions that cut across all sectors of society:

> . . . a system of interrelated social structures through which men exploit women . . . The key sets of patriarchal relations are to be found in domestic work, paid work, the state and male violence and sexuality, while other practices in civil society have a limited significance.
>
> (Walby, 1986, pp. 50–1)

Although there are many problems as yet not overcome within the 'joint orientation', its comprehensiveness seems to me to be its great strength. Neither structural nor cultural, material or ideological factors *alone* can explain the origins and persistence of segregation and sex-typing. That is why I argue that we need the insights from all these orientations to further our understanding of the processes of sexual differentiation.

References

Allen, S. and Wolkowitz, C. (1987), *Homeworking: Myths and Realities*, London: Macmillan.

Ardener, S. (1975), *Perceiving Women*, London: Malaby Press.

Barrett, M. (1980), *Women's Oppression Today*, London: Verso.

Barron, R. D. and Norris, G. M. (1976), 'Sexual Divisions and the Dual Labour Market' in Allen, S. and Barker, D. L. (eds), *Dependence and Exploitation in Work and Marriage*, London: Longman.

Beauvoir, S. de (1953), *The Second Sex*, London: Jonathan Cape.

Becker, G. (1957), *The Economics of Discrimination*, Chicago: University Press.

Beechey, V. (1977), 'Some Notes on Female Wage Labour in Capitalist Production', *Capital and Class* **3**, pp. 45–66.

Braverman, H. (1974), *Labor and Monopoly Capital*, New York: Monthly Review Press.

Cockburn, C. (1985), *Machinery of Dominance*, London: Pluto.

Dalla Costa, M. and James, S. (1972), *The Power of Women and the Subversion of the Community*, Bristol: Falling Wall Press.

Davidoff, L. and Hall, C. (1987), *Family Fortunes*, London: Hutchinson.

Delphy, C. (1977), *The Main Enemy*, London: Women's Research and Resources Centre.

Dex, S. (1985), *The Sexual Division of Labour*, Brighton: Harvester.

Firestone, S. (1979), *The Dialectic of Sex*, London: Women's Press.

Game, A. and Pringle, R. (1983), *Gender at Work*, Sydney: George Allen and Unwin.

Hakim, C. (1979), *Occupational Segregation by Sex*, Department of Employment Research Paper No. 9.

Hall, C. (1979), 'The early Formation of Victorian Domestic Ideology' in Burman, S. (ed.), *Fit Work for Women*, London: Croom Helm.

Hall, C. (1980), 'The History of the Housewife' in Malos, E. (ed.), *The Politics of Housework*, London: Allison and Busby.

Huws, U. (1982), *Your Job in the Eighties*, London: Pluto.

Jonung, C. (1984). 'Patterns of Occupational Segregation by Sex in the Labour Market' in Schmid, G. and Weitzel, R. (eds), *Sex Discrimination and Equal Opportunity* Aldershot: Gower.

Kreckel, R. (1980), 'Unequal Opportunity Structure and Labour Market Segmentation', *Sociology* **4**, pp. 525–50.

Lockwood, Baroness and Knowles, W. (1984), 'Women at Work in Great Britain' in Davidson, M. and Cooper, C. (eds), *Working Women: An International Survey*, Chichester: John Wiley.

Lyotard, J. P. (1984), *The Postmodern Condition*, Manchester: University Press.

Mackintosh, M. (1977), 'Reproduction and Property: A Critique of Meillassoux,' *Capital and Class* **2**, pp. 119–27.

Mallier, A. and Rosser, M. (1987), *Women and the Economy*, London: Macmillan.

Mies, M. (1986), *Patriarchy and Accumulation on a World Scale*, London: Zed.

Mitchell, J. (1975), *Psychoanalysis and Feminism*, Harmondsworth: Penguin.

Murdock, G. and Provost, C. (1973), 'Factors in the Division of Labour by Sex: A Cross-Cultural Analysis', *Ethnology* **12**: **2**, pp. 203–35.

Murgatroyd, L. (1985), 'The Production of People and Domestic Labour Revisited' in Close, P. and Collins, R. (eds), *Family and Economy in Modern Society*, London: Macmillan.

Novarra, V. (1980), *Women's Work, Men's Work*, London: Marion Boyars.

Ortner, S. (1974), 'Is Female to Male as Nature is to Culture?' in Rosaldo, M. and Lamphere, L. (eds), *Women, Culture and Society*, Stanford: University Press.
Parmar, P. (1982), 'Gender, Race and Class: Asian Women in Resistance' in Centre for Contemporary Cultural Studies, *The Empire Strikes Back*, London: Hutchinson.
Rosaldo, M. (1974), 'Women, Culture and Society: A Theoretical Overview' in Rosaldo, M. and Lamphere, L. (eds), *Women, Culture and Society*, Stanford: University Press.
Rubery, J. (1980), 'Structured Labour Markets, Worker Organisation and Low Pay', in Amsden, A. (ed.), *The Economics of Women and Work*, Harmondsworth: Penguin.
Walby, S. (1986), *Patriarchy at Work*, Cambridge: Polity.
Wilson, A. (1978), *Finding a Voice*, London: Virago.

26 Doreen Massey

'Industrial Restructuring as Class Restructuring: Production Decentralization and Local Uniqueness'

Excerpts from: *Regional Studies* **17**, 73–89 (1983)

Introduction

A good deal is now known about processes of industrial reorganization in the United Kingdom in the 1960s and early 1970s: about the wholesale desertion of areas, the changes in skill-requirements consequent upon reorganization of the labour process, the use of locational change within this overall reconstruction, and the ability of capital – in times of high unemployment nationally – to play off different groups of workers against each other.[1]

The aim of this paper is to examine some of the implications of this reorganization, and in particular to focus on the impact of industrial change on social change, and especially on class structure. If industrial reorganization is necessary in order to drag the British economy out of the long-term decline in which it is locked, so also is a change, in one direction or another, in the balance of social forces. Class reconstruction (which in the UK means breaking the class stalemate) is central to 'getting out of the crisis', on whoever's terms this eventually happens. And 'industrial restructuring' is a process of class restructuring; it is one of the mechanisms by which the social structure is re-shaped, social relations changed and the basis for political action broken down or reconstructed.

Individual firms are not motivated to this end. They set out to lower costs, perhaps, or to find a workforce with less organizational strength. Nonetheless, their aggregate actions can lay the basis for considerable social change. Nor are workers and local communities the passive recipients of industrial change. The kinds of organization and resistance which have been built on and around the existing forms of industrial organization and the existing uses of space by industry are often part of the stimulus to industrial restructuring in the first place. And that strength and resistance is itself not some 'automatic response' to the existing forms of organization of industry and its use of space; it results from active organization. What industrial change does is to alter the preconditions for such responses (see, for instance, Foster, 1974). It may both undermine the effectiveness of existing forms of organization and present new opportunities. It is both an offensive and the presentation of new challenges.

This paper focuses on just one aspect of the industrial change of the last two decades – the well known, geographical decentralization of jobs 'for women'. It is just this aspect of restructuring (a spatial change inextricably intertwined with the changing requirements of production) which will be considered here. It is a shift which has been going on since the Second World War, but it has varied in importance between different periods, and in particular was quite strong between the mid-1960s and the mid-1970s. It has involved both intra-regional decentralization away from conurbations, and inter-regional shifts to peripheral areas. And it has involved a range of different industries – clothing and footwear, light engineering, electronics and 'services' are probably the best documented. There are many contrasts between the jobs decentralized in these different industries – between what is normally classified as 'mental' and 'manual' work, for instance, and between the industry-level spatial structures in which the plants are embedded.[2] But what is more important for this paper is that the jobs have a lot in common – they are frequently in branch plants and, as far as jobs in production are concerned, they are characteristically low-paid, they are most often classified as semi-skilled or unskilled and, above all, they are jobs which are done by women.[3]

What this paper does is look at the impact of this kind of industrial decentralization in two very different areas. The first and longer study examines some aspects of the impact on old industrial areas which have for long been heavily dominated by coal-mining. The second study looks at Cornwall, an area where previously the economic structure has been based around a mixture of agriculture and tourism. These area studies are sketches, attempts to characterize fairly simply what are in local detail far more complex economies and societies. Much work is already under way and available which examines these areas – and in particular the coalfields – in greater depth (see especially Hudson's work on the North East, e.g. Hudson, 1979, and the papers by the 'Wales Regionalism Group' at UWIST on South Wales). My present purpose is rather different. The aim here is to emphasize and analyse how what at one level may be perceived as 'national' changes (changes such as the growth in jobs done by women, shifts in class structure, geographical decentralization) may vary greatly in operation and in impact between different parts of the country.

There are a number of different threads to this argument. First, there is the question of how one analyses change at local level. I have argued elsewhere (Massey, 1979) that the structures of local economies can be seen as a product of the combination of 'layers', of the successive imposition of new 'rounds of investment', representing in turn the successive roles the areas have played within the wider national and international divisions of labour. This paper examines the entry of one particular kind of investment into two very different kinds of area, and its 'combination' with the already-established characteristics of those areas.

The fact that the two studies look at very different areas is important. For it means that the effects of this new 'round of investment' are different in each case. Thus, although both types of region are now being drawn into a similar place in an emerging wider division of labour, their roles in previous spatial divisions of labour have been very different; they have different histories. They bring with them very different class structures and social characteristics. And, as a result, the changes which they undergo, as they are drawn into a new division of labour, are also different. In the schematic and formal terms used above – this process of the combination of 'layers' – any one particular layer or 'round of investment' may produce very different effects in different areas as a result of its combination with different pre-existing structures.

Before moving on to the analyses, there are a couple of points which should be made. First, this paper is about the influence of industrial change on social change, of industrial restructuring on class restructuring. But it must be stressed that this is only one influence. It is not being argued that industrial structure can provide a complete explanation of local class characteristics. To begin with, no full explanation of the processes of class formation could confine itself to changes within the structure and relations of production. On the one hand, such processes are not solely economic; on the other hand, even the economic forces are not confined within production. Broader social structures of community, changing patterns of consumption (for instance, Cooke, 1981a), the restructuring of spatial forms (for instance, Hudson, 1979; and Rees and Rees, 1981), the changing national ideological and political climate and the marked patterns of geographical cultural differentiation – all of these will combine with changes in the social relations of production in determining both the overall pattern of class structure and the more detailed internal characteristics of classes.

Second, we are looking here at *local* social effects of national geographical reorganization. These sketches are not meant to characterize whole regions; intra-regional variety is far greater than is dealt with here. Rather, these should be seen as studies of kinds of labour-market areas which typify parts of the wider regions. But the situation is still more complicated. For one important element of social restructuring has been the internal spatial reorganization of these areas themselves. Class restructuring has not taken place on a given and continuing spatial base. As argued elsewhere (Massey, 1978), 'regions' are not pre-given to analysis, nor are they unchanging. Spatial structures, and coherent areas and regions within them, are themselves continually reproduced in changing form as part and parcel of the kinds of processes looked at here.

The old coalmining areas

The existing structure

The first example is that of those areas of the country which have for a century or more been dominated by the spatial division of labour which characterized the dominant industries of Imperial Britain, regions where production in the basic industries of that period has for long structured the overall pattern of economic life.

These mining areas have evolved, as a result of their role in the previously-dominant spatial division of labour, very specific and coherent structures of economic and social relations. They were dominated by, often developed by, the coal-mining industry. The decline of Britain's Imperial position, and the collapse of the coal-exporting industry therefore entailed the beginning of their long-term economic decline. The pattern on which the newly-emerging role of these areas, in a new complex of uses of space by industry, is being imposed is thus a relatively simple one.

The most characteristic element of the class structure of these areas, in terms both of its numerical importance and of its geographical distinctiveness, has been the working class. The very fact of single-industry dominance has been one of the conditions for creating a degree of coherence both in the internal structure of the working class and in its organization. In most of the small mining towns themselves the majority of those employed in capitalist wage relations has, for over a century, been employed in the coal-mining industry. This comparative homogeneity of economic structure was reinforced in its class and organizational effects by the nature of the industry itself. For, in mining, the work experience of the bulk of the workers is similar, the detail of hierarchical structures is relatively under-developed, and a number of the potentially divisive differentiations which do exist (such as between surface and face workers) are less effective in their divisiveness because individual workers may well experience both at different stages of their working lives.[4] This situation has had a number of effects. First, the working class itself, within the mines, was relatively undifferentiated. It was the kind of situation which Friedman, 1977, p. 53, refers to as producing 'resonance' between workers, a basis for solidarity. Second, there were relatively few within the industry in these areas who could *not* be defined as working class. The division of labour and the labour process were not here developed sufficiently to form the basis of any significant element of those strata whose functions include middle-range management, the maintenance of capitalist discipline, and mental as opposed to manual work.

Moreover the dominance, within these capitalist relations, of a single industry was further reinforced in its effects by the nature of trades-union organization within the industry. The history of sectoral unionization (which has included an important syndicalist component) together with a very high degree of union membership, meant that a high proportion of those in paid employment in the community would be in the same union. Moreover, even before the nationalization of the coal industry, this similarity of union membership was reinforced within individual communities by the fact that there would often be only a single employer to fight.

Of course, while these conditions may make organizing easier, there is no automatic response to the structure of economic conditions. Trade union coherence and militancy also require active political organization. All that is being examined here are the preconditions for such organization. The inter-war history of the South Wales Miners' Federation was in fact a constant battle against *dis*unity between workers of different companies and pits (Beacham, 1958) and between the Federation and the 'scab' union. And there have, too, been long periods of relative quiescence and submission (Allen, 1982; Cooke, 1981a). Coherence of industrial structure in itself guarantees nothing. Nevertheless, these geographical areas and the unions associated with them have, both symbolically and actually, been centres of strength for the trades-union movement in the UK.

It has, however, been an organizational strength of a particular kind. It is economically militant. In 1974, of course, it led to the resignation of the government but the record over the longer period is equally notable. Beacham, 1958, reckons that from the thirties to the fifties, miners (a mere 4% of the economically active) accounted for over one third of the losses caused by industrial disputes in Great Britain.[5] But this is also a unionism which is defensive of status within the working class, a status based not so much on craft skills and higher wages, as say with engineers, but on closed-shop solidarity (the National Union of Mineworkers was a leading union in the fight for the closed shop, and was one of the first to win recognition of this principle from employers), clear identity of job, and a pride in the toughness of their work. This is a pride which refers itself to masculinity, a characteristic based not only on the nature of the work but also on the complementary lack of paid employment for women.

It should not be thought that any of these characteristics are simply a product of what might be called 'purely economic' mechanisms. Even the single-industry nature of these areas, and consequently their coherence, was also the result of social and political processes. First, the lack of paid work for women. All these areas had, and in varying degrees still have, very low female activity rates (with the exception of the large towns and the conurbations, which are anyway off the coalfields). In the colliery towns the rate was sometimes extremely low. The increase in paid employment for women began in the Second World War, but even in the mid-1960s in some towns it was as low as 20–25% (Department of Employment and Productivity, 1970). This seems more than anything to have been a result of the nature of the work done by the men (in particular the fact that it demanded large inputs of domestic labour, and the fact that it was often shift-work) and of the primacy accorded to that work. As Humphrys, 1972, points out, 'traditionally, too, the labour force is predominantly male. There were few jobs for women in the coal and steel industries, and their place was clearly seen as being in the home looking after the breadwinner' (p. 30). Second, there seems in a number of cases to have been an active policy of keeping out alternative employment for men. The dominance of these local labour markets by single industries was not simply a result of 'location factors'. The National Coal Board, for instance, seems to have guarded rather jealously the monopoly of the male labour force which it held in parts of South Wales. Third, very little competition for this labour was generated locally by the rise of indigenous firms. This, again, is a characteristic which these older coalfields have in common. It is also in part a product of the class structure associated with the dominant industries.

The impact, or 'the combination of layers'

What have been the effects on the social structure of these areas of the combination of the continuing decline of this old use of space with the increasing importance of their insertion into a new spatial division of labour?

The incorporation of these areas into different spatial divisions of labour is having its most important effects on the structure and organization, the lives and the politics, of the working-class. In particular, the old form of coherence, organization and strength of this class are being undermined.

First, the size and structure of the class are undergoing change. Most obviously, over the longer period there has been a considerable decline in its size, as a result both of quite massive outmigration and of deliberate policies of transfer, particularly for miners. To the extent that this migration was 'selective' it also entailed some change in the composition of the working-class that remained. Moreover, the internal structure of the class has been changed in other ways too. Capitalist relations of production are being continually extended to people who were previously not wage-earners, primarily married women. One aspect of this change has been the expansion of wage relations in clerical and service work. Such relations involve both unproductive sectors and what is usually termed mental (as opposed to manual) labour. Without arguing that low level clerical and service workers are therefore not part of the working class, it is true that the growing importance of wage relations in these sectors adds another degree of differentiation within the working class in these regions. It is perhaps best seen as another aspect of the changing internal structure of the economically active working class, another way in which its previous relative coherence and homogeneity is being disturbed.

Moreover, although for capitalism this has represented expansion, for the local working class it represents a more dramatic change. Not only are women now increasingly absorbed into capitalist wage relations, but with the continued decline of the basic industries older men are increasingly excluded from them. Socially, this has produced dramatic changes. It is common to hear that what such areas need is jobs for men, and *not* jobs for women. It is a call with which a number of academics and policy-makers appear to concur. In fact, the availability of work for women, and the female activity rate, are still lower in these regions than they are nationally. Further, when new jobs are made available to men we hear, as though it were patently funny, that Welsh ex-miners cannot be expected to turn their attention to making marshmallows, or underwear. What is at stake is the maintenance not just of a social structure in which the men are 'the breadwinners', but also of a long-held self-conception of a role within the working class – the uniqueness, the status, and the masculinity of working down the mine.

But while it is difficult to concur with such defensive postures, there are very real issues. An attack on traditional patriarchal structures does involve enormous personal dislocation. A study of the closure of a colliery in the North East of England reported, 'one man said he thought a partial solution would be to give older men "girls' jobs", such as selling men's wear in chain-stores. He thought it unjust that young women could be taken on while men were unemployed' (Department of Employment and Productivity, 1970, p. 73).

Given the virtual impossibility of these older miners getting new jobs, and the 'breadwinner' ideology to which they are accustomed, this kind of resentment is understandable. The same study also reported that out of some 26,000 men made redundant as a result of the colliery closures in 1967 and 1968, less than 7,000 were placed in employment or were known to have found work two years later. Moreover an extremely high proportion of these long-term unemployed were registered disabled, and nearly 90% of the remainder were aged 55 and over. Increasing numbers of men, now in middle-age, are facing the prospect of never again being able to get work in these regions. The problem is that the quotation about girls taking men's jobs is also mirrored in the emphasis on demands for male employment made by trades union organizations, and of their criticism of the incoming jobs as being 'only for women'. It is an outstandingly clear example of the conflicts – in this case within the working class – brought about by the changing use by capital of a particular area and the political confusions consequent upon the confrontation of very different spatial divisions of labour as they replace each other in dominant importance. In regions such as these it would seem that the changing use of the area is more likely to provoke conflicts within the working class than within capital.

This shift between male and female within the employed working class is also producing changes in the relation between unemployment and the reserve of labour. Because new groups of workers are being brought into the workforce the level of unemployment is not decreasing in proportion to the arrival of new jobs. Indeed, with the present closures in the older industries, it is increasing dramatically. Further, to the extent that a higher proportion of those not working will now register as unemployed, what was once a hidden supply of labour, not appearing at all amongst the ranks of the economically active (what Marx called a latent reserve army) is now increasingly transformed into explicit unemployment, an active reserve army. This means in turn that its existence is more likely to become a political issue; differentials in female activity rates have never been an important part of the demands from the regions for regional policy. Moreover the social characteristics of the unemployed are undergoing some change. As already indicated there is a considerable growth of long-term male unemployment. The nature of redundancy payments from the steel industry indeed recognises this fact.

Finally, of course, what is really wrong in an immediate sense with the new jobs is not that they are for women but that they pay low wages and involve little interest or job-control, and this applies also for many of the jobs which are taken by men (Department of Employment and Productivity, 1970). There is, as might be expected, in some areas some evidence of an increasing dichotomization in the wage structure even within the working class within manufacturing (Cooke, 1980).

There are also considerable changes to be faced by the men who are still employed. All the basic industries of these regions, and not only coalmining, have since the mid-1960s been subject to intense pressure to increase productivity. So those who have kept their jobs have often seen considerable changes in the nature of the labour process and in work practices – changes which in turn have often involved a shift in the skill composition of the workforce. But the changes experienced by those who have found jobs in other industries have been even greater. Few of the previously-dominant industries, even in the wider regions, employed flow-line and mass-production forms of organization of the labour process. This is obvious in coalmining and also in steel, but it is true also of the heavy engineering industry, which has been based around the production of small numbers of large-scale, often individually-specified commodities (Massey and Meegan, 1979). Such forms of production involve far greater individual control over the job than do those of the newer, incoming industries where production is more often based on conveyorbelt or related principles, and with a fairly high degree of subdivision of tasks.[6] There is, therefore, under way a process of conversion of the workforce of these regions to the greater discipline, the reduced autonomy and job-control of the modern factory. Jobs are increasingly broken down, and therefore people increasingly divided, into the disciplined and the maintainers of discipline. Report after report discusses the 'problems' of 'attitude to work', of absenteeism and high turnover, the difficulties of the 'acclimatization' of ex-miners.

These, then, are some of the changes in the structure and composition of the working-class to which the changing balance of spatial divisions of labour has contributed. This changing balance has also disrupted the level and type of labour movement organization. First, instead of the focus of the economically active community being around one industry, and often one company, there is now a disparity both of branches of production and, even more particularly, of employers. The individual plants are often smaller. The basis for coherence and solidarity which existed in the single industry/single owner mining towns is thus disappearing. Friedman, 1977, has pointed to the difficulties for working-class organization which such increased disparity may produce. The growth of long-term unemployment is a further element contributing to this disparity of fortunes within individual communities. Further, the very fact of the entry of these new forms of capital is partly a result of the lack of organization of the workers they wish to employ. It has already been pointed out that the female activity rate in these areas was previously extremely low. A large proportion of the new workforce has thus had no previous experience of capitalist wage relations and, at least in recent decades, male dominance has excluded women from many forms of public and political activity. No basis of union organization exists therefore, and the fact that these new workers are female compounds the difficulties, both because of the problems for women of doing two jobs and attending union meetings (for while women may be doing an increasing share of the breadwinning, men do not seem to be doing much more domestic labour) and because of the lack of commitment to organizing women workers traditional in the trades union movement as a whole.

Cornwall

The existing structure

The old coalfields are, however, by no means the only areas which have been experiencing the growth through decentralization of these kinds of economic activity and these kinds of jobs. Cornwall, too, has been on the receiving end of the process. And here the social structure into which the new industries have been implanted could hardly be more of a contrast with the coalfields. Instead of the combination of state ownership, big industry and working class, Cornwall has much more significant proportions of small industry and self-employment, and a much lower importance of large industry and highly-organized, relatively well-paid male working class. All this, of course, in some measure again results from the very different role which this part of the country has played so far in the UK economy, in other words its previous location in a wider territorial division of labour.

The class structure of Cornwall is not dominated by an industrial proletariat. In sectoral terms, its historically-important industries have been agriculture, mining and tourism. Only the second of these, together with some small presence of shipbuilding and engineering, form a basis for the kind of unionization which exists in the coalfields, and even then these industries are on a far smaller scale.

Moreover, while Cornwall is like the old coalfield areas in having had historically a very low female activity rate, the reasons for that low participation were very different. The Cornish County Structure Planners (Cornwall County Council, 1976b) analyse this in detail and argue that the most important explanations are; first, a straightforward lack of work opportunities; second, the availability of seasonal work during the summer (which they argue is probably a factor reducing the willingness of married women to seek year-round employment); third, the fairly high overall level of unemployment; and fourth, the geographical nature of the County. This last, the distribution of much of the population in villages and small towns, is argued to be 'perhaps the most important factor affecting married female activity rates' (p. 99). It lengthens the average journey-to-work, and this in turn particularly affects women as a result of the restricted public transport facilities and of their having less access to the 'family' car. The Cornish analysis points out that it is therefore probably often more feasible 'to help on the farm, take in summer guests or to be less well off and stay at home' (p. 99). It also points out that this spatial structure reduces not only the number but also the range of jobs within easy access. The continuing influence on activity rates of these problems of accessibility is confirmed by the variations within Cornwall between urban and rural areas: 'rates were up to 8–9% lower in rural areas than in the towns in 1971'. It is a clear example of the joint operation of spatial and patriarchal structures. It is of course probable that, precisely because of the higher proportion than nationally of small capital and self-employment, particularly in agriculture, more women were involved in economic productive activity than the statistics show. But this kind of work has been declining dramatically – between 1961 and 1971 the number of female workers actually registered in agriculture declined by 12%, while nationally there was an 11% increase (Cornwall County Council, 1976b). So the labour-reserve, which was one of the attractions to in-coming industry, was 'produced' by a very different social history from the apparently similar labour-reserve not that many miles away in South Wales.

It was, moreover, a reserve with different characteristics. A married woman in this region was less likely to be 'just' a wife and housekeeper to a 'breadwinner' and more likely to be involved in some form of non-domestic labour, keeping a bed-and-breakfast boarding-house maybe, or doing (paid or unpaid) work on the 'family' farm. What this means in turn, of course, is that attempts to produce (or refute) correlations between the in-migration of industry and low female activity rates miss the real point. The attraction to industry is a reserve of labour which – so management hopes – will be relatively cheap but above all non-combative. In Britain in the second half of the twentieth century, such reserves are often, though by no means always, composed of women. But the vast variety of ways in which patriarchy works, and in particular its geographical variations, means that the processes by which such characteristics are produced, and therefore the 'indices' on which they can be measured, also vary widely. While many married women in regions such as Cornwall may not have been confined to the home and to domestic labour, but worked in tourism, agriculture or small-scale services and may even have appeared in official statistics as 'economically active', they have nonetheless rarely been working in fully capitalist wage-relations and even more rarely within any labour process likely to provide the basis for solidarity and workplace organization. In the Cornish economy, especially that part in which women were directly and economically involved, a kind of petty-bourgeois self-sufficiency and independence had been the order of the day.

Cornwall also differs from the old coalfield areas in having been historically a low-wage economy. It has over the last ten years usually come bottom of the league of counties in the *New Earnings Survey*. This appears to be the result both of industrial structure and of low relative wages within that structure, and of the availability of seasonal work. This last provides extremely low (annual) wages to those who have to rely on it altogether, but it can also have the effect of subsidizing other wages so that people can supplement the income from their regular employment by second seasonal or part-time jobs, although if the pay from their main job were higher, the need for this might not arise (Cornwall County Council, 1976b, p. 40). Such employment may in other words enable wages in other sectors to be held down. It is an interesting parallel to the subsidization of wages in capitalist firms by subsistence farming in peasant areas.[7]

Finally the structure of the ownership of production is different in Cornwall. Neither locally-based big capital nor state ownership is of major significance. There is on the other hand a considerable proportionate presence of small capital and self-employment. As a proportion of the population, self-employed people are two and a half times more important in Cornwall than nationally. And Cornwall is much more dependent than the economy as a whole on profits from such sources and from professional earnings (Cornwall County Council, 1976b, p. 41), much of it coming from agriculture and from tourist-related activities.

The impact, or 'the combination of layers'

Up until the 1950s, the manufacturing basis of the Cornish economy was in decline but in the mid-1960s this long-term trend was reversed and manufacturing firms began moving into the County. The immigration reached a peak in 1969 and 1970 but in the mid-1970s was still well above the figure for the mid-1960s. The new employment is very similar to that arriving in the coalfield areas. Two-thirds of it is in branch-plants which in turn are the element most heavily biased towards female employment (about 60%). And much of this employment is low-paid and classified as semi-skilled or unskilled. The largest factories (in employment terms) have gone to the towns (themselves small), but the biggest proportionate impact has been in rural areas.[8]

So what, then, of the effects of the entry of these new forms of employment, and how do they compare with the impact in the old coalfield areas? Obviously some of the implications have been similar: the size of the population engaged in wage labour has increased, again primarily because of the incorporation of more women (almost entirely married women) in the workforce, and unemployment in some areas has increased and its structure changed.[9]

But it is the differences which are striking. In the first place, the effects on the working class are far less dramatic. There has been no existing and well-organized working class. The extension of the work force, and particularly its extension to more women and in low-paid jobs, has therefore not had the same effects. It has not produced the same basis for conflict, it has not been associated with such a relative lowering of wages, and it has not meant the undermining of a major base of trades-union strength.

But the main difference in impact between the two kinds of region is that, while in the heavy-industry areas capital did not seem to have been adversely affected by the entry of the new spatial divisions of labour, in Cornwall it seems to have witnessed the process with some horror. The most important aspect of this concerns wages.[10] Cornwall has for a long time been a low-wage area, and the Structure Plan reports bemoan the fact that the jobs being created there in the 1960s have done too little to change this. The Structure Plan recommendations include the aims of creating the conditions to make it easier for women to work, encouraging the provision of opportunities for people to increase their earnings, and, quite explicitly, increasing the pressure of demand for labour. It is the last point which goes down badly with existing local industry. The Structure Plan findings are based on a detailed analysis of the causes of low pay and in particular of the relation between pay and activity rates, as it might operate in Cornwall. The analysis concludes that in Cornwall the female activity rate and average female wages are both relatively low. It is possible for employers to offer lower wages than in many other places because there is a higher proportion of women potentially able to work but not actually working and hence less pressure of demand for labour (Cornwall County Council, 1976b, p. 45). In reply, the Cornwall Industrial Development Association's document on regional policy (CIDA, 1977) and the Cornwall Structure Plan mounts a sustained attack on branch plants employing women and owned by big capital. For while it argues strongly for more high-level technical and administrative employment, more unskilled and semi-skilled jobs represent only more competition for labour. Unlike the dominant elements of capital in the heavy-industry regions, these industries are still locally owned and based. They have not been nationalized, they have not diversified geographically, they are more likely to overlap in their labour requirements with the incoming industry, and they are just not large enough to compete. They write 'the provision of more jobs and/or higher wages and salaries can only add to the already strong pressures arising from inwards migration and are not therefore prime objectives. Indeed, it is doubtful whether Cornwall could *ever* have a high wage economy: it would lead to a loss in profitability and additional unemployment, and hence would be self-defeating' (p. 119). Their own recommendations concentrate on policies to encourage the development of existing and locally-owned businesses.

Conclusions

National processes and local change

It is clear from these two studies that the national process of industrial decentralization has been very locally varied in its operation and effects. While what regional capital there is in the old coalfield regions remains intact, if it does not actually benefit from the process, existing local capital in Cornwall clearly feels distinctly threatened.

Another 'national process' is the increasing incorporation of married women into the waged labour force. The availability of an alternative, relatively cheap, and relatively 'green' source of new labour has been enormously important to the kinds of industrial change we have seen in the post-war period. The kinds of changes in the production process, and the spatial patterns adopted, have been a response to, and moulded by, the fact that such a reserve (one important element of which was female) existed. What we have seen in the studies here are hints at how such reserves are constructed. At a general (national) level, analyses propose a variety of (frequently unquestioned) reasons for this 'availability' and 'suitability'. However, when the changes are examined at sub-national level it is clear that the processes vary. The reserves of female labour in the two types of areas looked at here were produced by different sets of interactions between patriarchal and class relations over space. Further, this difference between areas helps highlight the fact that such reserves are the product of social relations, and not the inevitable outcome of biological difference. It is not the fact that these people are biologically female which explains their characteristics and their new availability for the wage-labour force in emerging forms of organization of production, but the fact that certain kinds of (patriarchal) social relations construct them as having all these characteristics. Beechey, 1978, shows:

> how the existence of the sexual division of labour which consigns women to the family and the patriarchal ideology embodied in it must be presupposed in order that female labour can constitute these advantages to capital. This suggests that it is the sexual division of labour and the family rather than women's 'natural' lesser physical strength (the explanation which Marx actually resorts to) [or, one might add, the 'dexterity' which so many regional analysts so unquestioningly rely on – D.M.] whose existence must be assumed if the specificity of the position of female wage labour in [capitalist societies] is to be understood.
>
> (p. 192)

Of course, regional geographers do refer to such social characteristics – the 'dual role' of women, their lack of access to private transport, etc. But as we have seen here these characteristics too, are social constructs. They are not 'necessary', and they should not be left by analysts unquestioned. The fact that such relations vary geographically is not only interesting in itself, it also makes it plain that such 'female characteristics' should not be taken as given. They, too, need to be explained. Without the construction of such a reserve – i.e. such divisions within the working class – the changes within production might not have been the same. And the 'sphere of production' is not a passive element in these processes, either. Such jobs, when done by women, are labelled 'unskilled'. They are said to require 'dexterity', a characteristic which, unlike male 'strength', goes unrewarded financially. Relations within production, in other words, do not just take advantage of an externally-constructed ideology, they further reinforce it.

Notes

1 Much of this knowledge has been produced outside academe, strictly defined. It includes the publications of the Community Development Projects, the work of numerous local groups, of which the Merseyside Socialist Research Group and the Joint Docklands Action Group are good examples, and papers and discussions in gatherings such as the Regionalism Group of the Conference of Socialist Economists. Other references are cited elsewhere in this article.

2 It should be pointed out that these different industries also exhibit different uses of space. 'Monopoly capital' is more varied than is sometimes implied. Thus, although

the sectors represent very similar kinds of new jobs for the areas under discussion, they have different implications for the form of differentiation between these areas and the rest of the country. Thus in the case of electronics the regions are part of the well-known 'hierarchy of production', being the classic 'bottom-rung' regions of assembly-work. In the case of services the structure is similarly hierarchical, but it is a different hierarchy. This does not involve skilled craft work or research and development workers in other parts of the country. The division of labour and associated labour processes are different. The form of intra-sectoral and inter-regional differentiation in this case is between clerical workers on the one hand and high level administrators, managers and state functionaries on the other. In the case of clothing and footwear the situation is different again. Here there is no hierarchy within production. Inter-regional differences are confined to those which result from the different effects of internal and external control. The sectors also differ in terms of the conditions under which they have established production in these regions. The electronics and service industries have done so on the basis of developments of new internal divisions of labour and labour processes. In the clothing sector, on the whole, no such developments have taken place; the arrival of this sector in these regions has been predicted much more on a mixture of cost pressures in the old central areas and the attractions of regional policy. In a sense, then, this represents an influx into these areas of a whole complex of new uses of space, a complexity which is in itself a contrast with the previous relative simplicity of industrial structure.

3 These two characteristics are often related. 'Skill' classification may be as much a matter of history and industrial strength as of any intrinsic nature of the job. It is a real question to what extent the technical changes in production over this period actually led to *simpler labour processes*, and to what extent they were the occasion for the introduction of a *category of labour called unskilled*. It should also be noted that the industries discussed here are not the only sources of new jobs in those areas. But they *are* among the most important elements of industrial change, particularly in terms of their social impact.

4 It must be stressed that all this is comparative, with other industries and other kinds of community. It would be wrong to imply that there are no bases for sectionalism within mining. There are winding-enginemen, colliers, craftsmen and so on, and the history of union organization in the industry has often revolved about their inter-relation (see, for instance, Francis and Smith, 1980; and this paper). But the relative homogeneity of the coal industry is nonetheless clear, even in comparison with the other basic industries – such as steel – in the wider regions (Cooke, 1981b).

5 Carney, Lewis and Hudson, 1977, refer to this history in the specific context of the North East region; and Francis and Smith, 1980, document in detail the history of the miners' union in South Wales in the 20th century.

6 These kinds of changes apply not just to the range of incoming sectors considered here, but also to others such as the car industry.

7 Neither is it without its parallels in the regions discussed previously. Here the sources of alternative income are state transfer payments of various sorts – including redundancy pay, miners' pensions, and forms of EEC and ECSC income maintenance (mostly in steel areas). These are all temporary. Longer-lasting but at lower levels is unemployment benefit, a form of state transfer which, in a 'standard married couple' situation, is far more likely to cheapen female wages than male (since a higher proportion of unemployed men are eligible for unemployment benefit). These factors are small, but not insignificant. Similar situations occur on a rather larger scale in Ireland (Stanton, 1979) and Southern Italy (Wade, 1979).

8 Much of the information in this section has been drawn from the County Structure Plan and associated documents (Cornwall County Council, 1976a and 1976b).

9 And even these descriptively similar effects are different in detail and the result of rather different combinations of causes.

10 Other locally expressed reservations do exist particularly concerning environmental impact.

References

Allen V. (1982) *The Militancy of British Miners*, The Moor Press, Shipley, Yorks.

Beacham A. (1958) The coal industry, in Burn D. (Ed.) *The Structure of British Industry: A Symposium*, vol. 7, pp. 108–55. Cambridge University Press, Cambridge.

Beechey V. (1978) Women and production: a critical analysis of some sociological theories of women's work, in Kuhn A. and Wolpe A-M. (Eds) *Feminism and Materialism*, pp. 155–97. Routledge and Kegan Paul, London.

Carney J., Lewis J. and Hudson R. (1977) Goal combines and interregional uneven development in the UK, in Massey D.B. and Batey P.W.J. (Eds) *Alternative Frameworks for Analysis*, pp. 52–67. Pion, London.

Cooke P.N. (1980) Dependency and the regional problem: economic disintegration and reintegration in Wales, UWIST, Cardiff (mimeo).

Cooke, P.N. (1981a) Interregional class relations and the redevelopment process, Papers in Planning Research 36, UWIST, Cardiff.

Cooke, P.N. (1981b) Local class structure in Wales, Papers in Planning Research 31, UWIST, Cardiff.

Cornwall County Council (1976a) County Structure Plan: policy choice consultation document, Cornwall C.C., Truro.

Cornwall County Council (1976b) County Structure Plan: Topic Report: Employment, income and industry, Cornwall C.C., Truro.

Cornwall Industrial Development Association (1977) The economy of Cornwall: a discussion document prepared with particular reference to the Cornwall Structure Plan, 2nd edition, CIDA.

Department of Employment and Productivity (1970) *Ryhope: A Pit Closes: A Study in Redeployment*. HMSO, London.

Foster J. (1974) *Class Struggle and the Industrial Revolution: Early Industrial Capitalism in Three English Towns*. Methuen, London.

Francis H. and and Smith D. (1980) *The Fed: a History of South Wales Miners in the Twentieth Century*. Lawrence and Wishart.

Friedman A. (1977) *Industry and Labour: Class Struggle at Work and Monopoly Capitalism*. Macmillan, London.

Hudson R. (1979) New Towns and spatial policy: the case of Washington New Town, Department of Geography, University of Durham (mimeo).

Humphrys G. (1972) *South Wales*. David and Charles, Newton Abbott.

Massey D.B. (1978) Regionalism: some current issues, *Capital and Class*, no. 6, 106–25.

Massey D.B. (1979) In what sense a regional problem? *Reg. Studies* 13, 233–43.

Massey D.B. and Meegan R.A. (1979) The geography of industrial reorganisation: the spatial effects of the restructuring of the electrical engineering sector under the Industrial Reorganization Corporation, *Prog. Plann.* 10, part 3.

Rees G. and Rees T. (1981) Migration, industrial restructuring and class relations: the case of South Wales, Papers in Planning Research 22, UWIST, Cardiff.

Stanton R. (1979) Foreign investment and host country politics: the Irish cases, in Seers D., Schaffer B. and Kiljunen M-L. (Eds) *Underdeveloped Europe: Studies in Core-Periphery Relations*, pp. 103–24. Harvester, Brighton, Sussex.

Wade R. (1979) Fast growth and slow development in Southern Italy, in Seers D., Schaffer B. and Kiljunen M-L. (Eds) *Underdeveloped Europe: Studies in Core-Periphery Relations*, pp. 197–221. Harvester, Brighton, Sussex.

27 Linda McDowell and Gill Court

'Missing Subjects: Gender, Sexuality and Power in Merchant Banking'

Excerpts from: *Economic Geography* **70**, 229–51 (1994)

The economic structure of advanced industrial nations in the postwar decades has undergone fundamental changes, as the long debate among geographers about its nature and spatial form attests. Probably the most significant change has been the shift to a service-based economy, such that more than two-thirds of all waged workers in countries like the United States and the United Kingdom are now employed in service occupations. Associated with this transition has been a far-reaching transformation in the gender division of labor. The feminization of the labor force has been profound, particularly in the last decade. In Great Britain, for example, no less than eight out of ten new jobs created in the 1980s have been for women. While parts of the service sector have long been recognized as a feminized job ghetto, increasingly these "new" occupations rely on the marketing of attributes conventionally associated with the "natural" attributes of femininity – sociability, caring, and, indeed, servicing – which are marketed as an integral part of the product for sale (Hochschild 1983; Jenson 1989; Leidner 1991).

One consequence of the expansion of the service sector, however, is that increasing numbers of men are also employed in these jobs, which Leidner (1991) has termed interactive service occupations, in which personal bodily attributes and character traits are a significant part of the job. These jobs involve selling the worker as part of the overall service. As Leidner argues, "these jobs differ from other types of work in that the distinctions among product, work process, and worker are blurred or non-existent, since the quality of the interaction may itself be part of the service offered" (1991, 155). These jobs, she argues,

> have several distinctive features that make them especially revealing for the investigation of the interrelation of work, gender and identity as . . . workers' identities are not incidental to the work but are an integral part of it. Interactive jobs make use

> of workers' looks, personalities, and emotions, as well as their physical and intellectual capacities, sometimes forcing them to manipulate their identities more self-consciously than do workers in other kinds of jobs.
>
> (Leidner 1991, 155–56)

As these jobs are increasingly central to the economy, and will no doubt become even more significant (Christopherson 1989; Sayer and Walker 1992; Smith, Knights, and Willmott 1991), a set of new and interesting questions is posed for economic geographers interested in the causes and consequences of new social and spatial divisions of labor. In particular, it seems clear that a new set of issues about subjectivity and gendered identities should become central to economic geography. We should begin to consider how the characteristics of service sector jobs are connected to the gendered attributes of workers and how this varies across space and time. Are the familiar patterns of sex segregation in the labor market being restructured and, if so, how and to whose advantage? (McDowell 1991) How do jobs become gendered in the first place? What flexibility is there in this process? And how is the gender encodement of tasks achieved and maintained or contested and challenged? While some of these questions have been addressed with reference to manufacturing, only a few studies have raised similar questions in relation to the expansion of service sector occupations in advanced economies. In the main, these studies have focused on the bottom end of the sector, on what might perhaps be termed "servicing" occupations, such as secretarial work (Pringle 1989), selling fast food (Crang 1992; Gabriel 1988; Leidner 1991), and selling insurance (Leidner 1991; Knights and Morgan 1991; Morgan and Knights 1991). Little investigation has been carried out of the ways in which the construction and manipulation of gendered identities have been important in the expansion of that set of high-powered, high-status occupations, particularly in the financial services sector, that have been portrayed as the epitome of success in 1980s Britain and North America. This paper focuses on a subset of such occupations: high-status employment within merchant banking.

In the 1980s in Britain, the financial services sector was one of the most rapidly expanding employment sectors. Retail and merchant banking, broking, insurance, and associated legal services all exhibited record growth rates. Financial services thus out-performed virtually all sectors of the economy, with annual growth rates exceeding 7 percent in the boom years of the mid-1980s. The spatially concentrated growth of the wholesale banking sector, in particular, in the South East of England was a key element in the restructuring of the space economy that occurred during the decade and significantly deepened the north-south divide (Court and McDowell 1993; Thrift and Leyshon 1992; Peck 1993). Unprecedented rates of growth took place, for example, in the traditional area for banking, the Square Mile in the City of London, where in the mid-1980s employment growth occurred at rates eight times faster than those of any other small area in the United Kingdom.

This expansion of financial services throughout the 1980s has had a profound effect not only on the occupational structure of the City of London but also on its built environment (Pryke 1991). There has been a huge boom in new construction and renovation (Zukin 1991), not only of office buildings but also of spaces for recreation, play, and pleasure. In areas like the new Broadgate development, for example, the developers have carefully constructed a total landscape of work and leisure: a new global landscape of financial power complete with upmarket restaurants, the obligatory atrium or two, a circular open space for staging spectacles, and plenty of outdoor statuary.

In a stimulating collection of papers, Budd and Whimster (1992) argue that the globalization of financial services, their dominance in a small area of London, and the construction of new spaces of work and play have effected an "interpenetration of areas of life previously separated by hierarchies and boundaries," allowing "a repatterning of personality, lifestyle, neighborhoods and the metropolis" (p. 3). This interpenetration results in a blurring of the boundaries of work and pleasure and in new notions of the ideal worker. Social and cultural aspects of the construction of work are becoming increasingly important. Hence, Budd and Whimster argue, greater attention must be paid to the cultural aspects of current economic restructuring. Sayer and Walker (1992) develop related arguments when they suggest the theoretical centrality of the social division of labor for geographic analysis. The new social division of labor, they argue, extends over a vast geographic scale to "encompass new sectors, new jobs and new ways of working" (Sayer and Walker 1992, 2). This complex division of labor, in their view, is "an active force in social ordering, economic development and the lived experience of the participant" (1992, 1), thus placing questions of daily experience and social identity of workers on the geographic agenda. They also suggest that one of the most important aspects of these new divisions and new ways of working is the growing numbers of women entering waged labor in an array of occupations. They argue that this feminization of the labor force is at the heart of new questions about the service economy, and yet, as Sayer and Walker recognize, "scholars have barely begun to grasp this nettle" (1992, 103, fn.).

The purpose of this paper is to grasp the nettle and examine a set of questions about men and women's lived experience at work, looking at ways in which economic geographers might begin to focus on cultural aspects of economic change in investigating the social construction of labor power and its gendered attributes. While a considerable set of questions arising from feminization demand investigation – from regional variations in growth rates to the impact of women's labor force entry on spatial patterns of demand for housing and other consumer goods – this paper will focus on social practices within the workplace itself. Our aim is to uncover the ways in which gendered identities are constructed at work to produce and reproduce the worker as a subject. We argue that not only are everyday social relations in the workplace imbued with notions of power and domination in a general sense, but also that particular ideas about sexuality are key mechanisms in the maintenance of women's occupational segregation. So far, the significance of the construction of gender identities at work – the development of particular versions of masculinity and femininity appropriate to success in particular occupations – has been neglected by geographers.

What geographers have not yet explored is the social construction of occupations themselves as gendered. In their work, they tend to represent occupations as empty slots to be filled and workers as already socially constituted men and women with fixed gender attributes. For most geographers, the focus of analysis has been on explanations of why women enter particular sectors or certain occupations in the labor market in growing numbers, rather than on the logically preceding question of how these occupational ghettos arise. The gendering of jobs and workers has been taken for granted. As Scott (1988, 47) has argued.

> if we write the history of women's work by gathering data that describe the activities, needs, interests, and culture of 'women workers,' we leave in place the naturalized contrast and reify a fixed categorical difference between women and men. We start the story, in other words, too late, by uncritically accepting a gendered category (the 'woman worker') that itself needs investigation because its meaning is relative to its history.

Thus the process of *occupational sex typing* (or better, *stereotyping*), as distinct from occupational segregation, tends to be taken for granted. But jobs are not gender-neutral; rather, they are created as appropriate for either men or women. Jobs and occupations themselves, and the set of social practices that constitute them, are constructed so as to embody socially sanctioned but *variable* characteristics of masculinity and femininity. This association is apparently self-evident in the analysis of classically "masculine" occupations; consider, for example, the heroic struggle and camaraderie involved in male heavy manual labor compared with the characteristics of feminized occupations such as secretarial work, although even these have changed their gender associations over the century (Bradley 1989).

Sexing jobs

Suggesting that many studies of occupational segregation have neglected the processes by which jobs become gendered, that they have focused on occupational segregation rather than on occupational sex stereotyping, is not to imply that they have ignored the association of, for example, skill designation with gender or the embodiment of gender attributes in job definitions and workplace practices. This clearly is not so. A large body of work has revealed how the supposedly natural attributes of femininity (for example, docility, dexterity, or "caring") have been set up in opposition to masculine attributes to organize and reorganize labor processes and to reward workers differentially on the basis of gender (see, for example, Beechey and Perkins 1987; Cockburn 1983; Crompton and Jones 1984; Crompton and Sanderson 1990; Milkman 1987; Phillips and Taylor 1980; Walby 1986).

Little attention has been paid, however, to the ways in which new jobs are stereotyped initially and to the ways in which everyday social practices reaffirm or challenge these gender attributions over time. Formal organizational structures and informal workplace practices are not gender-neutral but are saturated with gendered meanings and practices that construct gendered subjectivities at work. Indeed, Pratt and Hanson have recently argued for an exploration of gendered subjectivities, suggesting that "women's subjectivity is structured partially at home, within the family household, and partially at work, but the actual dynamics of these processes lie unexplored. For men, we assume a simpler formula: their subjectivity is shaped through their work experience" (Pratt and Hanson 1991, 245). While we disagree with their assumption about men's subjectivity, we concur with them in their suggestion that geographers explore the construction of gendered subjectivities at work.

Transforming gender relations in the new service economy?

Clearly, then, a question needs to be asked about how jobs become gendered and what relations of power and domination maintain or transform these gender associations. How, in a period of rapid economic transformation like the 1980s, when a range of new service sector jobs became dominant, was gender encodement established? Among those "top-end" service sector jobs in financial services, in the wide range of managerial and white-collar service sector jobs that have become an increasingly important part of the labor market in recent years, the association with gender is much less apparent than in the "old" jobs or in many of the expanding "bottom-end" servicing jobs (Filby 1992). These high-status occupations would seem not to embody gendered attributes, not to require those "natural" gendered characteristics, be they strength or docility, but to be gender-neutral and so, theoretically, open to suitably qualified applicants of either sex.

Connell (1987) has suggested that in periods of rapid change, what he calls the gender regime or gender order may be disrupted. He maintains that the history of gender relations is an uneven, or lumpy, one, in which hegemonic notions of masculinity and femininity may be suddenly transformed. Further, Connell points to the workplace as one of the crucial milieux in which gendered subjectivities are constructed. As he argues, contrasting and competing femininities within and between different milieux are common. In a period of rapid economic change it may be that gendered identities in the home and the workplace are being transformed and disrupted.

McDowell (1991) has suggested that part of the "flexible" restructuring strategies of contemporary economic change has been the creation of specifically "female" or feminized jobs. Women's growing labor force participation has not been solely a *consequence* of the expansion of jobs particularly suited to them (in the service sector or the caring professions or in part-time employment) but also part of the deliberate creation of particular job categories and working practices to draw women in as cheap labor. It therefore seems that the gendering, and possibly the regendering, of occupations may be an important strategy in economic restructuring and labor market change, although how widespread and explicit a management strategy this is remains to be ascertained. Although a number of geographers have begun to investigate aspects of the relationship between the feminization of the labor market and economic restructuring (Christopherson 1989; Villeneuve and Rose 1988), few geographers have yet directly addressed questions about power and domination and the ways in which gendered social practices in everyday social relations are part of the production and renegotiation of sex segregation. A number of scholars in other disciplines have proved a helpful stimulus to our own theorizing about the sort of ways in which occupational gendering is accomplished and regendering occurs. Therefore, before turning to the details of our own empirical work, we briefly outline some of this literature.

Gendered organizations: sexing and resexing jobs

An important stimulus for our own work on the construction and maintenance of gendered occupations in the financial services came from within organization theory, especially from recent studies that have drawn attention to the ubiquity of sexuality in organizational processes and the ways in which it is related to the structures of power (Acker 1990; Hearn and Parkin 1987; Pringle 1989). This work develops a very different notion of organizations and firms from that which is conventional within economic geography. Sexuality in these studies is defined as a socially constructed set of processes that includes patterns of desire, fantasy, pleasure, and self-image. Hence it is not restricted solely, nor indeed mainly, to sexual relations and the associated policy implications around the issue of sexual harassment. Rather the focus is on power and domination and the way in which assumptions about gender-appropriate behavior and sexuality as broadly defined influence management practices and the everyday social relations between workers.

Dominant notions of sexuality in contemporary Western culture are based on a set of gendered power relations in which men's dominance over women is expressed and re-created. Images of hostility and domination, including fantasies of humiliation and revenge, have been shown to be a central part of masculine sexual identity. (A cursory glance at the images of women used in advertising or in many popular films should bring home the validity of this assertion.) As Stoller (1979, 94) has argued, central to the construction of sexuality in discourse, symbolism, and in social practice is an image of the phallus as "aggressive, unfettered, unsympathetic, humiliating." Recent work on organizational structures in a range of industries has begun to demonstrate the centrality of this image in everyday social relations and in interactions in many workplaces. It is a particularly significant image in the world of merchant banking, as we demonstrate below.

The growing recognition of the ways in which male sexuality structures organizational practices counters commonly held views that sexuality at work is a defining characteristic of *women* workers. According to Acker (1990, 139), organizations'

> gendered nature is partly masked through obscuring the embodied nature of work. Abstract jobs and hierarchies . . . assume a disembodied and universal worker. This worker is actually a man: men's bodies, sexuality and relationships to procreation

> and waged work are subsumed in the image of the worker. Images of men's bodies and masculinity pervade organisational processes, marginalising women and contributing to the maintenance of gender segregation in organizations.

Pringle (1989), in her study of the relationships between secretaries and bosses, made a similar observation, noting that the association between masculinity and rationality allowed male sexuality to remain invisible yet dominant, positioning women as the inferior "other" at work.

Epidermalizing the world

In her stimulating book, *Justice and the Politics of Difference*, Young (1990) has also singled out aspects of embodiment as a key element in the structure of oppression in contemporary capitalist societies. Like Acker, she too focuses on deviations from the contemporary hegemonic version of an idealized body, which is not only male but also slim and light-skinned. This idealization establishes as "the other" not only women but also people of color and those who are not the perfect shape or size. Young also emphasizes the significance of lived social experience in the maintenance of oppression through what she calls cultural imperialism and violence.

The net result is what Young refers to as "epidermalizing the world," or the "scaling of bodies." As Young explains, this epidermalization is a particularly significant part of the explanation of women's inferior position:

> Women's oppression [is] clearly structured by the interactive dynamics of desire, the pulses of attraction and aversion, and people's experiences of bodies and embodiment. While a certain cultural space is reserved for revering feminine beauty and desirability, in part that very cameo ideal renders most women drab, ugly, loathsome, or fearful bodies.
>
> (Young 1990, 123)

Gender performance and regulatory fictions

Both Young and Butler, in her stimulating analysis of the acquisition of gendered identities, *Gender Trouble* (1990), draw on Foucauldian notions to show how dominant discourses construct some groups as ugly or degenerate bodies in contrast to the purity or respectability of neutral or rational subjects. Butler extends these ideas to suggest that gendered identities are multiple and fluid, indeed that gender itself is an imitative structure and is contingent. In her view gender is a parody, a disciplinary production, a fabrication, or a performance, and so is open to resignification and recontextualization. This idea of a gender performance is particularly helpful in understanding the construction and maintenance of gender identities and gendered power relations in the workplace.

A number of examples should suffice to demonstrate the general relevance of this argument. As Acker suggested, women may have to become honorary men to be successful. Others, including Pringle (1989) in her study of secretaries, have pointed out the limited number of roles or performances available to women working in subordinate occupations. Pringle (1989, 3) distinguished "wives, mothers, spinster aunts, mistresses and femme fatales" in her work. Davidson and Cooper (1992), in their study of women in management positions, *Shattering the Glass Ceiling*, argued that the available scripts, what Butler termed "regulatory fictions," are restricted to the earth mother, the pet, the seductress, and the honorary man. We might add the bitch or the ball breaker, a male designation of professional women rather than a role that women themselves actively embrace.

Doing gender on the job: male performance

This idea of gender as performance is not, however, restricted to women. Interactive service work demands a performance of men, too. Leidner (1991), in her fine study of the fast food industry and door-to-door insurance selling, demonstrated the ways in which employees in these industries reconciled the work they did with a gendered identity they could accept. She showed how low-status, often demeaning work was imbued with socially defined masculine attributes in ways that excluded women, reinforced male power, and reconstructed the work as heroic, even glorious, for the men who did it. The insurance selling example makes her point most clearly.

At first sight it might seem that selling insurance on the doorstep – cold calling, as it is known in the trade – is a job that might draw on quintessentially feminine attributes. It involves the presentation of self as pleasant, sympathetic, and non-threatening, a congenial attitude, eagerness to please, and an ability to talk people into doing something they might not want to do. As Leidner argues, "deferential behavior and forced amiability are often associated with servility. . . . Such behavior is not easy to reconcile with the autonomy and assertiveness that are considered central to 'acting like a man'" (1991, 165). Leidner found, however, that the insurance sales force that she interviewed was almost entirely male and that these men were able to interpret their jobs as congruent with proper gender encodement. They emphasized the "manly" parts of the job and stressed control of their clients rather than deference to them. "They assigned a heroic character to the job, framing interactions with customers as contests of will. To succeed, they emphasized, required determination, aggressiveness, persistence and stoicism" (Leidner 1991, 166). The men interviewed regarded women as lacking "the killer instinct" needed to succeed, to control, win, or conquer a customer – language that, as Leidner points out, resembles the sexual seduction of an initially unwilling partner. This language of power and seduction also saturates very different types of work in high-status financial services occupations in the City of London, as we shall demonstrate.

Management and corporate strategy play a central role in the construction of gendered identities in the workplace. Organizations invent and reproduce cultural images of gender, and, through recruitment strategies, job classifications, and the regulation of workplace behavior, they create occupations that embody gendered images of the people who should occupy them. These images are not always stable: they are sometimes challenged by workers who are excluded and, on occasion, reshaped by management as part of the restructuring process or as a way of expanding the pool of potential recruits to an occupation. On a daily basis, relations of power and domination and the ways in which these are associated with men, male bodies, and masculine performances play a central role in the subordination of women. These work practices are imbued with gendered meanings, which play a crucial part in the construction of a gendered subjectivity in the workplace. It was these processes that we set out to investigate in a study of merchant banking in the City of London in the early 1990s.

It might be argued that financial services jobs in the City – in the world of high finance and the trading floor in particular – represent all that is socially defined as masculine. The cultural capital and associated social characteristics traditionally valued by City bankers – private sector schooling, a good university, the right accent and class background – still dominate recruitment and selection practices for professional positions, especially in corporate finance, despite some relaxation of these criteria in recent decades. Masculinity seems an almost inevitable corollary of these attributes.

An interesting difference emerged from the analysis of employees' educational and class backgrounds between men in the corporate finance divisions and men employed in dealing and trading occupations. The cultural capital possessed by these men varied. Dealers were less likely to have degrees, having entered the banking profession usually at the age of 18. As we delved deeper into their everyday workplace practices and behaviors, we found that these educational differences were mirrored by two contrasting versions of masculinity. Both male and female respondents remarked on the differences between them. As one of our respondents, a senior woman in corporate finance, suggested, in a remark that is representative of a range of comments that we recorded: "There's a cultural divide between them, between traders and what I might call small 'e' executives."

Dual masculinities in merchant banking

In the upper echelons of corporate finance the valued masculinity more closely reflected the staid and sober paternalism of managers in the retail banking sector, whereas in the heady world of sales and trading the male culture that is established valorizes a more macho masculinity of "guts," "iron balls," and the "killer instinct" necessary to overcome clients' resistance, to make sales, and to conclude deals. It is this latter version of masculinity that has become more dominant in media representations of City life in the 1980s. It also appears to have become a more dominant way of behaving in City institutions during the 1980s, particularly since deregulation, as expanded opportunities enabled men from a wider range of class backgrounds to enter the City. In popular parlance, a category of men known in London as "barrow boys" gained well-paid employment. This trend, in the words of a representative of a more traditional masculinity, stems from a time when "American practices" began to dominate social exchange in the City (City Lives Project 1990). "A gentleman's word" and trust between men who knew each other socially, often from their schooldays, was replaced by more formal rules of exchange, as well as by opportunities to break them. The 1980s, in Britain and in the United States, certainly seem to have been distinguished by a number of scandals, by insider trading, malpractice, and other problems.

Macho Masculinity. An indication of "American" practices and the type of attitudes and behavior that define what we have termed "macho masculinity" in US-owned institutions may be gained from Lewis's book *Liar's Poker* (1989). This is a racy account of his own experience as a salesman for Salomon Brothers in New York and London during the mid-1980s. The extent to which a particular construction of macho masculinity based on notions of power, seduction, and domination structured everyday social practices is clearly revealed in his text. This masculinity clearly operated to exclude women from key positions by emphasizing bodily "otherness." Lewis documented the construction of a male culture and camaraderie in Salomon's that was something between the worlds of a boys' private school and a street gang. The imagery of the trading floor was that of "a jungle of chest-pounding males": "a trader is a savage and a great trader a great savage" (p. 20).

In London, as our interviews revealed, the world of trading and selling was similarly constructed. As many of our respondents, both men and women, noted, the atmosphere in the dealing rooms and on the trading floors is masculine. One respondent suggested that "foreign exchange has always been the lads' floor . . . they kick a paper football round the office and things like that." Another reported that "they throw the secretary's soft toys around and make sexist remarks when they mess about," and a third remarked on her own exclusion because "it's all a good laugh with the lads and then go out and be sleazy together."

Sexual metaphor was commonly used to describe the "thrill of conquest" in concluding a deal, or "consummating a deal," as one of our interviewees put it. Everyday social relations habitually included ribald language, sexualized "horseplay," and a degree of sexual harassment of women. Thus, it was reported to us that "in one room, they had a blow-up woman they used to throw around and things like that." And, as another of our respondents wryly concluded, in this part of banking "I think the worse men behave, the faster they get promoted; obviously the same does not apply to women." Another commented that "the bolshiest people make the most money."

The atmosphere in the dealing rooms makes it hard for women, perhaps unused to such vigorous or unruly social relations at work, to survive. As one respondent suggested, "I think when you start you are at a bit of a disadvantage because you are not used to shouting down phones at men and giving people orders and being really tough."

A small number of women whom we interviewed had experienced unwelcome sexual advances from both colleagues and clients. Most women, however, regarded the sexual innuendo and repartee that was commonplace in the banks, especially among traders, analysts, and the sales force, as generally good natured and not directed either at them or at any individual in particular. As one respondent, an analyst in a merchant bank, reported, "there is harassment of women as a group, but not of women individually." She was able to ignore the foul language, blue jokes, and posters on the trading floor (and to "take it like a man"?). Indeed, several of the small number of women working in this area whom we interviewed emphasized that their male colleagues saw them as "one of the boys." They explained to us that one of the reasons why the excessively masculinist behavior did not get too far out of hand was because of the underlying relations of interdependence that bind analysts, traders, and salesmen (usually men) together as a team. Another respondent, who had found the harassment more intimidating, especially when directed at her (with comments about her underwear, for example), also commented on how the ties of interdependence that bound people together enabled her to dismiss unwelcome attention. This woman, a dealer, suggested that "brokers will try and sort of score with you," but "if it is a broker, you can just tell them to piss off because basically they need you to get the money."

One striking fact about the ribaldry and sexual innuendo was its overwhelmingly heterosexual nature. As Lewis argued about Salomon's, "the firm tolerated sexual harassment but not sexual deviance" (1989, 57). Gay men, along with women, whether straight or lesbian, were marginalized. In our interviews we also found an overwhelming emphasis on heterosexual discourse and practices. As a female dealer commented about her co-workers, "most of them are completely homophobic." Although she was aware of a single gay man in her department, he kept his sexual preferences hidden from his male co-workers.

Traditional Masculinity. In the corporate finance divisions, an alternative male gender performance was dominant. Among the bank directors, treasury specialists, managers, and directors in capital development divisions, a different set of social attributes and acceptable behaviors was valued. Here, we found the more traditional image of bankers preeminent. The right school and university, the correct accent and tailor were important, and the "old boy network" operated in recruitment and mentoring practices.

A significant number of our male respondents (over half of them) who conformed to a variant of what we might term "formal" or "traditional" masculinity, quite self-consciously referred to their work as a performance. A commonly used phrase, for example, was "we are selling ourselves as well as a product," and many of them discussed, unprompted, the sort of image that they tried to create.

Multiple femininities

What about women? Are they able to "fit in" to the world of corporate finance by becoming Acker's honorary men? Clearly, in most parts of a merchant bank, apart from the dealing and trading areas, being "one of the boys" would be the wrong script. For the majority of female employees, in fact, this question was not appropriate; they worked in female-dominated occupations and in areas of the bank where their investment in a particular version of femininity was taken for granted, both by the women themselves and by their male colleagues, or, more usually, male superiors. Thus, for example, for the secretaries we interviewed, the adoption of a masculinized role or performance was not an option.

Foucault's analysis of the way in which subjectivity is constituted through different disciplinary mechanisms, by techniques of surveillance, and by power-knowledge strategies so that we discipline ourselves is a helpful way of understanding women's investment in their sense of self as feminine, even when their participation in that set of power relations rationally seems to be disadvantageous. For secretaries, all of whom were women, being female was part of the job. A wide range of social practices reinforced the association of femininity with inferior status. The very layout of space in merchant banks, the size of rooms, the way in which they are partitioned, the organization of entrances and exits, the informal separation of workers by status and gender that we so clearly observed in the dining room and in open public spaces outside the banks, creates and reinforces gendered subjectivities and their relationship to power and status (Colomina 1992; Spain 1992). For the secretarial staff, the fact that they were female was not an issue in everyday social interaction in the workplace, as their own gender, social behavior, and the gender of the job coincided. There are, as Pringle (1989) has shown, however, a number of alternative femininities available to secretarial staff that structure the boss-secretary relationship in particular ways. The "mother," "mistress or office wife," or "dolly bird" models are available, and the selection of one often depends on the age of the secretary and her physical appearance. The particular stereotype also influences the range of tasks that a secretary is routinely expected to perform.

The honorary male – or not?

The relationship between a female body and lack of power is, of course, part of the reason why women have been forced to act as if they were men to achieve success. For women in senior positions in merchant banks, their gender and appearance is at odds with the masculinist nature of the occupation that they fill and the tasks they perform. Similarly, the masculine ethos that dominates everyday interactions in merchant banking, perhaps especially on the trading floor and in the dealing rooms, constructs women as "the other." Thus, a common option that many women take at work in order to minimize their difference is to become "an honorary man" (Acker 1990), adopting masculinist norms and behavior and, often, a variant of the male business dress. Many of our respondents explained their strategy in similar terms: "I always wear a suit. I like to look as male as I can or at least neutral."

The analysis of our interviews with women working in the most senior positions, however, revealed that the honorary man was not the only performance on offer. For women, like men, more than one gender role was appropriate. Women bankers were also able to choose between different regulatory fictions at work. Indeed, it was suggested to us by most of the senior women we interviewed that playing male was in fact an undesirable option that could not be sustained. As one woman who had "made it" to director level remarked about the masculine strategy she had originally adopted, "Over the years I've come to the conclusion why should I try to be a little man? I'm not a little man. It's not going to work. I'll never be a man as well as a man is." As feminist theorists Threadgold and Cranny-Francis have pointed out, "masculine and feminine behaviours have different personal and social significances when acted out by male and female subjects. What is valorized in patriarchy is not masculinity but male masculinity" (1990, 31).

There are strategies of resistance, however, to the dominance of masculinity and its association with power. It became evident that there are ways in which femininity might be constructed as an advantage in the workplace. A number of our respondents who had achieved seniority argued that they were able to use and play on their femininity to achieve visibility. And, further, that the availability of multiple images of femininity might be used to advantage by women bankers.

Masquerade? The consequences of these games, of a masquerade at work, of the adoption of a masculinized version of femininity by some women and a deliberately feminine version by others as a workplace strategy, are important for women's sense of self. The French feminist theorist Irigaray (1985) has written about womanliness or femininity in similar terms to those Butler uses. Thus she has defined femininity as a mask or a performance, describing the mimicry that she sees as constitutive of feminine subjectivity as a masquerade of femininity in which a woman loses herself by playing on her femininity.

In a provocative paper, originally published in 1929 but recently influential again, Joan Riviere (1986) documents the implications for women who take on a masculine identity in order to compete in the professions. She suggests that women who take on a masculine identity may also put on "a mask of womanliness" or masquerade in a feminine guise in order to "avert anxiety and the retribution feared from men" (p. 35). Thus, in Riviere's case notes about a woman university lecturer, it is suggested that she copes with her male colleagues by being flippant or frivolous, that "she has to treat the situation of displaying her masculinity to men as a 'game', as something not real, as a 'joke' " (p. 39).

Although we have not yet explored these ideas in the context of merchant banking, they seem a potential way forward in exploring why so many women do not achieve the really powerful positions. There are also fruitful lines of comparison between women who parody masculinity and those who parody femininity. Something more is at work than seeing women as straightforwardly oppressed by the structures of power and practices of representation. Rather it is important to begin to explore "the problem of the investments that subjects have in complying with the practices of representation" (Threadgold and Cranny-Francis 1990, 7). As Coward (1984) has asked, "What is the lure of these discourses which causes us to take up and inhabit the female position?" It also seems clear from our results that there is no single female position, but rather a range of performances open to subjects who are women, and that these performances have different implications for success in the world of waged labor. Important questions about embodiment and performance remain to be raised. Thus we argue that the processes of being and becoming a woman (or a man) in the world of waged work should not be ignored by geographers, especially in the context of the shift toward service sector employment and a feminized labor force. As Threadgold and Cranny-Francis have pointed out, "the feminist story that rewrites the liberal humanist and capitalist narrative of individualism sees subjectivities, too, as a function of their discursive and bodily histories in a signifying network of meaning and representation" (1990, 3).

Conclusions

We have attempted to indicate some of the directions that work on the sexual division of labor might take in the future. As service sector occupations expand in advanced industrial societies, questions about the construction of subjectivity at work will become increasingly important. The lines between the worker and the product are blurred in jobs in which a personal relationship is part of the product. A fruitful line of further research might therefore explore the relationships between economic restructuring, changes in competitive conditions, and the social construction of occupations, including their gendering. At a large geographic scale, in the context of the general shift toward a low-wage, feminized economy in both Britain and the United States, the social construction of occupations clearly has immense consequences.

At a smaller spatial scale, it is apparent that masculine and feminine subjectivities are both constructed and changed within the workplace. Particular gender identities are embedded in everyday social practices and in what we might term occupational cultures. This recognition enables a more complex notion of the significance of sexual identity and gender relations in the workplace to be developed and so moves away from static notions of men and women seeking slots in an already gender-differentiated labor market. It also opens up space to investigate the ways in which men and women are able to resist the apparent gender appropriateness of particular occupations and realign them in ways that more closely match their own sense of self. Thus the apparently "female" attributes of certain occupations are able to be reinterpreted by male workers to embody socially designated attributes of masculinity, and vice versa. Here, perhaps, geographers might reach outside their disciplinary boundaries and ask broader questions about economic restructuring, which involves not only struggles over the gendering of new jobs but also the reallocation of existing occupations in a period of intense economic change.

References

Acker, J. 1990. Hierarchies, jobs, bodies: A theory of gendered organisation. *Gender and Society* 4: 139–58.

Beechey, V., and Perkins, T. 1987. *A matter of hours: Women, part-time work and the labour market.* Oxford: Polity Press.

Bradley, H. 1989. *Men's work: Women's work.* Oxford: Polity Press.

Budd, L., and Whimster, S., eds. 1992. *Global finance and urban living: A study of metropolitan change.* London: Routledge.

Butler, J. 1990. *Gender trouble: Feminism and the subversion of identity.* London: Routledge.

Christopherson, S. 1989. Flexibility in the US service economy and the emerging spatial division of labour. *Transactions of the Institute of British Geographers* 14: 131–43.

City Lives Project. 1989–92. Unpublished interviews with key decision makers in the City of London, National Sound Archives, London SW7.

Cockburn, C. 1983. *Brothers: Male dominance and technological change.* London: Pluto Press.

Colomina, B. 1992. The split wall: Domestic voyeurism. In *Sexuality and space*, ed. B. Colomina, 73–130. Princeton, NJ: Princeton Architectural Press.

Connell, R. W. 1987. *Gender and power.* Cambridge: Polity Press.

Court, G., and McDowell, L. 1993. Serious trouble?: Financial services and structural change. Working Paper No. 3, The South East Research Programme, The Open University and Cambridge University. (Available from Faculty of Social Sciences, The Open University, Milton Keynes, MK7 6AA.)

Coward, R. 1984. *Female desire, Women's sexuality today.* London: Granada Publishing.

Crang, P. 1992. A new service society? On the geographies of service employment. Ph.D. thesis, University of Cambridge.

Crompton, R., and Jones, G. 1984. *White collar proletariat.* London: Macmillan.

Crompton, R., and Sanderson, K. 1990. *Gendered jobs and social change.* London: Unwin Hyman.

Davidson, M., and Cooper, C. 1992. *Shattering the glass ceiling: The woman manager.* London: Paul Chapman.

Filby, M. P. 1992. "The figures, the personality and the bums": Service work and sexuality. *Work, Employment and Society* 6: 23–42.

Gabriel, Y. 1988. *Working lives in catering.* London: Routledge.

Hearn, J., and Parkin, P. W. 1987. *Sex at work.* Brighton: Wheatsheaf.

Hochschild, A. 1983. *The managed heart: Commercialization of human feeling.* Berkeley: University of California Press.

Irigaray, L. 1985. *The sex which is not one.* Trans. Catherine Porter with Caroline Burke. Ithaca, N.Y. Cornell University Press.

Jenson, J. 1989. The talents of women and the skills of men: Flexible specialisation and women. In *The tranformation of work*, ed. S. Wood, 141–55. London: Unwin Hyman.

Knights, D., and Morgan, G. 1991. Selling oneself: Subjectivity and the labour process in the sale of life insurance. In *White collar work: The non-manual labour process*, ed. C. Smith, D. Knights, and H. Willmott. London: Macmillan.

Leidner, R. 1991. Selling hamburgers, selling insurance: Gender, work and identity. *Gender and Society* 5: 154–77.

Lewis, M. 1989. *Liar's poker: Two cities, true greed.* London: Hodder and Stoughton.

McDowell, L. 1991. Life without father and Ford: The new gender order of postfordism. *Transactions of the Institute of British Geographers* 16: 400–419.

Milkman, R. 1987. *Gender at work: The dynamics of job segregation by sex during World War II*. Chicago: University of Chicago Press.

Morgan, G., and Knights, D. 1991. Gendering jobs: Corporate strategy, managerial control and the dynamics of job segregation. *Work, Employment and Society* 5: 181–200.

Peck, J. 1993. End of great divide as south-east is dragged into the jobless mire. *The Guardian*, 1 April, 11.

Phillips, A., and Taylor, B. 1980. Sex and skill: Notes towards a feminist economics. *Feminist Review* 6: 79–88.

Pratt, G., and Hanson, S. 1991. Time, space and the occupational segregation of women: A critique of human capital theory. *Geoforum* 22: 149–58.

Pringle, R. 1989. *Secretaries talk*. London: Verso.

Pryke, M. 1991. An international city going "global": Spatial change in the City of London. *Environment and Planning D: Society and Space* 9: 197–220.

Riviere, J. 1986. Womanliness as a masquerade. In *Formations of Fantasy*, ed. V. Burgin, J, Donald, and C. Kaplan, 35–45. London: Methuen.

Sayer, A., and Walker, D. 1992. *The new social economy: Reworking the division of labour*. Oxford: Basil Blackwell.

Scott, J. 1988. Deconstructing equality versus difference: Or, the uses of post-structuralist theory for feminism. *Feminist Studies* 14: 33–50.

Smith, C.; Knights, D.; and Wilmott, H., eds. 1991. *White collar work: The non-manual labour process*. London: Macmillan.

Spain, D. 1992. *Gendered spaces*. London: University of North Carolina Press.

Stoller, R. 1979. *Sexual excitement: The dynamics of erotic life*. New York: Pantheon.

Threadgold, T., and Cranny-Francis, A. 1990. *Feminine, masculine and representation*. London: Allen and Unwin.

Thrift, N., and Leyshon, A. 1992. In the wake of money: The City of London and the accumulation of value. In *Global finance and urban living*. ed. L. Budd and S. Whimster, 282–311. London: Routledge.

Villeneuve, P., and Rose, D. 1988. Gender and the separation of employment from home in metropolitan Montreal, 1971–1981. *Urban Geography* 9: 155–79.

Walby, S. 1986. *Patriarchy at work*. Cambridge: Polity.

Young, I. M. 1990. *Justice and the politics of difference*. Princeton, NJ: Princeton University Press.

Zukin, S. 1991. *Landscapes of power: From Detroit to Disney Land*. Berkeley and Los Angeles: University of California Press.

28 S. Baxter and G. Raw

'Fast Food, Fettered Work: Chinese Women in the Ethnic Catering Industry'

Excerpts from: S. Westwood and P. Bhachu (eds) *Enterprising Women: Ethnicity and Gender Relations*, pp. 58–75. London: Routledge (1988)

For a quarter of a century, Chinese restaurants, chip shops, and 'takeaways' have been a common feature of towns and cities throughout the country. Yet although they are the third largest ethnic community in Britain, the Chinese have received relatively little attention from politicians and researchers. Partly a result of their demographic dispersal, the low profile of Britain's Chinese has been due more to their isolation from the mainstream economy. For not only is the Chinese community concentrated in the fast food catering industry,[1] in which workplaces are atomized and workers' organization difficult, but also ownership as well as the staffing of firms tends to lie in Chinese hands. This was confirmed by a Home Affairs Committee report in 1985 which estimated that about 90 per cent of Britain's Chinese were employed in the catering industry and that of these, perhaps 60 per cent were employed in small, family shops (Home Affairs Committee 1985: xi). However, so little is known about the lives of those who work in this insular trade that there has developed a popular misconception that Chinese people are culturally inclined towards economic and social detachment (see, for example, Jones, 1979; Watson 1974). Less still is known about Chinese women in the ethnic economy, yet it is their experiences in particular that can provide the most penetrating insight into the Chinese catering niche.

Because so little is known about the structural factors that to a large extent dictate the lives of Chinese people in the fast food industry, it is necessary to establish a wider explanatory framework within which to understand their specific experiences. In relation to Chinese women, this means first apprehending the way in which gender roles have been and continue to be transformed as a consequence of the erratic development of capitalism in southeast China. Imposed with the expansion of British and other imperial powers, rapid capitalist industrialization entailed the destruction of the local agricultural economy and prompted massive demographic movements which stretched far beyond the borders of China to the developing colonies and to Britain. Second, the lives of Chinese women must be seen not only within the context of racist and sexist immigration laws affecting female workers and dependants but also with a critical appreciation of employment conditions within the fast food industry. From this perspective it is evident that the concentration of Chinese women in an ethnic sub-economy is a direct result of the postwar economic demand for cheap, colonial labour to provide inexpensive, ready-cooked meals. In this sense, Chinese women in the catering trade share a common historical oppression with other women from New Commonwealth countries, despite appearing often to enjoy a petit bourgeois class position. Such findings challenge the conclusions of a growing body of research on exclusively ethnic economic 'enclaves' which maintain that migrants employed in these sectors fare considerably better than those who are not, and that the underlying impetus for such differences is rooted in cultural practice (e.g. Waldinger 1984; Wilson and Portes 1980; Model 1986; Light 1972).

Fast food

During the Second World War both migrants and women were drawn into industrial catering on a massive scale, a response to the sudden need for publically provided, ready-cooked meals caused by the new demands of a society at war. As women were drafted into waged work, families separated through the evacuation of children, and men conscripted into the armed forces, new methods of satisfying the basic needs of housing, clothing, and eating had to be found. One measure taken towards meeting those needs which was introduced by the Government was the setting up of British Restaurants and Factory Canteens. Factory Canteens (instituted under the Factory Canteens Order of 1940 No. 1993) catered for employees in factories where the workforce exceeded 250, whereas British Restaurants provided subsidized and unrationed hot meals to the community as a whole. For it was observed that 'If women are to enter industry . . . they must be freed from the necessity of providing meals for husbands and children. The extension of canteens in schools and factories accomplishes this purpose' (Labour Research Department 1943: 4). Similarly, it was noted, 'Hotels, restaurants and pubs have had to fall back on refugee and Irish labour to a great extent' (Labour Research Department 1943: 24). Doubtless, Chinese in Britain during this period entered the burgeoning catering trade along with other European migrants. By 1942 over 108 million meals per week were being consumed outside the home, over half of these being provided through private rather than state catering outlets (Labour Research Department 1943: 10).

During the same period, the economy of Hong Kong (formerly China's best trade port) was also undergoing considerable change. Its transition from an *entrepôt* and trading foothold for Britain in southeast Asia into a lucrative industrial and financial centre accelerated after the Japanese retreat and the rise to power of Mao in 1949. The latter propelled the majority of remaining capitalist interests from China to the colony, where they met with a ready supply of refugee labour. (By 1972 one American source calculated that Hong Kong was providing as much as half all the backing for the pound Hong Kong Research Project 1974: 30).

Rapid industrialization in Hong Kong soon outstripped the food supplies produced by its rural hinterlands, the New Territories. Agriculture was transformed into cash cropping of specialized rice grains and vegetables, whilst the main bulk of the colony's rice was imported cheaply from Thailand. This led to the swift demise of previous forms of rice production as a viable source of income throughout the New Territories and the hastened disintegration of the local economy, as pressure for industrial, commercial, and housing land encroached into rural areas. This was at a time when Britain was beckoning cheap, colonial workers to staff new jobs in the postwar 'boom'.

Postwar economic restructuring and expansion in Britain gave rise to the rapid growth of public and private 'service' industries. Subsequently the need for ready-cooked meals was kept alive. Founded upon labour-intensive production techniques, the new service jobs necessitated the employment of a flexible 'reserve army of workers' (whose labour power could be utilized and shed according to the rate of labour-saving technological development) in order to secure the fastest rate of capital accumulation.[2] As cheapness and docility of labour was a prime consideration in this process, women and migrant workers – who lacked a strong organizational power base and thus were most vulnerable – were a preferred source for the development of the service sector.[3] The evidence for this is well documented. Women, who in 1951 constituted less than one-third of the labour force (Bruegel 1979: 16) now account for roughly half of all employed workers (1981 Census: Economic Activity, Table 1). Moreover, their patterns of employment reveal a substantial shift towards service work (see Foord 1984). Hence, the expansion of the fast food market was not merely a source of profit in itself; it also potentially released women from time-consuming home cookery (i.e. simply producing use values) in order to sell their labour power (i.e. to produce surplus values). In Hong Kong, the response to Britain's brief period of active encouragement of New Commonwealth immigration resulted in many Chinese men departing for the colonial metropolis (most of them from the New Territories). Following in the tradition of their forefathers who had arrived as seamen, they used established contacts – by now mostly concentrated in the restaurant industry – to find work.

Chinese fast food

By the time the majority of men began to leave Hong Kong, relying upon Chinese sponsors already established in Britain had become a statutorily codified condition of emigration, as those arriving during the 1960s were subject to the work voucher system imposed by the 1962 Commonwealth Immigrants Act. The legislation, designed to tailor immigration more specifically to the demands of the British economy, served to channel incoming Chinese migrants into the ethnic fast food industry. As highlighted by Rees, by the latter part of the 1960s and early 1970s, 'the majority of permits were issued to workers in the hotel and catering industry and in hospital employment' (Rees 1982: 85). Lacking industrial skills and speaking little or no English, most of those who came from the New Territories were therefore forced to rely on Chinese sponsors who would guarantee them specific jobs in Britain. Moreover, under the terms of the Act, men were allowed to bring their wives and dependants under the age of eighteen to join them, whereas women were not. Migration under these conditions was thus inevitably male-led, whilst other members of the family remained in Hong Kong and lived off the money sent home from abroad. In this way, it was not uncommon for entire villages to be dependant upon remittances sent from kinsmen in Britain (see Watson 1975): returns from farming and industrial work performed by women left behind in the New Territories during this period were so scarce that they could but supplement the frail and declining village economies.

> When my daughter was born in 1974, her father went to England after she was six months old. So it was me and my mother-in-law who brought her up. He used to send money to us for our living expenses – about £50 a month – and we could earn about £25 from farming. He had some land, you see, about one or two acres, and I worked on that. Everyone was the same then, not just my family.

Confined to a system of economic patronage based on kinship networks, and due to their linguistic and technical differences in the context of a racist society, the migrants were obliged to take restaurant jobs. However, the proliferation of Chinese restaurants during the 1960s and 1970s to a large extent masked the conditions upon which their success was founded. Total reliance on Chinese employment channels put workers completely at the mercy of their employers. Long hours of split shift work, flexible duties, and lack of statutory entitlements (such as sick pay, holiday pay, overtime pay, payslips, and National Insurance payments) all too frequently accompanied tied housing and arbitrary management. Nevertheless, despite being below average British rates, remuneration levels were very much higher than any of the workers could have expected to receive back home.

> 1968 I came to England because you couldn't get a job in Hong Kong to put food in your mouth. My uncle worked in a restaurant, you see – The White Lotus in Richmond, just outside London. He signed the form for me to come. In those days, someone I knew would be coming over every week. My uncle came first and I came later. He got me a job working in the restaurant. No wages – just food and somewhere to live – no money. It wasn't supposed to be a real job, you see. But once I found a job with pay, it was so much better than Hong Kong. I got £17 for the week. In Hong Kong I got £15 for the month!

The universality of such conditions of service mitigated their perpetuation, as did the 'sojourner' orientation[4] of workers who remained separated from their families.

Whatever rights the Chinese restaurant workers had been accorded under the 1962 Act were subsequently eroded with the Immigration Act of 1971. Despite the introduction of concessionary Rules to administer the 1971 Act, it officially removed the automatic right of entry to wives and children of men already established in Britain. This prompted a 'beat the ban' wave of dependants' immigration, such that between 1971 and 1973 (when the Act came into force) dependants accounted for upwards of 90 per cent of total immigration from Hong Kong (Baxter 1986: 14). Dependants, however, were obliged to demonstrate that they could be supported by their sponsors, which rendered them in a far more subordinate position than women who had emigrated independently from urban Hong Kong and Singapore and who, on the whole, filled jobs in other areas of the service sector.

Chinese women in Britain

When I came in 1968 I went to live with my husband above his uncle's takeaway shop for a year. We both worked there in the evenings and he gave us free food and never asked for any rent . . . Oh no, we didn't get paid. Then we rented a fish and chip shop for a couple of years but it didn't do too well, so we moved to Chinese Street. I had two kids by then. In those days, all the Chinese used to live there in the big houses. There were five families in ours – well, one was only a single man. But the rent was only £3 a week, which was all we could afford. Only three families used to cook there – the others ate at work. So things weren't too bad.

Housing was the immediate problem confronting women and children upon joining their male relatives in Britain. The dormitory work system suited families even less than it had done single men, and delapidated, overcrowded staff houses became standard accommodation for the Chinese community.

I came here in 1972 but I just couldn't get used to it. My nephew was here already with his wife but I brought their three children with me so we could all live together again. My nephew and his wife slept in one room and I slept with the children in another room. The youngest one was so young, she had to use a potty all the time. We had to cook and eat in the same room. It was really horrible. There were mice everywhere. There was only two good things about it: the rent was only about £6 or £7 a week for all of us and it was easy to make friends. Everyone was Chinese so we could all look after each other.

For some wives, reunification with their husbands in itself was a bewildering experience.

My husband has been here twenty years now. His father before him was here for many, many years. I never met him because he died over ten years ago. He had an English wife as well, you know, and had two kids but they lost contact. I know when he sent for his son, it was a lot of hard work. I hadn't seen him for about six years before he came to England because he had already gone to Hong Kong to work in a clothing factory. But the wages here were good, I think. I stayed on our farm in China and grew vegetables and my husband would send us quite a lot of money. The only time I went to Hong Kong was to catch the aeroplane to England. It was very different, so modern and so busy. When I came to London Airport, I didn't even recognize him! I waited and waited for about two hours and still nobody came for me. In the end I talked to someone in Chinese and they announced me on the loud speaker. Then it turned out he had been there all the time in the same place as me!

Many women (and children) were absorbed into Chinese restaurants as kitchen hands and cleaners, often for no wages at all since it was assumed by employers that their labours were spent in part payment for their accommodation.

My sister can't speak English so she just washes dishes in a restaurant. I was the same when I first came to this country. My husband got me the job in the same place as he was learning to be a cook but it's hard work and I didn't get much money.

I remember when I first came to this country we lived above a Chinese restaurant in Wales. I had to work there on Thursday and Friday nights and every weekend because my parents had to work so hard on those days, they couldn't keep up with the business. Sometimes the boss would give me 50p pocket money for the week but that's nothing, is it?

From the mid-1970s onwards, Chinese restaurants increasingly met with competition from fast food 'chains' such as Kentucky Fried Chicken (introduced to Britain during the 1960s) and McDonalds (first established on this side of the Atlantic as early as 1974) (Jones 1985: 56). The advantage of the fast food chains lay in their heavy investment in cost-cutting technology which obviated the need for labour-intensive production, a saving passed on to customers. Since, by this period, rapidly diminishing new work permits kept wages reasonably stable for Chinese restaurant workers, the relative profitability of restaurants began to decline as economic competition began to bite. Together with the availability of family workers upon which to draw and the total unsuitability of dormitory, tied accommodation to family life, the falling profitability of restaurants transformed the Chinese catering economy into one composed of smaller capital units operating on lower running costs but with a similar rate of profitability to restaurants. In short, the mid- to late 1970s witnessed the simultaneous decline of Chinese restaurants and the rapid spread of Chinese 'takeaway' and fish and chip shops, the majority run as family businesses.

Running a restaurant involves a lot of money. But to open up a takeaway only costs about a third of what it costs to open up a restaurant. The profit margin is about the same – in fact it's more without the heavy outgoings. All you need is one chef in the kitchen and one waiter at the counter.

Twenty years ago there wasn't any takeaways. It was all restaurants. Me and my husband used to work in one of the big ones in the city centre. He was just learning then, like me. It was no good, though. Then we got a job in this place here for two years. He worked in the kitchen and I worked at the counter. There was another cook as well but he got the sack soon after because business wasn't very good then – not enough customers. After two years, we bought the business. That was ten years ago. The old owner, he's only got one business now. That's where he works with his family. I work seven days a week – but only in the evening, not in the daytime. We don't open for lunch. Some of our friends do, though. We both have to work at night but we can have a rest during the day, although we've still got to prepare the food for 5 o'clock, when we open. Thursday, Friday, and Saturday we work quite hard in the day – cutting the meat, getting it ready, chips, vegetables – things like that. Some of the stuff we get delivered but other stuff we've got to get from the wholesale market. You have to get up early for that. It's very difficult sometimes because we can't go to bed until about 2 o'clock (a.m.) and the market closes at 11 (a.m.). Weekends, it must be 2 or 3 (a.m.) before we're even finished in the kitchen.

> You have to clean everything after you've closed at night, you know. There's a lot of cleaning.

Whilst it is true that the lives of both Chinese women and men are dictated by the opening hours of the takeaway shops, it is nevertheless women who generally bear the brunt of the social and economic marginality such a living imposes. Time and again the sentiment, 'It's his business, he makes the decisions' is reiterated and this is noticeably reflected in community life.

John is a waiter with a dependent wife and three children. A description of how he spends his spare time reveals that he and his wife have little in common socially.

> This is my second job. The first is in the casino or down the bookies. I go out with all the others in the restaurant all the time. In this business, there's nothing – nothing. Last weekend I played Mah Jeung for seventy-two hours – seventy-two hours and no sleep but I couldn't stop. We finished at 5 o'clock in the morning and I got in the car and drove straight back to work. Last night, quarter to four (a.m.) I'd lost £90. By four o'clock I won £230. That's about £300 in fifteen minutes. The other night, we all put money in the kitty – about £50. In four spins we'd made £500. But what happened? We lost it again. Easy come, easy go. If I was in Hong Kong, I would save all my money for my children's future and buy my son and his family a house but it's not the same here. When they grow up, they say, 'bye bye Daddy, I'm off'. So what's the point? I might as well enjoy myself – as long as they've got enough to eat.

John's social life contrasts sharply with that of many Chinese women, not least his wife:

> There's nothing for me here. That's the truth. Everybody's too busy in England to make real friends. If you want to be friends with the English, you have to make an appointment to see them. Back East you can go and see people any time you like and do anything you like. My husband doesn't feel the same as me. He likes it here. But he's got more friends and he can go out more than me. I've got to look after the kids.

Social isolation is a far more acute problem for Chinese women than for men. John's wife is not alone in her frustration. Whether as dependent wives of workers or themselves working in the family takeaway, the monotony and alienation of life in the Chinese fast food industry for all women is striking.

> My husband gives me pocket money whenever we can afford it. Most of the time I'm too busy or too tired to spend it, though. Sometimes I go to play Mah Jeung at my friend's house. A lot of Chinese like to do that – or go to the casino. That's where my husband goes. That's all there is to do at 1 o'clock in the morning when we finish work. Catering makes you very cut off from other people. The only time you have off is when most people are in bed and when most other people are off, that's when you're working hardest. Some of my friends just go to the casino to watch the Chinese films. They're not even interested in the gambling. It's a very boring life really, but what else can I do?

I used to cry a lot – not really for any particular reason. There was just no-one I could really talk to. I've been trying to learn English ever since I came but it's very difficult when you've got nobody to talk to.

A second generation of fast food workers

The demanding hours of Chinese restaurant and 'takeaway' work has meant that for many women, rearing young children has been a virtual impossibility. As a consequence, a new generation of British-born Chinese children has not been automatically brought up with their parents in their country of birth. Instead, many were sent back to Hong Kong to be looked after by relatives and friends, paid and unpaid. Thus, even before 1973, whilst Chinese people already established in Britain enjoyed the same *de jure* legal and political rights as the indigenous population, their concentration in the ethnic fast food industry meant that they were *de facto* relegated to a position similar to migrant workers admitted after 1973 under the much less secure work permit system with regard to their families.[5] The situation has arisen, therefore, whereby many Chinese children born in this country have nevertheless entered it as much strangers as their parents did before them.

I was actually born here, you know. But you wouldn't think so ten years ago because I didn't actually come here till I was eight. I lived with my auntie in Hong Kong until then, from when I was a baby. I really wanted to come here, though. I remember it very clearly. All I knew was my parents were over here and I'd always expected to come some time or other. So when the day came for me and my sister, I didn't think nothing of it. It was almost automatic. But my parents, they were like strangers. I thought I knew my mother, but I hadn't seen her for four years and she was really different to how I remembered.

My brother – he grew up here; he spoke English – he ganged up on me with his classmates when I came to England from Hong Kong. He never could identify with me.

By the time me and my sister arrived they'd settled down in their own shop. It was good business then. We helped sometimes, chopping potatoes for chips and things like that. We helped quite a lot really, especially at weekends. I didn't get any homework, so it didn't disturb my schooling. I didn't like school anyway. When I came, I only knew simple English like 'pen' and 'pencil', so I didn't learn a lot. I just sat round and watched what was going on. I was in a class where everyone was slow, you see, so how could I learn much? There were only three girls including me and the other two didn't like me, so that was that. My parents were too busy to teach me anything – my Dad speaks English you see – so I never learned to speak it, not having any friends or anything. All my friends now are from my village in Hong Kong. It's still good business. There's five of us working. We also get paid now because me and my husband, we do most of the cooking. Actually we get over £100 a week. That works out about £10 a day each and we get every Monday off. I'd like to go to Hong Kong for a holiday. It's just so boring here. All we do is watch a video and go to bed. In Hong Kong you could just step outside the front door and all your family and friends would be around to talk to. In the future I'd like us to get our own

> shop. My parents will probably give this one to my brothers, you see, even though me and my sister are older. No, I don't resent it. It's not really unfair because it's always been like that, right from my great grandfather. Everyone does it; not just my parents – so I know what to expect.

New horizons

For Chinese women the isolation of the family workplace mitigates against conscious, collective struggle. However, it would be incorrect to conclude that Chinese women are merely the passive recipients of the adversities imposed upon them through migration. Working in a family business where profitability depends as much upon the efforts of the wife as upon those of the husband, actually means that there is a material basis for Chinese women to assert a certain degree of control, albeit on an individualized level. Indeed, an increasing minority of women are becoming joint partners in their husbands' businesses and others are running businesses themselves. This is tempered however, by a growing awareness of the economic and social trap that running an ethnic takeaway shop or restaurant holds in store for future generations of Chinese women. Consequently, it is accompanied often by a desire to see children leave the fast food trade through educational attainment:

> I'll tell you one thing, I don't want this life for my daughter. I want better for her. I want her to speak English and get a good job.

This attitude finds resonance with a younger generation of Chinese women and girls:

> A few years ago my Dad kept asking us if we would help him run a takeaway and we all said 'no way'. I don't want to work like a slave. I know what it's like because I've got friends who work in their parents' takeaway shops and they have to work like dogs. My Mum saw it from our point of view. She said it would be better for us to carry on at school so we could choose what we wanted to do when we finished.

Conclusion

The problems presented by precarious employment, and long and unsociable working hours for relatively few material rewards compounded by a lack of organizational support, exist as much for Chinese people in the ethnic fast food industry as for other catering workers. However, the Chinese are at the sharp end of this postwar development in that they have been pushed by immigration laws and by a historical legacy of indigenous racism towards a form of self-exploitation which has forged new ground for the development of the fast food market. Increasingly, they are forced to pitch the cheapness of their labour against the unabated, 'high tech' competition of multinational firms. For Chinese women, the situation is doubly oppressive. Whereas a formally paid female kitchen hand might earn a weekly £82.70, many Chinese women working in family takeaway shops or restaurants receive no wage at all, except for that which they might personally negotiate from husbands, fathers, or sons on a weekly basis. This confounding of economic and domestic roles has rendered them even more vulnerable to marginalization and isolation. When these experiences are placed within a broad historical context, it becomes apparent how the contemporary predicament of many Chinese women falls within the mould cast over a century ago with the imperialist penetration of southeast China. However, whilst the self-exploitation and oppression within the family remain at a high level, Chinese women are no longer content to accept the narrow horizons of the 'take-away'.

Notes

1 The term 'fast food' denotes a technological and organizational trend within food catering whereby standardized meals are produced at a rapid rate for immediate consumption, either on or off the premises where they are purchased. Production characteristically comprises extensive use of highly developed, labour saving cooking equipment combined with simplified cooking techniques making for quick, 'component assembly' of meals from standardized ingredients. Whilst the major hamburger, fried chicken, and pizza outlets have taken this trend to its most advanced form by capital investment on a massive scale in labour-saving technology, 'fast food' is a term which also describes the trade of 'fish and chip shops', cafés, snack bars, 'takeaways', and many restaurants. The term is least applicable to haute cuisine restaurants, where labour-intensive meals are produced on an individual basis by specialists.

2 'Marx (1867) saw the expansion of a reserve army of labour as an inevitable outcome of the process of capital accumulation (*Capital* Vol. 1). As capital accumulated, it threw certain workers out of employment into a reserve army; conversely, in order to accumulate, capital needed a reserve army of labour. Without such a reserve, capital accumulation would cause wages to rise, and the process of accumulation would itself be threatened as surplus value was squeezed' (Bruegel 1979: 12).

3 For example, see Baudouin *et al.* (1978), 'Women and immigrants: marginal workers', in C. Crouch, and A. Pizzorno. *The Resurgence of Class Conflict*, Vol. 2. London: Macmillan.

4 This concept is discussed at length in relation to Chinese migrants in Siu 1952–3.

5 The removal of reproduction and all its attendant costs to the country of origin facilitates optimum use of migrant labour power in the accumulation of capital in the receiving society.

All excerpts are taken from interviews conducted with Chinese respondents in Birmingham during 1985/6 as part of fieldwork for a forthcoming PhD thesis entitled 'A Political Economy of the Ethnic Chinese Catering Industry', to be submitted by Susan Baxter to Aston University, Strategic Management and Policy Studies Division, 1988. The research was supported by the Economic and Social Research Council.

References

Baxter, S. (1986) *The Chinese and Vietnamese in Birmingham*, Race Relations and Equal Opportunities Unit, Birmingham City Council.

Bruegel, I. (1979) 'Women as a reserve army of labour: a note on recent British experience', *Feminist Review* **3**: 12–23.

Foord, J. (1984) 'New technology and gender relations', Discussion Paper No. 58, Centre for Urban Regional Studies (CURS): University of Newcastle upon Tyne.

Home Affairs Committee (1985) *Second Report and Proceedings on the Chinese Community in Britain*, Vol. 1., London: HMSO.

Hong Kong Research Project (1974) *Hong Kong: A Case to Answer*, London and Nottingham: Hong Kong Research Project & Spokesman Books.

Jones, D. (1979) 'The Chinese in Britain: origins and development of a community', *New Community* **7**: 397–402.

Jones, P. (1985) 'Fast food operations in Britain', *Service Industries Journal* **5**(1): 55–63.

Labour Research Department (1943) *Works Canteens and the Catering Trade*, London: Labour Research Dept.

Light, I. (1972) *Ethnic Enterprise in America*, Berkeley and Los Angeles: University of California Press.

Model, S. (1986) 'A comparative perspective on the ethnic enclave', *International Migration Review* **XIX**(1): 64–81.

Rees, T. (1982) 'Immigration policies in the United Kingdom', in C. Husband (ed.) *'Race' in Britain*. London: Hutchinson.

Waldinger, R. (1984) 'Immigrant enterprise and the structure of the labour market', in R. Finnegan and D. Gallie (eds) *New Approaches to Economic Life*. Manchester: Manchester University Press.

Watson, J. (1974) 'Restaurants and remittances: Chinese emigrant workers in London', in G. Foster and R. Kemper (eds) *Anthropologists in Cities*. Boston: Little, Brown & Co.

Watson, J. (1975) *Emigration and the Chinese Lineage*, Berkeley and Los Angeles: University of California Press.

Wilson, K. and Portes, A. (1980) 'Immigrant enclaves: an analysis of the labor market experiences of Cubans in Miami', *American Journal of Sociology* **86**: 295–319.

SECTION SEVEN

GENDER, NATION AND INTERNATIONAL RELATIONS

Editors' introduction

> As a woman I have a country; as a woman I cannot divest myself of that country merely by condemning its government or by saying three times 'As a woman my country is the whole world.' Tribal loyalties aside, and even if nation-states are now just pretexts used by multinational conglomerates to serve their interests, I need to understand how a place on the map is also a place in history within which as a woman, a Jew, a lesbian, a feminist I am created and trying to create.
>
> Adrienne Rich, Notes towards a politics of location (p. 212)

In 'Three Guineas' Virginia Woolf stated famously, 'as a woman I have no country. As a woman I want no country. As a woman my country is the whole world.' Adrienne Rich's critical reflection on Woolf's global sisterhood shows how problematic such universalist claims can be. Rich insists that womanhood is constructed specifically in different locations, as a result of many geographies – and historical geographies – playing out local and global relationships, of colonialism, trade, exploration, struggle. Rich's opportunities, experiences, expectations and actions are both constrained and made possible by her multiple positionings within different power 'containers,' perhaps most significantly, the nation-state within which she is a citizen.

Our argument throughout this collection has been that gendered identities, roles and powers are actively constituted at a number of different spatial scales and in relation to a number of different social, cultural, economic and political processes. In this section the interrelationships between gendered identities and those produced through processes of state formation, nationalisms and international relations are examined.

Clearly political identity is not just something that is experienced at the national level. As we have suggested in earlier sections of the reader, the division between political and non-political identities is merely a social construction. Feminists have been active in bringing to the foreground the silenced politics of the self, the body, the private sphere, of nature, work practices and so on. The silencing of these in traditional theories, which locate politics solely in the formal arena of the state, has helped to reinforce a political map that has left women out. Yet institutions of statehood and processes of nation building are of great significance in contemporary society – it is important not to lose track of this within moves to recognise every act as political. Feminists have looked to the workings of the apparently universal assumptions of nationalist rhetoric, state citizenship and international relations and have found gendered constructions lying behind. In a sense, these projects of rendering the private political, and challenging the definition of (nation-)state political agency are not dissimilar, as Simon Dalby (1994 p. 531) suggests:

> Just as some feminists challenge the ideology of the family in suggesting that private spaces are 'safe' because of the presence of a male protector, whereas public spaces are dangerous to women [. . .] it is a simple extension of these arguments to argue that states do not really protect all their (domestic) citizens while providing protection from the perils of the anarchy beyond the bounds of the state.

These feminist analyses challenge the dominant (neo)realist world-view that sees the world-order as composed of a series of independent states, whose primary role is to provide its citizens protection from the threat of political chaos emanating from the unregulated international sphere (see also Peterson, 1992; Enloe, 1989, 1993).

State sponsored nationalism reached its peak in Europe at the close of the nineteenth century at which point, George Mosse (1985) has argued, the virtues of manliness were projected to become virtues of nation. The values of national citizenship have thus been derived from gendered origins despite the apparent universalism of nationalist rhetoric. Symbols of national identification have become naturalised into landscapes of everyday life, through repeated performances as mundane as sports spectatorship, or as overtly nationalist as paying tribute to national heroes (including those who have literally become part of the landscape in statue form). Along with this has been the naturalisation of a particular national citizen. The influential theorist of nationalism Benedict Anderson has suggested that the Unknown Soldier is the architypal emblem for nations, as *he* could be any one of the national community. Yet with significant exceptions, this is not a role generally offered to women. Peterson (1992 p. 23) has suggested that the '"idea and ideal of sacrificial identity" goes back to the hoplite warfare of classical Greece; on tombstones, men were honored only if they died in war, women only if they died in childbirth.'

This helps to explain the resistance to women joining the military in many societies today: they are seen to be weakening the institution of the family in America (see Sparke, 1994) or threatening the moral of (heterosexual) men 'who had been recruited partly with the promise that joining the military would confirm their manliness' (Enloe 1993 p. 214).

However, this is not to argue that the project of nationalism is nothing more than a part of a project of patriarchal domination. But simply put, national agency is not constructed in blindness to gender (nor for that matter, race or sexuality): the reproduction of national norms has tended to reinforce certain gender relations. Indeed, in times of threat to the nation – the all-too-familiar cry for protection of the 'national interest' – women's issues are often marginalised. Women are accused of being divisive in a time of need, privileging one internal division over the community as a whole:

> Women who have called for a more genuine equality between the sexes – in the movement, in the home – have been told that now is not the time, the nation is too fragile, the enemy is too near. Women must be patient, they must wait until the national goal is achieved; *then* relations between women and men can be addressed.
>
> (Enloe, 1989 p. 62)

The general situation described by Enloe is no exaggeration. Partha Chatterjee (1993) discusses the manner in which the 'women's question' was removed from the public sphere of political activism in India during the struggle for independence, with the promise that after this was achieved it would once again become a political issue. Sally Marston (1990) similarly shows how even the radicalism of the American Revolution failed to change the status of women, as it depended heavily upon the long established division between the public and private spheres, and the assumption that women were located in the private arena, not in the arena of political agency.

However the politics described by Enloe, Chatterjee and Marston replicate the domestic–international politics of exclusion that produce an Other – an enemy outside the nation-state borders – upon which all evils can be projected in order to maintain an image of a coherent population and a coherent national identity. This drive to produce a homogenous community is at the expense of internal difference, and assumes that leaving women's issues until 'later' will not make it more difficult for women to achieve their aims. The nature of this future national society will be irrevocably altered if it does or does not engage with feminist politics at the onset.

The first two excerpts in this section discuss the nature of the linkages and dependencies of national and gendered identities. Nira **Yuval-Davis**'s paper 'Gender and nation' discusses the blindness of theories of nationalism to gender differences. Using examples from around the world, she demonstrates the ways in which nationalist discourse – the national interest – has been used to discipline women and their political demands. She also highlights the blindness of traditional theorists of nationalism to questions of gender, and the concomitant limitations on their understandings of national citizenship.

Anne **McClintock** similarly discusses the role of women in the rhetoric of nation. Drawing upon the work of Benedict Anderson, she explores the integration of women into the 'imagined community' of nation-state. She observes that in this symbolic community citizenship is differentiated by gender so that women 'are typically construed as the symbolic bearers of the nation but are denied any direct relation to national agency' (McClintock, 1993 p. 62). It is the metonymic bond of male citizens who must act to save or promote the female nation. Women passively represent the nation, symbolically as Marianne or Mother Russia, materially as they are protected from 'front line' exposure, and in their capacity as literal mothers of the national populace. McClintock delves into Fanon's work on nationalism to discuss the psychoanalytic dimensions of this political identity which draws it closely to a familiar structure.

Other feminists have sought to illustrate the differential access to citizenship available to men and women in specific cases. The recent changes to post-Soviet societies in Eastern Europe where state-socialism has been replaced by nationalist, market-based societies as a normative goal, has brought tensions between gender and nationalism to the surface. With changes to the public sphere, come changes to dominant conceptions of women's roles. Often the first state-sponsored services to be terminated with the fall of communism were childcare provision and maternity leave. Women have also been expected to return to the home as full employment becomes part of history, and in reaction to defunct communist values, gender equality is being challenged by calls for a return to 'traditional' family values and religious morality (see Sharp, 1996). Most powerful of all is the symbolism of the health of the nation in the figure of the woman's body. The issue of a woman's right to choose abortion – politically debated as an issue of 'national interest' – is described by a number of the contributors to Nanette Funk and Magda Mueller's collection 'Gender politics and post-communism' to illustrate women's role as symbolic bearers of the nation, but not of national agency.

Simona **Sharoni** turns to the Middle East to examine the relationship between gender and citizenship, in the third excerpt included in this section. Her paper illustrates the convergence of sexual, national and orientalist stereotypes in the placing of women in national political life, and also in feminist theorisations of them. Sharoni tackles the problem of how the struggles of women against these dominant representations can be articulated so as to give voice to those involved, how to 'theorise from women's struggles' in her terms. In other words, where feminist beliefs may lead some commentators to outrage at particular situations, they (we) must remain aware of the importance of certain gender roles to the maintenance of dignity or agency. It is important then, as Section Two of this reader suggested, to let in the voices of the women in question to understand their agendas, fears, and hopes.

Sharoni's paper also indicates the permeability of political boundaries. The nation-state is presented as a coherent and unified identity, and yet the prospects of women in the Middle East are influenced not only by the events in their own countries but also by the continued influence of colonial orientalist images. Other feminist writers have also pointed to the fallacy of unitary national identity.

It is not only the agency offered by nationalism that has come under feminist scrutiny. Others have sought to challenge the very concept of a territorial state, and identity created within boundaries that sharply define one identity from something quite different. Gloria **Anzaldúa** writes of her experience of identity formed over the lines imposed by political identification. Her sense of belonging is not a centered community defined by cartographic divisions but of a community that is centered along this border, a community of the borderland. Her experience of Chicana identity is shaped by its existence over the international border between the USA and Mexico. Her identity is further destabilised by her positionality as a woman and as a lesbian. She relates her sense of division, hybridity and mixedness not only in the topic of her writing but also in the style of communication: a style which ranges from prose to poetry, autobiography, mythology, children's stories and history. In addition, she narrates in a mixture of languages – Spanish, Chicano, English – sometimes translated, sometimes not. As monolingual English speakers, sometimes we are let in to her community of meaning if the translation is provided – at other times we are left out, wondering what a passage might mean. Anzaldúa subverts traditional inclusive narratives of national mythologising by allowing us to feel that we both do and do not belong.

International relations are similarly constructed around gender differences, rather than being a gender neutral affair. The final piece, by Cynthia **Enloe**, challenges the national–international boundary more directly than in the preceding papers. The story of international politics has traditionally been one of the spectacular confrontation of mighty states led by powerful statesmen, of the speeches and heroic acts of the elite, and the specialist knowledge of 'intellectuals of statecraft'. Enloe refuses to accept this story as the extent of the workings of international relations, and instead focuses on those elements that the traditional story excludes and silences: the role of international labour migration, the availability of cheap female labour for transnational corporate investment, the availability of sex workers for the tourist industry in southeast Asia and so on. Enloe's is a very different account of international politics than the traditional story and certainly one that lacks its glamour. Instead the international is linked intimately to everyday events of gender relations, as well as to the grand narratives of international trade, diplomacy and war. Her account perhaps ultimately links the personal and the political – not simply a micropolitics of the body, or a local politics of struggle – but a politics of international linkages and exploitations.

References and further reading

Anderson, B 1991: *Imagined Communities*. Second edition. London: Verso.

Antić, M 1991: Democracy between tyranny and liberty: women in post-'Socialist' Slovenia. *Feminist Review* **39**, 149–54.

Bell, M 1995: A woman's place in 'a white man's country'. *Ecumene* **2**(2), 129–48.

Browen, W 1995: Irishness, gender and space. *Environment and Planning D: Society and Space* **13**, 35–50.

Chatterjee, P 1993: *The Nation and Its Fragments*. Princeton University Press: Princeton, NJ.

Dalby, S 1994: Gender and geopolitics: reading security discourse in the new world order. *Environment and Planning D: Society and Space* **12**(5), 525–46.

Einhorn, B 1991: Where have all the women gone? Women and the women's movement in East Central Europe. *Feminist Review* **39**, 16–36.

Enloe, C 1989: *Bananas, Beaches and Bases: Making Feminist Sense of International Relations*. University of California Press: Berkeley.

Enloe, C 1993: *The Morning After: Sexual Politics at the End of the Cold War*. University of California Press: Berkeley.

Funk, N and Mueller, M (eds) 1993: *Gender Politics and Post-Communism: Reflections from Eastern Europe and the Former Soviet Union*. Routledge: New York.

Jackson, P and Penrose, J 1994: *Constructions of 'Race', Place and Nation*. UCL Press: London.

Johnson, N 1995: Cast in stone: monuments, geography and nationalism. *Environment and Planning D: Society and Space* **13**, 51–65.

Kandiyoti, D 1991: Identity and its discontents: women and the nation. *Millennium* **20**(3), 429–43.

Marston, S 1990: Who are 'the people'? Gender, citizenship and the making of the American nation. *Environment and Planning D: Society and Space* **8**(4), 449–58.

McClintock, A 1993: Family Feuds: gender, nationalism and the family. *Feminist Review* **44**, 61–80.

Menchú, R 1992: *I, Rigoberta Menchú*. Routledge: London.

Mosse, G 1985: *Nationalism and Sexuality*. University of Wisconsin Press: Madison.

Mouffe, C 1992: Feminism, citizenship and radical democratic politics. In J. Butler and J. Scott (eds) *Feminists Theorize the Political*. 369–84. Routledge: New York.

Parker, A M, Russo, D, Sommer, and P Yaeger 1992: *Nationalisms and Sexualities*. Routledge: New York.

Peterson, V S 1992: Introduction. *Gendered States: Feminist (Re)visions of International Relations Theory*. Lynne Rienner: London.

Radcliffe, S 1996: Gendered nations: nostalgia, development and territory in Ecuador. *Gender, Place and Culture* **3**(1), 5–21.

Rich, A 1986: Notes towards a politics of location. *Blood, Bread and Poetry: Selected Prose 1979–1985*. New York: Norton.

Sharp, J 1996: Gendering nationhood: a feminist engagement with national identity. In N. Duncan (ed.) *BodySpace*. Routledge: London, pp. 97–108.

Sparke, M 1994: Writing on patriarchal missiles: the chauvinism of the 'Gulf War' and the limits of critique. *Environment and Planning A* **26**, 1061–89.

Tickner, J A 1992: *Gender and International Relations*. Columbia University Press: New York.

Warner, M 1985: *Monuments and Maidens: the Allegory of the Female Form*. Picador: London.

Yuval-Davis, N 1991: The citizenship debate: women, ethnic processes and the state. *Feminist Review* **39**, 58–68.

29 Nira Yuval-Davis

'Gender and Nation'

Excerpt from: *Ethnic and Racial Studies* **16**(4), 621–32 (1993)

Gender relations, citizenship and membership in the national collectivity

The 'universalistic' nature of citizenship which emanates, out of the traditional liberal and social democratic discourses is very deceptive (Balibar 1989; Yuval-Davis 1991a, 1991b; Yeatman 1992). The expression 'nation-state' camouflages the only-partial overlap between the boundaries of the hegemonic national collectivity and the settled residents or even citizens of the state. Nevertheless, even beyond this, the integrity and viability of the 'community of citizens' thus defined is very much dependent on clear-cut definitions of who belongs and who does not belong to it – hence continuous fears and debates about immigration as well as systemic exclusions of many who are situated within the boundaries of the state, such as indigenous peoples and other minorities. The exclusions become much clearer if we take into account the three dimensions of citizenship as defined by Marshall (1950; 1975; 1981) – civil, political, and social.

The exclusion of women is of particular importance for this article. The whole social philosophy which was at the base of the rise of the notion of state citizenship was constructed in terms of the 'Rights of Man', a social contract based on the 'fraternity of men' (as one of the slogans of the French revolution states – and not incidentally) (Carol Pateman 1988). Women were not simply late comers to citizenship rights, as in Marshall's evolutionary model of the development of citizenship rights. Their exclusion was part and parcel of the construction of the entitlement of men to democratic participation which conferred citizen status not upon individuals as such, but upon men in their capacity as members and representatives of a family (i.e., a group of non-citizens)' (Vogel 1989, p. 2).

Unlike Marshall's scheme, where political rights followed civil rights, married women have still not been given full civil and social rights. The image of the Thatcherite 'Active Citizen' of the late 1980s in Britain, has still been personified in the image of the man as responsible head of his family. The construction by the state of relationships in the private domain, namely, marriage and the family, is what has determined women's status as citizens within the public domain. In some non-European countries, especially those ruled by Muslim laws, the right of women even to work and travel in the public domain is dependent on formal permission of her 'responsible' male relative (Kandiyoti 1991).

Given the recent changes in eastern and central Europe, some have attempted to formulate them in terms of the reconstruction of civil society.

By this they mean a presence of a social sphere which is independent of the state. Many Western feminist analyses of the relationship between women and the state have shown this 'independence' to be largely illusory, as it is the state which constructs, and often keeps surveillance of, the private domain, especially of the lower classes (e.g., Wilson 1977; Showstack Sasoon 1987). However, in Third World societies there is often only partial penetration of the state into civil society, especially in its rural and other peripheral sections. In such cases, gender and other social relations are determined by cultural and religious customs of the national collectivity. This may also happen in 'private domains' of ethnic and national minorities in the state.

It is not only in the 'private domain', however, that gender relations are differential in the state. Often the citizenship rights and duties of women from different ethnic and racial groupings are different. They could have different legal positions and entitlements; sometimes they could be under the jurisdiction of different religious courts; they could be under different residential regulations, including rights of re-entry when leaving the country; they might or might not be allowed to confer citizenship rights to their children, or – in the case of women migrant workers who had to leave their children behind – may or may not receive child and other welfare benefits as part of their social rights (Women, Immigration and Nationality Group [WING] 1985).

With all these differences, there is one characteristic which specifies women's citizenship. That is its dualistic nature: on the one hand, women are always included, at least to some extent, in the general body of citizens of the state and its social, political and legal policies; on the other hand, there is always, more or less developed, a separate body of legislation which relates to them specifically as women. These policies can express different ideological constructions of gender, such as requiring different retirement ages for women and men. They might discriminate against women – as in cases where women might be forbidden to vote, forbidden to be elected to certain public posts, etc.; or they might favour women as in cases where they are granted maternity leave, or special 'privileges' in labour legislation, etc.

Marshall's definition of citizenship has been 'full membership in a community', which encompasses civil, political and social rights and responsibilities. This has led some feminists to think that the only way women could gain full equality would be if they were to share equally all citizenship responsibilities and duties. This has been the debate especially in relation to women's participation in the military (Enloe 1983; Yuval-Davis 1986; 1991b). In many ways this debate is similar to earlier debates on the entry of women into the waged labour market – especially in modern highly technological armies which are professional, rather than based on national draft. As in the civil labour market, the entrance of women to the military has usually resulted in introducing a new arena rather than changing the principle of the sexual division of labour and power. This can change only when men as well as women are defined in a dualistic manner as reproducers as well as producers of the nation – a project which has only begun in a few western countries and even there generally in a virtually purely symbolic way. Nevertheless, the participation of women in the military can erode one of the most powerful cultural constructions of national collectives – that of "womenandchildren' (Enloe 1990) as the reason men go to war.

Gender relations and cultural constructions of collectivities

The mythical unity of national 'imagined communities' which divides the world between 'us' and 'them', is maintained and ideologically reproduced by a whole system of what Armstrong (1982) calls symbolic 'border guards'. These 'border guards' can identify people as members or non-members of a specific collectivity. They are closely linked to particular cultural codes of style of dress and behaviour as well as to more elaborate bodies of customs, literary and artistic modes of production, and, of course, language. Gender symbols play an especially significant role in this.

Just beyond Cyprus airport there is a large poster of a mother mourning her child – Greek Cyprus mourning and commemorating the Turkish invasion. In France, it was *La Patrie*, a figure of a woman giving birth which personified the revolution. Women often come to symbolize the national collectivity, its roots, its spirit, its national project. Moreover, women often symbolize national and collective 'honour'. Shaving the heads of women who 'dared' to fraternize, or even to fall in love with 'the enemy' is but one expression of this.

As we have written elsewhere (Anthias and Yuval-Davis 1983; 1989), women are often the ones who are given the social role of intergenerational transmitters of cultural traditions, customs, songs, cuisine, and, of course, the mother tongue (*sic*!). The actual behaviour of women can also signify ethnic and cultural boundaries:

> Often the distinction between one ethnic group and another is constituted centrally by the sexual behaviour of women. For example, a 'true' Sikh or Cypriot girl should behave in sexually appropriate ways. If she does not then neither her children nor herself may be constituted part of the 'community'.
>
> (Anthias and Yuval-Davis 1989, p. 10)

The importance of women's culturally 'appropriate behaviour' can gain special significance in 'multicultural societies'. A basic problem in the construction of multiculturalism is the assumption that all members of a specific cultural collectivity are equally committed to that culture. It tends to construct the members of minority collectivities as basically homogeneous, speaking with a unified cultural voice. These cultural voices have to be as distinguishable as possible from the majority culture in order for them to be perceived as being 'different'; thus, the more traditional and distanced from the majority culture the voice of the 'community representatives' is, the more 'authentic' it will be seen to be within such a construction (Sahgal and Yuval-Davis 1992).

Such a construction therefore would not have space for internal power conflicts and interest differences within the minority collectivity, conflicts along the lines of class and gender as well as politics and culture, for instance.

> It becomes clear that the liberal conception of the group requires the group to assume an authoritarian character: there has to be headship of the group which represents its homogeneity of purpose by speaking with the one, authoritative voice. For this to occur, the politics of voice and representation latent within the heterogeneity of perspectives and interests must be suppressed.
>
> (Yeatman 1992, p. 4)

This liberal construction of group voice therefore can collude with fundamentalist leaderships who claim to represent the true 'essence' of their collectivity's culture and religion, and who have high on their agenda the control of women and their behaviour – as campaigns like the enforced veiling of women by Muslim fundamentalists and the major anti-abortion campaigns by Christian fundamentalists demonstrate.

As a general rule, collectivities are composed of family units. A central link between the place of women as national reproducers and women's subjugation can be found in the different regulations – customary, religious or legal – that determine the family units within the boundaries of the collectivity, and the ways they come into existence (marriage), end (divorce and widowhood) and which children are considered legitimate members of the family.

The question of legitimacy of children relates to the ideologically constructed boundaries of families and collectivities. However, a major part of the control of women as national reproducers relate to their actual biological role as bearers of children.

Gender relations and the biological reproduction of 'the nation'

If membership in the national collectivity depends on one's being born into it, then those who do not share the myth of common origin are completely excluded. The only way 'outsiders' can conceivably join the national collectivity is by intermarriage. Not incidentally, those who are preoccupied with the 'purity' of the race would also be preoccupied with the sexual relationships between members of different collectivities. Typically, the first (and only) law proposal that Rabbi Kahana, the leader of the Israeli fascist party Kach, raised in the Israeli parliament was to forbid sexual relationships between Jews and Arabs. The legal permission for people of different 'races' to have sexual intercourse and to marry has been one of the more significant first steps that the South African government has taken in its slow but inevitable journey towards the abolition of apartheid.

In different religious and customary laws, the membership of a child in a national collectivity might depend exclusively on the father's membership, the mother's membership, or it might be open for a dual, or voluntary choice membership. The inclusion of the collectivity is far from being only a biological issue. There are always rules and regulations governing the cases where children born to 'mixed parenthood' would be part of the collectivity and cases where they would not; about when they would be considered a separate social category, as in South Africa; part of the 'inferior' collectivity, as during slavery; or – although this is rarer – part of the 'superior' collectivity, as was the case in marriages between Spanish settlers and aristocratic Indians in Mexico (Gutierrez 1994). When a man from Ghana a few years ago tried to claim his British origin under the patriality clause of the British Immigration Act, arguing that his African grandmother was legally married to his British grandfather, the judge rejected his claim, arguing that at that period no British man would genuinely have married an African woman (WING 1985).

The quality of the 'stock' has been a major worry in the British empire and its Settler Societies. The Royal Commission on Population declared in its 1949 report that

> British traditions, manners, and ideas in the world have to be borne in mind. Immigration is thus not a desirable means of keeping the population at a replacement level as it would in effect reduce the proportion of home-bred stock in the population.
>
> (quoted in Riley 1981)

It was concern for the 'British race' which Beveridge describes in his famous report as the motivation for establishing the British welfare state system (Beveridge 1942).

The control of women as producers of 'national stocks' starts with pre-natal policies. A variety of techniques and technologies, used by various social agencies, exist for controlling rates of birth. These can include allowances for maternity leave and child care facilities for working mothers; availability and encouragement of contraception as a means of family planning; availability and legality of abortion; infertility clinics; and, on the other hand, compulsory contraception and sterilization. The encouragement or discouragement of women to bear children is determined, to a great extent, by the specific historical situation of the collectivity, and by no means exists as a 'laissez-faire' institution even in the most permissive societies. Notions like 'population explosion', 'demographic balance' (or 'holocaust' or 'race'), or 'children as a national asset' are expressions of various ideologies that can lead controllers of national reproduction towards different population control policies. These policies are rarely, if ever, applied in a similar manner to all members of the civil society. While class differences often play a major role in this, membership in different racial, ethnic and national collectivities is usually the most important determinant in being subject to differential population policies, and can affect differently, but as effectively, women of both hegemonic majorities and subjugated minorities. These policies, however, are not used only by national collectivities who have control over states, but can be used also as a mode of resistance. A common Palestinian saying a few years ago, for instance, was 'the Israelis beat us on the borders but we beat them in the bedrooms' (Yuval-Davis 1987, p. 80).

Nevertheless, it would be a mistake to see women as passive victims in such 'national/biological warfare', whether pro- or anti-natal. Older women will often play an important part in controlling younger women, and all women may be parties to these ideologies, as the active participation of women in various religious fundamentalist and fascist movements clearly show.

References

Anthias, Floya and Yuval-Davis, Nira 1983 'Contextualizing feminism: ethnic, gender and class division', *Feminist Review*, no. 15, pp. 62–75.

—— 1989 'Introduction' in Nira Yuval-Davis and Floya Anthias (eds), *Woman–Nation–State*, Basingstoke: Macmillan.

Armstrong, John 1982 *Nations before Nationalism*, Chapel Hill: University of North Carolina Press.

Balibar, Étienne 1989 'Y-a-til un "neo-racisme"?', in Etienne Balibar and Emannuel Wallerstein, *Race, classe, nation: les identités ambigues*, Paris: Editions La Découverte.

Beveridge, William 1942 *Report on Social Insurance and Allied Services*, London: HMSO.

Enloe, Cynthia 1983 *Does Khaki Become You? The Militarization of Women's Lives*, London: Pluto Press.

—— 1990 'Womenandchildren: making feminist sense of the Persian Gulf crisis', *The Village Voice*, 25 September, New York.

Gutierrez, Natividad 1994 'Mixing races for nation building: native and settler women in Mexico', in Daiva Stasiulis and Nira Yuval-Davis, (eds), *Beyond Dichotomies: Gender, Race, Ethnicity and Class in Settler Societies*, London: Sage.

Kandiyoti, Denis (ed.) 1991 *Women, Islam and the State*, Basingstoke: Macmillan.

Marshall, Thomas H. 1950 *Citizenship and Social Class*, Cambridge: Cambridge University Press.

—— 1975 *Social Policy in the Twentieth Century*, London: Hutchinson.

—— 1981 *The Right To Welfare and Other Essays*, London: Heinemann Educational Books.

Pateman, Carol 1988 *The Sexual Contract*, Cambridge: Polity Press.

Riley, Denise 1981 'The free mothers', *History Workshop Journal*, pp. 59–119.

Saghal, Gita and Yuval-Davis, Nira 1992 *Refusing Holy Orders: Women and Fundamentalism in Contemporary Britain*, London: Virago.

Showstack Sasoon, Ann (ed.) 1987 *Women and the State*, London: Hutchinson.

Vogel, Ursula 1989 'Is citizenship gender specific?', paper presented at the Political Science Association Annual Conference, April.

Wilson, Elizabeth 1977 *Women and the Welfare State*, London: Tavistock.

Wing (Women and Immigration and Nationality Group) 1985 *Worlds Apart, Women Under Immigration and Nationality Law*, London: Pluto Press.

Yeatman, Anna 1992 'Minorities and the politics of difference', *Political Theory Newsletter*, Canberra, March.

Yuval-Davis, Nira 1986 'Front and rear: sexual division of labour in the Israeli military' in Haleh Afshar (ed.) *Women, State and Ideology*, Basingstoke: Macmillan Press, pp. 185–203.

—— 1987, 'The Jewish collectivity and national reproduction in Israel', *Khamsin*, special issue on *Women in the Middle East*, London: Zed Books.

—— and Anthias, Floya (eds) 1989 *Woman–Nation–State*, Basingstoke: Macmillan.

—— 1991a 'The citizenship debate: women, the state and ethnic processes', *Feminist Review*, no. 39, pp. 56–68.

—— 1991b 'The gendered Gulf War: women's citizenship and modern warfare'. in *World Order*, London: Zed Books, pp. 219–25.

30 Anne McClintock

'No Longer in a Future Heaven: Gender, Race and Nationalism'

Excerpt from: *Imperial Leather: Race, Gender and Sexuality in the Colonial Contest*, pp. 352–90. New York: Routledge (1994)

> The tribes of the Blackfoot confederacy, living along what is now known as the United States/Canadian border, fleeing northward after a raiding attack, watched with growing amazement as the soldiers of the United States army came to an sudden, magical stop. Fleeing southwards, they saw the same thing happen, as the Canadian mounties reined to an abrupt halt. They came to call this invisible demarcation the 'medicine line.'

All nationalisms are gendered, all are invented and all are dangerous – dangerous, not in Eric Hobsbawm's sense of having to be opposed but in the sense that they represent relations to political power and to the technologies of violence.[1] As such, nations are not simply phantasmagoria of the mind; as systems of cultural representation whereby people come to imagine a shared experience of identification with an extended community, they are historical practices through which social difference is both invented and performed.[2] Nationalism becomes in this way constitutive of people's identities through social contests that are frequently violent and always gendered. Yet, if following Benedict Anderson, the invented nature of nationalism has recently found wide theoretical currency, explorations of the gendering of the national imaginary have been conspicuously paltry.

Nations are contested systems of cultural representation that limit and legitimize peoples' access to the resources of the nation-state, but despite many nationalists ideological investment in the idea of popular *unity*, nations have historically amounted to the sanctioned institutionalization of gender *difference*. No nation in the world grants women and men the same access to the rights and resources of the nation-state. Yet, with the notable exception of Frantz Fanon, male theorists have seldom felt moved to explore how nationalism is implicated in gender power. As a result, as Cynthia Enloe remarks, nationalisms have 'typically sprung from masculinized memory, masculinized humiliation and masculinized hope'.[3]

Not only are the needs of the nation typically identified with the frustrations and aspirations of men, but the representation of male *national* power depends on the prior construction of *gender* difference. All too often in male nationalisms, gender difference between women and men serves to symbolically define the limits of national difference and power between *men*. Even Fanon, who at other moments knew better, writes: 'The look that the native turns on the settler town is a look of lust . . . to sit at the settler's table, to sleep in the settler's bed, with his wife if possible. The colonized man is an envious man.'[4] For Fanon, both colonizer and colonized are here unthinkingly male and the Manichean agon of decolonization is waged over the territoriality of female, domestic space.

Excluded from direct action as national citizens, women are subsumed symbolically into the national body politic as its boundary and metaphoric limit: 'Singapore girl, you're a great way to fly.' Women are typically constructed as the symbolic bearers of the nation, but are denied any direct relation to national agency. As Elleke Boehmer notes in her fine essay, the 'motherland' of male nationalism thus may 'not signify "home" and "source" to women.'[5] Boehmer notes that the male role in the nationalist scenario is typically 'metonymic;' that is, men are contiguous with each other and with the national whole. Women, by contrast, appear 'in a metaphoric or symbolic role.'[6] Yet it is also crucial to note that not all men enjoy the privilege of political contiguity with each other in the national community.

In an important intervention, Nira Yuval-Davis and Floya Anthias identify five major ways in which women have been implicated in nationalism:

- as biological reproducers of the members of national collectivities
- as reproducers of the boundaries of national groups (through restrictions on sexual or marital relations)
- as active transmitters and producers of the national culture
- as symbolic signifiers of national difference
- as active participants in national struggles.[7]

Nationalism is thus constituted from the very beginning as a gendered discourse and cannot be understood without a theory of gender power. Nonetheless, theories of nationalism reveal a double disavowal. If male theorists are typically indifferent to the gendering of nations, feminist analyses of nationalism have been lamentably few and far between. White feminists, in particular, have been slow to recognize nationalism as a feminist issue. In much Western, socialist feminism, as Yuval-Davis and Anthias point out, 'Issues of ethnicity and nationality have tended to be ignored.'[8]

A feminist theory of nationalism might thus be strategically fourfold: investigating the gendered formation of sanctioned male theories; bringing into historical visibility women's active cultural and political participation in national formations; bringing nationalist institutions into critical relation with other social structures and institutions; and at the same time paying scrupulous attention to the structures of racial, ethnic and class power that continue to bedevil privileged forms of feminism.

The national family of man: a domestic genealogy

A paradox lies at the heart of most national narratives. Nations are frequently figured through the iconography of familial and domestic space. The term nation derives from *natio*: to be born. We speak of nations as 'motherlands' and 'fatherlands.' Foreigners 'adopt' countries that are not their native homes and are naturalized into the national 'family'. We talk of the 'Family of Nations,' of 'homelands' and 'native' lands. In Britain, immigration matters are dealt with at the Home Office; in the United States, the president and his wife are called the First Family. Winnie Mandela was, until her recent fall from grace, honored as South Africa's 'Mother of the Nation.' In this way, despite their myriad differences, nations are symbolically figured as domestic genealogies. Yet, as I argued in the earlier chapters of this book, since the mid nineteenth century in the West at least, the family itself has been figured as the antithesis of history.

The family trope is important for nationalism in at least two ways. First, it offers a 'natural' figure for sanctioning national *hierarchy* within a putative organic *unity* of interests. Second, it offers a 'natural' trope for figuring national time. After 1859 and the advent of social Darwinism, Britain's emergent national narrative took increasing shape around the image of the evolutionary Family of Man. The family offered an indispensable metaphoric figure by which national difference could be shaped into a single historical genesis narrative. Yet a curious paradox emerged. The family as a *metaphor* offered a single genesis narrative for national history while, at the same time, the family as an *institution* became void of history and excluded from national power. The family became, at one and the same time, both the *organizing figure* for national history and its *antithesis*.

In the course of the nineteenth century, the social function of the great service families were displaced onto the national bureaucracies, while the image of the family was projected onto these nationalisms as their shadowy, naturalized form. Since the subordination of woman to man and child to adult, was deemed a natural fact, hierarchies within the nation could be depicted in familial terms to guarantee social difference as a category of nature. The metaphoric depiction of social hierarchy as natural and familial – the 'national family,' the global 'family of nations,' the colony as a 'family of black children ruled over by a white father' – depended in this way on the prior naturalizing of the social subordination of women and children within the domestic sphere.

In modern Europe, citizenship is the legal representation of a person's relationship to the rights and resources of the nation-state. But the putatively universalist concept of national citizenship becomes unstable when seen from the position of women. In post-French revolution Europe, women were incorporated directly into the nation-state not directly as citizens, but only indirectly, through men, as dependent members of the family in private and public law. The Code Napoleon was the first modern statute to decree that the wife's nationality should follow her husband's, an example other European countries briskly followed. A woman's *political* relation to the nation was thus submerged as a *social* relation to a man through marriage. For women, citizenship in the nation was mediated by the marriage relation within the family. This chapter is directly concerned with the consequences for women of this uneven gendering of the national citizen.

The gendering of nation time

A number of critics have followed Tom Nairn in naming the nation 'the modern Janus.'[9] For Nairn, the nation takes shape as a contradictory figure of time: one face gazing back into the primordial mists of the past, the other into an infinite future. Deniz Kandiyoti expresses the temporal contradiction with clarity: '[Nationalism] presents itself both as a modern project that melts and transforms traditional attachments in favour of new identities and as a reflection of authentic cultural values culled from the depths of a presumed communal past.'[10] Bhabha, following Nairn and Anderson, writes: 'Nations, like narratives, lose their origins in the myths of time and only fully realize their horizons in the mind's eye.'[11] Bhabha and Anderson borrow here on Walter Benjamin's crucial insight into the temporal paradox of modernity. For Benjamin, a central feature of nineteenth-century industrial capitalism was the 'use of archaic images to identify what was historically new about the "nature" of commodities.'[12] In Benjamin's insight, the mapping of Progress depends on systematically inventing images of archaic time to identify what is historically new about enlightened, national progress. Anderson can thus ask: 'Supposing "antiquity" were, at a certain historical juncture, the *necessary consequence* of "novelty?"'[13]

What is less often noticed, however, is that the temporal anomaly within nationalism – veering between nostalgia for the past and the impatient, progressive sloughing off of the past – is typically resolved by figuring the contradiction in the representation of *time* as a natural division of *gender*. Women are represented as the atavistic and authentic body of national tradition (inert, backward-looking and natural), embodying nationalism's conservative principle of continuity. Men, by contrast, represent the progressive agent of national modernity (forward-thrusting, potent and historic), embodying nationalism's progressive, or revolutionary principle of discontinuity. Nationalism's anomalous relation to time is thus managed as a natural relation to gender.

In the nineteenth century, the social evolutionists secularized time and placed it at the disposal of the national, imperial project. The axis of *time*, was projected onto the axis of *space* and history became global. Now not only natural space but also historical time was collected, measured and mapped onto a global science of the surface. In the process history, especially national and imperial history, took on the character of a spectacle.

Secularizing time has a threefold significance for nationalism. First figured in the evolutionists' global Family Tree, the world's discontinuous nations appear to be marshalled within a single, hierarchical European Ur-narrative. Second, national history is imaged as naturally teleogical, an organic process of upward growth, with the European nation as the apogee of world progress. Third, inconvenient discontinuities are ranked and subordinated into a hierarchical structure of branching time – the progress of 'racially' different nations mapped against the tree's self-evident boughs, with 'lesser nations' destined, by nature, to perch on its lower branches.

National time is thus not only *secularized*, it is also *domesticated*. Social evolutionism and anthropology gave to national politics a concept of natural time as familial. In the image of the Family Tree, evolutionary progress was represented as a series of anatomically distinct family types, organized into a linear procession, from the 'childhood' of 'primitive' races to the enlightened 'adulthood' of European imperial nationalism. Violent national change takes on the character of an evolving spectacle under the organizing rubric of the family. The merging of the racial evolutionary Tree and the gendered family into the Family Tree of Man provided scientific racism with a simultaneously gendered and racial image through which it could popularize the idea of linear national Progress.

Britain's emerging national narrative gendered time by figuring women (like the colonized and the working class) as inherently atavistic – the conservative repository of the national archaic. Women were not seen as inhabiting history proper but existing, like colonized peoples, in a permanently anterior time within the modern nation. White, middle-class men, by contrast, were seen to embody the forward-thrusting agency of national progress. Thus the figure of the national Family of Man reveals a persistent paradox. National Progress (conventionally the invented domain of male, public space) was figured as familial, while the family itself (conventionally the domain of private, female space) was figured as beyond history.

One can safely say, at this point, that there is no single narrative of the nation. Different groups (genders, classes, ethnicities, generations and so on) do not experience the myriad national formations in the same way. Nationalisms are invented, performed and consumed in ways that do not follow a universal blueprint. At the very least, the breathtaking Eurocentricism of Hobsbawm's dismissal of Third World nationalisms warrants sustained criticism. In a gesture of sweeping condescension, Hobsbawm nominates Europe as nationalism's 'original home,' while 'all the anti-imperial movements of any significance' are unceremoniously dumped into three categories: mimicry of Europe, anti-Western xenophobia and the 'natural high spirits of martial tribes.'[14] By way of contrast, it might be useful to turn at this point to Frantz Fanon's quite different analysis of the gendering of the national formation.

Fanon and gender agency

As male theorists of nationalism go, Frantz Fanon is exemplary, not only for recognizing gender as a formative dimension of nationalism but also for recognizing – and immediately rejecting – the Western metaphor of the nation as a family. 'There are close connections,' he observes in *Black Skin, White Masks* 'between the structure of the family and the structure of the nation.'[15] Refusing, however, to collude with the notion of the familial metaphor as natural and normative, Fanon instead understands it as a cultural projection ('the characteristics of the family are projected onto the social environment') that has very different consequences for families placed discrepantly within the colonial hierarchy.[16] 'A normal Negro child, having grown up within a normal family, will become abnormal on the slightest contact with the white world.'[17]

The challenge of Fanon's insight is threefold. He throws radically into question the naturalness of nationalism as a domestic genealogy. At the same time, he reads familial normality as a product of social power – indeed, of social violence. Fanon is remarkable for recognizing, in this early text, how military violence and the authority of a centralized state borrow on and enlarge the domestication of gender power within the family: 'Militarization and the centralization of authority in a country automatically entail a resurgence of the authority of the father.'[18]

Perhaps one of Fanon's most provocative ideas is his challenge to any easy relation of identity between the psychodynamics of the unconscious and the psychodynamics of political life. The audacity of his insight is that it allows one to ask whether the psychodynamics of colonial power and of anti-colonial subversion can be interpreted by deploying (without mediation) the same concepts and techniques used to interpret the psychodynamics of the unconscious. If the family is not 'a miniature of the nation,' are metaphoric projections from family life (the Lacanian 'Law of the Father,' say) adequate for an understanding of colonial or anticolonial power? Fanon himself seems to say no. Relations between the individual unconscious and political life are, I argue, neither separable from each other nor reducible to each other. Instead, they comprise crisscrossing and dynamic mediations, reciprocally and untidily transforming each other, rather than duplicating a relation of structural analogy.

Even in *Black Skins, White Masks*, the most psychological of Fanon's texts, he insists that racial alienation is a 'double process.'[19] First, it 'entails an immediate recognition of social and economic realities.' Then, it entails the 'internalization' of inferiority. Racial alienation, in other words, is not only an 'individual question' but also involves what Fanon calls a 'sociodiagnostic.'[20] Reducing Fanon to a purely formal psychoanalysis, or a purely structural Marxism, risks foreclosing precisely those suggestive tensions that animate, in my view, the most subversive elements of his work. These tensions are nowhere more marked than in his tentative exploration of the gendering of national agency.

Gender runs like a multiple fissure through Fanon's work, splitting and displacing the 'Manichean delirium' to which he repeatedly returns. For Fanon, the colonial agon appears, at first, to be fundamentally Manichean. In *Black Skins, White Masks*, he sees colonial space as divided into 'two camps: the white and the black.'[21] Nearly a decade later, writing from the crucible of the Algerian resistance in *The Wretched of the Earth*, Fanon once again sees anticolonial nationalism as erupting from the violent Manicheanism of a colonial world 'cut in two,' its boundaries walled by barracks and police stations.[22] Colonial space is split by a pathological geography of power, separating the bright, well-fed settler's town from the hungry, crouching casbah: 'This world . . . cut in two is inhabited by two different species.'[23] As Edward Said puts it: 'From this Manichean and physically grounded statement Fanon's entire work follows, set in motion, so to speak, by the native's violence, a force intended to bridge the gap between white and nonwhite.'[24] Yet the fateful chiaroscuro of race is at almost every turn disrupted by the crisscrossings of gender.

Fanon's Manichean agon appears at first to be fundamentally male: 'There can be no further doubt that the real Other for the white man is and will continue to be the black man.' As Homi Bhabha writes: 'It is always in relation to the place of the Other that colonial desire is articulated.'[25] But Fanon's anguished musings on race and sexuality disclose that 'colonial desire' is not the same for men and women:

> Since he is the master and more simply the male, the white man can allow himself the luxury of sleeping with many women . . . But when a white woman accepts a black man there is automatically a romantic aspect. It is a giving, not a seizing.[26]

Leaving aside, for the moment, Fanon's complicity with the stereotype of women as romantically rather than sexually inclined, as giving rather than taking, Fanon opens race to a problematics of sexuality that reveals far more intricate entanglements than a mere doubling of 'the Otherness of the Self.' The psychological Manicheanism of *Black Skins, White Masks* and the more political Manicheanism of *The Wretched of the Earth* are persistently inflected by gender in such a way as to radically disrupt the binary dialectic.

For Fanon, the envy of the black man takes the form of a fantasy of territorial displacement: 'The fantasy of the native is precisely to occupy the master's place.'[27] This fantasy can be called a *politics of substitution*. Fanon knows, however, that the relation to the white woman is altogether different: 'When my restless hands caress those white breasts, they grasp white civilization and dignity and make them mine.'[28] The white woman is seized, possessed and taken hold of, not as an act of *substitution*, but as an act of *appropriation*. However, Fanon does not bring this critical distinction between a politics of substitution and a politics of appropriation into explicit elaboration as a theory of gender power.

As Bhabha astutely observes, Fanon's *Black Skins, White Masks* is inflected by a 'palpable pressure of division and displacement' – though gender is a form of self-division that Bhabha himself fastidiously declines to explore.[29] Bhabha would have us believe that 'Fanon's use of the word "man" usually connotes a phenomenological quality of humanness, inclusive of man and woman.'[30] But this claim is not borne out by Fanon's texts. Potentially generic terms like 'the Negro' or 'the Native' – syntactically unmarked for gender – are almost everywhere immediately contextually marked as male: 'Sometimes people wonder that the native, rather than giving his wife a dress, buys instead a transistor radio;'[31] '. . . the Negro who wants to go to bed with a white woman;'[32] '. . . the Negro who is viewed as a penis symbol.'[33] The generic category 'native' does not include women; women are merely possessed by the (male) native as an appendage: 'When the native is tortured, when his wife is killed or raped, he complains to no one.'[34]

For Fanon, colonized men inhabit 'two places at once.' If so, how many places do colonized women inhabit? Certainly, Bhabha's text is not one of them. Except for a cursory appearance in one paragraph, women haunt

Bhabha's analysis as an elided shadow – deferred, displaced and dis-remembered. Bhabha concludes his eloquent meditation on Fanon with the overarching question: 'How can the human world live its difference? How can a human being live Other-wise?'[35] Yet immediately appended to his foreword appears a peculiar Note. In it Bhabha announces, without apology, that the 'crucial issue' of the woman of color 'goes well beyond the scope' of his foreword. Yet its scope, as he himself insists, is bounded by nothing less than the question of *humanity*: 'How can the human world live its difference? how can a human being live Other-wise?' Apparently, the question of the woman of color falls beyond the question of human difference and Bhabha is content simply to 'note the importance of the problem' and leave it at that. Bhabha's belated note on gender appears after his authorial signature, after the time and date of his essay. Women are thus effectively deferred to a no-where land, beyond time and place, outside theory. If, indeed, 'the state of emergency is also a state of emergence,' the question remains whether the national state of emergency turns out to be a state of emergence for women at all.[36]

To ask 'the question of the subject' ('What does a man want? What does the black man want?'), while postponing a theory of gender, presumes that subjectivity itself is neutral with respect to gender.[37] From the limbo of the male afterthought, however, gender returns to challenge the male question not as women's 'lack,' but as that excess that the masculine '"Otherness" of the Self' can neither admit nor fully elide. This presumption is perhaps nowhere more evident than in Fanon's remarkable meditations on the gendering of the national revolution.

At least two concepts of national agency shape Fanon's vision. His anticolonial project is split between a Hegelian vision of colonizer and colonized locked in a life-and-death conflict and an altogether more complex and unsteady view of agency. These paradigms slide discrepantly against each other throughout his work, giving rise to a number of internal fissures. These fissures appear most visibly in his analysis of gender as a category of social power.

On one hand, Fanon draws on a Hegelian metaphysics of agency inherited, by and large, through Jean-Paul Sartre and the French academy. In this view, anticolonial nationalism erupts violently and irrevocably into history as the logical counterpart to colonial power. This nationalism is, as Edward Said puts it, 'cadenced and stressed from beginning to end with the accents and inflections of liberation.'[38] It is a liberation, moreover, that is structurally guaranteed, immanent in the binary logic of the Manichean dialectic. This metaphysics speaks, as Terrence Eagleton nicely phrases it, 'of the entry into full self-realization of a unitary subject known as the people.'[39] Nonetheless, the privileged national agents are urban, male, vanguardist and violent. The progressive nature of the violence is preordained and sanctioned by the structural logic of Hegelian progress.

This kind of nationalism can be called an *anticipatory nationalism*. Eagleton calls it nationalism 'in the subjunctive mood,' a premature utopianism that 'grabs instinctively for a future, projecting itself by an act of will or imagination beyond the compromised political structures of the present.'[40] Yet, ironically, anticipatory nationalism often claims legitimacy by appealing precisely to the august figure of inevitable progress inherited from the Western societies it seeks to dismantle.

Alongside this Manichean, mechanical nationalism, however, appears an altogether more open-ended and strategically difficult view of national agency. This nationalism stems not from the inexorable machinery of Hegelian dialectics but from the messy and disobliging circumstances of Fanon's own activism, as well as from the often dispiriting lessons of the anticolonial revolutions that preceded him. In this view, agency is multiple rather than unitary, unpredictable rather than immanent, bereft of dialectical guarantees and animated by an unsteady and nonlinear relation to time. There is no preordained rendevous with victory; no single, undivided national subject; no immanent historical logic. The national project must be laboriously and sometimes catastrophically invented, with unforeseen results. Time is dispersed and agency is heterogeneous. Here, in the unsteady, sliding interstices between conflicting national narratives, women's national agency makes its uncertain appearance.

In 'Algeria Unveiled,' Fanon ventriloquizes – only to refute – the long Western dream of colonial conquest as an erotics of ravishment. Under the hallucinations of empire, the Algerian woman is seen as the living flesh of the national body, unveiled and laid bare for the colonials' lascivious grip, revealing 'piece by piece, the flesh of Algeria laid bare.'[41] In this remarkable essay, Fanon recognizes the colonial gendering of women as symbolic mediators, the boundary markers of an agon that is fundamentally male. The Algerian woman is 'an intermediary between obscure forces and the group.'[42] 'The young Algerian woman . . . establishes a link,' he writes.[43]

Fanon understands brilliantly how colonialism inflicts itself as a *domestication* of the colony, a reordering of the labor and sexual economy of the people, so as to divert female power into colonial hands and disrupt the patriarchal power of colonized men. Fanon ventriloquizes colonial thinking: 'If we want to destroy the structure of Algerian society, its capacity for resistance, we must first of all conquer the women.'[44] His insight here is that the dynamics of colonial power are fundamentally, though not solely, the dynamics of gender: 'It is the situation of women that was accordingly taken as the theme of action.'[45] Yet, in his work as a whole, Fanon fails to bring these insights into theoretical focus.

Long before Anderson, Fanon recognizes the inventedness of national community. He also recognizes the power of nationalism as a *scopic* politics, most visibly embodied in the power of sumptuary customs to fabricate a sense of national unity: 'It is by their apparel that types of society first become known.'[46] Fanon perceives, moreover, that nationalism, as a politics of visibility, implicates women and men in different ways. Because, for male nationalists, women serve as the visible markers of national homogeneity, they become subjected to especially vigilant and violent discipline. Hence the intense emotive politics of dress.

Yet a curious rupture opens in Fanon's text over the question of women's agency. At first, Fanon recognizes the historical meaning of the veil as open to the subtlest shifts and subversions. From the outset, colonials tried to grant Algerian women a traitorous agency, affecting to rescue them from the sadistic thrall of Algerian men. But, as Fanon knows, the colonial masquerade of giving women power by unveiling them was merely a ruse for achieving 'a real power over the man.'[47] Mimicking the colonial masquerade, militant Algerian women deliberately began to unveil themselves. Believing their own ruse, colonials at first misread the unveiled Algerian women as pieces of 'sound currency' circulating between the casbah and the white city, mistaking them for the visible coinage of cultural conversion.[48] For the *fidai*, however, the militant woman was 'his arsenal,' a technique of counterinfiltration, duplicitously penetrating the body of the enemy with the armaments of death.

So eager is Fanon to deny the colonial rescue fantasy that he refuses to grant the veil any prior role at all in the gender dynamics of Algerian society. Having refused the colonial's desire to invest the veil with an essentialist meaning (the sign of women's servitude), he bends over backward to insist on the veil's semiotic innocence in Algerian society. The veil, Fanon writes, was no more than 'a formerly inert element of the native cultural configuration.'[49] At once the veil loses its historic mutability and becomes a fixed, 'inert' element in Algerian culture: 'an undifferentiated element in a homogeneous whole.'[50] Fanon denies the 'historic dynamism of the veil' and banishes its intricate history to a footnote, from where, however, it displaces the main text with the insistent force of self-division and denial.[51]

Fanon's thoughts on women's agency proceed through a series of contradictions. Where, for Fanon, does women's agency begin? He takes pains to point out that women's militancy does not precede the national revolution. Algerian women are not self-motivating agents, nor do they have prior histories or consciousness of revolt from which to draw. Their initiation in the revolution is learned, but it is not learned from other women or from other societies, nor is it transferred analogously from local feminist grievances. The revolutionary mission is 'without apprenticeship, without briefing.'[52] The Algerian woman learns her 'revolutionary mission instinctively.'[53] This theory is not, however, a theory of feminist spontaneity, for women learn their militancy only at men's invitation. Theirs is a *designated agency* – an agency by invitation only. Before the national uprising, women's agency was null, void, inert as the veil. Here Fanon colludes not only with the stereotype of women as bereft of historical motivation, but he also resorts, uncharacteristically, to a reproductive image of natural birthing: 'It is an authentic birth in a pure state.'[54]

Why were women invited into the revolution? Fanon resorts immediately to a mechanistic determinism. The ferocity of the war was such, the urgency so great, that sheer structural necessity dictated the move: 'The revolutionary wheels had assumed such proportions; the mechanism was running at a given rate. The machine would have to be complicated.'[55] Female militancy, in short, is simply a passive offspring of male agency and the structural necessity of the war. The problem of women's agency, so brilliantly raised as a question, is abruptly foreclosed.

Women's agency for Fanon is thus agency by designation. It makes its appearance not as a direct political relation to the revolution but as a mediated, domestic relation to a man: 'At the beginning, it was the married women who were contacted. Later, widows or divorced women were designated.'[56] Women's first relation to the revolution is constituted as a domestic one. But domesticity, here, also constitutes a relation of possession. The militant was, in the beginning, obliged to keep 'his woman' in 'absolute ignorance.'[57] As designated agents, moreover, women do not commit themselves: 'It is relatively easy to commit oneself . . . The matter is a little more difficult when it involves designating someone.'[58] Fanon does not consider the possibility of women committing themselves to action. He thus manages women's agency by resorting to contradictory frames: the authentic, instinctive birth of nationalist fervor; the mechanical logic of revolutionary necessity; male designation. In this way, the possibility of a distinctive feminist agency is never broached.

Once he has contained women's militancy in this way, Fanon applauds women for their 'exemplary constancy, self-mastery and success.'[59] Nonetheless, his descriptions of women teem with instrumentalist similes and metaphors. Women are not women, they are 'fish,' they are 'the group's lighthouse and barometer,' the *fidai's* 'women-arsenal.'[60] Most tellingly, Fanon resorts to a curiously eroticized image of militarized sexuality. Carrying the men's pistols, guns and grenades beneath her skirts, 'the Algerian woman penetrates a little further into the flesh of the Revolution.'[61] Here, the Algerian woman is not a victim of rape but a masculinized rapist. As if to contain the unmanning threat of armed women – in their dangerous crossings – Fanon masculinizes the female militant, turning her into a phallic substitute, detached from the male body but remaining, still, the man's 'woman-arsenal.' Most tellingly, however, Fanon describes the phallic woman as penetrating the flesh of the 'revolution,' not the flesh of the colonials. This odd image suggests an unbidden fear of emasculation, a dread that the arming of women might entail a fatal unmanning of Algerian men. A curious instability of gender power is here effected as the women are figured as masculinized and the male revolution is penetrated.

Fanon's vision of the political role of the Algerian family in the national uprising likewise proceeds through contradiction. Having brilliantly shown how the family constitutes the first ground of the colonial onslaught, Fanon seeks to reappropriate it as an arena of nationalist resistance. Yet the broader implications of the politicizing of family life are resolutely naturalized after the revolution. Having recognized that women 'constituted for a long time the fundamental strength of the occupied,' Fanon is reluctant to acknowledge any gender conflict or feminist grievance within the family prior to the anti-colonial struggle, or after the national revolution.[62] Although, on the one hand, he admits that in 'the Algerian family, the girl is always a notch behind the boy,' he quickly insists that she is assigned to this position 'without being humiliated or neglected.'[63] Although the men's words are 'Law,' women 'voluntarily' submit themselves to 'a form of existence limited in scope.'[64]

The revolution shakes the 'old paternal assurance' so that the father no longer knows 'how to keep his balance,' and the woman 'ceased to be a complement for man.'[65] It is telling, moreover, that in his analysis of the family, the category of mother does not exist. Women's liberation is credited entirely to national liberation and it is only with nationalism that women 'enter into history.' Prior to nationalism, women have no history, no resistance, no independent agency.[66] And since the national revolution automatically revolutionizes the family, gender conflict naturally vanishes after the revolution. Feminist agency, then, is contained by and subordinated to national agency and the heterosexual family is preserved as the 'truth' of society – its organic, authentic form. The family is revolutionalized, taken to a higher plane through a Hegelian vision of transcendence, but the rupturing force of gender is firmly foreclosed: 'The family emerges strengthened from this ordeal.'[67] Women's militancy is contained within the postrevolutionary frame of the reformed, heterosexual family, as the natural image of national life.

In the post-revolutionary period, moreover, the tenacity of the father's 'unchallengeable and massive authority' is not raised as one of the 'pitfalls' of the national consciousness.[68] The Manichean dialectic – as generating an inherent resistant agency – does not, it seems, apply to gender. Deeply reluctant as he is to see women's agency apart from national agency, Fanon does not foresee the degree to which the Algerian National Liberation Front (FLN) will seek to co-opt and control women, subordinating them unequivocally once the revolution is won.

A feminist investigation of national difference might, by contrast, take into account the dynamic social and historical contexts of national struggles; their strategic mobilizing of popular forces; their myriad, varied trajectories; and their relation to other social institutions. We might do well to develop a more theoretically complex and strategically subtle genealogy of nationalisms.

Feminism and nationalism

For many decades, African women have been loath to talk of women's emancipation outside the terms of the national liberation movement.[69] During the 1960s and 1970s, black women were understandably wary of the middle-class feminism that was sputtering fitfully to life in the white universities and suburbs. African women raised justifiably skeptical eyebrows at a white feminism that vaunted itself as giving tongue to a universal sisterhood in suffering. At the same time, women's position within the nationalist movement was still precarious and women could ill afford to antagonize men so embattled, who were already reluctant to surrender whatever patriarchal power they still enjoyed.

In recent years, however, a transformed African discourse on feminism has emerged, with black women demanding the right to fashion the terms of nationalist feminism to meet their own needs and situations.[70] On May 2, 1990, the National Executive of the ANC issued a historic 'Statement on the Emancipation of Women,' which forthrightly proclaimed:

> The experience of other societies has shown that the emancipation of women is not a by-product of a struggle for democracy, national liberation or socialism. It has to be addressed within our own organisation, the mass democratic movement and in the society as a whole.

The document is unprecedented in placing South African women's resistance in an international context; in granting feminism independent historic agency; and in declaring, into the bargain, that all 'laws, customs, traditions and practices which discriminate against women shall be held to be unconstitutional.' If the ANC remains faithful to this document, virtually all existing practices in South Africa's legal, political and social life will be rendered unconstitutional.

A few months later, on June 17 1990, the leaders of the ANC Women's Section, recently returned to South Africa from exile, insisted on the strategic validity of the term feminism:

> Feminism has been misinterpreted in most third world countries . . . there is nothing wrong with feminism. It is as progressive or reactionary as nationalism. Nationalism can be reactionary or progressive. We have not got rid of the term nationalism. And with feminism it is the same.

Feminism, they believed, should be tailored to meet local needs and concerns.

Yet very real uncertainties for women remain. So far, theoretical and strategic analyses of South Africa's gender imbalances have not run deep. There has been little strategic rethinking of how, in particular, to transform labor relations within the household and women are not given the same political visibility as men. At a Congress of South African Trade Unions (COSATU) convention, trade union women called for attention to sexual harrassment in the unions, but their demand was brusquely flicked aside by male unionists as a decadent symptom of 'bourgeois imperialist feminism.' Lesbian and gay activists have been similarly condemned as supporting life-styles that are no more than invidious imports of empire.[71]

There is not only one feminism, nor is there one patriarchy. Feminism is imperialist when it puts the interests and needs of privileged women in imperialist countries above the local needs of disempowered women and men, borrowing from patriarchal privilege. In the last decade, women of color have been vehement in challenging privileged feminists who don't recognize their own racial and class power. In an important article, Chandra Mohanty challenges the appropriation of women of color's struggles by white women, specifically through the use of the category 'Third World Woman' as a singular, monolithic and paradigmatically victimized subject.[72]

Denouncing all feminisms as imperialist, however, erases from memory the long histories of women's resistance to local and imperialist patriarchies. As Kumari Jayawardena notes, many women's mutinies around the world pre-dated Western feminism or occurred without any contact with Western feminists.[73] Moreover, if all feminisms are derided as a pathology of the West, there is a very real danger that Western, white feminists will remain hegemonic, for the simple reason that such women have comparatively privileged access to publishing, the international media, education and money.

A good deal of this kind of feminism may well be inappropriate to women living under very different situations. Instead, women of color are calling for the right to fashion feminism to suit their own worlds. The singular contribution of nationalist feminism has been its insistence on relating feminist struggles to other liberation movements.

All too frequently, male nationalists have condemned feminism as divisive, bidding women hold their tongues until after the revolution. Yet feminism is a political response to gender conflict, not its cause. To insist on silence about gender conflict when it already exists is to cover and thereby ratify, women's disempowerment. Asking women to wait until after the revolution serves merely as a strategic tactic to defer women's demands. Not only does it conceal the fact that nationalisms are from the outset constituted in gender power, but, as the lessons of international history portend, women who are not empowered to organize during the struggle will not be empowered to organize after the struggle. If nationalism is not transformed by an analysis of gender power, the nation-state will remain a repository of male hopes, male aspirations and male privilege.

All too often, the doors of tradition are slammed in women's faces. Yet traditions are both the outcome and the record of past political contests as well as the sites of present contest. In a nationalist revolution, both women and men should be empowered to decide which traditions are outmoded, which should be transformed and which should be preserved. Male nationalists frequently argue that colonialism or capitalism has been women's ruin, with patriarchy merely a nasty second cousin destined to wither away when the real villain expires. Yet nowhere has a national or socialist revolution brought a full feminist revolution in its train. In many nationalist or socialist countries, women's concerns are at best paid lip service, at worst greeted with hilarity. If women have come to do men's work, men have not come to share women's work. Nowhere has feminism in its own right been allowed to be more than the maidservant to nationalism.

A crucial question thus remains for progressive nationalism: Can the iconography of the family be retained as the figure for national unity, or must an alternative, radical iconography be developed? In South Africa currently, critical questions for women remain. The Freedom Charter promises that the land will be given to those that work it. Since, in South Africa, women do much of the farming, will the land be given to them? Or, as in so many other post-independence countries, will the property rights, the technology, the loans and aid, be given to men? Will men become the principal beneficiaries of the rights and resources of the new nation state? When these questions are answered, perhaps we can begin to talk about a new South Africa.

Frantz Fanon's prescient warnings against the pitfalls of the national consciousness were never more urgent than now. For Fanon, nationalism gives vital expression to popular memory and is strategically essential for mobilizing the populace. At the same time, no one was more aware than Fanon of the attendant risks of projecting a fetishistic denial of difference onto a conveniently abstracted 'collective will.' In South Africa, to borrow Fanon's phrase, national transformation is 'no longer in a future heaven.' Yet the current situation gives sober poignancy, especially for women, to the lines from Giles Pontecorvo's famous film on the Algerian national war of liberation, *The Battle of Algiers*: 'It is difficult to start a revolution, more difficult to sustain it. But it's later, when we've won, that the real difficulties will begin.'

Notes

1 See Eric Hobsbawm's critique of nationalism in *Nations and Nationalism Since 1780* (Cambridge: Cambridge University Press, 1990). Ernest Gellner, *Thought and Change* (London: Weidenfeld and Nicholson, 1964) and *Nations and Nationalism* (Oxford: Blackwell, 1983).
2 Benedict Anderson, *Imagined Communities* (London: Verso, 1983, 1991), p. 6.
3 Cynthia Enloe, *Bananas, Beaches and Bases: Making Feminist Sense of International Politics* (Berkeley: University of California Press, 1989), p. 44.
4 Fanon, *The Wretched of the Earth*, trans. Constance Farrington (London: Penguin, 1963), p. 30.
5 Elleke Boehmer, 'Stories of Women and Mothers: Gender and Nationalism in the Early Fiction of Flora Nwapa,' in Susheila Nasta, ed., *Motherlands: Black Women's Writings from Africa, the Caribbean and South Asia* (London: The Women's Press, 1991), p. 5.
6 Boehmer, 'Stories of Women and Mothers,' p. 6.
7 Nira Yuval-Davis and Floya Anthias, eds., *Women–Nation–State* (London: Macmillan, 1989), p. 7.
8 Yuval-Davis and Anthias, *Women–Nation–State*, p. 1.
9 Tom Nairn, *The Break-up of Britain* (London: New Left Books, 1977).
10 Deniz Kandiyoti, 'Identity and Its Discontents: Women and the Nation' *Millenium: Journal of International Studies* 20, 3 (1991): 431.
11 Homi K. Bhabha, ed., *Nation and Narration* (London: Routledge, 1991), p. 1.
12 Susan Buck-Morss, *The Dialectics of Seeing: Walter Benjamin and the Arcades Project* (Cambridge, Mass.: MIT Press, 1990), p. 67.
13 Buck-Morss, *The Dialectics of Seeing*, p. xiv.
14 Hobsbawm, *Nations and Nationalism*, p. 151.
15 Frantz Fanon, *Black Skin, White Masks*, trans. Charles Lam Markmann (London: Pluto Press, 1986), p. 141.
16 Fanon, *Black Skin, White Masks*, p. 142.
17 Fanon, *Black Skin, White Masks*, p. 143.
18 Fanon, *Black Skin, White Masks*, pp. 141–142.
19 Fanon, *Black Skin, White Masks*, p. 13.
20 Fanon, *Black Skin, White Masks*, p. 13.
21 Fanon, *Black Skin, White Masks*, p. 10.
22 Fanon, *The Wretched of the Earth*, p. 29.
23 Fanon, *The Wretched of the Earth*, p. 30.
24 Edward Said, *Culture and Imperialism* (London: Chatto and Windus, 1993), p. 326.
25 Bhabha, introduction to Fanon, *Black Skin, White Masks*, p. ix.
26 Fanon, *Black Skin, White Masks*, p. 46.
27 Fanon, *Black Skin, White Masks*, p. 46.
28 Fanon, *Black Skin, White Masks*, p. 63.

29 Bhabha, intro. to Fanon, *Black Skin, White Masks*, p. ix.
30 Bhabha, intro. to Fanon, *Black Skin, White Masks*, p. xxvi.
31 Fanon, *Black Skin, White Masks*, p. 81.
32 Black Skin, White Masks, p. 16.
33 Fanon, *Black Skin, White Masks*, p. 159.
34 Fanon, *Black Skin, White Masks*, p. 92.
35 Bhabha, intro. to Fanon, *Black Skin, White Masks*, p. xxv.
36 Bhabha, intro. to Fanon, *Black Skin, White Masks*, p. xi.
37 Bhabha, intro. to Fanon, *Black Skin, White Masks*, p. xi.
38 Said, *Culture and Imperialism*, p. 89.
39 Terrence Eagleton, 'Nationalism, Irony and Commitment,' in Terrence Eagleton, Fredric Jameson and Edward Said, *Nationalism, Colonialism and Literature* (Minneapolis: University of Minnesota Press, 1990), p. 28.
40 Eagleton, 'Nationalism,' p. 25.
41 Frantz Fanon, 'Algeria Unveiled,' in *A Dying Colonialism*, trans. Haakon Chevalier (New York: Grove Press, 1965), p. 42.
42 Fanon, 'Algeria Unveiled,' p. 37.
43 Fanon, 'Algeria Unveiled,' p. 53.
44 Fanon, 'Algeria Unveiled,' pp. 37–38.
45 Fanon, 'Algeria Unveiled,' p. 38.
46 Fanon, 'Algeria Unveiled,' p. 35.
47 Fanon, 'Algeria Unveiled,' p. 39.
48 Fanon, 'Algeria Unveiled,' p. 42.
49 Fanon, 'Algeria Unveiled,' p. 46.
50 Fanon, 'Algeria Unveiled,' p. 47.
51 Fanon, 'Algeria Unveiled,' p. 63.
52 Fanon, 'Algeria Unveiled,' p. 50.
53 Fanon, 'Algeria Unveiled,' p. 50.
54 Fanon, 'Algeria Unveiled,' p. 50.
55 Fanon, 'Algeria Unveiled,' p. 48.
56 Fanon, 'Algeria Unveiled,' p. 51.
57 Fanon, 'Algeria Unveiled,' p. 48.
58 Fanon, 'Algeria Unveiled,' p. 49.
59 Fanon, 'Algeria Unveiled,' p. 54.
60 Fanon, 'Algeria Unveiled,' pp. 58, 54.
61 Fanon, 'Algeria Unveiled,' p. 54.
62 Fanon, 'Algeria Unveiled,' p. 66.
63 Fanon, 'Algeria Unveiled,' p. 105.
64 Fanon, 'Algeria Unveiled,' p. 66.
65 Fanon, 'Algeria Unveiled,' p. 109.
66 Fanon, 'Algeria Unveiled,' p. 107.
67 Fanon, 'Algeria Unveiled,' p. 116.
68 Fanon, 'Algeria Unveiled,' p. 115.
69 The ANC delegation to the Nairobi Conference on Women in 1985 declared: 'It would be suicidal for us to adopt feminist ideas. Our enemy is the system and we cannot exhaust our energies on women's issues.'
70 At a seminar titled 'Feminism and National Liberation,' convened by the Woman's Section of the ANC in London in 1989, a representative from the South African Youth Congress (SAYCO) exclaimed: 'How good it feels that feminism is finally accepted as a legitimate school of thought in our struggles and is not seen as a foreign ideology.'

71 For a groundbreaking book on the history, politics and culture of lesbian and gay life in South Africa, see Edwin Cameron and Mark Gewisser, eds., *Defiant Desire: Gay and Lesbian Lives in South Africa* (New York: Routledge, 1994).
72 Chandra T. Mohanty, 'Under Western Eyes: Feminist Scholarship and Colonial Discourses' in Chandra T. Mohanty, Ann Russo and Lourdes Torres, eds. *Third World Women and the Politics of Feminism* (Bloomington: Indiana University Press, 1991), p. 52.
73 Kumari Jayawardena, *Feminism and Nationalism in the Third World* (London: Zed Press, 1986).

31 Simona Sharoni

'Middle East Politics through Feminist Lenses: Toward Theorizing International Relations from Women's Struggles'

From: *Alternatives: Social Transformation and Humane Governance* **18**(1), 5–28 (1993)

> Less partial and distorted descriptions and explanations of nature and social relations tend to result when research starts from the lives of women of Third World descent rather than only from the lives of men or of women of European descent.
> Sandra Harding[1]

> The very practice of remembering against the grain of "public" or hegemonic history, of locating the silences and the struggle to assert knowledge which is outside the parameters of the dominant, suggests a rethinking of sociality itself.
> Chandra Talpade Mohanty[2]

That Middle Eastern women have been leading opposition movements and taking positions on international politics would not come as a surprise to those who are familiar with histories of women's struggles in the region. Widespread women's resistance to the recent Gulf War is only the latest chapter of an unwritten genealogy of Middle Eastern women's resistance to encroaching imperial powers.[3] In the late nineteenth and early twentieth centuries, women's movements in Egypt and Palestine used feminist and nationalist discourses to articulate positions against imperial and neocolonial practices.[4] Almost a century later, in the aftermath of the Gulf War, the Egyptian government dissolved the Arab Women's Solidarity Association (AWSA) following a September 1990 conference AWSA had organized on Arab women and journalism in which the participants opposed both Iraq's invasion of Kuwait and the burgeoning US-organized military intervention. Egyptian feminists argue that this act was not simply a "domestic" political backlash or regime pandering to Islamist positions on women's rights; they are convinced that their intervention into the international politics of the Gulf War was taken quite seriously and perceived as a threat to the "stability" of the region as defined by privileged male elites in Washington, DC, and in Cairo.[5]

The above examples are indicative of the complex (and usually concealed) relationships between gender, Middle East politics, and international relations that could be explored and theorized through detailed studies of women's struggles in the Middle East. I owe some credit to both Sandra Harding and Chandra Mohanty for the phrase "theorizing from women's struggles." The phrase is a synthesis of Mohanty's exploration of the "links among the histories and struggles of third world women against racism, sexism, colonialism, imperialism and monopoly capital" and of Harding's most recent work, which argues that "if feminist research and scholarship were to start from women's lives, they would have to start from all women's lives."[6] The disruption of old forms of thinking called for by Harding and Mohanty is creating space in which women's initiatives and struggles in the Middle East and elsewhere may elicit new interpretations regarding the gendered nature of local, regional, and international politics.

In this space, for example, one may begin to read against the grain of hegemonic interpretations of the Israeli-Palestinian conflict by paying attention to feminist interventions in the international policymaking arena. Long before there was any official acquiescence to the 1991 Madrid peace conference, and unnoticed by the gender-blind international media, Palestinian and Israeli women engaged in a series of international conferences to develop feminist frameworks for Middle East peace. They alerted the international community to the serious need for an international initiative to address the broader Middle Eastern disputes, of which the Israeli-Palestinian conflict is at the forefront.[7] One might also recognize that the participation of prominent Palestinian women such as Hanan Mikhail-Ashrawi and Zahira Kamal at the Middle East peace talks in 1991–1992 was not simply decorative affirmative action. It marked the entry of women articulating within the arena of Middle East and international politics explicit feminist agendas grounded in the complex ongoing struggles of Palestinian women in the West Bank and Gaza Strip.[8] Demonstrating the conflicts between feminist and nationalist identities, some Israeli feminist peace activists, whose own voices have been excluded from the political process, went to express solidarity and congratulate Mikhail-Ashrawi and Kamal for their significant role in the process.[9]

Working in the space opened by critiques such as those of Harding and Mohanty, this article seeks to demonstrate how theory and research in international relations (IR) and related fields can benefit from looking more carefully at women's struggles and at the relationship of gender and international politics. My intention is not to provide a comprehensive theory or a complete historical account of the struggles of women in the Middle East, but rather to introduce a different location from which Middle East politics could be engaged and theorized. The term "struggles," which is used extensively in this article, includes a broad range of feminist interventions into local regional and international politics. I prefer the term "women's struggles" to "women's lives" because it evokes assertive and dynamic images of women as political actors and thus contributes to the critiques of stereotypical depictions of Middle Eastern women.

This article outlines three major aspects of the move toward theorizing IR from women's struggles in the Middle East: (1) the need to situate ourselves and our projects in the sociopolitical contexts within which we work; (2) the need to critically examine previous work on the subject; and (3) the need to listen to women's struggles in order to articulate new questions and conceptual frameworks. By situating myself vis-à-vis the "object of study," by critiquing sterotypical representations of Middle Eastern women and the political implications of those stereotypes, and by calling attention to questions, contradictions, and complexities embedded in the discourses that inform the lives and struggles of women in the region, I intend to open up space for theoretical frameworks that will not only make feminist sense of Middle East politics, but also engage in writing feminist IR theories.[10]

Situated knowledge: authorship, identity, and responsibility

> Knowledge production in literary and social-scientific disciplines is clearly an important discursive site for struggle. The practice of scholarship is also a form of rule and of resistance, and constitutes an increasingly important arena for third world feminisms.
>
> Chandra Talpade Mohanty[11]

> The concern to develop socially responsible research has forced us to question continually the relationship between investigation and the needs and rights of the people. It has also forced us to rethink our research practices and our own motives for engaging in this activity.
>
> Rina Benmayor[12]

Throughout the past decade, feminists who challenge traditional disciplinary paradigms of social science research have joined feminists of Third World descent in attempts to develop socially responsible projects grounded in the relationship between research and the needs and rights of the people in the complex sociopolitical context within which they unfold.[13] Although she is aware of new trends of self-reflexive feminist theorizing, research, and practice, Daphne Patai stresses that "because 'women,' gender notwithstanding, are not a monolithic block, ethical questions about our actions and implications of those actions are especially appropriate if such research is [to become] for women and not merely by or about them."[14]

Can feminists theorize and locate the links between the histories, multiple identities, and experiences of women without perpetuating the global power relations that have been reinforced through the institutionalization of dichotomies such as self/other, researcher/researched, insider/outsider, and theory/practice? Can we contribute to dominance in spite of our liberatory intentions? Because both these questions could be answered in the affirmative, we are left with a host of challenges and unresolved tensions, which are constant reminders that our academic projects do not unfold in a controlled laboratory environment but within a complex web of global and local power relations. The calls for ethically responsible research, and the emphasis on the situatedness of "the author" in the processes of knowledge production have recently led feminist scholars and activists "to construct projects that more directly and immediately benefit the women who are researched."[15] By calling attention to the lens through which women's lives and struggles are interpreted and by taking into account the location(s) and political contexts that inform the writing and the political implications of particular interventions, feminists have been trying to rearticulate the relationship between the researcher and researched and the author and text. In this context, Middle Eastern feminists have recently engaged in projects of studying their own societies and reflecting on the problems that might be involved in carrying out such projects.

A significant contribution to this emerging body of research is the anthology, *Arab Women in the Field: Studying Your Own Society*, edited by Soraya Altorki and Camillia Fawzi El-Solh.[16] The contributors challenge the power asymmetries between researcher and researched and examine the political implications of their academic work. For example, Soheir Morsy, an Egyptian-born feminist who came to the United States with her family at age fifteen, reflects on the transformation she underwent in the United States and then during her field research in the Middle East:

> It was my Arab identity that proved to be of greater import, gaining unprecedented significance with the outbreak of the 1967 June War between Israel and the Arabs. Watching television coverage of the suffering of Arab civilians and the humiliating defeat of the Arab armies was a most agonizing experience. . . . My readings in search of explanations of my homeland's underdevelopment and answers as to how it can be transformed, promoted me to focus on the lives of the poor, exploited and powerless.[17]

In what she describes as an attempt to reconcile her personal motivation with academic scholarship, Morsy reflects on the politics of her academic practice:

> I reasoned that unlike Western researchers, my study of "my people" would be to our benefit. Furthermore, I convinced myself that anthropological knowledge, as a form of power, has the potential for use not only as a means for exploitation, but also as an instrument of liberation, the critical factor being the political framework within which it is applied. . . . I now recognize that while academic training does not offer us strategies for liberating "our people," it empowers us to articulate and defend their interests, if we so choose.[18]

The significance of Morsy's self-reflexive work lies in her political commitment and sensitivity. She is well aware of how the different aspects of her fractured identity are perceived by the people in the village and how their perceptions of her have been politically constructed in accordance with significant events:

> I had arrived in Fatiha less than a year after the 1973 October War between Egypt and Israel, when many Egyptians had still not been made to forget (through the

> mass media), the role of the US in supporting the Zionist state. Unfortunately, I became the target of the anger of some of the villagers. For example, during a household census-related interview with an old woman, I asked her about the composition of her family. I was unaware of the fact that one of her sons had been reported missing, possibly dead or imprisoned by the Zionists. She screamed at me and told me to leave her alone, ending our conversation thus: "Go ask your Americans, where did my son go?"[19]

At the same time, on the opposite side of the border in Israel, I lived through the 1967 and 1973 wars without yet being aware of their significant role in the construction of my identity. I was only six years old in 1967 in a country that was in a state of euphoria, celebrating a "victory" that included the occupation of the West Bank, Gaza Strip, and the Golan Heights. The myths about the omnipotence of the Israeli military that the 1967 war invoked resulted in the further militarization of Israeli society and the institutionalization of "national security" as the top priority in Israel. Only years later, I realized that this fact played a significant role in the social and political construction of gender (as well as race and class) relations in Israel, and has contributed a great deal to the militarization of the region and to the escalation of domestic, regional, and global conflicts.[20]

It was not until the Israeli invasion of and war in Lebanon in 1982 that I became politically active in opposition to my country's policies. The development of my feminist identity, my commitment to the struggle to end the Israeli occupation of the West Bank and Gaza Strip, and the pursuit of a doctoral degree in conflict analysis and resolution have enabled me to write and speak up against injustices committed in my name. My feminist identity was shaped primarily in a political climate that prevented me from learning about the struggles of my sisters across the border and from coming to terms with my Middle Eastern identity. Therefore, by introducing myself as a Middle Eastern feminist and expressing solidarity with the struggles of women in the Middle East to affect regional and international politics, I defy attempts to control my identity through the imposition of sovereign national boundaries. Recognizing the interconnectedness of militarism and sexism in Israel, understanding that my identity as an Israeli feminist is inseparable from the rest of my political views and especially from my positions on the Israeli-Palestinian conflict, and enjoying the privilege of collaborating with other Middle Eastern feminists (particularly Palestinian feminists) on several projects, I feel part of an "imagined community" of Middle Eastern feminists. It is with this sense of commitment and responsibility to our separate and joint struggles that I write this article, recognizing that being able to use my voice to share my perspective is in itself a privilege for which many of my sisters in Israel as in other parts of the Middle East are still struggling.

My choice to draw linkages between Morsy's feminist scholarship and political commitment and my Middle Eastern feminist identity, political beliefs, and academic work is grounded in the realization that it is impossible to produce "neutral" scholarship on Middle East politics in general, or on women in the Middle East in particular. I am working within the space opened by a new genre in feminist critique, which insists on experience as a potential location for the construction of feminist theory, emphasizing that "the confessional can lead to the conceptual and the political."[21] The space opened by this new mode of feminist scholarship provides significant examples of self-reflexive and explicitly political academic projects that are grounded in the sociopolitical context that they seek to change.[22]

Because studies of women's life experiences and political struggles in the Middle East are, without doubt, political projects, they require careful examination of the methods, ethics, and political implications that are involved. For example, Julie Peteet, who lived and worked among Palestinian women in refugee camps in Lebanon, points out that

> Palestinian women had a specific intent when participating in researching or talking to outsiders. The women I worked with viewed me as a foreigner to whom by telling their story they would be conveying it to the West. To them the ethnographic experience was an experience in dialogue in international politics.[23]

Peteet's remark is of particular relevance to the present project, for it highlights the interplay between the lives and struggles of Palestinian women in Lebanon and the broader political context of global power relations. Peteet's point substantiates the contention that scholars and researchers who engage in projects such as theorizing Middle East politics from women's struggles have a responsibility to listen closely, grasp the messages, and deliver them to their audiences as they constantly ask themselves if and how women in the Middle East can benefit from such projects.

The politics of re-presentation: challenging stereotypes, contextualizing women's struggles

Despite the fact that Middle Eastern women have been raising their voices and organizing for social emancipation and social and political change since the beginning of this century, the stories of their struggles have been marginalized and written out of conventional IR scholarship on Middle East politics. These women are portrayed in academic studies and in the media usually as passive victims with little or no attention paid to their struggles to achieve control over their lives and over social and political changes in their societies. There are three interlocking stereotypical images of Middle Eastern women: orientalist, sexist, and nationalist.

Orientalist and sexist seem to go together in the scholarship and media reports of foreign men and women. There is a tendency, first, to establish the essential otherness of Middle Eastern women by juxtaposing Western and Middle Eastern cultures and images of Occidental women with Oriental women. When the author is a man, the interplay between orientalist and sexist images is unequivocal.[24] Sexist images then reconstruct a traditional division of labor based on rigid distinctions between the so-called "public" and "private" spheres: Men are portrayed as being in charge of the public-political sphere whereas women are confined to the domestic sphere. Middle Eastern feminists argue that both such images can be and are utilized to reinforce women's subordination, to silence women's voices, and to complicate their ongoing struggles to affect politics.[25]

Nationalist stereotypical images of Middle Eastern women, by contrast, have emerged in the context of women's participation in anticolonial struggles and national liberation movements throughout the region. Two common representations are of women as fighters and of women as mothers of the nation.[26] Marnia Lazreg challenges the political agendas behind the orientalist and sexist stereotypical portrayals of Middle Eastern women by linking her critique of the representation of Middle Eastern women to the broader context of international relations in the region, emphasizing that "the intense current interest in 'Middle Eastern women' is occurring at a time when the 'Middle East' has been neutralized as a self-sustaining political and economic force."[27] The stereotypical depictions of Middle Eastern women reinforce the existing power disparities in the international political arena. Feminist examinations of US media coverage, such as that of the recent Gulf War by Cynthia Enloe, have underscored the political consequences of the juxtaposition of an archetypal veiled Arab woman with a "liberated" US woman soldier: "by contrasting the allegedly liberated US woman tank mechanic with the Saudi woman deprived of a driver's license, US reporters are implying that the United States is the advanced civilized country whose duty it is to take the lead in resolving the Persian Gulf crisis."[28]

The same powerful subtext was at play during the Iranian revolution of 1979. During that time, millions of Iranian women stepped into the political arena. Nayereh Tohidi points out that women's

> involvement in the revolution took many forms: some collected and disseminated news or distributed leaflets; others gave shelter to the wounded or to political activists under attack. Many actively marched and demonstrated in the streets, some went so far as to help erect barricades against the police, and a few even took up arms and went underground as members of a guerrilla movement.[29]

The Western media, however, did not display the range of political activities carried out by Iranian women during the revolution. Instead, reporters used the images of veiled Iranian women demonstrating against the Shah to reinforce the dichotomies of "Us" (the "West," especially the United States) versus "Them" (Iran) in order to mobilize public opinion in the United States and in Europe against the Islamic revolution and in support of continued US intervention in Iranian politics. Nevertheless, the media as well as Western policymakers failed to notice that Iranian women did not show obeisance to the stereotypes. Tohidi calls attention to the fact that

> many Iranian activists at the time, both women and men, considered the veil part of the superstructure or a secondary phenomenon which bothered only Western feminists or a few Iranian women intellectuals. The immediate concern was to rid the country of the Shah's regime and his imperialist supporters. . . . many women, even

nonreligious, nontraditional, and highly educated women, took up the veil as symbol of solidarity and opposition to the Shah.[30]

The complex and detailed accounts that are required in order to challenge crude dichotomies such as veiled Middle Eastern women versus liberated Western women call attention to another example: the political crisis in Algeria following the national elections in December 1991. Shafla Jemame, an Algerian socialist feminist, highlights the direct relationship between global intervention in Algeria, state policies, and women's daily battles: "women do not only have to deal with democratic questions. Another issue is the economic crisis, which has worsened in recent months. The agreement made by the government with the International Monetary Fund (IMF) has meant massive job losses."[31] She points out the direct implications of Algeria's economic dependency on the West – the rise in unemployment, the decline in purchasing power resulting from the government's decision to abolish subsidies on basic necessities, the shortage in housing, and the cuts in health care programs that used to be free – and emphasizes that "the subjection of the government to the IMF's diktats will only serve to make reactionary ideology of the fundamentalists more popular."[32] To ensure the "stability" of the region, which is essential for unchallenged US hegemony in the region, the Western media has been mobilized again: discourses of democracy, human rights, and women's rights are suddenly invoked, rendering invisible the destructive effects of the global capitalist economy. Understanding the political economy of foreign interventions in Middle East politics is crucial to feminist attempts to alter the gendered politics of the region and intervene in the broader arena of international politics.

To counter Western orientalist and sexist depictions of women in the Middle East, which reinforce the image of women as victims, newly published works on women in the Middle East record the previously unknown stories of women's struggles in the region through biographies, oral histories, testimonies, and interviews, as well as through critical analyses of the political and economic contexts within which they unfold.[33] By taking into consideration Middle Eastern women's perspectives on political developments that impact their lives and struggles, we may gain insights into the subtexts and unwritten arrangements of international politics in the Middle East. That is, detailed accounts of women's resistance and struggles to influence the course of politics in the Middle East hold the potential to map alternative feminist interpretations of both regional and international politics. Thus, demonstrating how Orientalist and sexist images of Middle Eastern women contribute to the construction of US and European superiority and the legitimization of further political and economic exploitation is an example of feminist interventions into the making of international politics.

Despite the growing body of feminist scholarship that challenges orientalist and sexist stereotypical representations of women in the Middle East, the project of challenging nationalist images, which is far more complex and politically risky, is only in its early stages. One problem that arises from the new body of scholarship by Middle Eastern feminists and their allies is a tendency to replace women as victims with a new archetype of women as heroines. The latter image is empowering, but, because it is confined by nationalist discourses it subsumes too much diversity and complexity and it falls short of addressing the dynamic relationship between different experiences and struggles of Middle Eastern women and the historical and sociopolitical contexts within which they unfold. The image of women as mothers of the nation is a particular example of the archetypes of women as heroines. Palestinian sociologist Nahla Abdo elaborates on the centrality of this image among Palestinians, especially since the beginning of the Palestinian uprising, known as the intifada. In December 1987, according to Abdo:

> The construction of motherhood equals nationhood within the Palestinian context emerged as an expression of Palestinian lived reality. Expulsion from the homeland and refugeeism in foreign territories provided the impetus for the mother-nation relationship. . . . It is not surprising, therefore, that the discourse on women and the intifada focused largely on one particular image of social relations, that is, "the heroic mother." She is the middle-aged woman, usually in her national embroidered dress, who is often in demonstrations. She has invented effective tactics for saving children in danger of being arrested or beaten by Israeli soldiers. By expanding her mothering role to encompass all other children, she has dissolved herself into the wider nation.[34]

In this context, to challenge the image of motherhood could be interpreted as lack of support for the national struggle. On the other hand, Palestinian women activists have addressed the short- and long-term ramifications of the portrayal of women in traditional static auxiliary roles such as the mothers of the nation. Many Palestinian women activists make references to the disillusions of Algerian women following their participation in their nation's struggle against the French colonial rule, stressing that they are determined not to let the erasure of women's contributions happen in their case. Some, such as Sahar Khalifa, a prominent Palestinian feminist, activist, and novelist, explicitly challenge nationalist images. In a recent interview, Khalifa said:

> I am not at all happy when I see and know women who have sacrificed so much for the national struggle and in the end are left with nothing. . . . I don't want to be a heroine for the sake of the nation; I don't want my body to be the bridge for the state; I want to undergo a transformation, to have the experience of changing, to reap the benefits of the struggle, not to serve as the fuel for the national flame.[35]

Voices such as Khalifa's open up space for feminist theorizing that will not only challenge the simplistic juxtaposition of stereotypical images of Middle Eastern women with those of Western women, but will also call attention to the limitations of contrasting stereotypical images of women as victims with those of women as heroines. In order to make feminist sense of Middle East politics, it is not enough to record a wide range of images and representations of women in struggle. One also needs to examine the social construction of such images, listen to the multiplicity of voices and silences, identify common themes and contradictory meanings; and address differences, tensions, and changes in political interventions carried out by women in the Middle East. For example, in Israel, women's voices have been ignored, marginalized, or silenced unless they echo the "national consensus" and unconditionally support the political agenda set by privileged men of the Zionist Ashkenazi ruling elites.[36] Yet Israel is portrayed in the West as a country with gender equality. This distorted myth has been constructed through references to the facts that Israeli women must serve in the army and that Israel once had a female prime minister, and through the juxtaposition of veiled Arab women with "liberated" Israeli women.

It is important to note that Palestinian women and Israeli women have been stereotyped through similar and contrasted images. Whereas Palestinian women have to challenge orientalist, sexist, and nationalist modes of representation, Israeli women do not have to confront orientalist images, but do confront a hybrid of sexist and nationalist images.[37] As in the Palestinian case, the image of women as the mothers of the nation fuses both sexist and nationalist discourses. For example, Ben Gurion, the founding father of the Jewish state, raised the issue of women's fertility to a level of national duty by stressing that "increasing the Jewish birthrate is a vital need for the existence of Israel, and a Jewish woman who does not bring at least four children into the world is defrauding the Jewish mission."[38] By turning the diverse and complex life stories of Palestinian, Israeli, and other women in the Middle East and elsewhere into locations from which international politics could be engaged and theorized, one not only challenges stereotypical depictions, but also responds to the need for new innovative theorizing in IR and in related fields.

Theorizing from women's struggles: politics, discourse, and resistance

In this section, I explore several examples of women's struggles in different parts of the Middle East as locations from which crucial political phenomena can be theorized. I choose to highlight primarily the colonial and neocolonial projects in Egypt and Palestine in the late nineteenth and early twentieth centuries, the Palestinian resistance movement in the refugee camps in Lebanon in the early 1980s, and the Palestinian intifada because they present rich, nonstereotypical, and complex accounts of women's experiences and political interventions. Also, these particular examples highlight the tensions between nationalism and feminism, the interconnectedness of militarism and sexism, the social construction of gender, and the necessity and complexity of solidarity movements that transcend national sovereignty.

Julie Peteet's research among Palestinian women in Lebanon between 1980 and 1982 represents an important step toward theorizing international relations from women's struggles in the Middle East. Peteet traces dynamic changes in Palestinian women's participation in the resistance movement vis-à-vis patterns of male-female relations within one particular community. Peteet notes that "women's participation in the Palestinian national movement has been uneven in form, content and meaning" and illustrates her contention with a multiplicity of voices and perspectives. Peteet's detailed accounts of dynamic and unsystematic patterns of change in the roles of women as well as the differences that exist among women activists go far beyond replacing the stereotype of women as victims with the stereotype of women as heroines.[39]

Rita Giacaman and Penny Johnson present a gendered examination of the first year of the intifada through the story of Umm Ruquyya (Arabic for "Ruquyya's mother"), the mother of a young Palestinian woman activist in the West Bank.[40] Giacaman and Johnson introduce the reader to the reality of life under occupation by weaving information concerning house demolitions, imprisonment, deportations, the Palestinian women's movement, and the sociopolitical implications of the intifada with the story of a particular Palestinian family. They describe one of their visits to Umm Ruquyya's house:

> we went to visit her on November 6, 1988 when we heard that the family house had been demolished by the Israeli army. Hers was one of more than 100 houses destroyed in the northern Jordan Valley, leaving about 1000 persons homeless and devastated.[41]

Ruquyya was the first in her community to mobilize in response to the demolition of her family's house by the Israeli military, yet Giacaman and Johnson do not turn her into a symbol of the Palestinian woman heroine. Instead, they situate Ruquyya's story in the broader political context and point out that she "is one of many women forging a new chapter in the history of the Palestinian women's movement."[42]

Another example of making feminist sense of Middle East politics that takes into account the specific historical and contextual meanings of women's struggles in the region, as well as the multiplicity of voices, silences, disruptions, and changes in the discourses that inform their resistance, records how Egyptian women in the late nineteenth and early twentieth centuries appropriated nationalist discourses to articulate their positions on domestic, regional, and international politics. Situating her critique in the broader historical and sociopolitical context of Middle East politics at the time, Margot Badran identifies the economic conditions, ideologies, and institutional practices that facilitated and informed the crystalization and utilization of feminist-nationalist discourses in Egypt between the 1870s and 1925. She explains that

> feminism and nationalism emerged in late nineteenth-century Egypt several decades after the rise of a capitalist economy and the entry of the country into the world market system dominated by Eruope. During this period vast social, economic, political, and technological transformation occurred.[43]

The appropriation of such discourses centered around upper- and middle-class women who were associated with the Egyptian Feminist Union (founded by Huda Sha'rawi in 1923) and those who participated in an international women's conference organized by the Alliance of Women for Suffrage and Equal Citizenship that took place in Rome in 1923. Mervat Hatem demonstrates how Egyptian women used international conferences as forums to publicize their social and political concerns. She points out that "support for the Palestinian national struggle and its women became one of these concerns in the 1930s."[44]

But this example is not representative of other struggles carried out by Middle Eastern women at the time. For some Palestinian women, for example, the nationalist-feminist discourses of Egyptian women did not convey strong enough anticolonialist sentiments, perhaps because Egypt enjoyed some measure of independence under neocolonial rule whereas Palestine was under direct British colonial rule. In a clear attempt to encourage an exchange between the strategies of struggles utilized by women in different sociopolitical contexts in the Middle East, the Egyptian feminist magazine, *Al-Misriya*, published a memorandum that was sent to the Egyptian Feminist Union by the Society of Arab Women of Jerusalem in 1938. The message described the violation of Palestinian women's rights and the disrupted lives of men and women as they resisted the British-administered transformation of Palestine into a Jewish homeland, and also called on the Egyptian Feminist Union to take a more explicit public stand against British colonial rule.[45] Huda Sha'rawi responded with a strong denunciation of British rule in Palestine and expressed solidarity with these Palestinian women: "the women of Egypt are distressed to see Britain which calls for world peace tarnish this call by using force in a sacred land (Palestine) whose people are only requesting that Britain protect their legitimate (national) rights."[46] These examples demonstrate complex and dynamic relationships in the Middle East at different times and under different political circumstances and social arrangements. Thus, discourses of struggles cannot be fully understood in isolation from the time, place, and sociopolitical structures that inform them; they are shaped and transformed within networks and alliances – such as the international women's conference and the exchange between Palestinian and Egyptian women – that transcend local and national boundaries.

Today, in a more explicit manner than ever before, Palestinian women in the West Bank and Gaza Strip insist that the problems they face as women are connected to the hardship suffered by Palestinian men and women due to the Israeli occupation. The discourses through which Palestinian women in the West Bank and Gaza Strip have been articulating their struggles and concerns, especially since the beginning of the intifada, challenge the primacy of "political issues" over "women's issues" and the limitations of nationalism as the superior discourse to inform their struggles.[47]

Thus, although recent critiques by Palestinian women address the tensions between nationalist projects and women's liberation, most Palestinian women continue to argue that these projects are interdependent; their struggles for women's liberation and against the Israeli occupation of West Bank and Gaza Strip are linked.[48] But they challenge the separation of "women's issues" and "politics," and thereby indicate that multifaceted struggles cannot be hidden anymore between the lines of nationalist discourses. In rendering the discourse more complex, they open up spaces for the emergence of even more complex "discourses of liberation" that can address linkages, contradictions, and changes embedded in the political struggles of women in different contexts. Furthermore, discourses of liberation have the potential to open up space for alliances and solidarities that transcend national sovereignty. For this potential to materialize depends on women on both sides of the national divide.

On the other side of the Israeli-Palestinian conflict, in Israel, Israeli-Jewish women rarely organized on a large scale around political issues until the intifada. Nevertheless, there was some organizing following the Israeli invasion of Lebanon in 1982 in the form of groups such as Women Against the War and Parents Against Silence. These groups utilized different strategies against the Israeli invasion and the war. Women Against the War involved women who had been active in the feminist movement in Israel, and worked in solidarity with Palestinian women on the West Bank and Gaza Strip. On the other hand, Parents Against Silence, which the media and the Israeli public referred to as Mothers Against Silence, publicly disassociated themselves from feminism and tried to project an image that would not be threatening to most Israelis; they were mothers (and fathers) who were worried about their sons in combat.[49] Israeli society and media were by and large sympathetic in the case of Parents Against Silence, which centered on the image of the victimized yet patriotic mother who has suffered enough war; they did not, however, tolerate the discourse used by Women Against the War. After the Israeli withdrawal from Lebanon, this latter group changed its name to Women Against the Occupation and called attention both to the subordination of Israeli women and the oppression of Palestinians in the West Bank and Gaza Strip.

Parents Against Silence dispersed soon after the Israeli army pulled out of Lebanon (in 1985); the women who were active in the group did not join feminist and/or other political organizations. Despite political marginalization and the systematic attempts to silence their voices, Women Against the Occupation laid the groundwork, in my opinion, for some of the strategies of struggle triggered by the intifada. Specifically, they articulated the linkages between the oppression of Palestinians in the Israeli-Occupied Territories and the subordination and discrimination of Israeli women, and they expressed solidarity with Palestinian women in the West Bank and Gaza Strip through words and political actions.[50] The outbreak of the intifada represented a crucial turning point in the political consciousness and activism of Israeli women. It triggered three unprecedented processes in the political lives of women in Israel, resulting in: (1) the establishment of women's groups such as Women in Black, Women for Women Political Prisoners, Women Against the Occupation (Shani), and the Women and Peace Coalition, among others, that took clear positions against the Israeli occupation of the West Bank and Gaza Strip; (2) manifold attempts to build alliances with Palestinian women; and (3) the emergence of new discourses, addressing the interconnectedness of militarism and sexism and linking the oppression of Palestinians in the Occupied Territories to the oppression of women in Israel.

Whereas the majority of Israeli society, including large segments of the nominal male-dominated peace movement, failed to grasp the message and challenges of the intifada, Israeli women participated in large-scale activities with two major goals: (1) to mobilize public opinion against the occupation in Israel and abroad, and (2) to build bridges of solidarity with Palestinian women in the West Bank and Gaza Strip.[51] The new women's groups that emerged during the intifada provided a context for the demarginalization of feminist discourses by appropriating national principles of self-determination, justice, and equality for women's political activism.[52] The intifada also created better conditions for cooperation between Israeli Jewish women and Palestinian women in Israel and in the West Bank and Gaza Strip; these new and fragile alliances take the shape of joint protests, solidarity visits, and conferences.[53] Most alliances and encounters between Palestinian and Israeli-Jewish women challenge the primacy of national identities and the rigidity of sovereign boundaries. For example, Hana Safran, an Israeli-Jewish feminist and peace activist and a founding member of Women in Black in Haifa, alludes to the political significance of collaborative solidarity work:

> In the sociopolitical context there are many forces that work to keep us separate and the joint protest enables the building of a common ground, opens a dialogue. There is a process of trust building. They [Palestinian women] know that we are their allies, that we do not represent the Israeli government. Our persistence in the vigils makes us more credible to Palestinian women.[54]

But the emerging alliances between Palestinian and Israeli-Jewish women face constant challenges. As Palestinian feminist scholar and activist Rita Giacaman reminds us: "Bridges cannot be built in a vacuum. Not every woman will agree with another woman. Sisterhood is not necessarily global. For sisterhood to be global certain predispositions need to be met. . . . Bridges are very difficult to build."[55]

To begin even to imagine how we might construct bridges between Israeli-Jewish and Palestinian women, women on each side of the divide need to develop critical consciousness of where we stand in relation to the divisions between us. For example, Safran points out:

> Before the intifada there was a fear amongst feminists that were interested in building bridges of solidarity with Palestinian women; those of us who were aware that feminism as we practice it emerged in a Western context, we were afraid of imposing our "way of life" on Palestinians who carry a struggle which emerged in a completely different context than ours. There wasn't enough knowledge about their struggles and their local unique definitions of feminism. Our fear was of patronizing in cooperation.[56]

This example alludes to the differences between Western feminist frameworks and Third world feminist frameworks and to disparities in power and privilege between Israeli-Jewish women and Palestinian women.

The intifada also triggered the emergence of feminist voices in Israel linking narrowly defined "women's issues," such as violence against women, discrimination in the workplace, and women's health and reproductive rights, to the sociopolitical context within which they occur. For example, following her participation with Palestinian women at an international women's conference in Brussels entitled "Give Peace a Chance – Women Speak Out," Rachel Ostrowitz, a long-time feminist, peace activist, and editor of the only feminist magazine in Israel, linked the oppression of Israeli women with the oppression of Palestinians:

> Our oppression is not acceptable nor is the oppression of others. . . . We, the women of the Israeli peace movement will not allow our senses to be numbed by the daily killings. We will not accept oppression, discrimination, or exploitation as part of our political system. . . . We won't give up, or shut up, or put up with the current version of reality.[57]

With the unprecedented increase in violence against women and children since the outbreak of the intifada, and especially in the aftermath of the Gulf War, more feminists and peace activists in Israel realize that the struggle for women's liberation in Israel is directly connected to the Israeli-Palestinian conflict and that militarism and sexism are interconnected. Furthermore, some Israeli feminists argue that the construction of Israeli masculinity is linked to the militarized political climate in Israel and in the region. They demonstrate how the institutionalization of "national security" as a top priority in Israel contributes to gender inequalities on the one hand, and legitimizes violence against Palestinians and against women on the other. The major argument is that the social construction of masculinity in Israel has to be addressed in its historical context, especially in light of the holocaust and the creation of the Jewish state. Accordingly, the state of Israel can be seen as a reassertion of manhood, justified by the need to end a history of weakness and suffering by creating an image of an Israeli man who is exceedingly masculine, pragmatic, protective, assertive, and emotionally tough. Utilizing images such as "a nation under siege" surrounded by enemies that threaten to throw the entire population into the sea, the Zionist ideology of the state made "national security" a top priority and gave rise to the centrality of the army and its practices in all spheres of Israeli life, thus offering the "new" Israeli men a privileged status in Israeli society.[58]

Although conventional scholarship on the Israeli-Palestinian conflict tends to ignore contentions such as those presented above, Middle Eastern scholars should not overlook the fact that in a context where every man is a soldier, every woman becomes an occupied territory. Furthermore, they should view as significant the fact that there are Palestinian and Israeli women who, despite fundamental differences separating them (especially since the outbreak of the intifada), have begun to seriously challenge the primacy of national identity and the confines of sovereign boundaries and link their struggles for women's liberation with their broader political struggles. The emergence of transnational discourses of liberation opens up space for new politics of resistance and new avenues for conflict resolution that have the potential to rewrite conventional international politics.

Conclusions

In a recent Middle East report on gender and politics in the Middle East, Julie Peteet and Barbara Harlow point out that "in the Middle East and in Middle East Studies, the disruption of old forms of thinking is creating space in which women's independent initiatives are mapping new paths of social, cultural and political transformation."[59] Women in different parts of the Middle East are gradually becoming aware of the explicit and implicit ways in which gender is embedded in the politics of the region as in international relations. However, as Hanan Mikhail-Ashrawi pointed out in an interview that took place before she appeared on the stage of the international political arena, there is an urgent and concentrated need to crystallize feminist perspectives on regional and international politics because "unless women's spontaneous activity is given a context of theory and ideology, it will not fulfil its potential of transforming women's position in society."[60]

In order to articulate the kind of feminist perspective that Mikhail-Ashrawi calls for, we need to situate ourselves in the broader political context within which we work and examine the language and conceptual frameworks that we use to articulate our ideas. To do so without creating new essentialist categories and/or stereotypical images of Middle Eastern women, one needs to move beyond the confines of positivist theoretical frameworks that search for causal patterns and distinct categories. By focusing on the struggles of women in the Middle East, challenging the stereotypical depiction of Middle Eastern women, and exploring different discourses that inform women's struggles, we find that women's interpretations of domestic, regional, and international politics, and the different meanings of identity, community, and solidarity that they forge, present an alternative site for feminist theorizing on international politics.

The conceptual frameworks needed for feminist IR theorizing cannot emerge in academic settings and then be applied to the Middle East. They have to be grounded in the complex and multidimensional daily struggles of women in the region. As we move toward theorizing from women's struggles in the Middle East, we have to go beyond feminist critiques of existing scholarship and engage in writing the unwritten prefaces of conventional IR textbooks on Middle East politics. Projects of exploring Middle East politics through feminist lenses could begin with women's struggles for voice and visibility informed by the larger discursive and historical contexts within which they unfold.[61] To do so, we should integrate feminist critiques of conventional IR scholarship – which uncover local, regional, and global gender arrangements that underlie the complex relationships between and within nations and people – with different detailed examples of Middle Eastern women's life experiences, resistance struggles, and solidarities that transcend sovereign boundaries.[62] Finally, we need to explode the artificial distinctions between narrowly defined "women's issues" and "international politics" by making topics such as the social construction of gender identities and roles, the interconnectedness of militarism and sexism, and the complex relationship of colonialism, nationalism, and feminism, integral parts of IR scholarship.

Notes

1 Sandra Harding, *Whose Science? Whose Knowledge? Thinking from Women's Lives* (New York: Cornell University Press, 1991), p. 211.
2 Chandra Talpade Mohanty, "Cartographies of Struggle: Third World Women and the Politics of Feminism," in Chandra T. Mohanty, Ann Russo, and Lourdes Torres, eds., *Third World Women and the Politics of Feminism* (Bloomington: Indiana University Press, 1991), p. 39.
3 In the Middle East, as in the United States, information concerning women's resistance to the Gulf War has been systematically suppressed and not well documented. For example, a significantly unreported event was an urgent meeting of seventy-five women from twenty-six countries in Geneva on February 4, 1991, at which women across the globe were asked to use International Women's Day to express horror and opposition to the war. Also, the Tunisian Women's Union took upon itself the coordination of a women's peace march on March 8, 1991, from Jordan to Baghdad.
4 Mervat Hatem, "Egyptian Upper- and Middle-Class Early Nationalist Discourses on National Liberation and Peace in Palestine (1922–1944)," *Women & Politics* 9, no. 3 (1989): 49–56.
5 From personal conversations with the prominent Egyptian feminist, Nawal el-Saadawi, January 9–11, 1992.
6 Mohanty, note 2, p. 4; Harding, note 1, p. 268.
7 For more on the joint conferences of Israeli and Palestinian women, see Naomi Chazan, "Israeli Women and Peace Activism," in Barbara Swirsky and Marilyn Safir, eds., *Calling the Equality Bluff: Women in Israel* (New York: Pergamon Press, 1991); and Yvonne Deutsch, "Israeli Women: From Protest to a Culture of Peace," *Connections* (Fall 1991).
8 For an articulation of Hanan Mikhail-Ashrawi's feminist agenda for Middle East politics, see the recent interview she gave to Rabab Hadi, "*Ms.* Exclusive: The Feminist Behind the Spokeswoman: A Candid Talk with Hanan Ashrawi," in *Ms. Magazine*, April/March 1992, pp. 14–18.
9 The admiration and appreciation of Israeli feminists toward Hanan Mikhail-Ashrawi is exemplified in the editorial of the Israeli feminist magazine *Noga*, which reads (in free translation from Hebrew): "as this issue goes to press, Hanan Ashrawi is gradually finding her place among the leadership of the Palestinian delegation to the peace process. We should be envious that she doesn't have a feminist counterpart on the Israeli side of the negotiation table." *Noga* 22 (Fall 1991): 5.
10 Although the terms "Middle East" and "Middle Eastern women" are used extensively throughout this article, I agree with Marnia Lazreg that "North African and Middle Eastern societies are more complex and more diverse than is admitted, and cannot be understood in terms of monolithic, unitary concepts." Marnia Lazreg, "Feminism and Difference: The Perils of Writing as a Woman on Women in Algeria," in Marianne Hirsch and Evelyn Fox-Keller, eds., *Conflicts in Feminism* (New York: Routledge, 1990), p. 337. When I use the terms "Middle East" and "Middle Eastern women" as analytical and political categories, I do so – and not without hesitation – in order to subvert their dominant meaning both in academic

scholarship and in international politics. Mohanty acknowledges a similar dilemma regarding the use of the term "Third World women." Although she is well aware that the term can be used (and has been used) as a monolithic essentializing construct, she chooses to subvert the conventional use and its attached meaning by using "third world women" to refer to an "imagined community" of Third World oppositional struggles. In appropriating what Benedict Anderson called "horizontal comradeships," Mohanty emphasizes "the potential alliances and collaborations across divisive boundaries." See Mohanty, note 2, p. 4; and Benedict Anderson, *Imagined Communities: Reflections on the Origin and Spread of Nationalism* (New York: Verso Books, 1983), esp. pp. 11–16.

11 Mohanty, note 2, p. 32.

12 Rina Benmayor, "Testimony, Action Research, and Empowerment: Puerto Rican Women and Popular Education," in Sherna Berger Gluck and Daphne Patai, eds., *Women's Words: The Feminist Practice of Oral History* (New York: Pergamon, 1991), p. 159.

13 For example, see the following anthologies: Teresa de Lauretis, *Feminist Studies/Critical Studies* (Bloomington: Indiana University Press, 1986); Elizabeth Weed, *Coming To Terms: Feminism, Theory, Politics* (New York: Routledge, 1989); Hirsch and Fox-Keller, *Conflicts in Feminism*, note 10; Patti Lather, *Getting Smart: Feminist Research and Pedagogy with/in the Postmodern* (New York, Routledge, 1991); see also the third section, entitled "Others," in Harding, note 1.

14 Daphne Patai, "US Academics and Third World Women: Is Ethical Research Possible?" in Gluck and Patai, note, 12, p. 138.

15 Sherna Berger Gluck, "Advocacy Oral History: Palestinian Women in Resistance," in Gluck and Patai, note 12, p. 205.

16 Soraya Altorki and Camillia Fawzi El-Solh, eds., *Arab Women in the Field: Studying Your Own Society* (New York: Syracuse University Press, 1988).

17 Soheir Morsy, "Fieldwork in My Egyptian Homeland: Toward the Demise of Anthropology's Distinctive-Other Hegemonic Tradition," in Altorki and El-Solh, note 16, p. 71.

18 Altorki and El-Solh, p. 72–73.

19 Altorki and El-Solh, p. 77.

20 For discussions on the militarization of Israeli society and on the relationship of militarism and sexism, see, for example, Regine Waintrater, "Living in a State of Siege," in Swirsky and Safir, note 7. See also Simona Sharoni, "To Be a Man in the Jewish State: The Sociopolitical Context of Violence and Oppression," *Challenge* 2, no. 5 (September/October 1991): 26–28.

21 Mary Childers and bell hooks, "A Conversation About Race and Class," in Hirsch and Fox-Keller, note 13, p. 77. For more contributions to this genre see Minnie Bruce Pratt, "Identity: Skin Blood Heart" in Elly Bulkin, Minnie Bruce Pratt, and Barbara Smith, *Yours in Struggle: Three Feminist Perspectives on Anti-Semitism and Racism* (New York: Firebrand, 1988); and the reading of this essay by Biddy Martin and Chandra Talpade Mohanty, "Feminist Politics: What's Home Got to Do with It?" in Teresa de Lauretis, ed., *Feminist Studies/Critical Studies* (Bloomington: Indiana University Press, 1986). See also Ann Russo, "We Cannot Live Without Our Lives: White Women, Antiracism and Feminism" in Mohanty, Russo, and Torres, note 2, pp. 295–313.

22 For example, see Sherna Berger Gluck, "Advocacy Oral History: Palestinian Women in Resistance," in Gluck and Patai, note 12, and Rina Benmayor's and Daphne Patai's essays in the same anthology. Also see the chapter by Rey Chow,

"Violence in the Other Country: China in Crisis," in Mohanty, Russo, and Smith, note 2.

23 Julie Peteet, *Gender in Crisis: Women and the Palestinian Resistance Movement* (New York: Columbia University Press, 1991), p. 16–17.

24 "Orientalism" refers to the view according to which the "Orient" is antithetical to and radically different from the "West." For an in-depth examination of the social and political implications of orientalism, see Edward W. Said, *Orientalism* (New York: Vintage Books Edition, 1979). For more on the representation of Middle Eastern women, see, for example, Zjaleh Hajibashi, "Feminism or Ventriloquism: Western Presentations of Middle Eastern Women," *Middle East Report* 172 (September/October 1991): 43–45. See also Judy Mabro, *Veiled Half Truths: Western Travellers' Perceptions of Middle Eastern Women* (New York: St. Martin's Press, 1991).

25 For examples of critiques of sexist images of Middle Eastern women and of the public-private split, see Suad Joseph, "Women and Politics in the Middle East," *Middle East Report* 138 (January/February 1986): 3–7; as well as other contributions to this special issue of women and politics. See also Julie Peteet's work on the Palestinian women's resistance movement in the refugee camps in Lebanon, note 23.

26 Palestinian sociologist Nahla Abdo points out that "most studies on women and national liberation have focused on two major images: that of the woman-victim who participated, gave and sacrificed during the movement, but who after liberation was victimized by the patriarchal structure of the state; and that of the strong woman fighter in the forefront of the armed struggle." As for the image of women as mothers of the nation, it is a later construction that I argue, and Nahla Abdo agrees, works to resolve the tension between the first two contrasting images. See Nahla Abdo, "Women of the Intifada: Gender, Class and National Liberation," *Race & Class* 32, no. 4. (1991): 19–34.

27 Lazreg, note 10, p. 343.

28 Cynthia Enloe, "Womenandchildren: Making Feminist Sense of the Persian Gulf Crisis," *The Village Voice*, September 25, 1990, p. 30.

29 Nayereh Tohidi, "Gender and Islamic Fundamentalism: Feminist Politics in Iran," in Mohanty, Russo, and Torres, note 2, p. 251.

30 Nayereh Tohidi, p. 251–252. "The Women's Struggle in Iran," *Monthly Review* (March 1981): 22–30.

31 Shafla Jemame, "Women Fight Exclusion and Poverty," *International Viewpoint* 220 (January 20, 1992): 5.

32 Shafla Jemame, p. 5.

33 For example, see Evelyne Accad, *Sexuality and War: Literary Masks of the Middle East* (New York: New York University Press, 1990); Deniz Kandiyoti, ed., *Women, Islam and the State* (London: Macmillan, and Philadelphia: Temple University Press, 1991); Bouthaina Shaaban, *Both Right and Left Handed: Arab Women Talk About Their Lives* (Bloomington: Indiana University Press, 1991); Fadwa Tuqan, *A Mountainous Journey: An Autobiography*, translated from Arabic by Olive Kenny (London: The Women's Press, 1990); Kitty Warnock, *Land Before Honour: Palestinian Women in the Occupied Territories* (New York: Monthly Review Press and London: Macmillan, 1990). See also Peteet, *Gender in Crisis*, note 23.

34 Abdo, note 26, p. 25.

35 Excerpt from an interview with Sahar Khalifa. For the entire interview, see Maya Rosenfeld, "I Don't Want My Body To Be a Bridge for the State: Palestinian

Novelist Sahar Khalifa on Her Work and Issues of Women in the Intifada," *Challenge* 2, no. 6 (November–December, 1991): 24–27.

36 The Ashkenazim, Israel's ruling minority, constitute only 28 percent of the citizenry. Of the 4.3 million Israeli citizens, about 20 percent are Palestinians who hold Israeli citizenship and 54 percent are Mizrahim (literally meaning "Orientals" in Hebrew). Recently, critical analyses of Israeli society (not yet gendered) began to challenge the power and privilege held by the Ashkenazi demographic minority that constitutes the Eurocenter of Israeli culture and politics. For example, see Ella Shohat, *Israeli Cinema: East/West and the Politics of Representation* (Austin: University of Texas Press, 1989); and Smadar Lavie, "Blow-Ups in the Borderzones: Third World Israeli Authors' Gropings for Home" in Smadar Lavie and Ted Swedenburg, eds., *Displacement, Diaspora & Geographies of Identity* (Durham, NC: Duke University Press, 1993).

37 For numerous examples of the hybrid of nationalist and sexist depictions of Israeli women, see Elise G. Young, *Keepers of the History: Women and the Israeli-Palestinian Conflict* (New York and London: Teachers College Press, 1992); Nira Yuval-Davis, "The Jewish Collectivity," in The Khamsin Collective, ed., *Women in the Middle East* (London: Zed Books Ltd., 1987); and Simona Sharoni, "Feminist Reflections on the Interplay of Racism and Sexism in Israel," in Betty Rosoff and Ethel Tobach, eds., *Racism and Sexism: A Challenge to Genetic Explanations* (New York: The Feminist Press, 1993).

38 Quoted in Lesley Hazleton, *Israeli Women: The Reality Behind the Myths*, (New York: Simon and Schuster, 1977), p. 63.

39 See Julie Peteet, "No Going Back: Women and the Palestinian Movement," *Middle East Report* 138 (January–February 1986): 20–24. See also Peteet, note 23.

40 Rita Giacaman and Penny Johnson, "Palestinian Women: Building Barricades and Breaking Barriers," in Zachary Lockman and Joel Beinin, eds., *Intifada: The Palestinian Uprising Against Israeli Occupation* (Boston: South End Press, 1989). In Middle Eastern cultures, parents may be referred to as the father or the mother of their eldest son. However, Giacaman and Johnson emphasize that "Umm 'Uthman is to us Umm Ruquyya, because we came to know her through her radical and activist daughter." The linguistic construct "Umm Ruquyya" is used subversively to call attention to the active political participation of Palestinian women during the intifada, and its potential transformative impact on Palestinian society in the West Bank and Gaza Strip (p. 155).

41 Giacaman and Johnson, p. 155.

42 Giacaman and Johnson, p. 157.

43 Margot Badran, "Dual Liberation: Feminism and Nationalism in Egypt 1870–1925," *Feminist Issues* (Spring 1988): 17.

44 Hatem, note 4, p. 56. Huda Sha'rawi (1879–1974), a pathbreaking upper-class Egyptian feminist and nationalist, was one of the prominent figures of the first wave of feminist consciousness and activism in late nineteenth and early twentieth century Egypt. This is an excerpt from her reply (written in May 1938) to a memorandum sent to the feminist magazine *Al-Misriya* by women in Palestinine.

45 Hatem, note 4, pp. 59–60.

46 Hatem, p. 49.

47 For discussions concerning the inseparability of the struggle for women's liberation from the struggle against the Israeli occupation, see, for example, Warnock, *Land Before Honour*, note 33; Rita Giacaman, *Life and Health in Three Palestinian Villages* (London: Ithaca Press, 1988); and Rosemary Sayigh, "Encounters with Palestinian Women Under Occupation," in Nasir Aruri, ed., *Occupation:*

Israel over Palestine (London: Zed Books, 1984). See also Young, *Keepers of the History*, note 37.

48 Nahla Abdo identifies additional examples of projects that represent a shift toward different discourses of struggle. For example, she mentions the project of feminist deconstruction of Palestinian revolutionary national literature and the establishment of three women's research and resource centers in the Israeli-Occupied Territories. See Abdo, note 26, pp. 28–34.

49 Nurith Gillath, "Women Against War: Parents Against Silence," in Swirsky and Safir, note 7, pp. 142–146.

50 For example, see Ghada Talhami's account of the solidarity work of Women Against the Occupation and the warm welcome they received from Palestinian women. Ghada Talhami, "Women Under Occupation: The Great Transformation," in Suha Sabbagh and Ghada Talhami, eds., *Images and Realities: Palestinian Women Under Occupation* (Washington, DC: The Institute for Arab Women's Studies, Inc., 1990), pp. 15–27.

51 For more information on the various groups, see, for example, the special issue of *New Outlook* entitled "Women in Action" (June/July 1989); Naomi Chazan, "Israeli Women and Peace Activism," in Swirsky and Safir, note 7. On Women in Black, see Simona Sharoni, "Women in Black," *Palestine Perspectives* (May/June, 1990): 3–4; and Gila Svirsky, "Women in Black," in Rita Falbel, Irena Klepfisz, and Donna Nevel, eds., *Jewish Women's Call for Peace: A Handbook for Jewish Women on the Israeli/Palestinian Conflict* (New York: Firebrand Books, 1990).

52 For example, see Chazan, "Israeli Women and Peace Activism," note 7; Deutsch, "Israeli Women," note 7.

53 Simona Sharoni, "Women's Alliances and Middle East Politics: Conflict Resolution Through Feminist Lenses," paper presented at the annual meeting of the Association for Israel Studies, Milwaukee, Wisconsin, May 1992.

54 Excerpts from an interview with Hana Safran, May 1990, Haifa, Israel.

55 Excerpts from a conversation with Rita Giacaman, Maha Nasar, and Eileen Kuttab, Ramalla, Occupied West Bank.

56 See note 54.

57 Rachel Ostrowitz, "The Israeli Women's Peace Movement," *New Outlook* 32, nos. 6/7 (June–July 1989): 14–15.

58 See Sharoni, "To Be a Man," note 20. See also Sharoni, "Every Woman is an Occupied Territory: The Politics of Militarism and Sexism and the Israeli-Palestinian Conflict," paper presented at the First International Conference of the Ethnic Studies Network, Northern Ireland, June 1992.

59 Julie Peteet and Barbara Harlow, "Gender and Political Change," *Middle East Report* 173 (November/December 1991): 4–8.

60 From an interview with Hanan Mikhail-Ashrawi. See Warnock, *Land Before Honour*, note 33, p. 188.

61 I am referring to discourse in a roughly "Foucauldian" manner as more than ways of thinking and producing meaning. According to Chris Weedon, "discourses, in Foucault's work, are ways of constituting knowledge, together with social practices, forms of subjectivity and power relations which inhere in such knowledges and the relations between them." See Chris Weedon, *Feminist Practice and Poststructuralist Theory* (New York: Basil Blackwell, 1987). See also Michel Foucault, "The Order of Discourse," in Robert Young, ed., *Untying the Text* (Boston and London: Routledge & Kegan Paul, 1981).

62 Because the primary purpose of this article is to highlight the struggles of women

in the Middle East as a potential site of IR theorizing, I have not discussed here the emerging body of feminist scholarship in IR, which uncovers the gendered aspects of politics. For example, see Christine Sylvester, "The Emperors' Theories and Transformations: Looking at the Field Through Feminist Lenses," in Dennis Pirages and Christine Sylvester, eds., *Transformations in the Global Political Economy* (London: Macmillan, 1990); Anne Sisson Runyan and V. Spike Peterson, "The Radical Future of Realism: Feminist Subversions of IR Theory," *Alternatives* 16 (1991): 67–106; V. Spike Peterson, ed., *Gendered States: Feminist (Re)visions of International Relations Theory* (Boulder, Colo.: Lynne Rienner Publishers, 1992); and Ann Tickner, *Gender in International Relations: Feminist Perspectives on Achieving Global Security* (New York: Columbia University Press, 1992).

32 Gloria Anzaldúa

'To Live in the Borderlands Means You . . .'

From: *Borderlands/La Frontera: the New Mestiza*, pp. 194–195.
San Francisco: Spinsters/Aunt Lute (1987)

To live in the Borderlands means you
are neither *hispana india negra española*
ni gabacha, eres mestiza, mulata, half-breed
caught in the crossfire between camps
while carrying all five races on your back
not knowing which side to turn to, run from;

To live in the Borderlands means knowing
that the *india* in you, betrayed for 500 years,
is no longer speaking to you,
that *mexicanas* call you *rajetas*,
that denying the Anglo inside you
is as bad as having denied the Indian or Black;

Cuando vives en la frontera
people walk through you, the wind steals your voice,
you're a *burra*, *buey*, scapegoat,
forerunner of a new race,
half and half – both woman and man, neither –
a new gender;

To live in the Borderlands means to
put *chile* in the borscht,
eat whole wheat *tortillas*,
speak Tex-Mex with a Brooklyn accent;
be stopped by *la migra* at the border checkpoints;

Living in the Borderlands means you fight hard to
resist the gold elixer beckoning from the bottle,
the pull of the gun barrel,
the rope crushing the hollow of your throat;

In the Borderlands
you are the battleground
where enemies are kin to each other;
you are at home, a stranger,
the border disputes have been settled
the volley of shots have shattered the truce
you are wounded, lost in action
dead, fighting back;

To live in the Borderlands means
the mill with the razor white teeth wants to shred off
your olive-red skin, crush out the kernel, your heart
pound you pinch you roll you out
smelling like white bread but dead;

To survive the Borderlands
you must live *sin fronteras*
be a crossroads.

gabacha a Chicano term for a white woman
rajetas literally, "split," that is, having betrayed your word
burra donkey
buey oxen
sin fronteras without borders

---

33 Cynthia Enloe

'Gender Makes the World Go Round'

Excerpts from: *Bananas, Beaches and Bases: Making Feminist Sense of International Politics*, pp. 1–17. Berkeley: University of California Press (1989)

Ambassadors cabling their home ministries, legislators passing laws to restrict foreign imports, bank executives negotiating overseas loans, soldiers landing on foreign hillsides – these are some of the sites from which one can watch the international political system being made. But if we employ only the conventional, ungendered compass to chart international politics, we are likely to end up mapping a landscape peopled only by men, mostly élite men. The real landscape of international politics is less exclusively male.

A European woman decides to take her holiday in Jamaica because the weather is warm, it is cheap and safe for tourists. In choosing this form of pleasure, she is playing her part in creating the current international political system. She is helping the Jamaican government earn badly needed foreign currency to repay overseas debts. She is transforming 'chambermaid' into a major job category. And, unwittingly, if she travels on holiday with a white man, she may make some Jamaican men, seeing every day the privileges – economic and sexual – garnered by white men, feel humiliated and so nourish nationalist identities rooted in injured masculinity.

A school teacher plans a lesson around the life of Pocahontas, the brave Powhantan 'princess' who saved Captain John Smith from execution at Jamestown and so cleared the way for English colonization of America. The students come away from the lesson believing the convenient myth that local women are likely to be charmed by their own people's conquerors.

In the 1930s Hollywood moguls turned Brazilian singer Carmen Miranda into an American movie star. They were trying to aid President Franklin Roosevelt's efforts to promote friendlier relations between the US and are relegated to the 'human interest' column. Women's roles in creating and sustaining international politics have been treated as if they were 'natural' and thus not worthy of investigation. Consequently, how the conduct of international politics has *depended* on men's control of women has been left unexamined. This has meant that those wielding influence over foreign policy have escaped responsibility for how women have been affected by international politics.

Perhaps international politics has been impervious to feminist ideas precisely because for so many centuries in so many cultures it has been thought of as a typically 'masculine' sphere of life. Only men, not women or children, have been imagined capable of the sort of public decisiveness international politics is presumed to require. Foreign affairs are written about with a total disregard for feminist revelations about how power depends on sustaining notions about masculinity and femininity. Local housing officials, the assumption goes, may have to take women's experiences into account now and then. Social workers may have to pay some attention to feminist theorizing about poverty. Trade-union leaders and economists have to give at least a nod in the direction of feminist explanations of wage inequalities. Yet officials making international policy and their professional critics are freed from even a token consideration of women's experiences and feminist understandings of those experiences.

. . . It's difficult to imagine just what feminist questioning would sound like in the area of international politics. Some women have come to believe that there is a fundamental difference between men and women. 'Virtually everyone at the top of the foreign-policy bureaucracies is male,' they argue, 'so how could the outcome be other than violent international conflict?' That is, men are men, and men seem almost inherently prone to violence; so violence is bound to come about if men are allowed to dominate international politics. At times this sweeping assertion has the unsettling ring of truth. There's scarcely a woman who on a dark day hasn't had a suspicion this just might be so. Yet most of the women from various cultures who have created the theories and practices which add up to feminism have not found this 'essentialist' argument convincing. Digging into the past and present has made them reluctant to accept explanations that rest on an assertion that men and women are inherently different.

Men trying to invalidate any discussion of gender in international politics tend to quote a litany of militaristic women leaders: 'Well, if you think it's men who are causing all the international violence, what about Margaret Thatcher, Indira Gandhi and Jeanne Kirkpatrick?' Most women – or men – who have been treating feminist analyses seriously have little trouble in responding to this now ritualistic jibe. It's quite clear to them that a woman isn't inherently or irreversibly anti-militaristic or anti-authoritarian. It's not a matter of her chromosomes or her menstrual cycle. It's a matter of social processes and structures that have been created and sustained over the generations – sometimes coercively – to keep most women out of any political position with influence over state force. On occasion, élite men *may* let in a woman here or a woman there, but these women aren't randomly selected.

Most of the time we scarcely notice that governments look like men's clubs. We see a photo of members of the Soviet Union's Politburo, or the US Cabinet's sub-committee on national security, of negotiators at a Geneva textile bargaining sesson, and it's easy to miss the fact that all the people in these photographs are men. One of the most useful functions that Margaret Thatcher has served is to break through our numbness. When Margaret Thatcher stood in Venice with Mitterand, Nakasone, Reagan and the other heads of state, we suddenly noticed that everyone else was male. One woman in a photo makes it harder to ignore that the men are men.

However, when a woman is let in by the men who control the political élite it usually is precisely because that woman has learned the lessons of masculinized political behaviour well enough not to threaten male political privilege. Indeed she may even entrench that privilege, for when a Margaret Thatcher or Jeanne Kirkpatrick uses her state office to foment international conflict, that conflict looks less man-made, more people-made, and thus more legitimate and harder to reverse.

Still, being able to counter the 'What about Margaret Thatcher?' taunt isn't by itself a satisfactory basis for a full feminist analysis of international politics. We have to push further, open up new political terrain, listen carefully to new voices.

A fictional James Bond may have an energetic sex life, but neither sexuality nor notions of manhood nor roles of women are taken seriously by most commentators in the 'real' world of power relations between societies and their governments. What really matters, conventional international observers imply, are money, guns and the personalities of leaders – of the men who make up the political élite. The processes holding sway in most societies have been designed so that it is mainly men who have the opportunities to accumulate money, control weaponry and become public personalities. As a consequence, any investigation that treats money, guns and personalities as the key ingredients in relations between societies is almost guaranteed to obliterate women from the picture.

Where are the women? Clues from the Iran/Contra affair

In July 1987 I turn on my television to watch the congressional hearings on the Iran/Contra affair. Senior members of the Reagan administration are accused of selling weapons to the Iranian government and funnelling the proceeds to the anti-Sandinista rebels in Nicaragua in violation of congressional policy. All of the congressional representatives sitting at their tiers of desks under the media's bright lights are men. All but one of the congressional committees' lawyers asking questions are men. All but two of the scores of witnesses subpoenaed to answer their questions are men. All of their attorneys are men. All of the men have been told that dark blue suits and red ties look best on television. Everyone wears a dark blue suit and a red tie.

The Iran/Contra hearings are heralded as the event of the decade in international politics. Now we, the ordinary folk, are going to see how foreign policy actually gets made. Some of my friends become hooked on the congressional hearings: watching or listening from morning to evening, arranging their work and social schedules so as not to miss a word. In Britain, Canada and Australia TV viewers see excerpts every evening. As much as Europe's endless drizzle, the Iran/Contra hearings seem to define the summer of 1987. Information from the hearings is woven into popular culture. There are 'instant books', songs and jokes, 'I luv Ollie' T-shirts, even Ollie North and Betsy North haircuts.

Women do appear during the hearings, though their appearances confirm rather than disturb the implied naturalness of the otherwise all-male cast. Maybe it's because all the women captured by the media's eye are so marginalized.

Ellen Garwood is one of the few women called to the witness table. She is a wealthy conservative who has donated over $2 million to the Contras after being appealed to by Colonel North and other American Contra fund raisers. Congressmen and their attorneys ask her about how her donation was solicited. They aren't interested in her views on US foreign policy. She is not a retired general or a former CIA agent. A public opinion survey comes out at about this time showing that American women are significantly less enthusiastic about US aid to the Contras than are American men, especially American white men.[1] But such revelations do not prompt any of the legislators to ask Ellen Garwood for her foreign policy ideas.

Women appear so infrequently that their very appearance in any authoritative role becomes 'news'. One day a woman assistant-attorney for the congressional committee appears on TV. She asks a minor witness questions. Feminist viewers sit up and take notice. One viewer counts seven young women sitting on the chairs arranged awkwardly just behind the congressmen.

They don't speak in public. They are staff aides, ready to serve their male bosses.

Men comprise the majority of the media people assigned to tell us what each day's revelations 'mean'. The women reporters covering the hearings for radio and television do take extra care with their gender pronouns, yet shy away from posing any feminist questions. They haven't climbed this high on the news media ladder by questioning how masculinity and femininity might be shaping foreign policy. They must take care to look feminine while still sounding as though gender were irrelevant to their commentary.[2]

The one woman witness who becomes front page news is Fawn Hall. She is the 27-year-old who worked in the National Security Council as civil-service secretary to Oliver North and who admits assisting her boss in shredding important government documents. Fawn Hall is routinely described, even by the most low-key media commentators, as 'the beautiful Fawn Hall'. It's as if Fawn Hall is meant to represent the feminine side of High Politics of the 1980s: worldly, stylish, exciting, sexy. Beauty, secrecy and state power: they all enhance one another. In the élite politics of the present era, the 'beautiful secretary', the 'handsome, can-do military officer', and the bureaucratic shredding machine make an almost irresistible combination.[3]

There are at least two sorts of feminized beauty, however: the revealed and the hidden. Fawn Hall is set up as beauty revealed. She stands in stark contrast to the popularly constructed image of beauty hidden: the veiled Muslim woman. Until it began selling weapons to Iran, the Reagan administration liked to emphasize the Iranian regime's wrong-headed regressiveness by pointing to its anti-modern confinement of women. Reagan aides thought their arms sales were giving them access to a moderate 'second channel' in the Teheran political élite. Does the Iranian 'second channel' insist his secretary wear a veil? I try to imagine what Fawn Hall and the 'second channel's' secretary, if they ever had a chance to meet, would find they had in common, as government secretaries to male bosses carrying out secret operations.

Some Republicans deem Fawn Hall worthy of imitation, if not emulation. In Arkansas Republican party activists hold a gathering to celebrate Oliver North with '"Ollie Dogs" on the grill, tough-talking T-shirts, water-melons and 95° heat', according to the press report. There is also a Fawn Hall look-alike contest. Women entering the contest have to perform dramatic readings of Fawn Hall's congressional testimony and act her feeding documents into a paper shredder. The winner is sixteen-year-old Renee Kumpee, who, when asked about her attitude to Oliver North, replies, 'I like him OK.'[4]

Women supply most of the clerical labor force that has made the complex communications, money transfers and arms shipments possible. They handle the procedures and technology, and more importantly, they provide many male officials with on-the-job encouragement. In today's international political system, large bureaucracies are vehicles for making, implementing and remembering decisions. Since the deliberate feminization of clerical work in the early twentieth century, every government has required women to acquire certain skills and attitudes towards their work, their superiors and themselves.

Even in small states without the huge bureaucratic machines the public agencies rely on women for their smooth running. If secretaries went out on strike, foreign affairs might grind to a standstill. Without women's willingness to fill these positions in acceptably feminine ways, many men in posts of international influence might be less able to convince themselves of their own rationality, courage and seriousness.

Other Washington secretaries felt ambivalent towards Fawn Hall. They resented the media for treating Fawn Hall as the quintessential government secretary. 'I guess the media wouldn't be making such a big deal out of it if she had been fifty years old and not blond.' Still, they also saw Fawn Hall's dilemma as their own. 'You develop personal relationships when you work in a high-pressure operation like that . . . She was more than a receptionist or a typist, and she was expected to keep things confidential.' Patricia Holmes, a Black woman working as a secretary in the Department of the Interior, summed up many Washington secretaries' feelings: Fawn Hall was caught 'between a rock and a hard place.'[5]

Each woman who appears in person during the Iran/Contra proceedings is considered peripheral to the 'real' political story. None of their stories is interpreted in a way that could transform the masculinized meaning of complex international political relationships. Most of us see them as marginal characters who simply add 'color' to the all-male, blue-suited, red-tied political proceedings.

Several of the male witnesses assure their congressional interrogators that they took their state jobs so seriously that they didn't tell 'even their wives' about the secrets they were guarding. On the other hand, they expected to receive from their wives an automatic stamp of moral approval. This is the kind of marriage on which the national-security state depends.

Thousands of women today tailor their marriages to fit the peculiar demands of states operating in a trust-starved international system. Some of those women are married to men who work as national-security advisors; others have husbands who are civilian weapons-engineers working on classified contracts; still others are married to foreign-service careerists. Most of these men would not be deemed trustworthy if they were not in 'stable' marriages. Being a reliable husband and a man the state can trust with its secrets appear to be connected.

And yet it is precisely that elevation to a position of state confidence which can shake the foundations of a marriage. Patriotic marriages may serve the husbands, giving them a greater sense of public importance and less of a sense of guilt for damaging the lives of people in other countries. And they serve the national-security state. But they don't necessarily provide the women in those marriages with satisfaction or self-esteem. Typically, it is left up to the wife to cope with the tensions and disappointments. She may respond by trying to cultivate interests of her own outside her marriage, investing her relationship with her distracted husband with less importance than she once did. Or she may continue to see her relationship with her husband as her most important friendship, but adjust her notion of it so that it becomes a marriage of unequals: she will continue to confide in him all her hopes and worries, while resigning herself to hearing from him only what is 'unclassified'. In such cases some women express admiration for their husbands' patriotism, a patriotism they believe, as wives, they cannot match. This is the stance taken by Betsy North. She is praised. Her haircut becomes the new fashion.

Marriages between élite men and patriotic wives are a building block holding up the international political system. It can continue to work the way it now does, dependent on secrecy, risk-taking and state loyalty, only if men can convince women to accept the sorts of marriages that not only sustain, but also legitimize, that system. And it isn't just marriages at the pinnacles of power that must be made to fit. As we will see, marriages up and down the international pyramid can jeopardize power relations between governments if the women refuse to play their parts. They must be willing to see their husbands leave home for long periods of time – as multinational plantation workers, as migrant workers on Middle East construction projects, as soldiers posted to foreign bases. Women working as domestic servants must be willing to leave their husband and children to service other families and, in the process, their country's foreign debt. But no one asked either Betsy North or the wife of a Honduran banana-plantation worker what her analysis was of the international political system that produced the Iran/Contra affair.

One of the beliefs that informs this book is this: if we listened to women more carefully – to those trying to break out of the strait-jacket of conventional femininity *and* to those who find security and satisfaction in those very conventions – and if we made concepts such as 'wife', 'mother', 'sexy broad' central to our investigations, we might find that the Iran/Contra affair and international politics generally looked different. It's not that we would abandon our curiosity about arms dealers, presidents' men and concepts such as 'covert operations'. Rather, we would no longer find them sufficient to understand how the international political system works.

Masculinity and international politics

Making women invisible hides the workings of both femininity and masculinity in international politics. Some women watching the Iran/Contra hearings found it useful to speculate about how the politics of masculinity shape foreign-policy debates. They considered the verbal rituals that public men use to blunt the edges of their mutual antagonism. A congressman would, for instance, preface a devastating attack on Admiral Poindexter's rationale for destroying a document by reassuring the admiral – and his male colleagues – that he believed the admiral was 'honourable' and 'a gentleman'. Another congressman would insist that, despite his differences with Reagan officials Robert McFarlane and Oliver North, he considered them to be 'patriots'. Would these same male members of Congress, selected for this special committee partly because they had experience of dealing with military officers and foreign-policy administrators, have used the word 'honorable' if the witness had been a woman? Would 'patriot' have been the term of respect if these men had been commending a woman? There appeared to be a platform of trust holding up these investigations of US foreign policy. It was a platform that was supported by pillars of masculinity, pillars that were never subjected to political scrutiny, but which had to be maintained by daily personal exchanges, memos and formal policy.

A theme that surfaced repeatedly during the weeks of the Iran/Contra hearings was 'We live in a dangerous world'. Critics as well as supporters of selling arms to Iran and using the profits to fund the Contras were in agreement on this view of the world in 1987. No one chimed in with, 'Well, I don't know; it doesn't feel so dangerous to me.' No one questioned this portrayal of the world as permeated by risk and violence. No one even attempted to redefine 'danger' by suggesting that the world may indeed be dangerous, but especially so for those people who are losing access to land or being subjected to unsafe contraceptives. Instead, the vision that informed these male officials' foreign-policy choices was of a world in which two super-powers were eyeball-to-eyeball, where small risks were justified in the name of staving off bigger risks – the risk of Soviet expansion, the risk of nuclear war. It was a world in which taking risks was proof of one's manliness and therefore of one's qualification to govern. Listening to these officials, I was struck by the similarity to the 'manliness' now said to be necessary for success in the international financial markets. With Britain's 'Big Bang', which deregulated its financial industry, and with the French and Japanese deregulators following close behind, financial observers began to warn that the era of gentlemanliness in banking was over. British, European and Japanese bankers and stockbrokers would now have to adopt the more robust, competitive form of manliness associated with American bankers. It wouldn't necessarily be easy. There might even be some resistance. Thus international finance and international diplomacy seem to be converging in their notions of the world and the kind of masculinity required to wield power in that world in the 1990s.[6]

At first glance, this portrayal of danger and risk is a familiar one, rooted in capitalist and Cold War ideology. But when it's a patriarchal world that is 'dangerous', masculine men and feminine women are expected to react in opposite but complementary ways. A 'real man' will become the protector in such a world. He will suppress his own fears, brace himself and step forward to defend the weak, women and children. In the same 'dangerous world' women will turn gratefully and expectantly to their fathers and husbands, real or surrogate. If a woman is a mother, then she will think first of her children, protecting them not in a manly way, but as a self-sacrificing mother. In this fashion, the 'dangerous world' evoked repeatedly in the Iran/Contra hearings is upheld by unspoken notions about masculinity. Ideas of masculinity have to be perpetuated to justify foreign-policy risk-taking. To accept the Cold War interpretation of living in a 'dangerous' world also confirms the segregation of politics into national and international. The national political arena is dominated by men but allows women some select access; the international political arena is a sphere for men only, or for those rare women who can successfully play at being men, or at least not shake masculine presumptions.

Notions of masculinity aren't necessarily identical across generations or across cultural boundaries. An Oliver North may be a peculiarly American phenomenon. He doesn't have a carbon copy in current British or Japanese politics. Even the Hollywood character 'Rambo', to whom so many likened Oliver North, may take on rather different meanings in America, Britain and Japan.[7] A Lebanese Shiite militiaman may be fulfilling an explicitly masculinist mandate, but it would be a mistake to collapse the values he represents into those of a British SAS officer or an American 'Rambo'. Introducing masculinity into a discussion of international politics, and thereby making men visible as men, should prompt us to explore differences in the politics of masculinity between countries – and between ethnic groups in the same country.

These differences have ignited nationalist movements which have challenged the existing international order, dismantling empires, ousting foreign bases, expropriating foreign mines and factories. But there have been nationalist movements which have engaged in such world challenges without upsetting patriarchal relationships within that nation. It is important, I think, to understand which kinds of nationalist movement rely on the perpetuation of patriarchal ideas of masculinity for their international political campaigns and which kinds see redefining masculinity as integral to re-establishing national sovereignty. Women do not benefit automatically every time the international system is re-ordered by a successful nationalist movement. It has taken awareness, questioning and organizing by women inside those nationalist movements to turn nationalism into something good for women.

In conventional commentaries men who wield influence in international politics are analyzed in terms of their national identities, their class origins and their paid work. Rarely are they analyzed as men who have been taught how to be manly, how to size up the trustworthiness or competence of other men in terms of their manliness. If international commentators do find masculinity interesting, it is typically when they try to make sense of 'great men' – Teddy Roosevelt, Winston Churchill, Mao Tse-t'ung – not when they seek to understand humdrum plantation workers or foreign tourists. Such men's presumptions about how to be masculine in doing their jobs, exercising influence, or seeking relief from stress are made invisible.

Beyond the global victim

Some men and women active in campaigns to influence their country's foreign policy – on the right as well as the left – have called on women to become more involved in international issues, to learn more about 'what's going on in the world': 'You have to take more interest in international affairs because it affects how you live.' The gist of the argument is that women need to devote precious time and energy to learning about events outside their own country because as women they are the objects of those events. For instance, a woman working in a garment factory in Ireland should learn more about the European Economic Community because what the EEC commissioners do in Brussels is going to help determine her wages and maybe even the hazards she faces on the job. An American woman will be encouraged to learn the difference between a cruise and Pershing missile because international nuclear strategies are shaping her and her children's chances of a safe future.

Two things are striking about this line of argument. First, the activists who are trying to persuade women to 'get involved' are not inviting women to reinterpret international politics by drawing on their own experiences as women. If the explanations of how the EEC or nuclear rivalry works don't already include any concepts of femininity, masculinity or patriarchy, they are unlikely to after the women join the movement. Because organizers aren't curious about what women's experiences could lend to an understanding of international politics, many women, especially those whose energies are already stretched to the limit, are wary of becoming involved in an international campaign. It can seem like one more attempt by privileged outsiders – women and men – to dilute their political efforts. If women are asked to join an international campaign – for peace, against communism, for refugees, against apartheid, for religious evangelism, against hunger – but are not allowed to define the problem, it looks to many locally engaged women like abstract do-gooding with minimal connection to the battles for a decent life in their households and in their communities.

Second, the typical 'women-need-to-learn-more-about-foreign-affairs' approach usually portrays women as victims of the international political system. Women should learn about the EEC, the United Nations, the CIA, the IMF, NATO, the Warsaw Pact, the 'greenhouse effect' because each has an impact on them. In this world view, women are forever being acted upon; rarely are they seen to be actors.

It's true that in international politics women historically have not had access to the resources enabling them to wield influence. Today women are at the bottom of most international hierarchies: women are routinely paid less than even the lowest-paid men in multinational companies; women are two thirds of all refugees. Women activists have a harder time influencing struggling ethnic nationalist movements than do men; women get less of the ideological and job rewards from fighting in foreign wars than do men. Though a pretty dismal picture, it can tell us a lot about how the international political system has been designed and how it is maintained every day: some men at the top, most women at the bottom.

But in many arenas of power feminists have been uncovering a reality that is less simple. First, they have discovered that some women's class aspirations and their racist fears lured them into the role of controlling other women for the sake of imperial rule. British, American, Dutch, French, Spanish, Portuguese women may not have been the architects of their countries' colonial policies, but many of them took on the roles of colonial administrators' wives, missionaries, travel writers and anthropologists in ways that tightened the noose of colonial rule around the necks of African, Latin American and Asian women. To describe colonization as a process that has been carried on solely by men overlooks the ways in which male colonizers' success depended on some women's complicity. Without the willingness of 'respectable' women to see that colonization offered them an opportunity for adventure, or a new chance of financial security or moral commitment, colonization would have been even more problematic.[8]

Second, feminists who listen to women working for multinational corporations have heard these women articulate their own strategies for coping with their husbands' resentment, their foremen's sexual harassment and the paternalism of male union leaders. To depict these women merely as passive victims in the international politics of the banana or garment industries doesn't do them justice. It also produces an inaccurate picture of how these global systems operate. Corporate executives and development technocrats need some women to depend on cash wages; they need some women to see a factory or plantation job as a means of delaying marriage or fulfilling daughterly obligations. Without women's own needs, values and worries, the global assembly line would grind to a halt. But many of those needs, values and worries are defined by patriarchal structures and strictures. If fathers, brothers, husbands didn't gain some privilege, however small in global terms, from women's acquiescence to those confining notions of femininity, it might be much harder for the foreign executives and their local élite allies to recruit the cheap labor they desire. Consequently, women's capacity to challenge the men in their families, their communities or their political movements, will be a key to remaking the world.

Notes

1 *New York Times*/CBS poll, reported in the *New York Times*, July 18, 1987. Only 25 per cent of the American women sampled said that they approved of the US government giving aid to the anti-Sandinista Contra forces; 37 per cent of the American men polled approved of US aid to the Contras, though proportionately fewer Black men approved than did white men.

2 For an interesting exploration of the contradictions behind women as television-news reporters, see Patricia Holland, 'When a Woman Reads the News', in Helen Baehr and Gillian Dyers, editors, *Boxed In: Women and Television*, London and Winchester, MA, Pandora Press, 1988, pp. 133–50.

3 For more on the politics of beauty, see Wendy Chapkis, *Beauty Secrets*, Boston, South End Press, 1986, London, Women's Press, 1987.

4 *Boston Globe*, August 10, 1987.

5 Barbara Gamarekian, 'Consequences of Fawn Hall', *New York Times*, February 28, 1987. For Fawn Hall's own views on a secretary's professional relationships, see Mary Sit, 'Hall Tells Secretaries: "Stand by Your Boss" ', *Boston Globe*, September 30, 1988. See also Beatrix Campbell's analysis of the Sara Keayes–Cecil Parkinson scandal in Britain in her *Iron Ladies*, London, Virago Press, 1987, pp. 274–5.

6 For a suggestive report on the changing images of male bankers in the era of internationalized and deregularized banking, see 'City of London, Survey', *The Economist*, June 25, 1988, pp. 25–9; Sebastian Kinsman, 'Confessions of a Commodity Broker', *New Statesman*, 19 February 1988, pp. 10–11; Steve Lohr, 'London's Resurgent Markets', *New York Times*, September 22, 1986; 'The Risk Game: A Survey of International Banking', *The Economist*, March 21, 1987; 'A Survey of Wall Street', *The Economist*, July 11, 1987; Barbara Rogers, *Men Only*, London and Winchester, MA, Pandora Press, 1988.

7 Cynthia Enloe, 'Beyond Rambo', in Eva Isaksson, editor, *Women and the Military*

System, Brighton, Wheatsheaf Books, and New York, St Martins Press, 1988; a somewhat different version appears as 'Beyond Rambo and Steve Canyon', in John Gillis, editor, *The Militarization of the World*, New Brunswick, NJ, Rutgers University Press, 1989.

8 See, for instance, 'Western Women and Imperialism', special issue of *Women's Studies International Forum*, edited by Margaret Strobel and Nupur Chaudhuri, vol. 13, no. 2, 1990.

INDEX